Multilevel Modeling Methods With Introductory and Advanced Applications

A volume in
Quantitative Methods in Education and the Behavioral Sciences: Issues, Research, and Teaching
Jeffrey R. Harring, *Series Editor*

Quantitative Methods in Education and the Behavioral Sciences: Issues, Research, and Teaching

Jeffrey R. Harring, *Series Editor*

Structural Equation Modeling: A Second Course (2nd ed.) (2013)
Gregory R. Hancock and Ralph O. Mueller

Multilevel Modeling of Educational Data (2008)
Ann A. O'Connell and D. Betsy McCoach

Real Data Analysis (2007)
Shlomo S. Sawilowsky

Structural Equation Modeling: A Second Course (2006)
Gregory R. Hancock and Ralph O. Mueller

Multilevel Modeling Methods With Introductory and Advanced Applications

edited by

Ann A. O'Connell
The Ohio State University

D. Betsy McCoach
University of Connecticut

Bethany A. Bell
University of Virginia

IAP
INFORMATION AGE PUBLISHING, INC.
Charlotte, NC • www.infoagepub.com

Library of Congress Cataloging-in-Publication Data

A CIP record for this book is available from the Library of Congress
http://www.loc.gov

ISBN: 978-1-64802-871-7 (Paperback)
978-1-64802-872-4 (Hardcover)
978-1-64802-873-1 (E-Book)

Printed and bound by CPI Group (UK) Ltd, Croydon, CR0 4YY

To the teachers and health care workers of 2020
—AOC

To the Siegles
—DBM

To my supportive and loving husband Daniel O'Connell
—BAB

CONTENTS

SECTION II

LONGITUDINAL DATA

SECTION III

DESIGN AND SPECIAL ISSUES

ACKNOWLEDGMENTS

First, we would like to thank all the chapter authors for their contributions and for their patience. Finishing the book during a global pandemic proved challenging, but we believe that the final product is worth the wait. Thanks to the editorial and publications staff at Information Age Publishing, especially George Johnson. We would also like to thank the Educational Statisticians Special Interest Group of AERA. This book is part of their series, Quantitative Methods in Education and the Behavioral Sciences: Issues, Research, and Teaching.

There are many people who provided assistance at various stages of this project. First, thanks are due to James Uanhoro who assisted with reviewing each chapter's "Try This" data and code/syntax provided by the authors. In addition, we would like to thank Susie Mauck, who provided major editorial assistance for several of the chapters. Also, thanks to Sarah Newton, Pamela Peters, Joselyn Perez, and Anthony Gambino who helped proofread and format several of the chapters and Raymond Smith for data management support.

Thanks also to all of our current and former graduate students in our multilevel modeling classes. Working with you inspired us to undertake this project, and helped shape the content and structure of the volume. Thanks to our exceptional faculty colleagues at The Ohio State University, University of Connecticut, and University of Virginia. Finally, we owe a special debt of gratitude to our families, who were patient with us through this entire process.

Multilevel Modeling Methods with Introductory and Advanced Applications, page xi

CHAPTER 1

INTRODUCTION TO MULTILEVEL MODELING METHODS

Pedagogy and Context

Ann A. O'Connell
The Ohio State University

D. Betsy McCoach
University of Connecticut

Bethany A. Bell
University of Virginia

"Hierarchies abound." That is how we started our first edited volume for this series, more than a decade ago (O'Connell & McCoach, 2008). Since that time, the science of multilevel modeling has continued to advance, with development of new theory and applications, and improved algorithms and access to software capable of fitting multilevel models. This new volume is a collection of work by researchers dedicated to furthering the

Multilevel Modeling Methods with Introductory and Advanced Appiications, pages 1–10

use of multilevel analyses in education, health, and the social/behavioral sciences. Underpinning these efforts is the recognition that data are often correlated in ways that affect our inferences about relationships among variables of interest, and accommodating this correlation assures better understanding of educational and social phenomena. This correlation can arise with hierarchical or clustered data, such as observations on students within different schools, or longitudinal data, where repeated measurements are clustered within individuals.

Once you start to view the world through a multilevel lens, the importance of context or place and its impact on individuals becomes apparent in nearly every human setting. The use of multilevel analyses to examine similarities or differences across groups or contexts—such as classrooms, schools, neighborhoods, hospitals, or community settings—on individual outcomes continues to expand. The widespread adoption of multilevel models to represent these hierarchical systems and address questions of critical social and educational significance presents complex challenges to the applied research community, particularly to those of us who teach in this field. Our goal in this new volume is to provide a comprehensive and instructional resource text on multilevel analysis for quantitative and applied researchers, and for graduate students and professors of quantitative methods courses.

To achieve this goal, most chapters included in this text use datasets that are accessible to all readers in an online appendix (available at http://modeling.uconn.edu). The online materials include software code and supplemental material to help guide readers through the presented material. Each chapter offers a "Try This!" section with data that can be used for classroom practice and for individuals to check their own learning. We also asked contributing authors to provide a brief collection of current applied and technical readings, in addition to their detailed references for each chapter. This approach makes the volume unique among other options for classroom instruction and guided learning. Although one volume cannot possibly cover all applications and new directions for multilevel data, the emphasis placed here on introductory and advanced treatment of organizational and longitudinal data and related design issues provides a strong foundation for the theory and use of multilevel modeling to further capacity for continuing work in this area.

BRIEF BACKGROUND ON MULTILEVEL MODELING

Historically, the treatment of multilevel data has its roots in sampling designs where information is collected from clusters or groups of individuals who experience some common phenomena or event. This "common phenomena" could arise from simply living in the same neighborhood or attending the same class, school, or community health clinic. Alternatively, group membership could be manipulated, which might occur when students are assigned to work in small collaborative groups to examine the effects of increasingly difficult puzzle tasks on individual solution strategies. Of central importance in these scenarios is that information obtained from individuals who are clustered within a population—either in naturally occurring or intentionally formed groups—tend to be more similar within the groups in comparison to information obtained from people if they had been randomly assigned to the groups and did not experience any common or shared events. The degree to which individuals within a group or cluster are similar is measured through the intraclass correlation coefficient (ICC). The ICC, which decomposes total variance into its within-group and between-group components, represents the proportion of variance that is between clusters or groups (Raudenbush & Bryk, 2002). As a measure of similarity or homogeneity, the ICC also represents the correlation between pairs of individuals randomly selected from within the same randomly selected group or cluster (Snijders & Bosker, 2012). Outcome data that are wholly independent and contain no within-cluster similarity yield an ICC of zero. Thus, the ICC indicates the degree of dependency within the dataset.

Just as observations taken from individuals within a group or cluster tend to be similar to each other, so too are repeated observations from the same individual. Thus, repeated measures designs or longitudinal studies exhibit the same intra-unit dependency problems as data collected from clustered-sample designs. Multilevel models applied to longitudinal data accommodate this clustering or intra-individual correlation. Conceptually, each person in the longitudinal design simply functions as his or her own "cluster."

The effects of a lack of independence on statistical analyses and subsequent conclusions have been known for some time (see Kenny & Judd, 1986, for a review), and sampling statisticians have long accommodated the presence of intraclass correlation in analyses using cluster or multi-stage data collection designs (Donner et al., 1981; Fowler, 2001; Kish, 1965/1995; Murray & Hannon, 1990; Sudman, 1985). If the hierarchical structure is ignored for clustered data, the person-level (or lowest level) variance is exaggerated relative to the (assumed homogeneous) variance within the existing groups or clusters—and potential differences *between* clusters are not captured. Thus, when attempting to estimate the variance of an estimate, such as the overall mean, obtained from a clustered sample *when the clustering is ignored*, the resulting variance is much smaller than what would realistically be expected if the clustered structure of the data were incorporated into the calculation. The variance inflation factor (VIF), also referred to as the design effect, provides information regarding the degree to which the variance of an estimate needs to increase (for clustered samples) over what would be obtained if a simple random sample of the same size were conducted in the same population (Donner et al., 1981; Kish, 1965/1995; Murray, 1998; Snijders & Bosker, 2012). The amount of variance inflation is directly related to the intraclass correlation—as the ICC increases, the VIF also increases. The amount of variance inflation is also related to the average cluster size. As the size of the average cluster increases, the VIF increases. Therefore, treating cluster-sampled data as if it were obtained using simple random sampling results in variance estimates (and standard errors) that are too low.

This situation, well-known from the sampling literature, has grave implications for how hierarchical data should best be analyzed. In short, when considering variance from data obtained through grouped or clustered designs, the hierarchical sources of variability cannot be ignored without seriously contributing to errors of inference, compromising the validity of results and research conclusions. Statisticians use variance estimates in nearly all statistical tests and analyses, so ignoring the structure of a research setting or the sampling design adversely affects these variance estimates and increases the likelihood of Type I errors for any comparisons, contrasts, or effects we wish to assess. Multilevel modeling accommodates the structure of many familiar research settings—such as data collected from children within multiple schools or classrooms, or data collected from patients within multiple health clinics—and helps to protect the validity of statistical conclusions drawn from our analyses.

Often, however, the existence of clustering is not simply a nuisance to be overcome. In fact, differences in relationships among individual-level variables may exist between the clusters or contexts themselves, representing an additional source of variability over and above that present at the individual level. Multilevel analyses are uniquely designed to capture and model this variability across multiple levels, allowing for a better understanding of the influence that contexts—such as classrooms, schools, or neighborhoods—may have on individual outcomes.

Perhaps there is no more obvious setting for consideration of multilevel systems than schools. As researchers, educators and policymakers strive to learn more about factors affecting student outcomes, there is particular interest in the effects of classroom, school, or district-level variables on students' academic achievement. Multilevel modeling techniques lend themselves to the exploration of these and other educational issues, but as with all observational or non-experimental studies where random assignment is constrained, causal conclusions need to be tempered. Causality requires far more than a sophisticated statistical model, and the use of multilevel modeling does not imply that causality can be readily assumed or inferred. Multilevel modeling does provide a valuable tool for analyzing and understanding data that arise from non-independent or clustered samples; however, the use of this technique in and of itself does not lead to identification of direct causal links. When the data are collected from cluster-randomized designs, that is, where entire clusters are the randomized elements, multilevel modeling can be used to make inferences about the efficacy of a treatment. However, when the data result from non-randomized designs, any causal claims attributed to the variables of interest may be far more speculative. The focus for this book is on methodology and applications of multilevel modeling for hierarchical data structures; these structures can occur within experimental or non-experimental settings.

STRUCTURE OF THE BOOK

This book is pedagogical in nature and is intended to provide a cogent and connected collection of methodologies and advancements of multilevel theory and practice for applied researchers as well as for advanced graduate students. It is our hope that readers and professors of educational statistics courses will use this book as a guide to both basic and advanced multilevel modeling techniques. The content covered in the different chapters is presented in such a way to allow each chapter to stand alone as an instructional unit on its respective topic, with emphasis on application and interpretation, while assuring theoretical and practical connections throughout the book.

The 'Try This!" section for each chapter poses a structured data problem for the reader. The website to which our book is linked (http://modeling.uconn.edu) contains data for the "Try This!" section, creating an opportunity for readers to learn by doing. The inclusion of the "Try This!" problems and data is meant to aid methodology instructors, who must continually search for class examples and homework problems. Consequently, this book provides an accessible and practical treatment of methods appropriate for use in a first and/or second course in multilevel analysis.

A Note on Software

It is no longer the case that a single software option may be sufficient for new learners or for advanced methodologists. Rather than constrain contributors to a specific software choice, we asked authors to use the best available software options for the chapters they authored. This purposeful choice exposes readers to a variety of potential alternatives for multilevel analyses, and encourages mindful recognition of the strengths and limitations of existing software—all of which have different default analysis options and estimation settings. Most chapters used R, SAS, HLM, Stata, or author-developed tools and packages. The online appendix for each chapter provides code to facilitate understanding of the material, thus exposing readers to multiple alternatives and encouraging analysts to build their own capacity for multilevel modeling that is not limited to a single software choice.

SUMMARY OF THE CHAPTERS

The first section in the book, "Organizational Data," provides foundational information on some of the most common applications of multilevel modeling. Chapter 2 by Bell and Schoeneberger introduces multilevel modeling in the context of organizational data. Bell and Schoeneberger frame their discussion of models for organizational data around an example examining academic self-efficacy of students nested within schools, highlighting essential features of multilevel models with random intercepts and slopes. In Chapter 3, McCoach, Newton, and Gambino, review methods for the evaluation of model fit and model adequacy within the multilevel framework. Through a series of questions posed to the reader, McCoach, Newton, and Gambino cover issues such as deviance and model selection criteria, the impact of full versus restricted maximum likelihood estimation on assessment of model fit, how nestedness affects decisions on model fit and adequacy, and determining variance explained measures within a multilevel framework.

Chapter 4 by Rhoads and Li discusses many of the issues related to causal inference within multilevel settings. Their chapter examines the use of randomized experiments versus observational studies and how design features contribute to an understanding of causal effects for multilevel data. Implications of Rubin's (1986) stable-unit treatment variable assumption, confounding and balance, differences between the average treatment effect on the treated (ATT) and the average treatment effect (ATE), and other aspects of inference-making in causal settings are presented and discussed in depth by Rhoades and Li.

Related to the discussion of causal claims from multilevel data are the twin issues of sample size and power. Chapter 5 provides an introduction to statistical power analysis within the multilevel framework. Spybrook, Kelcey, and Dong focus on randomized experiments and illustrate the use of the Power Up! software to determine the power or necessary sample sizes for several different kinds of multilevel designs. The authors provide a series of worked examples highlighting interpretation and value of power analysis to real world research problems. In particular, the examples by Spybrook, Kelcey, and Dong include estimations of power for moderation and mediation effects in a variety of multilevel designs.

Sometimes data are simultaneously nested within two or more crossed hierarchies. For example, in a cross-sectional design students are nested within teachers at a single grade level, however, in a longitudinal design students can be cross-classified by teachers across two or more grade levels. In Chapter 6, Leroux and Beretvas contrast cross-classified structures with purely hierarchical structures and introduce cross-classified random effects models through detailed examples, descriptions of data/model set-ups, and model interpretations.

The next two chapters in this section deal with dependent variables that are categorical, ordinal, or count in nature. In Chapter 7, O'Connell, Lo, Goldstein, Rogers, and Peng extend generalized linear models to the multilevel framework and provide readers with details on estimation and interpretation. The authors provide in-depth examples of generalized linear mixed models (GLMMs) for dichotomous and ordinal data, some details on unit-specific versus population-averaged models, a discussion on extensions to other educational contexts, and a review in the online appendix of options in six software packages for fitting logistic and ordinal GLMMs. In Chapter 8, O'Connell, Bhaktha, and Zhang introduce both single level and multilevel models for the analysis of count data, focusing on the two most commonly applied approaches: poisson models and negative binomial models. Detailed description and examples of both kinds of models are covered, including parameter interpretation, estimation of the ICC, and a set of guiding questions designed to assist in the choice among these alternatives when fitting multilevel models to count data.

Our next section, "Longitudinal Data," contains four chapters focusing on treatment and aspects of multilevel models within a longitudinal context. In Chapter 9, McCoach, Bell, and Bellara introduce individual growth curve modeling with detailed examples and comparisons between linear growth, piecewise linear growth, and polynomial growth models. The nature of time-structured versus time-unstructured data, the value of centering, and the use of time-varying and time-invariant covariates are discussed and illustrated in depth. Harring and Blozis, in Chapter 10, expand on the treatment of longitudinal data using nonlinear mixed-effects models (NLME), a flexible analytic framework for describing individuals' continuous repeated measures data that exhibit complex, curvilinear patterns of change. The authors cover interpretation and notation issues, covariance structures for summarizing level-1 variability, the importance of graphical displays of data, and lead the reader through a series of analytic issues and decision points encountered when employing NLME analyses. Chapter 11 by Kim, Hsu, and Kwok explores alternative within-subjects variance-covariance structures for two-level models. Kim, Hsu, and Kwok note that misspecifying or oversimplifying the covariance structure can lead to bias in standard errors of parameters, affecting inferences from statistical significance tests and confidence intervals. They illustrate a series of models with alternative covariance structures and provide guidance for researchers on approaches to specifying an optimal and meaningful structure. In Chapter 12, Hedeker and Mermelstein examine methods for modeling variability in intensive longitudinal data (ILD). Adequately capturing the within-subject or intra-individual variation in ILD becomes increasingly more complex as the number of observations contributed per individual increases. The authors detail the use of the mixed-effects location scale (MELS) model for variance modeling and provide an example based on ecological momentary assessments (EMA) data from a natural history study of adolescent smoking.

The final section of the book, "Design and Special Issues," covers several critical topics within multilevel research. Chapter 13 serves as an introduction to sampling methods in relation to multilevel analysis designs and provides an overview of several existing national survey databases housed within the National Center for Education Statistics. Stapleton and Thomas describe how multilevel modeling could be used to explore these datasets and provide recommendations and suggested avenues for research utilizing these extensive public-access databases. In Chapter 14 French, Finch, and Vo extend traditional measurement models and procedures to a multilevel framework. Their chapter includes discussion of multilevel factor analysis, multilevel differential item functioning (MDIF), and multilevel reliability estimates. In Chapter 15, Enders and T. Hayes provide crucial guidance on how to conceptualize and handle missing data within a multilevel study. In Chapter 16, Rockwood and A. Hayes provide a comprehensive introduction to mediation within a multilevel context. In Chapter 17, Ferron, Wang, Yi, Yin, Kim, and Dedrick provide guidelines for interpreting results and writing about multilevel models. Without being overly prescriptive, the authors offer helpful suggestions for what results to present and how to use text, tables, and figures to best present these results.

A NOTE ON NOTATION

For most chapters in this volume, the notation used follows the conventions contained in Raudenbush and Bryk, 2002. For simplicity, and to introduce some consistency in language, a null or "empty" model may also be referred to as an unconditional means model (i.e., a one-way analysis of variance model with random effects), in which there are no level-1 or level-2 predictors. Each chapter presents the notation used for multilevel models containing level-1 and level-2 predictors. In the null model, and assuming organizational data, the outcome variable, Y, for person i in cluster j, Y_{ij}, is a function of the cluster-level intercept, which randomly varies across clusters, and a residual term, r_{ij} (sometimes also indicated as e_{ij}) which captures the deviation of the score for person$_{ij}$ from his or her cluster's intercept. Then, the collection of cluster level intercepts, β_{0j}, are modeled at level 2 through the grand mean intercept in the population, γ_{00}, where u_{0j} is the random effect or the deviation of cluster j's intercept from the grand intercept. At level 1, the within-cluster variability is represented as $\text{Var}(r_{ij}) = \sigma^2$, with variability assumed to be homogeneous across clusters—and Section II of this volume covers situations in which this homogeneity assumption is relaxed. At level 2, the between-cluster variability is represented as $\text{Var}(u_{0j}) = \tau_{00}$. In some chapters by necessity, the notation for this variability at level 2 may vary (e.g., τ_{00}^2), but the underlying concept of between-cluster variation is the same.

Thus, the unconditional model can be expressed as two separate equations, as shown below.

$$Y_{ij} = \beta_{0j} + r_{ij} \tag{1.1}$$

$$\beta_{0j} = \gamma_{00} + u_{0j} \tag{1.2}$$

Alternatively, the two separate equations can be combined into one equation:

$$Y_{ij} = \gamma_{00} + u_{0j} + r_{ij} \tag{1.3}$$

Although the combined equation accurately portrays the computational process being used when estimating multilevel models, the multiple equation layout is often simpler for novice multilevel modelers to grasp. Therefore, throughout much of this volume the multiple equation layout is presented, but the combined format is used in many examples and presentations of material. Departures from the notational conventions used by Raudenbush and Bryk (2002) are noted and explained within each chapter as necessary.

REFERENCES

Donner, A., Birkett, N., & Buck. C. (1981). Randomization by cluster: Sample size requirements and analysis. *American Journal of Epidemiology, 114,* 283–286.

Fowler, F. J. (2001). *Survey research methods* (3rd ed.). SAGE.

Kenny, D. A., & Judd, C. M. (1986). Consequences of violating the independence assumption in analysis of variance. *Psychological Bulletin, 99,* 422–431.

Kish, L. (1995). *Survey sampling.* John Wiley & Sons. (Originally published in 1965)

Murray, D. M. (1998). *Design and analysis of group-randomized trials.* Oxford.

Murray, D. M., & Hannan, P. J. (1990). Planning for the appropriate analysis in school-based drug-use prevention studies. *Journal of Consulting and Clinical Psychology, 58,* 458–468.

O'Connell, A. A., & McCoach, D. B. (Eds.). (2008). *Multilevel modeling of educational data.* Information Age Publishing.

Raudenbush, S. W., & Bryk, A. S. (2002). *Hierarchical linear models* (2nd ed.). SAGE Publications.

Rubin, D. B. (1986). Statistics and causal inference: Comment: Which ifs have causal answers. *Journal of the American Statistical Association, 81*(396), 961–962.

Snijders, T., & Bosker, R. (2012). *Multilevel analysis* (2nd ed). SAGE Publications.

Sudman, S. (1985). Efficient screening methods for the sampling of geographically clustered special populations. *Journal of Marketing Research, 22,* 20–29.

SECTION I

ORGANIZATIONAL DATA

CHAPTER 2

INTRODUCTION TO MULTILEVEL MODELS FOR ORGANIZATIONAL DATA

Bethany A. Bell
University of Virginia

Jason A. Schoeneberger
ICF

Hierarchically organized data are commonplace in a variety of research settings. For example, data on students are nested within classrooms or teachers, teachers are nested within schools, and schools are nested within districts. Alternatively, patients are nested within doctors providing services, who may in turn be nested within clinics or hospitals. In these examples, the independence assumption required for correctly estimating single-level regression models is violated. Independence implies there is no correlation between observations, but in a hierarchical system when observations are collected within multiple clusters or contexts such as classrooms or clinics, observations from within the same cluster tend to share some similarity to each other relative to what would be expected in a non-hierarchical

Multilevel Modeling Methods with Introductory and Advanced Applications, pages 13–49

sample. Statistically, if the nested structure of the data were ignored, standard errors for parameter estimates, power to detect effects, and Type I error rates could all be impacted (Donner & Klar, 2000; Julian, 2001; Moerbeek, 2004; Murray, 1998; Shadish et al., 2002; Wampold & Serlin, 2000). Multilevel models (MLMs) have been developed to properly account for the lack of independence that occurs with nested data.

MLMs can differ in terms of the number of levels (e.g., students nested within schools [two levels], students nested within schools nested within districts [three levels]), type of design (e.g., organizational, longitudinal with repeated measures, cross-classified), scale of the outcome variable (e.g., continuous, categorical), and number of outcomes (e.g., univariate, multivariate). Regardless of the type of model being estimated, MLMs can be conceptualized as regression models applied at different levels of a hierarchical system. Thus, regression serves as a valuable base from which to examine similarities as well as important differences between single-level regression models and MLMs. For example, as with single-level regression models, MLMs are used to examine relationships between predictors and criterion variables. Both single and multilevel models estimate regression coefficients for the predictors (referred to as fixed effects), and provide information on the errors or residuals between the model and the data. In the MLM, however, predictors can be included at any level of the model. The inclusion of error terms and what they represent in MLMs is different compared to single-level models where all predictors (and the criterion) are at the same level (i.e., the lowest level such as student level). Both provide a single error term at the lowest level of the model but in MLMs there are multiple possible error terms that may be included at higher levels of the model (i.e., the school level). Given the inclusion of multiple error terms at level 2—also referred to as random effects—the variances and covariances among these random effects can be estimated. The ability to capture multiple sources of variability in the data through these random effects is one of the key differences between single-level and multilevel models. When present, random effects provide unique information that cannot be obtained through a single-level regression model. We provide more details with interpretations on random effects in the examples throughout this chapter.

In this chapter we focus exclusively on organizational data structures for continuous outcome variables. We begin with an illustration of a simple two-level model to help define important concepts and terms of the MLM. We then emphasize several critical decisions that researchers must make when fitting MLMs to hierarchical data. These include methods of estimation, fixed versus random slopes, variance/covariance structures, centering of predictors, and determination of degrees of freedom for statistical tests. We then present an applied example based on data from the Early

Childhood Longitudinal Study (ECLS-K:2011; Tourangeau et al., 2015). We also review the extension of two-level models to a three-level context.

SIMPLE ILLUSTRATION OF A GENERIC TWO-LEVEL ORGANIZATIONAL MODEL

A simple example illustrates how to approach estimating a two-level organizational model. Imagine a researcher is interested in exploring the influence of school (level 2) and student (level 1) characteristics on the academic self-efficacy of individual students. The presence of a hierarchical structure requires an analytic method that accounts for the lack of independence among the data. We note that fitting a MLM is one approach to analyzing hierarchical data, and that other strategies exist, including generalized estimating equations (GEE; Liang & Zeger, 1986), fixed effects models (FEM; Allison, 2005), or single-level models with adjustment corrections to standard errors (McNeish & Stapleton, 2017). McNeish and Stapleton (2016, 2017) and Gardiner et al. (2009) provide reviews of these and other approaches for clustered data. MLMs allow for the exploration of varying coefficients across the level-2 units; variability in these coefficients between clusters is captured through the inclusion of random effects. In contrast, models estimated through GEE treat the random effects as a nuisance and this approach to model estimation is typically used when the random effects and their corresponding variances and covariances are not of substantive interest (Hardin & Hilbe, 2013). MLMs also allow for generalizability to the wider population of level-2 units from which the random sample of level-2 units was selected, which differs from the interpretation of FEM for which results are only generalizable to the units modeled (Allison, 2005).

Equations necessary for estimating a basic MLM are presented below. Because there are data at multiple levels, multiple equations are presented—one for each level of the model—followed by a combined equation capturing both levels. Equation 2.1 represents a simple level-1 model with one continuous student-level predictor, X_{ij}.

$$Y_{ij} = \beta_{0j} + \beta_{1j} X_{ij} + e_{ij} \tag{2.1}$$

In this model, Y_{ij} is the academic self-efficacy for student i in school j; likewise, X_{ij} is the value of the predictor for student i in school j. β_{0j} represents each schools' intercept and its estimate is the prediction for academic self-efficacy for school j when the student-level predictor $X_{ij} = 0$ (and conditional on any level-2 predictors). β_{1j} is the slope or regression coefficient associated with X_{ij} for each school. The slope β_{1j} depicts the relationship between the student-level variable and academic self-efficacy for each school, and just as intercepts may vary across schools, the slope may also vary across schools. Estimates for the level-1 regression coefficients are determined through empirical Bayes (EB) estimation and not via ordinary least squares (OLS). EB estimates are weighted based on parameter reliability and can be more informative than least-squares estimates when data are nested (Raudenbush & Bryk, 2002). The final term in Equation 2.1 is the student-level error term, e_{ij}; these residuals are assumed to be normally distributed and homoscedastic across the different level-2 schools, with a mean of 0 and common variance σ^2.

Next, we consider the model at level 2. Equation 2.2a corresponds to a simple two-level model that includes a single continuous school-level predictor W_j representing the value of W for the jth school. This example assumes the same student-level model shown in Equation 2.1.

$$\begin{aligned}\beta_{0j} &= \gamma_{00} + \gamma_{01}W_j + u_{0j} \\ \beta_{1j} &= \gamma_{10}\end{aligned} \tag{2.2a}$$

Note there are two level-2 equations, one for the level-1 intercept and one for the level-1 slope. Here, W_j represents the value of the level-2 predictor for the jth school. The intercept γ_{00} represents the predicted mean for all the intercepts from level 1 when $W_j = 0$. The gamma coefficients at level 2 are weighted estimates, determined through generalized least squares estimation (Raudenbush & Bryk, 2002). The gammas represent fixed effects for the model. Based on the definition for β_{0j} described in the preceding paragraph, we can interpret γ_{00} as the best estimate for a student's academic self-efficacy across all schools when both X_{ij} and W_j are 0. The parameter γ_{01} is the fixed-effect regression coefficient associated with W_j; as W_j increases by one unit, predictions for γ_{00} are expected to increase by γ_{01} units. Finally, u_{0j} are the random effects associated with the school-level intercepts and are unique for each school j. Thus, random effects are level-2 residuals or deviations from the fixed effect for the intercept, γ_{00}. If u_{0j} is positive, the predicted value for the intercept for school j is higher than γ_{00}; if u_{0j} is negative, the predicted value is lower. The variance in u_{0j} represents variability in the school-level intercepts at level 2. In general, it is the variance among the random effects that we are typically interested in estimating in MLM, not the residuals themselves. The level-2 random effects in Equation 2.2a are assumed to be normally distributed with a mean of 0 and variance of τ_{00}, typically written as $u_{0j} \sim N(0, \tau_{00})$.

Turning to the second level-2 equation in Equation 2.2a, γ_{10} represents the overall or weighted-average slope for the student-level predictor, X_{ij}, across all schools. The absence of an error term in the equation for β_{1j} indicates that the effect of the student-level predictor is fixed or held constant across schools. Whereas the intercepts could be uniquely estimated for schools based on the random effect, u_{0j}, the slopes are held constant for all schools. This type of two-level model is commonly referred to as a random intercept model (Raudenbush & Bryk, 2002).

Alternatively, the effect of the student-level predictor can be allowed to vary across level-2 units. The equation that includes a random effect for the student-level predictor, X_{ij}, is shown below in Equation 2.2b.

$$\begin{aligned} \beta_{0j} &= \gamma_{00} + \gamma_{01} W_j + u_{0j} \\ \beta_{1j} &= \gamma_{10} + u_{1j} \end{aligned} \tag{2.2b}$$

The only difference between Equation 2.2a and 2.2b is the inclusion of u_{1j} in the equation for β_{1j}. Including u_{1j} in the level-2 equation for the school slopes allows the relationship between the student-level predictor (X_{ij}) and the outcome (Y_{ij}) to vary across schools or level-2 units. This type of two-level model is commonly referred to as a "random-slopes and -intercepts model" (Raudenbush & Bryk, 2002). In these models, the level-2 residuals or random effects are assumed to be normally distributed with a mean of 0 and variances of τ_{00} and τ_{11}, respectively. The covariance between the random effects in the model, τ_{01}, is also estimated (although we note in a later section that some software sets this covariance to 0 by default). The variation and covariation for the random effects in a random-slopes and -intercepts model (assuming one random slope) can be collected and summarized in the Tau matrix, T, as:

$$T = \begin{pmatrix} \text{var}(u_{0j}) & \text{cov}(u_{0j}, u_{1j}) \\ & \text{var}(u_{1j}) \end{pmatrix} = \begin{pmatrix} \tau_{00} & \tau_{01} \\ & \tau_{11} \end{pmatrix}$$

Next, substituting the values of β_{0j} and β_{1j} from the level-2 model provided in Equation 2.2a into the level-1 Equation 2.1 produces the combined level-1 and level-2 random intercept model (Equation 2.3a). From this combined model, the regression framework underlying the MLM becomes more apparent. Specifically, the continuous academic self-efficacy outcome (Y_{ij}) is a function of the intercept (γ_{00}), the contribution of predictors based on their regression coefficients (γ_{10} and γ_{01}), and level-1 and level-2 error terms (e_{ij} and u_{0j}, respectively).

$$Y_{ij} = \gamma_{00} + \gamma_{01} W_j + \gamma_{10} X_{ij} + u_{0j} + e_{ij} \tag{2.3a}$$

Following the same logic, when substituting the values β_{0j} and β_{1j} from the level-2 equation in Equation 2.2b into level-1 Equation 2.1, the combined level-1 and level-2 random-slopes and -intercepts model can be expressed as (Equation 2.3b):

$$Y_{ij} = \gamma_{00} + \gamma_{01}W_j + \gamma_{10}X_{ij} + u_{0j} + u_{1j}X_{ij} + e_{ij} \tag{2.3b}$$

Equations 2.3a and 2.3b contain "fixed" components that are common across clusters (γ_{00}, γ_{01}, and γ_{10}) and "random" components that are allowed to vary from cluster to cluster (u_{0j} and u_{1j}). Variability of these random effects capture the differences in the intercepts and slopes across schools. The variance covariance matrix of the random effects ($\boldsymbol{T}$), together with the level-one variance σ^2, are referred to as the model's variance components.

In addition to estimating fixed and random effects, MLMs allow researchers to explore the influence of cluster-level variables on lower-level predictors. Not only can we estimate the main effects of W_j and X_{ij} (γ_{01} and γ_{10} respectively, as shown above in Equations 2.3a and 2.3b) but we can also estimate a cross-level interaction between W_j and X_{ij}. Specifically, in a multilevel model we can assess how the effect of a lower-level predictor such as X_{ij} can be moderated by a higher-level predictor, W_j. Researchers include cross-level interactions when they are interested in examining if a level-2 variable moderates (i.e., alters the strength or nature of) the relationship between a level-1 predictor and the outcome (see Aguinis et al., 2013). Working from Equation 2.2b specified above, a model with a cross-level interaction would appear as

$$\begin{aligned} \beta_{0j} &= \gamma_{00} + \gamma_{01}W_j + u_{0j} \\ \beta_{1j} &= \gamma_{10} + \gamma_{11}W_j + u_{1j} \end{aligned} \tag{2.4}$$

In Equation 2.4, the cross-level interaction appears in the equation for $\beta_{1j} : \gamma_{11}W_j$ which represents the effect of the level-2 predictor (W_j) on the slope of the level-1 predictor (X_{ij}; i.e., W_j predicts the estimated intercept and the slope of the level-1 predictor).

When equations are expanded to the combined model (Equation 2.5), the interaction between the predictors is readily apparent as $\gamma_{11}W_jX_{ij}$, that is, in the combined equation, it is easier to see that the model includes two main effects (γ_{10} for X_{ij} and γ_{01} for W_j) as well as the interaction between these two variables (γ_{11}).

$$Y_{ij} = \gamma_{00} + \gamma_{01}W_j + \gamma_{10}X_{ij} + \gamma_{11}W_jX_{ij} + u_{0j} + u_{1j}X_{ij} + e_{ij} \tag{2.5}$$

In addition to understanding the equations used to represent multilevel models, it is also important to understand the language that is used when estimating these models. In the remainder of this chapter, we define key terms as they are introduced, beginning with fixed effects and random effects. For a comprehensive list of key terminology used with multilevel model analyses, we recommend Diez Roux (2002).

Understanding of fixed and random effects is best clarified when we first consider how MLMs are estimated. Conceptually, when a researcher specifies a MLM the estimation process results in the establishment of a regression model based on empirical Bayes coefficients for each cluster. The estimates from these cluster-specific regressions feed into the estimates for the parameters at level 2, which are referred to as the fixed effects. In the level-2 model equations presented above, the value determined for γ_{00} would represent the overall value of the intercepts across clusters, γ_{10} would be the overall (weighted) average slope for the level-1 predictor across clusters, and γ_{01} would be the slope for the level-2 predictor at the cluster level. Figure 2.1 provides an illustration of this process assuming three clusters for a model with only one predictor at level 1 and no predictors at level 2. The intercept value for the overall regression line, γ_{00}, can be conceptualized as the average of the intercepts from the regression lines for clusters A, B, and C. A similar process is used to determine the overall slope, γ_{10}, based on the slopes for each cluster's regression line.

Random effects allow for the estimation of variance between cluster intercepts and coefficients at level 1. These level-2 variances are the variance components of the model and capture the degree of between-cluster differences. Within-cluster differences among observations are captured through the residual variance at level 1. Like single-level regression, all MLMs include a level-1 or within-cluster estimate of variance. A two-level linear organizational model includes at least one random effect; this is the key distinction between two-level linear models and single-level ordinary least

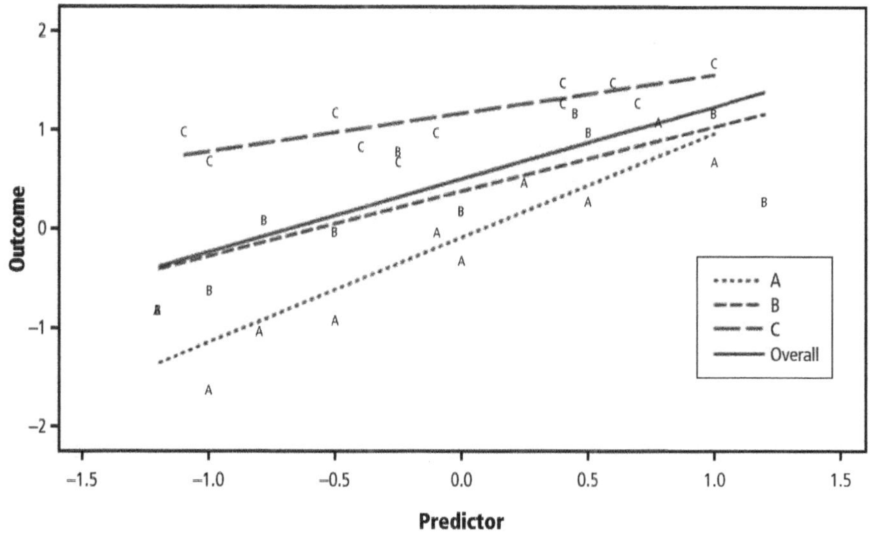

Figure 2.1 Conceptual illustration of random intercepts and slopes.

squares (OLS) models. For example, the random effect for the intercept is what makes the model depicted in Equation 2.3a a "random intercepts" model. In Equation 2.3a, if the random level-2 residual, u_{0j}, were removed, we would be left with a traditional OLS equation and there would be no deviations between clusters on the intercept. In the random-slopes and -intercepts model as shown in Equation 2.3b, we have an additional random effect, u_{1j}, which reflects the random slopes for our level-1 predictor variable X_{ij}. Inclusion of a random slope term indicates that the relationship between the student-level predictor (X_{ij}) and the outcome (Y_{ij}) are allowed to randomly vary across level-2 units (i.e., we are not forcing the relationship to be the same in each of the level-2 units). These random effects are key components of multilevel models that differentiate them from single-level models.

The model depicted in Figure 2.1 includes both random intercepts and random slopes. There are separate regression lines for each cluster. Each cluster's regression line can have a different intercept and a different slope—this is what is meant by allowing the intercept and slope of a level-1 predictor to vary as shown in Equation 2.3b. Cluster A has, on average, lower overall performance on the outcome, but the slope associated with the predictor is stronger in this cluster than any other. Cluster C, on the other hand, has the highest overall performance on the outcome, but the slope is the weakest. If we were to ignore the nested structure of these data and estimate a single-level OLS model, these differences in intercepts and slopes across the clusters would not be evident and would contribute to overall "error" in the model. However, in MLM, we are specifying unique group effects through random intercepts and random slopes. Remember, the fixed effects represent the average or overall intercepts and slopes across clusters (and controlling for any other predictors in the model). When we specify random effects in our models, statistically, we are capturing deviations from the overall intercept and slope for each cluster (u_{0j} and u_{1j}, respectively). The variability of these random effects are the parameter estimates of interest—τ_{00} and τ_{11}—and represent between-cluster variances of the intercepts and/or slopes in our model.

Intraclass Correlation Coefficient

Given this basic understanding of fixed and random effects and how random effects contribute to specific variance components of a MLM, the next key concept to clarify is the intraclass correlation coefficient (ICC). Using MLM, the variance in an outcome can be decomposed into variance attributable to differences between observations within-clusters at level 1 and between-clusters at level 2 using an unconditional (i.e., no predictors) model as

$$Y_{ij} = \gamma_{00} + u_{0j} + e_{ij}.$$

This model provides an estimate of the variance in the outcome of interest at level 1 $\left(\text{var}(e_{ij}) = \sigma^2\right)$ and at level-2 $\left(\text{var}(u_{0j}) = \tau_{00}\right)$. Using these variance estimates, the ICC can be calculated as:

$$\text{ICC} = \frac{\tau_{00}}{\sigma^2 + \tau_{00}}. \tag{2.6}$$

As shown in Equation 2.6, the intraclass correlation indicates the proportion of variance due to clustering in the population, or the proportion of total variance in the outcome attributable to differences between level-2 units. Additionally, the intraclass correlation can be interpreted as the correlation between two randomly-selected units from the same cluster. An ICC equal to 1 means that everyone in a given cluster had identical values for the outcome of interest and an ICC equal to 0 would indicate that people in a given cluster were no more similar to each other than people from different clusters were. In OLS regression, we are typically interested in the proportion of variance in the outcome explained by the predictors. In MLM, we first examine how much between cluster variability in the outcome exists. Then, after the statistical models are estimated, we examine how much variability is explained by predictors at both level 1 and level 2. By examining changes in the variances at different levels researchers can understand the predictive power of the variables included at each level of the model (e.g., does the inclusion of a school's proportion of low-SES students as a level-2 predictor reduce or explain the between-school variance of student academic self-efficacy). More information on understanding the amount of variance explained at each level of the model is covered in Chapter 3 of this volume ("Evaluation of Model Fit and Adequacy"). As shown in the applied example later in this chapter, the typical first step in building MLMs is to estimate an unconditional model with no predictors to calculate the ICC.

MODEL SPECIFICATION DECISIONS THAT NEED TO BE MADE BEFORE ESTIMATING MLMS

Before estimating MLMs, researchers need to make several model specification decisions including (a) deciding on what estimation method to use, (b) whether to include random slopes in the models, (c) what variance/covariance structure to estimate, (d) how to center predictors, and (e) which degrees of freedom to use. These decisions are not unique to organizational models—they apply to a variety of multilevel models. This section provides overviews of each of these areas along with references on each topic, including several topics that are covered in more detail in Chapter 3.

Estimation Method

One of the first decisions that researchers need to make before estimating their MLMs is deciding which estimation method they will use to fit their models. Different statistical software packages contain procedures for estimating MLMs using a variety of estimation methods (e.g., MLMs can be estimated using maximum likelihood [ML] methods, which is what we use throughout this chapter, or through a Bayesian framework [see Chapter 13 in Hox et al., 2018 and Chapter 13 in Raudenbush & Bryk, 2002 for more details on Bayesian approaches to MLM]). Sticking with ML-based options, the first decision that needs to be made is how to estimate the model—full-information maximum likelihood (FIML) or restricted maximum likelihood (REML). The difference between FIML and REML estimation involves the manner of computation for the variance components. With FIML, the estimates of the variances and covariances are conditional upon point estimates of the fixed effects, which has been shown to yield biased estimates of the variance components when samples are small (i.e., variances for the random effects; Raudenbush & Bryk, 2002). Specifically, with smaller level-2 sample sizes, the FIML estimates of the variance components are downwardly biased. REML, on the other hand, adjusts the estimates of the variance components for uncertainty in the fixed effects, which eliminates the bias problem (Raudenbush & Bryk, 2002; Snijders & Boskers, 2012). When the level-2 sample size is large, the difference in the variance component estimates will be trivial between FIML and REML estimation.

Random Slopes

Although MLM can handle random slopes, depending on sample size and the amount of variability that exists between higher level units (i.e., classrooms, schools, clinics, etc.), a researcher needs to decide which, if any, slopes should be allowed to randomly vary across clusters. When making this decision, researchers should balance theory (i.e., is it reasonable to assume that some relationships are fixed and do not vary across higher units) with model complexity (every additional random slope increases the size of the Tau matrix and thus estimation complexity). In addition, some approaches to assessing statistical significance of the variance components depend on the model estimation methods used. We discuss several of these considerations in our applied example and refer readers to Chapter 3 for more details on how inclusion of random slopes relates to other aspects of MLM and which methods can be used to examine model fit.

Variance/Covariance Structure

In addition to deciding which, if any slopes, will be allowed to vary across level-2 units, researchers also need to decide on the variance/covariance structure to use when estimating their models. Although there are a variety of choices for the form of the variance/covariance matrix that can be requested when estimating MLMs, the two most common options for organizational models are to (a) estimate only the variances of the random effects, without estimating their covariances—akin to assuming their covariances are 0 (referred to as a variance component (VC) structure) or (b) estimate both variances and covariances (referred to as an unstructured [UN] covariance matrix). In Chapter 11 ("Within-Subject Residual Variance/Covariance Structures in Longitudinal Data") additional structures are considered for longitudinal data.

The VC structure is relatively simple in terms of estimation; the covariance matrix of level-2 errors (T) is estimated to have separate variances but no covariances, and the level-1 errors are assumed to have a common variance, σ^2. When the VC structure is used, a variance component is estimated for each coefficient that was specified to randomly vary across level-2 units. Thus, in a model that has a random intercept and one random slope, two variance components would be estimated: τ_{00} and τ_{11} as shown below. Additionally, a single estimate for the variance of the level-1 errors, σ^2, is also estimated (i.e., the level-one residual variance component). Thus a total of three variances are estimated using this simple structure.

$$\text{Variance Component Matrix} = \begin{bmatrix} \tau_{00} & 0 \\ & \tau_{11} \end{bmatrix}.$$

The UN option is the most general covariance matrix that allows for both variances and covariances among the random effects to be estimated from the data. Thus, when estimating this matrix, many more parameters are estimated compared to the variance component option. Specifically, for models estimated based on an unstructured covariance structure, the Tau matrix contains estimates for $[r(r+1)/2]$ level-2 variances and covariances, where $r =$ the number of random effects specified in the model. Again, the pooled within-cluster variance of the level-1 errors (σ^2) is automatically estimated. If we expand the VC matrix above to a UN matrix, four variance/covariance components are estimated: (a) the intercept variance (τ_{00}),

(b) the slope variance (τ_{11}), (c) the covariance (τ_{10}) between the intercepts and slopes, and (d) the variance of the level-1 errors (σ^2).

$$\text{Unstructured Variance/Covariance Matrix} = \begin{bmatrix} \tau_{00} & \tau_{01} \\ & \tau_{11} \end{bmatrix}.$$

The addition of the single covariance in this example is not a large change in the number of estimated variance/covariance parameters. However, if we were to use an unstructured matrix for a model that included a random intercept and two random slopes (i.e., only changing the model by adding one additional random slope), now seven parameters would need to be estimated: τ_{00}, τ_{11}, τ_{22} (the variance for the second random slope), τ_{10}, τ_{20} (the covariance between the intercept and the second random slope), τ_{21} (the covariance between the two random slopes), and σ^2. This additional complexity is why it is important to thoughtfully consider how many random slopes to include in your models as well which variance/covariance structure to use.

Centering of Predictors

Centering of predictors is an established strategy for enhancing the interpretability of ordinary least squares (OLS) multiple regression main and interaction effect coefficients and reducing the effects of collinearity (Aiken & West, 1991; Blanton & Jaccard, 2006; Cronbach, 1987; Pedhazur, 1997). Choices for centering predictors in MLMs have greater implications and are considerably more complex as level-1 intercepts and slopes become outcome variables at higher levels (Enders & Tofighi, 2007; Raudenbush & Bryk, 2002). In addition to the advantages outlined above for OLS, centering in multilevel models can improve estimation stability and convergence (Hox et al., 2018; Longford, 1989), clarify the interpretation of intercept and slope variances (Hox et al., 2018), facilitate the interpretation of cross-level interactions present in the model (Hofmann & Gavin, 1998; Hox et al., 2018), and facilitate variance decomposition (Rights & Sterba, 2019).

Due to the hierarchical nature of the data, analysts can choose to either grand-mean or group-mean center covariates at the lowest levels in a model (i.e., for a 2-level model, level-1 covariates can be group-mean or grand-mean centered). At level two, only grand-mean centering is appropriate, as there is no within-group variability at level two. For level-1 predictors, the method of centering to use depends on the theoretical questions of interest and/or the desired substantive interpretation for a particular covariate's estimate (Enders & Tofighi, 2007; Kreft et al. 1995; Raudenbush & Bryk,

2002). As is the case with OLS, the meaning of the intercept depends on the coding of categorical level-1 predictors or the value of 0 for a predictor.

In Equation 2.1, the intercept β_{0j} is the expected value of the dependent variable in group j when the covariate X is 0. Centering X is particularly important if X in its raw form has no inherent, meaningful 0 value. For example, if we were predicting weight using height in its raw metric, the intercept would represent the expected weight for a student with a height of 0. Likewise, if we had a categorical predictor coded as 1 = does not receive free/reduced price lunch and 2 = receives free/reduced price lunch, then a 0 value is out of the range of plausible values. In that instance, a simple recode to set 0 = does not receive free/reduced price lunch and 1 = receives free/reduced price lunch allows the intercept to represent the mean outcome for students not receiving free/reduced price lunch as the reference group. In general, researchers should think carefully about how the coding of their data impacts the interpretation of their model results, and how different centering strategies can change those interpretations. We proceed now by reviewing in a bit more detail how to implement these strategies, when they might be used, and how results can be interpreted.

Grand-Mean Centering

One option for centering level-1 predictors is to center around the sample grand mean

$$Y_{ij} = \beta_{0j} + \beta_{ij}(X_{ij} = \bar{X}_{..}) + e_{ij} \tag{2.7}$$

where the overall mean ($\bar{X}_{..}$) across all level-1 units in the sample is subtracted from each value of the X covariate. Grand-mean centering does not impact the rank order of scores in the sample, does not change the mean outcome differences among clusters, and does not alter the correlation between predictors and the dependent variable. Note, however that assuming a non-zero intra-class correlation (ICC), level-1 predictors presumably contain both within and between-cluster variation that results in grand-mean centered (or uncentered) level-1 predictors potentially being correlated with other level-1 and level-2 predictors (Enders & Tofighi, 2007, p. 125).

Under grand-mean centering, the intercept for cluster j represents the expected value of the dependent variable for a level-1 unit i whose value on X is equal to the overall sample mean (i.e., an adjusted mean for group j). Correspondingly, because the intercept for group j is an adjusted mean, the intercept variance estimate quantifies the variation in these adjusted means after partialling out the effects of level-1 predictors (i.e., the expected variances for "average" subjects as defined by the predictors included in the model). Because the grand-mean centered predictors are a weighted combination of both within and between-cluster variance a clear interpretation of the slope estimate is not possible (Raudenbush & Bryk, 2002, p. 139). Similar complexities arise when models include random slopes since the magnitude of the adjustment to cluster means is dependent upon that cluster's slope coefficient(s). The shrinkage associated with empirical Bayes (EB; Raudenbush & Bryk, 2002, p. 157) intercept and slope estimates causes the dependency between intercepts and slopes to synthetically reduce slope variation, yielding negatively biased slope variance estimates (Enders & Tofighi, 2007, p. 126).

Researchers interested in exploring the influence of level-2 predictors after controlling for potential level-1 confounders would most likely want to employ grand-mean centering at level 1 because grand-mean centered level-1 predictors remain a composite of within and between-group variation and, as such, are correlated with level-2 predictors. Thus, level-2 coefficients from a model with grand-mean centered level-1 predictors reflect partial associations with the dependent variable controlling for the grand-mean centered level-1 predictors. As an example, assume a researcher conducts a cluster randomized controlled trial (C-RCT; Lohr et al., 2014) and randomly assigns schools to receive an intervention designed to improve reading achievement. In this context, the primary research question might be to examine whether students attending schools receiving the treatment performed differently than students attending schools that did not receive the treatment. A level-2 dummy variable reflecting school membership in the treatment group (or not) would be of primary interest, and the researcher might wish to control for the influence of student socioeconomic status (SES). In this scenario, the researcher would likely want to grand-mean center the level-1 SES predictor to obtain an estimate of the treatment effect controlling for student characteristics.

Group-Mean Centering

A second method is to center level-1 predictors around their group (i.e., level-2) means:

$$Y_{ij} = \beta_{0j} + \beta_{1j}(X_{ij} - \bar{X}_{.j}) + r_{ij} \tag{2.8}$$

where the cluster mean across all level-1 units in group j is subtracted from each value of the X covariate within that group. Unlike grand-mean centering, where relative standing on the variable is preserved, group-mean centering often impacts the rank order of scores in a sample, because each value of a level-1 predictor is now expressed relative to other cases in the group j to which it belongs (Hox et al., 2018). In addition, the means for every group on the group-mean centered predictor are now all equal to 0 (i.e., there are no mean differences among groups) because group-mean centering removes all between-group variation from the level-1 predictor.

Because between-group variation has been removed, group-mean centered level-1 predictors remain correlated with other level-1 predictors but are uncorrelated with any level-2 predictors (Bell et al., 2018; Enders & Tofighi, 2007). Correlations among group-mean centered level-1 predictors can differ compared to those obtained under grand-mean centering, particularly if a predictor is a blend of lower and group-level associations and the group-level influence is stronger than the lower-level influence. For example, the correlation between a student's prior test score and their socioeconomic status (SES) may change if SES is group-mean centered (Castellano et al., 2014; Ma & Klinger, 2000). If there is a relatively large amount of between-school mean SES variance, the removal of between-school variance from student SES can influence correlations with other level-1 predictors.

Group-mean centering now yields an overall intercept representing the average, unadjusted group mean when X is equal to its own group j's sample mean. Similarly, estimates of the intercept variance reflect the between-group variation in the outcome, or the variability of unadjusted group means; this estimate should be nearly equivalent to an estimate obtained from an unconditional model. Group-mean centered slope coefficients reflect the collective within-group relationship between the predictor and the dependent variable as all between-group variance has been removed from the predictor. Removal of the between-group variance yields estimates of slope variation that are less biased than their grand-mean centered counterparts due to the reduced influence of X on the intercept variance estimate (i.e., EB shrinkage), in turn reducing the impact of the intercept-slope dependency during estimation (Enders & Tofighi, 2007; Raudenbush & Bryk, 2002).

For the reasons outlined above, researchers most interested in level-1 predictor relationships should consider group-mean centering. Group-mean centered estimates reflect a within-group estimate devoid of between-group variation and yield a more accurate estimate of slope variances. For instance, a researcher might be interested to know whether students from lower SES families have different levels of academic performance relative to students from other SES levels within the same school. Given a sample of students from many different schools, the researcher would be most interested in obtaining an estimate of the level-1 SES predictor's relationship to achievement absent the impact of between-school differences.

A researcher may also be interested in whether the influence of a predictor(s) is the same at multiple levels. Contextual models are specified by entering individual values of a level-1 predictor *and* an aggregate of that level-1 variable at level 2 (Raudenbush & Bryk, 2002). Returning to our interest in SES as a predictor of student achievement, suppose a researcher has access to a student-level SES variable and standardized assessment scores nested within the school students attend. By including the mean SES calculated across all the students in a school as a predictor at level two, a researcher can learn both how a student's individual SES impacts their achievement and how the concentration of poverty/affluence at a school impacts that student's achievement (see Raudenbush & Bryk, 2002, p. 141). This modeling strategy decomposes the predictor into its within and between-level components as

$$Y_{ij} = \beta_{0j} + \beta_{1j}(X_{ij} - \bar{X}_{.j}) + \beta_2 \bar{X}_{.j} + r_{ij} \tag{2.9}$$

where the mean across all level-1 units within cluster $j(\bar{X}_{.j})$ is now included as a predictor at level 2. This yields two orthogonal coefficients β_{1j} and β_2, which could be compared for differences via custom contrasts. Technically, this type of analysis can also be accomplished using grand-mean centering (see Enders & Tofighi, 2007, p. 131 or Kreft et al., 1995 for details). Since grand-mean centering yields a level-1 predictor comprised of both within and between components, group-mean centering is the best strategy for investigating cross-level interactions as it removes the between-group variation in the level-1 predictor and provides a pure estimate of the moderating influence of the level-2 predictor (Hofmann & Gavin, 1998; Raudenbush & Bryk, 1989a, 1989b). Group-mean centering should also be used to obtain pure level-1 slope estimates when interest is focused on the interaction between two level-1 predictors. However, researchers should carefully consider their research questions, the construct being defined through aggregation, the sample sizes at each level and the sampling ratio of their hierarchical data structure, as these facets impact the expected reliability of the aggregate measures, which in turn influences bias, variability, and coverage in parameter estimates (Ludtke et al., 2008; Preacher et al., 2010).

Centering Dummy Variables

Dummy variables represent unit membership in groups, as is the case with commonly used biological sex/gender where a value of 1 might represent females and 0 represents non female (i.e., male). Though it seems odd to consider centering a binary variable, doing so has the same influence on the interpretation of parameters as when the predictor is continuous. Uncentered, an intercept at level 1 would represent the male outcome mean in group *j*, when gender is equal to 0, and the intercept variance represents variation in male outcome averages across the *j* level-2 units. If an uncentered gender dummy code were entered into a model along with a grand-mean centered student SES variable to predict student achievement, the level-1 intercept would be the male mean outcome adjusted for group (e.g., school) differences in student SES and the intercept variance would represent variation in adjusted achievement means *for males.*

However, grand-mean centering the gender variable (around the overall proportion of females) yields a level-1 intercept that represents the expected achievement for a student after adjusting schools on their proportion of female students (i.e., if the proportion of females were the same in each school). The overall model intercept (γ_{00}) represents the average mean outcome adjusted for the concentration of females in schools. Likewise, group-mean centering the gender variable involves subtracting the proportion of females in group *j* from each person in group *j*'s value on the variable. Now, the level-1 intercept is the predicted achievement for a student with a gender value equal to the school proportion of females (i.e., the unadjusted mean achievement for group *j*). A similar strategy could be used with a three-group categorical variable such as student race. Typically, when representing a variable with *k* groups using dummy-coded variables, $k - 1$ dummies are used. For example, assume we have White, African American, and Hispanic students represented in a data file. Using White students as the reference group, a dummy representing the African American and Hispanic students could each be centered (grand- or group-mean) on their respective proportions, yielding an intercept with an adjusted interpretation as above for the example using gender. Remember, group-mean centering is ideal when a researcher is interested in the pure relationship (and heterogeneity) between a level-1 predictor and an outcome because the strategy will remove between-group variation (e.g., the between-school variation in the concentration of females; Enders & Tofighi, 2007; Raudenbush & Bryk, 2002). In other words, group-mean centering gender suggests a student's relative position within their school is of importance (i.e., Does it matter if a student is female when attending a school that is comprised of 75% males?).

Degrees of Freedom

Degrees of freedom provide an approximate sample size based on the number of individuals and predictors at level 1 and the number of groups and predictors at level 2. Degrees of freedom are necessary for determining the statistical significance of the regression parameter estimates at each level, which are based on *t*-distributions. Some statistical software does not let users select the DF used when estimating MLMs. In those instances, users need to familiarize themselves with the default DF method employed and decide if that approach is acceptable given the nature of their sample sizes at level 1 and level 2 as well as model complexity. If it is not, researchers should consider using a software program that allows users to specify different DF approaches or familiarize themselves with options for adjusting degrees of freedom after the models have been estimated.

Because our applied example in this chapter uses SAS, in this section, we review the DF methods available when estimating organizational MLMs via PROC MIXED in SAS. Specifically, SAS provides four degrees of freedom methods to use within the PROC MIXED statement when estimating organizational multilevel models: residual, containment, Satterthwaite, and Kenward-Roger. The residual, Satterthwaite, and Kenward-Roger degrees of freedom methods are also available in Stata whereas the Satterthwaite and Kenward-Roger degrees of freedom methods are available in R through the lmerTest and pbkrtest packages, respectively. Both the Satterthwaite and Kenward-Roger DF methods are well-suited for most multilevel models as they both are intended to be used in models with unbalanced designs and both can handle complex covariance structures. Moreover, the Kenward-Roger DF method also adjusts for small sample bias, so it is ideal to use when estimating models with relatively small samples. Below is a brief summary of each of the four DF methods available in PROC MIXED that are applicable to estimating organizational models.

The residual degrees of freedom estimation method produces correct DFs only when the level-1 errors are independent and identically distributed and when there are no level-2 errors (i.e., a situation in which multilevel modeling is not needed). Given this information, there is no reason to ever use this DF method when using PROC MIXED for estimating MLMs. The containment degrees of freedom estimation method is the default method in PROC MIXED when a random statement is used. This option can lead to accurate degrees of freedom when the design is balanced and when the level-1 errors are independent and identically distributed. This choice becomes more questionable, however, as the design becomes less balanced or when a complex level-1 error structure is needed. Thus, we do not recommend this method given the likelihood that most models do not have balanced sample sizes across each level-2 unit.

The Satterthwaite method approximates the degrees of freedom and is designed for use with unbalanced designs and more complex covariance structures. Given that most data are unbalanced, this method is most applicable for researchers working with hierarchical data if the level-2 sample size is at least 30. However, the Type I error rates tend be conservative when there are less than 30 clusters and the cluster sizes are more unbalanced (Li & Redden, 2015). Thus, using the Satterthwaite DF method with less than 30 clusters can negatively impact statistical power. The Kenward-Roger method also approximates the degrees of freedom and again is designed for use with unbalanced designs and complex covariance structures. By adjusting the covariance matrix, this method also adjusts for small-sample bias. Previous research indicates the Kenward-Roger DF method yields unbiased parameter and standard error estimates when used with small sample sizes; therefore, we recommend this method be used for models with less than 30 level-2 units (Bell et al., 2014). When deciding between these two DF methods, there is one consideration: the Satterthwaite DF approach can be used with ML or REML whereas the Kenward-Roger DF should only be used in models estimated with REML (Li & Redden, 2015; Luke, 2017). Given that REML is the preferred estimator when the number of clusters is small, using Kenward Roger DF with REML is recommended when the number of clusters is less than 30.

APPLIED TWO-LEVEL ORGANIZATIONAL MODEL USING ECLS-K: 2011 DATA

To illustrate how to estimate a two-level linear multilevel model, we use the ECLS-K: 2011 (Tourangeau et al., 2015) public use data to examine the relationship between student race, school-level racial composition, and externalizing behavior among children enrolled in kindergarten classes during the 2011–2012 school year. This example is informed by both the continued existence of racial inequalities in a multitude of child outcomes as well as by Bronfenbrenner's (1979) ecological systems theory. Specifically, a core tenant of Bronfenbrenner's theory is that human development is influenced by the interrelations among settings in which a person actively participates (e.g., family, school, neighborhoods, religious institutions); thus, to study human development effectively, we need to look beyond a single environment and analyze the interactions among multiple environments. In examining the following research question: "Among kindergarten students, does the relationship between student race and externalizing behavior differ by school-level racial composition?"; we are looking at the interaction between individual-level and microsystem characteristics (Bronfenbrenner, 1979).

For our example, we created a reduced dataset from the larger ECLS-K master data file. A snippet of the dataset used in this section is shown in Figure 2.2, and Table 2.1 contains details on the variables included in our examples. As shown in Figure 2.2, the datafile is in wide format. In addition to the variables used in our predictive model and described in Table 2.1, the dataset also contains two ID variables—a student ID (child identification number) and school ID (Spring 2011 school identification number).

As outlined earlier, there are several decisions that need to be made before estimating the MLMs. Given the large level-2 sample size, we decided to estimate our models using FIML. REML is the preferred method when the level-2 sample size is not very large—in our example there are over 1,000 schools so there would not be substantive differences between the two estimation methods. In an effort to keep the example relatively simple for illustrative purposes, we included one random slope, allowing the slope for student SES to vary across clusters. Of the three level-1 slopes, the relationship

	Child Identification Number	Spring 2011 School Identification Number	X2 Teacher Report Extern Prob Behaviors	X12 Continuous SES Measure	X2 Percent Non-White Students in School	Male	Non-White	Private
1	10001062	1002	1	−0.07	34	0	0	0
2	10001485	1002	1	0.65	34	0	0	0
3	10001694	1002	2	0.9	34	0	0	0
4	10003310	1002	2	0	34	1	0	0
5	10004876	1002	1.4	0.7	34	1	0	0
6	10005561	1002	1	−0.76	34	1	1	0
7	10008202	1002	1.6	−0.62	34	1	1	0
8	10010154	1002	1	−0.29	34	0	0	0

Figure 2.2 Snippet of data file used for applied example of estimating two-level organizational linear models.

TABLE 2.1 Information on ECLS-K: 2011 Variables Included in Chapter Examples

Variable Name	Details
Student ID	Administrative variable provided in ECLS-K: 2011 data
School ID	Administrative variable provided in ECLS-K: 2011 data
Externalizing Behavior	Variable measuring teachers' reports on students externalizing behaviors. Values represent the mean score for the 6 items in this scale
SES	Family socioeconomic status variable provided in ECLS-K: 2011 data. Measured as a z-score
Male	Derived variable measuring self-reported sex, (0 = female, 1 = male)
Non_White	Derived variable measuring self-reported race, (0 = White, 1 = non_White)
Private	Derived variable measuring type of school student attended, (0 = public school, 1 = private school)
School_Non_White	ECLS-K: 2011 provided variable that represents the percent of non-White students in each school in the ECLS-K:2011 sample. For our analyses, we grand-mean centered this variable at 46.

between student SES and externalizing behavior is the one that is most likely to exhibit variation across schools; in some schools the relationship could be positive, whereas in others it could be negative. We decided to use the unstructured matrix as our selected variance/covariance matrix to provide readers with a detailed example using the more complex covariance structure. Once you understand how to interpret results from an unstructured matrix, you know how to interpret results from models estimated with a VC matrix. For centering decisions, the continuous level-1 predictor SES is in z-score form, and as such has a natural 0; thus we did not use centering.

We used grand-mean centering for the level-2 continuous predictor that was part of the primary interest for our research question (school racial composition). Lastly, we used the Satterthwaite degrees of freedom because when estimating MLMs via PROC MIXED this is the best DF option given its ability to handle unbalanced samples and complex covariance structures when the level-2 sample size is exceeds 30.

Traditionally, multilevel modeling textbooks have focused on model building approaches when introducing readers to the world of MLMs. Although different sources provide different guidelines on the model building process when estimating multilevel models, they all have the same goal—for researchers to estimate the most parsimonious models that best fit their data. Other common elements of the different model building strategies include (a) always starting with an unconditional model (i.e., a model that has no predictors) as this model is used to calculate the ICC which estimates how much variation in the outcome exists between level-2 units; and (b) gradually estimating more complex models while checking for improvement in model fit after each model is estimated. Until recently, this sequential process was necessary given the tools available for examining how different elements of the model fit, including how variance was explained by each part of the models (i.e., variance explained by fixed effects and random effects). However, Rights and Sterba (2019) developed an approach for calculating explained variance at different levels of a model, without relying on the traditional model building process. Rather than build models step by step in our example that follows, we estimate the unconditional model and then the full analytic model needed to answer our research question. We cover basic model interpretation through our demonstration, and remind readers that Chapter 3 expands on the model presented in this chapter and discusses how to examine explained variance as well as to determine the best fitting variance/covariance structure.

Estimated Models

Unconditional Model

With multilevel models, the first model typically estimated is the unconditional model with no predictors. This allows calculation of the ICC and assessment of between-cluster variation in the dependent variable. Thus, for this example, output from the unconditional model is used to calculate the ICC, which indicates how much variability in kindergartners' externalizing behavior varies across schools. Because the unconditional model has no predictors, the equations for the unconditional model in our applied example (Equation 2.10) looks like the unconditional model presented earlier in the chapter, with the exception that now we are using variable names and not generic equation notion; we use EB (for externalizing behavior) rather than a generic Y as the outcome variable.

$$\begin{aligned} &\text{Level 1:} && EB_{ij} = \beta_{0j} + e_{ij} \\ &\text{Level 2:} && \beta_{0j} = \gamma_{00} + u_{0j} \\ &\text{Combined:} && EB_{ij} = \gamma_{00} + u_{0j} + e_{ij} \end{aligned} \tag{2.10}$$

At level 1, EB_{ij} is the externalizing behavior for student i in school j and β_{0j} is the average externalizing behavior for school j, e_{ij} is the student-level error term. At level 2, γ_{00} represents the expected value on externalizing behavior (EB) in the population and u_{0j} is an error term representing a unique effect associated with school j. In the combined model, EB for person i in school j is predicted by overall EB (γ_{00}), school j's deviation from overall EB (u_{0j}) and student i's deviation from their school's predicted EB (e_{ij}; Raudenbush & Bryk, 2002). Using the output from the unconditional model (see Table 2.2), we compute the ICC, which indicates how much of the total variation in externalizing behavior is accounted for by the schools.

Equation 2.11 is the formula for calculating the ICC for a two-level linear model with students nested in schools. In this equation, τ_{00} refers to the variance estimate for the randomly varying intercepts (i.e., level-2 errors) and σ^2 refers to the variance estimate for the pooled within cluster residuals (i.e., level-1 errors).

$$ICC = \frac{\tau_{00}}{\sigma^2 + \tau_{00}}. \tag{2.11}$$

Based on the results in Table 2.2, the ICC is .03075/(.3732 + .03075) = .0761. This indicates that 7.61% of the variability in kindergartner's externalizing behavior is accounted for by the schools in the study, 92.39% of the variability in externalizing behavior is between students within schools.

Random-Slopes and -Intercepts Model

Next, we estimate the MLM needed to answer our research question. For this model, we include student-level (i.e., level-1) predictors (male, non_White, ses), school-level (i.e., level-2) predictors (private, school_non_White), and the cross-level interaction (non_White*school_non_White) as fixed effects and include a random slope for student SES (ses). Although the interaction between student race and school-level racial composition is the focus of our research question in this example, we wish to control for SES, gender, and school type, as would typically be the case in applied research studies. Equation 2.12 shows the level 1, level 2, and combined equations for the model estimated to answer our research question.

TABLE 2.2 Estimates From Two-level Organizational Linear Models Predicting Externalizing Behavior Among Using ECLS-K: 2011

	Unconditional Model			Cross-Level Interaction Model		
Fixed Effects	**Estimate**	**S.E.**	***p*-value**	**Estimate**	**S.E.**	***p*-value**
Intercept (γ_{00})	1.635	0.008	<.0001	1.513	0.013	<.0001
Male (γ_{10})				0.257	0.010	<.0001
Non-White (γ_{20})				–0.027	0.013	.0472
SES (γ_{30})				–0.090	0.008	<.0001
Private School (γ_{01})				0.080	0.021	.0002
Percent Non-White Students (γ_{02})				0.001	0.0004	<.0001
Non-White*Percent Non-White Students (γ_{22})				0.001	0.0004	.0146
Error Variance						
Level-1 (σ^2)	0.373	0.005	<.0001	0.352	0.004	<.0001
Level-2 Intercept (τ_{00})	0.031	0.003	<.0001	0.025	0.002	<.0001
SES (τ_{11})				0.005	0.004	.0076
Cov (τ_{00}, τ_{11})				–0.004	0.001	.0047

Note: Values based on SAS PROC MIXED. Estimation Method = ML; Satterthwaite degrees of freedom.

$$
\begin{aligned}
\text{Level 1:}\quad & EB_{ij} = \beta_{0j} + \beta_1 Male_{ij} + \beta_2 Non\_White_{ij} + \beta_3 SES_{ij} + e_{ij} \\
\text{Level 2:}\quad & \beta_{0j} = \gamma_{00} + \gamma_{01} Private_j + \gamma_{02} School\_Non\_White_j + u_{0j} \\
& \beta_{1j} = \gamma_{10} \\
& \beta_{2j} = \gamma_{20} + \gamma_{22} School\_Non\_White_j \\
& \beta_{3j} = \gamma_{30} + u_{1j}
\end{aligned}
\tag{2.12}
$$

Combined:

$$
\begin{aligned}
EB_{ij} = {} & \gamma_{00} + \gamma_{10} Male_{ij} + \gamma_{20} Non\_White_{ij} + \gamma_{30} SES_{ij} + \gamma_{01} Private_j \\
& + \gamma_{02} School\_Non\_White_j + \gamma_{22} School\_Non\_White_j Non\_White_{ij} \\
& + u_{0j} + u_{1j} SES_{ij} + e_{ij}
\end{aligned}
$$

The combined equation shows the (a) model intercept (γ_{00}), which represents the overall mean of externalizing behavior across students and across schools; (b) main effect fixed effects for student-level variables (γ_{10}, γ_{20}, and γ_{30}) which represent the overall or weighted-average slopes for each of student-level predictors ($Male_{ij}$, Non_White_{ij}, and SES_{ij}, respectively) across all schools; (c) main effect fixed effects (γ_{01} and γ_{02}) for the school-level variables

(*Private*$_j$ and *School_Non_White*$_j$) which represent the effects of these school characteristics on the distribution of outcomes within school j; (d) the cross-level interaction between the school percent of non-White students and student race ($\gamma_{22} School\_Non\_White_j Non\_White_{ij}$); and (e) the model random effects for the intercept, u_{0j} and the SES slope, $u_{1j} SES_{ij}$.

Interpreting Results. The regression coefficients are interpreted like OLS model results. For example, using the output provided in Table 2.2, after controlling for other variables in the model, males are predicted to have higher externalizing behavior scores than females ($b = 0.257$, $p < .0001$) whereas non-White students ($b = -0.027$, $p = .0472$) and those with higher SES ($b = -0.090$), $p < .0001$) are predicted to have slightly lower externalizing behavior score; given the coding of the level two variables, these estimates are for public schools with average proportions of non-White students. In addition, the relationship between student SES and externalizing behaviors varies across schools ($\tau_{11} = 0.005$, $p = .0076$). For our research question, the focal part of the output is the regression coefficient for the interaction between student race and school racial composition. These results suggest that the relationship between student race and externalizing behavior does differ based on school racial composition ($b = -0.001$, $p = .0146$), albeit the nature of the difference appears small.

As shown in Figure 2.3 (which is an interaction plot generated from the model results shown in Table 2.2), non-White students' externalizing behaviors do not substantively change based on school racial composition. This suggests that non-White students who attend schools where they are the minority (i.e., schools with lower percentages of non-White students)

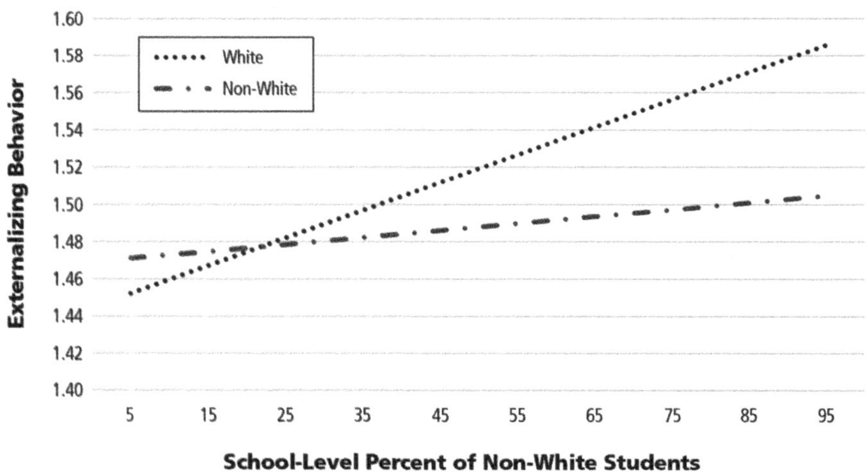

Figure 2.3 Kindergartners externalizing behavior as a function of child race and school-level racial composition.

exhibit relatively similar externalizing behaviors as non-White students who attend schools where they are the majority (i.e., schools with greater percentages of non-White students). Conversely, it appears that externalizing behaviors among White students does differ based on a school's racial composition. White students are predicted to have fewer externalizing behaviors when they attend schools where the percent of non-White students is low (i.e., schools where White students are the majority); however, their externalizing behaviors are higher when they attend schools with higher percentages of non-White students (i.e., schools where the White students are the minority).

Other Things to Consider When Estimating MLMs

As with single-level regression models, if the interaction examined in our example were not statistically significant, we would focus on the main effects for student race and school racial composition. Although there is not universal agreement across fields, in such a scenario, a researcher might consider removing the interaction and running a main effect model and interpreting those results. Also, although we based our interpretation of the random effects solely on the results of the default hypothesis test provided in the PROC MIXED output, this is not the best way to determine the statistical significance of variance and covariance parameters. As discussed in Chapter 3, best practice for assessing the significance of variance components is to compare the model fit of a model that includes the random effect to a model that does not.

Another important part of estimating multilevel models is to examine the distribution of lower level units across higher level units. In our example, that means that we need to examine the distribution of kindergarten students (level-1 units) across schools (level-2 units). This is often referred to as "level-2 density." In the ECLS-K: 2011 data used in our example, there are 14,176 students nested in 1029 schools—this yields an average of 14 students per school. Looking beyond the average, we also observed that the number of students per school ranges from 1 to 25 and that two-thirds of schools have 17 or less students. Syntax for generating this information is included in the SAS syntax file used to estimate the example two-level linear models provided in the online appendix available at http://modeling.uconn.edu.

EXAMINING DISTRIBUTIONAL ASSUMPTIONS

As with any other statistical procedures, it is important to examine and report the appropriate assumptions when using multilevel modeling techniques: There are certain distributional assumptions underlying the validity of the Type I error control. Therefore, the validity of inferences based on models estimated through these techniques depends on the degree to which assumptions are upheld about the structural and random parts of the model (Raudenbush & Bryk, 2002). Some assumptions associated with two-level linear models are similar to OLS model assumptions. These include residual normality, independence, and homoscedasticity. With two-level linear models, these assumptions need to be examined at both level 1 and level 2. In particular, the assumption of normally distributed errors applies to both level-1 and level-2 residuals, with violations adversely affecting level-2 estimated standard errors and inferential statistics (Raudenbush & Bryk, 2002). Violations at level 1 can also yield distorted random effect coefficients and variance-covariance components. In addition to examining these assumptions, detection of influential observations is also important when conducting multilevel analyses. Littell et al. (2006) recommend a "drill-down" approach to mixed model influence diagnostics, beginning with a global assessment of the influence on the overall model, followed by a more detailed exploration of the case-sets should they be warranted.

Different statistical software packages allow users to examine different elements of the distributional assumptions (see O'Connell et al., 2016 for an overview of several software packages). Researchers estimating MLMs need to review the software user's manual for details on the options available. SAS users can take advantage of the INFLUENCE and RESIDUAL options of the MODEL statement to generate a list of influence diagnostic values as well as visual plots depicting both influence and residual values. However, the amount of information generated by these options makes it difficult to actually examine if violations are occurring. Alternatively, Bell et al. (2010) developed the MIXED_DX macro to provide a comprehensive approach for conducting two-level linear model diagnostics, including examinations of residual normality, linearity, homogeneity of variance, and influential outliers. The macro produces both visual output (e.g., box-and-whisker plots, histograms, scatter plots) and summary tables of statistical output for both level-1 and level-2 residuals. In addition, MIXED_DX facilitates detection of influential observations in two-level linear models by creating a ranked summary table of the influence statistics automatically created in the SAS influence table. We encourage users to read more about assumptions and how to test them in the MIXED_DX paper, Bell et al. (2010). Likewise, Palmeri (n.d.) provides example code for examining model assumptions using R and Rabe-Hesketh and Skrondal (2022) provide information for examining assumptions when estimating MLMs in Stata. We provide the MIXED_DX macro that corresponds to the model estimated in our example in the online supplemental materials located at http://modeling.uconn.edu.

EXTENDING THE TWO-LEVEL ORGANIZATIONAL MODEL TO A THREE-LEVEL ORGANIZATIONAL MODEL

MLMs can easily be expanded to accommodate additional levels of hierarchy when data are available. In education, it is not uncommon for researchers to encounter hierarchical data at three levels—students (level 1) nested in classrooms/teachers (level 2) nested in schools (level 3). In health fields, patients are nested in wards/clinics, which are nested in hospitals. Although the expansion to 3-level models is relatively straightforward in terms of the fixed effects (i.e., the 3-level model can contain predictors, as both main effects and interactions, at each level), specification of the random effects is more complex.

For example, with a 2-level model, the intercept can vary across level-2 units and the slopes of level-1 predictors can also vary across level-2 units. In a 3-level model, the intercept can vary across level-2 units *and* across level-3 units, the slope of level-1 predictors can vary across level-2 units *and* level-3 units, and the slopes of level-2 predictors can vary across level-3 units. In addition, there are also multiple ICC values to calculate—one estimates the proportion of variance that exists between level-2 units (within level-3 units) and the other estimates the proportion of variance that exists between level-3 units. To illustrate these additional variance components and ICCs, in this portion of the chapter we walk through equations for an unconditional 3-level linear model, ICCs for a 3-level linear model, and equations for a 3-level random slopes and intercepts model.

3-Level Unconditional Model and Intraclass Correlation Coefficients

As in the 2-level linear MLM example, we first estimate the unconditional 3-level linear model with no predictors and random intercepts at level 2 and level 3. This allows us to estimate the variance in the continuous outcome attributable to level-2 units and level-3 units. First, we present how the data used in our 2-level linear model example could be extended to be a 3-level MLM with students (level 1) nested in classrooms (level 2) and schools (level 3). Specifically, the research question guiding our 3-level model in this section is: “What is the relative influence of students’ race, teachers’ race, and school racial composition on kindergarteners’ externalizing behaviors?” To answer this research question, we keep the same student-level and school-level fixed and random effects and we add teacher race (as a classroom-level predictor) as both a fixed and random effect.

3-Level Unconditional Model

Details for the unconditional 3-level model are illustrated in Equations 2.13–2.16. First, we model externalizing behavior for each student as a function of a classroom mean plus random error:

Level 1 (student-level model):

$$EB_{ijk} = \pi_{0jk} + e_{ijk} \tag{2.13}$$

where EB_{ijk} is externalizing behavior of child *i* in classroom *j* and school *k*. π_{0jk} is the mean externalizing behavior of classroom *j* in school *k*, and e_{ijk} is a random student effect (i.e., the deviation of child *ijk*'s score from the classroom mean). The random student effects are assumed to be normally distributed with a mean of 0 and variance σ^2 (Raudenbush & Bryk, 2002).

Next, we model each classroom mean, π_{0jk}, as an outcome that varies randomly around a school mean. This is depicted in Equation 2.14 where β_{00k} represents the mean externalizing behavior in school *k* and r_{0jk} is a random classroom effect (i.e., the deviation between classroom *jk*'s mean externalizing behavior from the school average externalizing behavior). These effects are assumed normally distributed with a mean of 0 and variance τ_π (Raudenbush & Bryk, 2002).

Level 2:

$$\pi_{0jk} = \beta_{00k} + r_{0jk} \tag{2.14}$$

The level-3 model in Equation 2.15 captures the variability of externalizing behavior among schools, where school means (β_{00k}) vary randomly around the grand mean (γ_{000}). Here, u_{00k} is the random school effect which represents the deviation of school *k*'s mean from the grand mean. These school effects are assumed normally distributed with a mean of 0 and variance τ_β (Raudenbush & Bryk, 2002).

Level 3:

$$\beta_{00k} = \gamma_{000} + u_{00k} \tag{2.15}$$

Equation 2.16 shows the combined 3-level model equation where EB_{ijk} is the externalizing behavior for student *i* in classroom *j* in school *k*, γ_{000} is the expected externalizing behavior across all schools, u_{00k} is the deviation of school *k* from γ_{000}, r_{0jk} is the deviation of classroom *j* within school *k* from school *k*'s expected EB, and e_{ijk} is the residual error term (aka, the random student effect). The random effects at each level and residual errors are assumed independent of one another and normally distributed with 0 means and constant variances (Raudenbush & Bryk, 2002).

Combined:

$$EB_{ijk} = \gamma_{000} + u_{00k} + r_{0jk} + e_{ijk} \tag{2.16}$$

3-Level Intraclass Correlation Coefficients

After estimating the unconditional 3-level model, researchers can calculate the ICCs. The formula for calculating classroom/teacher and school ICCs is similar to the formula used for two-level ICC calculations, however, the denominator now has three elements which form the total variance in the model. As shown in Equations 2.17 and 2.18, the three sources of variability are level 1 (σ^2) which represents the variability in externalizing behaviors among students within classrooms, level 2 (τ_π) which represents the variability in externalizing behaviors among classrooms within schools, and level 3 (τ_β) which represents variability in externalizing behaviors among schools.

$$\text{ICC}_{classroom} = \frac{\tau_\pi}{\tau_\pi + \tau_\beta + \sigma^2} \tag{2.17}$$

$$\text{ICC}_{school} = \frac{\tau_\beta}{\tau_\pi + \tau_\beta + \sigma^2} \tag{2.18}$$

In addition to being able to calculate the different ICC values, it is equally important to revisit what they represent. The $\text{ICC}_{\text{Classroom}}$ expresses the similarity (correlation between) of classrooms within the same school; it indicates the proportion of total variation in externalizing behaviors that exists between classrooms within schools. The $\text{ICC}_{\text{School}}$ indicates the proportion of variance in externalizing behaviors that exists between schools (Hox et al., 2018).

3-Level Random-Slopes and -Intercepts Model

The statistical model that corresponds to the research question, "What is the relative influence of student race, teacher race, and school racial composition on kindergarteners' externalizing behaviors?" is presented in Equation 2.19 and is then reduced to the combined mixed-model equation (Equation 2.20).

Level 1:

$$EB_{ijk} = \pi_{0jk} + \pi_{1jk}SES_{ijk} + \pi_{2jk}Non\_White_{ijk} + \pi_{3jk}Male_{ijk} + e_{ijk} \tag{2.19}$$

Level 2:

$$\pi_{0jk} = \beta_{00k} + \beta_{01k}TeacherRace\_Non\_White_{jk} + r_{0jk}$$
$$\pi_{1jk} = \beta_{10k} + r_{1jk}$$
$$\pi_{2jk} = \beta_{20k}$$
$$\pi_{3jk} = \beta_{30k}$$

Level 3:

$$\beta_{00k} = \gamma_{000} + \gamma_{001}Private_k + \gamma_{002}School\_Non\_White_k + u_{00k}$$
$$\beta_{01k} = \gamma_{010} + u_{01k}$$
$$\beta_{10k} = \gamma_{100} + u_{10k}$$
$$\beta_{20k} = \gamma_{200}$$
$$\beta_{30k} = \gamma_{300}$$

Combined:

$$\begin{aligned} EB_{ijk} = {} & \gamma_{000} + \gamma_{001}Private_k + \gamma_{002}Scholl\_Non\_White_k \\ & + \gamma_{010}TeacherRace\_Non\_White_{jk} + \gamma_{100}SES + \gamma_{200}Non\_White \\ & + \gamma_{300}Male + u_{00k} + u_{01k}TeacherRace\_Non\_White + u_{10k}SES \\ & + r_{0jk} + r_{1jk}SES + e_{ijk} \end{aligned} \tag{2.20}$$

Both Equations 2.19 and 2.20 predict externalizing behavior (EB) of the *i*th kindergartner in the *j*th classroom in the *k*th school as a function of three level-1 predictors (SES, child race [Non_White] and sex (Male)], one level-2 predictor [Teacher Race (TeacherRace_Non_White)], and two level-3 predictors [School Type (Private) and School Racial Composition (School_Non_White)]. In addition, the level-1 and level-2 intercepts and the slopes for student SES are allowed to randomly vary across classes and schools. The slope for teacher race is allowed to randomly vary across schools. More specifically, the model shown in Equations 2.19 and 2.20 contains seven fixed effects (γ_{000}, γ_{001}, γ_{002}, γ_{010}, γ_{100}, γ_{200}, and γ_{300}) and six random effects (u_{00k}, u_{01k}, u_{10k}, r_{0jk}, r_{1jk}, and e_{ijk}). Thus, using VC structure, this relatively simple model contains 13 parameters. However, if we fit the model with an unstructured variance/covariance structure at levels two and three, then the model contains 19 parameters (seven fixed effects, six variance/covariance components at level 2, and six variance/covariance components at level 3). Three-level models can quickly become very complex and challenging to estimate particularly with many random effects. Nonetheless, this brief overview shows that multilevel models can be easily extended beyond two levels. For more details on estimating 3-level linear MLMs, we recommend Chapter 7 in Raudenbush and Bryk (2002).

CHAPTER SUMMARY

In this chapter we provide an introduction to multilevel models for organizational data. We introduce the reader to basic MLM concepts including illustrating what random intercepts and slopes represent as well as how to calculate and interpret a model's ICC. We also provide information of key issues to consider when working with nested data structures—(a) estimation method, (b) inclusion of random slopes, (c) variance/covariance structure, (d) centering of predictors, and (e) degrees of freedoms. In addition, we share information on the importance of examining MLMs distributional assumption. Next, we apply the concepts presented in the chapter to examine the relationship between kindergartner's race, school racial composition, and externalizing behavior using the ECLS-K: 2011 public use data. We conclude the chapter with a brief overview of how 2-level MLMs can be expanded to 3 levels. Although introductory in nature, the conceptual knowledge provided in this chapter serves as a strong foundation for readers to engage with other types of multilevel models in this volume.

RECOMMENDED READINGS

Technical Readings

- Bell, B. A., Morgan, G. B., Schoeneberger, J. A., Kromrey, J. D., & Ferron, J. M. (2014). How low can you go? An investigation of the influence of sample size and model complexity on point and interval estimates in two-level linear models. *Methodology, 10,* 1–11. https://doi.org//10.1027/1614-2241/a000062
- Enders, C. K., & Tofighi, D. (2007). Centering predictor variables in cross-sectional multilevel models: A new look at an old issue. *Psychological Methods, 12,* 121–138. https://doi.org/10.1037/1082-989X.12.2.121

- González-Romá, V. (2017). Multilevel modeling: Research-based lessons for substantive researchers. *Annual Review of Organizational Psychology and Organizational Behavior, 4,* 183–210. https://doi.org/10.1146/annurev-orgpsych-041015-062407

Applied Readings

- Pituch, K. A., Murphy, D. L., & Tate, R. L. (2010). Three-level models for indirect effects in school- and class-randomized experiments in education. *Journal of Experimental Education, 78,* 60–95. https://doi.org/10.1080/00220970903224685
- Walsemann, K. M., Bell, B. A., & Maitra, D. (2011). The intersection of school racial composition and student race/ethnicity on adolescent depressive and somatic symptoms. *Social Science & Medicine, 72,* 1873–1883. https://doi.org/10.1016/j.socscimed.2011.03.033
- Richmond, T. K., Milliren, C., Walls, C. E., & Kawachi, I. (2014). School social capital and body mass index in the National Longitudinal Study of Adolescent Health. *Journal of School Health, 84,* 759–768. https://doi.org/10.1111/josh.12213

TRY THIS!

Our "Try This!" uses National Longitudinal Survey of Adolescent to Adult Health (Add Health; Harris, 2013) public use data to apply the skills and concepts covered in the chapter. Add Health is a longitudinal study that has followed a nationally representative sample of adolescents who were in grades 7 to 12 during the 1994-1995 academic year. Add Health data collection occurred in 132 middle and high schools across the United States and survey items cover a variety of important areas including demographics, social relations, familial and school context, and psychological and behavioral health (Harris, 2013). For this exercise we have created an analytic dataset from Wave I public use data (Add Health, n.d.). Whereas the complete Wave 1 public use data file contains 6504 adolescents, which is a subsample of the larger study, we applied listwise deletion to provide a cleaner data set for you to practice your MLM skills. The data for this exercise contain 5886 adolescents across 132 schools and is available at the online appendix located at http://modeling.uconn.edu

Sticking with Bronfenbrenner's (1979) ecological systems theory, use the Add Health data to answer the following research question: "What are the relative influences of individual, family, and school context variables on adolescent self-reported somatic symptoms?"

For this exercise, please complete the following:

1. Write the equations (level 1, level 2, and combined) for the unconditional model and the analytic model used to answer the research question.
2. Estimate the models (unconditional and analytic) making your own decisions on estimation method, random slopes, variance/covariance structure, how to center your variables, and which degrees of freedom to use.
3. Create a table that shows results from the two models, including the ICC.
4. Check model assumptions.
5. Calculate level-2 density.
6. Write a summary of methods and results for this exercise. In the methods, be sure to include information on the model estimation decisions that you made and why you made them.

Details on each of the variables are below. Use all of the predictors when estimating your model to answer the research question.

Administrative Variables

- AID—adolescent identification number
- Cluster2—school identification number

Dependent Variable

- Somatic symptoms—This is a sum of how often participants reported experiencing a range of conditions during the last 12 months. Responses ranged from never (coded 0) to everyday (coded 4). Conditions included in this variable are feeling hot, stomachache, cold sweats, weak, very sick, waking up tired, dizziness, chest pain, muscle pain, insomnia, and trouble relaxing. Higher values represent higher levels of somatic symptoms.

Level-1 Variables

- Individual
  - Male (0 = female, 1 = male)

    - Non-White (0 = White, 1 = non-White which includes Black, Native American, Asian, Hispanic, and other/mixed races)
    - Age (measured in years)
    - Adolescent feelings of school connectedness is a composite variable based on adolescents' responses to items asking about how close they feel to people at their school, if they feel part of their school, if they are happy at school, if they feel safe at school, and if teachers treat students fairly at school. Responses ranged from strongly agree (coded 1) to strongly disagree (coded 5). Higher values represent DISAGREEMENT with items so, higher values represent LOWER feelings of school connectedness.
- Family
    - Family SES is a z-score based on parental education, occupation, and income to poverty ratio. Higher values represent higher socioeconomic levels.
    - Relationship with mother is a composite variable that contains participants responses to items asking about how their mother is with them (e.g., mother is warm and loving towards you, mother encourages you to be independent, mother talks with you when you do something wrong) and how satisfied participants are with the way their mother and them communicate with each other and how satisfied they are with their relationship with their mother. Each of the items was coded where 1 = *strongly agree* and 5 = *strongly disagree*. So, HIGHER values represent a WEAKER maternal relationship.

Level-2 Variables

- School-level school connectedness—This is an aggregate measure based on participants feelings of school connectedness. So, higher values represent a school context with LOWER levels of overall connectedness.
- School-level prejudice—This is an aggregate measure based on participants response to the item "Students at your school are prejudice." Higher values represent AGREEMENT with this item thus higher values represent GREATER feelings of prejudice among students at school.

REFERENCES

Add Health (n.d.). *Add Health public use data.* https://data.cpc.unc.edu/projects/2/view#public_li

Aiken, L. S., & West, S. G. (1991). *Multiple regression: Testing and interpreting interactions.* SAGE.

Aguinis, H., Gottfredson, R. K., & Culpepper, S. A. (2013). Best practice recommendations for estimating cross-level interaction effects using multilevel modeling. *Journal of Management, 39, 6,* 1490–1528 https://doi.org/10.1177/0149206313478188

Allison, P. D. (2005). *Fixed effects regression methods for longitudinal data using SAS.* SAS Institute.

Bell, A., Jones, K., & Fairbrother, M. (2018). Understanding and misunderstanding group mean centering: A commentary on Kelley et al.'s dangerous practice. *Quality & Quantity, 52,* 2031–2036. https://doi.org/10.1007/s11135-017-0593-5

Bell, B. A., Morgan, G. B., Schoeneberger, J. A., Kromrey, J. D., & Ferron, J. M. (2014). How low can you go? An investigation of the influence of sample size and model complexity on point and interval estimates in two-level linear models. *Methodology: European Journal of Research Methods for the Behavioral and Social Sciences, 10*(1), 1–11. https://doi.org/10.1027/1614-2241/a000062

Bell, B. A., Schoeneberger, J. A., Morgan, G. B., Kromrey, J. D., & Ferron, J. M. (2010). *Fundamental diagnostics for two-level mixed models: The SAS® macro MIXED_DX.* SAS Global Forum 2010 Proceedings. http://support.sas.com/resources/papers/proceedings10/201-2010.pdf

Blanton, H., & Jaccard, J. (2006). Arbitrary metrics in psychology. *American Psychologist, 61,* 27–41. https://doi.org/10.1037/0003-066X.61.1.27

Bronfenbrenner, U. (1979). *The ecology of human development: Experiments by nature and design.* Harvard University Press.

Castellano, K. E., Rabe-Hesketh, R., & Skrondal, A. (2014). Composition, context and endogeneity in school and teacher comparisons. *Journal of Educational and Behavioral Statistics, 39*(5), 333–367. https://doi.org/10.3102/1076998614547576

Cronbach, L. (1987). Statistical tests for moderator variables: Flaws in analysis recently proposed. *Psychological Bulletin, 102,* 414–417. https://doi.org/10.1037/0033-2909.102.3.414

Diez Roux, A. V. (2002). A glossary for multilevel analysis. *Journal of Epidemiology and Community Health, 56,* 588–594. https://doi.org/10.1136/jech.56.8.588

Donner, A., & Klar, N. (2000). *Design and analysis of cluster randomization trials in health research.* Edward Arnold.

Enders, C. K., & Tofighi, D. (2007). Centering predictor variables in cross-sectional multilevel models: A new look at an old issue. *Psychological Methods, 12,* 121–138. https://doi.org/10.1037/1082-989X.12.2.121

Gardiner, J. C., Luo, Z., & Roman. L. E. (2009). Fixed effects, random effects and GEE: What are the differences? *Statistics in Medicine,* 28(2), 221–239. https://doi.org/10.1002/sim.3478

Harden, J. W., & Hilbe, J. M. (2013). *Generalized estimating equations* (3rd ed.). CRC Press.

Harris, K. M. (2013). The Add Health study: Design and accomplishments. Carolina Population Center. https://addhealth.cpc.unc.edu/wp-content/uploads/docs/user_guides/DesignPaperWave_I-IV.pdf

Hofmann, D. A., & Gavin, M. B. (1998). Centering decisions in hierarchical linear models: Implications for research in organizations. *Journal of Management, 24,* 623–641. https://doi.org/10.1016/S0149-2063(99)80077-4

Hox, J. J., Moerbeek, M., & van de Schoot, R. (2018). *Multilevel analysis: Techniques and applications* (3rd ed.). Routledge.

Julian, M. (2001). The consequences of ignoring multilevel data structures in nonhierarchical covariance modeling. *Structural Equation Modeling, 8*(3), 325–352. https://www.tandfonline.com/doi/abs/10.1207/S15328007SEM0803_1

Kreft, I. G. G., de Leeuw, J., & Aiken, L. S. (1995). The effect of different forms of centering in hierarchical linear models. *Multivariate Behavioral Research, 30,* 1–21. https://doi.org/10.1207/s15327906mbr3001_1

Li, P., & Redden, D. T. (2015). Comparing denominator degrees of freedom approximations for the generalized linear mixed model in analyzing binary outcome in small sample cluster-randomized trials. *BMC Medical Research Methodology, 15,* Article 38. https://doi.org/10.1186/s12874-015-0026-x

Liang, K. Y., & Zeger, S. L. (1986). Longitudinal data analysis using generalized linear models. *Biometrika, 73,* 1, 13–22. https://doi.org/10.1093/biomet/73.1.13

Littell, R. C., Milliken, G. A., Stroup, W. W., Wolfinger, R. D., & Schabenberger, O. (2006). *SAS for Mixed Models* (2nd ed.). SAS Institute.

Lohr, S., Schochet, P. Z., & Sanders, E. (2014). *Partially nested randomized controlled trials in education research: A guide to design and analysis* (NCER 2014-2000). National Center for Education Research, Institute of Education Sciences, U.S. Department of Education.

Longford, N. (1989). Contextual effects and group means. *Multilevel Modelling Newsletter, 1*(3), 5, 11.

Ludtke, O., Marsh, H., Robitzsch, A., Trautwein, U., Asparouhov, T., & Muthen B. (2008). The multilevel latent covariate model: A new, more reliable approach to group-level effects in contextual studies. *Psychological Methods, 13, 3,* 203–229. https://doi.org/10.1037/a0012869

Luke, S. G. (2017). Evaluating significance in linear mixed-effects models in R. *Behavior Research Methods, 49,* 1494–1502. https://doi.org/10.3758/s13428-016-0809-y

Ma, X., & Klinger, D. (2000). Hierarchical linear modelling of student and school effects on academic achievement. *Canadian Journal of Education, 25*(1), 41–55. https://doi.org/10.2307/1585867

McNeish, D., & Stapleton, L. M. (2016). Modeling clustered data with very few clusters. *Multivariate Behavioral Research, 51*(4), 495–518. https://doi.org/10.1080/00273171.2016.1137008

McNeish, D., & Stapleton, L. M. (2017). On the unnecessary ubiquity of hierarchical linear modeling. *Psychological Methods, 22*(1), 114–140.

Moerbeek, M. (2004). The consequences of ignoring a level of nesting in multilevel analysis. *Multivariate Behavioral Research, 39*, 129–149. https://doi.org/10.1207/s15327906mbr3901_5

Murray, D. M. (1998). *Design and analysis of group-randomized trials.* Oxford University Press.

O'Connell, A. A, Yeomans-Moldanado, G., & McCoach, D. B. (2016). Residual diagnostics and model assessment in a multilevel framework: Recommendations toward best practice. In J. Harring, L. Stapleton, & S. N. Beretvas (Eds.), *Advances in multilevel modeling* (pp. 97–137). Information Age Publishing.

Palmeri, M. (n.d.). Testing the assumptions of multilevel models. In *A language, not a letter: Learning statistics in R.* https://ademos.people.uic.edu/index.html

Pedhazur, E. (1997). *Multiple regression in behavioral research.* Harcourt Brace.

Preacher, K., Zyphur, M., & Zhang, Z. (2010). A general multilevel SEM framework for assessing multilevel mediation. *Psychological Methods, 15*(3), 209–233. https://doi.org/10.1037/a0020141

Rabe-Hesketh, S., & Skrondal, A. (2022). *Multilevel and longitudinal modeling using stata* (4th ed.). State Press.

Raudenbush, S. W. (1989a). "Centering" predictors in multilevel analysis: Choices and consequences. *Multilevel Modelling Newsletter, 1*(2), 10–12.

Raudenbush, S. W. (1989b). A response to Longford and Plewis. *Multilevel Modelling Newsletter, 1*(3), 8–11.

Raudenbush, S. W., & Bryk, A. S. (2002). *Hierarchical linear models: Applications and data analysis methods* (2nd ed.). SAGE Publications.

Rights, J. D., & Sterba, S. K. (2019). Quantifying explained variance in multilevel models: An integrative framework for defining R-squared measures. *Psychological Methods, 24*(3), 309–338. https://doi.org/10.1037/met0000184

Shadish, W., Cook, T., & Campbell, D. (2002). *Experimental and quasi-experimental designs for generalized causal inference.* Houghton Mifflin.

Snijders, T. A. B., & Bosker, R. J. (2012). *Multilevel analysis: An introduction to basic and advanced multilevel modeling* (2nd ed.). SAGE.

Tourangeau, K., Nord, C., Lê, T., Sorongon, A. G., Hagedorn, M. C., Daly, P., & Najarian, M. (2015). *Early childhood longitudinal study, Kindergarten class of 2010–11 (ECLS-K:2011): User's Manual for the ECLS-K:2011 Kindergarten Data File and Electronic Codebook, Public Version* (NCES 2015-074). U.S. Department of Education. National Center for Education Statistics.

Wampold, B. E., & Serlin, R. C. (2000). The consequences of ignoring a nested factor on measures of effect size in analysis of variance. *Psychological Methods, 5*, 425–433. https://doi.org/10.1037//1082-989X.5.4.425

CHAPTER 3

EVALUATION OF MODEL FIT AND ADEQUACY

D. Betsy McCoach
University of Connecticut

Sarah D. Newton
University of Connecticut

Anthony J. Gambino
University of Connecticut

How do researchers evaluate multilevel models (MLMs)? How should they choose among competing models? The utility of any model depends upon its ability to explain the phenomenon under investigation. Therefore, model evaluation should consider two aspects of the model: (a) model fit, or the use of model selection criteria to choose among competing models; and (b) the adequacy/explanatory power of the model, or the ability of the predictors to explain variability in the outcome variable. As such, model selection processes should assess both model fit and predictive utility.

In this chapter, we describe and demonstrate the model evaluation process, considering both model fit and model adequacy. After providing a

Multilevel Modeling Methods with Introductory and Advanced Applications, pages 51–94

brief conceptual overview of MLM estimation, we identify common measures of model fit and adequacy within the MLM literature; highlight several areas of controversy or confusion; and provide general recommendations for evaluating MLM fit and adequacy. We review the concept of deviance and explain how to use the chi-square difference test to compare the deviances of two nested models. We also describe index comparison approaches (e.g., the Akaike information criterion [AIC] and Bayesian information criterion [BIC]) to model evaluation. Then, we consider model adequacy and predictive utility, explaining use of proportion of variance explained-type measures to determine the predictive power of MLMs. Using the example data from Chapter 2 in this volume, "Introduction to Multilevel Models for Organizational Data," we compare the model fit and predictive utility of two competing models. Finally, we provide guidance for assessing model fit and model adequacy in MLM.

CONCEPTUAL OVERVIEW OF ESTIMATION IN MLM

Before discussing deviance and model fit, it is useful to have a rudimentary conceptual understanding of estimation in MLM.[1] To keep things simple, we contextualize this entire discussion using the unconditional random effects model, with a single, continuous outcome variable: $Y_{ij} = \gamma_{00} + u_{0j} + e_{ij}$. In this model: Y_{ij} represents the predicted value on the outcome for person i in cluster j; γ_{00} denotes the intercept—the expected person-level value on the outcome, absent any additional person- or cluster-level information; u_{0j} conveys unexplained variance in the outcome across clusters; and e_{ij} indicates the degree of person-level divergence from the overall intercept (γ_{00}).

MLM does not require balanced data: The number of units per cluster can vary across clusters. In fact, there is no minimum or maximum number of units per cluster, and MLMs can easily accommodate datasets that include some clusters with very few level-1 units and other clusters with very large numbers of level-1 units. A simple, but important, issue that arises when estimating parameters from unbalanced data is how to determine the expected value of the outcome variable $\left(E(Y_{ij}) = \gamma_{00}\right)$ from multiple clusters of unequal sizes. In non-clustered data, the sample mean provides our "best guess" about the population mean (the "expected" value in the population). But with clustered data, what is the expected value of the outcome variable, γ_{00}? Imagine randomly sampling students from within 100 schools. Further, the sample sizes within the schools vary widely: The smallest cluster size is 2 and the largest cluster size is 1,000. How should we determine expected achievement? One option would be to ignore clustering and to take the sample mean. In such a scenario, every person is weighted equally; however, the schools where larger numbers of students were sampled would have a much larger influence on the expected mean than the small schools. On the other hand, we could compute the mean of school means. Yet weighting all schools equally gives the school with only two students a disproportionately large influence on expected achievement.[2]

In MLM, larger clusters do have a larger influence on the expected mean. However, the intraclass correlation coefficient (ICC) tempers that influence. In the unconditional random effects ANOVA model, the ICC represents the proportion of between-cluster variance, indicating the degree of dependence within a cluster. Equation 3.1 shows this:

$$ICC = \frac{\tau_{00}}{\tau_{00} + \sigma^2} \quad (3.1)$$

where τ_{00} conveys between-cluster variability, σ^2 denotes within-cluster variability, and their sum (in the denominator of the ICC equation) represents total variance in the outcome variable.

An ICC of 0 indicates that two randomly selected people within a given cluster are no more similar to each other than two randomly selected people from different clusters. When the ICC is 0, ignoring clustering seems reasonable. On the other hand, if the ICC is 1, all people in a cluster are complete replicates of each other on the outcome variable of interest. In such a scenario, the mean of cluster means might be of greater interest, given the deterministic nature of within-cluster performance.

In MLM, when the ICC is 0, the influence of each cluster on γ_{00} is determined by the sample size of the cluster. When the ICC is 1, each cluster has an equal influence on γ_{00}, regardless of its size (Snijders & Bosker, 2012). In reality, the ICC is typically between 0 and 1. Therefore, γ_{00} is a compromise between the sample mean (as it would be when ICC = 0) and a mean of cluster means (as it would be when ICC = 1). The higher the ICC, the more γ_{00} approaches a mean of cluster means; the lower the ICC, the more γ_{00} approaches the sample mean.

Empirical Bayes

Continuing from our previous example, imagine we want to estimate academic achievement for a randomly selected set of schools, and we have an incomplete set of information. In most schools, we randomly sampled 1,000 or more students. However, in one school, we randomly sampled only two students. In schools where we sampled 1,000 students, the sample mean provides a reasonable estimate of the "true" level of achievement. But in the school where we selected only two students, the mean of these two students may provide a poor estimate of the true achievement level in that school. Imagine an even more-extreme example: What if we missed sampling one school entirely? What would be our best estimate of achievement in that school, in the absence of *any* school data?

There are two approaches to estimating this true school mean. First, we could compute the school's sample mean, $\bar{Y}_{\cdot j}$. The sample mean of the school ($\bar{Y}_{\cdot j}$) should provide a good estimate of the true mean in schools that have many sample observations (more information). However, in a school with no information, what is our best guess about school achievement? We could use the expected value of achievement, $\hat{\gamma}_{00}$, as an estimate of a given school's achievement. If we know nothing else about the school and have no other information from the school, the overall expected value of achievement provides our best estimate of the school's achievement. In this case, we "borrow" information from other schools to estimate the expected value for this school. But what do we do for schools with small samples of respondents? The school's sample data provide some information about achievement, but we cannot completely trust that the sample mean based on those students provides a reasonable estimate of the true school mean. In such a situation, we use a combination of the expected value of achievement and the school sample mean to derive an estimate of the true mean for the school. Such a strategy should place more weight on the cluster mean when we have more information within a cluster (more observations) and place more weight on the expected value when we have less information from the cluster. Conceptually, the empirical Bayes estimate of the randomly varying intercept (β^{*}_{0j}) balances these two sources of information to generate an estimate of the true cluster mean (McCoach & Cintron, 2021).

Reliability of Cluster *j*

Potential expected values of the *true* cluster mean include the estimate of γ_{00} and $\bar{Y}_{\cdot j}$. Empirical Bayes estimation combines these values, based on the reliability of cluster *j*, which incorporates three pieces of information: within-cluster variability (σ^2), between-cluster variability (τ_{00}), and the number of observations per cluster, n_j.

$$\text{Reliability of } \hat{\beta}_{0j} = \frac{\tau_{00}}{\tau_{00} + \frac{\sigma^2}{n_j}} \tag{3.2}$$

Theoretically, the reliability of cluster *j* can range from 0 (when there is no between-cluster variability in the outcome) to 1 (when all variability is between clusters and no variability is within clusters). However, within a given sample, the lower bound for reliability in any given cluster is the

ICC, and this occurs when there is only one unit in the cluster ($n_j = 1$). The ICC features prominently in this reliability formula (Raudenbush & Bryk, 2002). Larger ICCs, which indicate that within-cluster group variance is small relative to between-cluster variance, result in higher reliability.

Each cluster has its own estimate of reliability; however, variance estimates τ_{00} and σ^2 remain constant across clusters. Therefore, holding between- and within-school variance constant, larger cluster sizes (n_j's) result in higher reliability. Additionally, larger between-cluster variance (relative to within-cluster variance) also increases reliability. In other words, reliability is higher when the group means vary substantially across level-2 units (holding constant the sample size per group). So, increasing group size, increasing homogeneity within clusters, and increasing heterogeneity between clusters all increase reliability.

The reliability in cluster *j* is consequential: When the reliability in cluster *j* is higher, more weight is placed on the sample mean as the estimate of the true school mean. Conversely, when the reliability of the cluster *j* is lower, more weight is placed on the estimate of γ_{00} as the estimate of the true school mean.

Empirical Bayes Estimates

In the unconditional random effects ANOVA model, the empirical Bayes estimate of the true cluster mean (β^*_{0j}) weights the two potential estimates for each cluster as a function of the reliability for that cluster.

$$\hat{\beta}^*_{0j} = \lambda_j \bar{Y}_{\cdot j} + (1 - \lambda_j)\hat{\gamma}_{00} \tag{3.3}$$

where λ_j is the reliability of the sample mean, $\bar{Y}_{\cdot j}$ (Raudenbush & Bryk, 2002) and $\hat{\gamma}_{00}$ is the expected value of the intercept (the expected value of the outcome variable in the unconditional random effects ANOVA model).

The empirical Bayes estimate of the true cluster mean is a compromise between the sample mean ($\bar{Y}_{\cdot j}$) and the expected, or model-based mean ($\hat{\gamma}_{00}$). The sample mean ($\bar{Y}_{\cdot j}$) is weighted by the reliability for cluster *j* (λ_j) and the model-based mean ($\hat{\gamma}_{00}$) is weighted by 1 minus the reliability ($1-\lambda_j$). Thus, the higher the reliability of the estimate for cluster *j*, the more weight is placed on the sample mean ($\bar{Y}_{\cdot j}$) as the estimate of the true cluster mean, β_{0j}. In contrast, the lower the reliability of cluster *j* estimate, the more weight is placed on the model-based estimate $\hat{\gamma}_{00}$ as the estimate of the true cluster mean. In the extremes, if the reliability were 1, the unadjusted sample mean would provide the estimate of the true cluster mean. If the reliability were 0, $\hat{\gamma}_{00}$ would serve as the estimate of the true cluster mean. Sometimes, the empirical Bayes estimators are referred to as shrinkage estimators because empirical Bayes residuals are "shrunken" towards zero (Raudenbush & Bryk, 2002) as compared to OLS estimates, and the degree of shrinkage is a function of the reliability of the cluster.

Maximum Likelihood Estimation

MLM uses maximum likelihood (ML) estimation. The goal of ML estimation is to find the set of parameter values that maximizes the likelihood of observing the actual data. The most-common techniques for estimating variance components for MLMs with normal response variables are full information maximum likelihood (FIML) and restricted maximum likelihood (REML) estimation.

Deviance

Using ML to estimate model parameters also provides a likelihood, which can easily be transformed into a deviance statistic (Snijders & Bosker, 2012). The deviance is calculated as –2 multiplied by the difference between the log-likelihood of the specified model and the log-likelihood of a saturated model that fits the sample data perfectly. Therefore, the deviance is actually a measure of the badness of fit of a given model: higher deviances indicate greater model misfit (Singer & Willett, 2003). Although lower deviances indicate better model fit, we cannot interpret the deviance[3] in isolation. However, we can interpret differences in the deviances of competing models as long as both models: (a) are hierarchically nested; (b) analyze the same observations; and (c) use FIML to estimate the model parameters (if we wish to compare two models that differ in terms of their fixed effects).

Full Information Maximum Likelihood (FIML)

In FIML, the estimates of the variance and covariance components are conditional upon the point estimates of the fixed effects. FIML estimates of Γ (i.e., the fixed-effect variance/covariance matrix), **T** (i.e., the random-effect variance/covariance matrix; see Equation 3.4), and σ^2 "maximize the joint likelihood of these parameters for a fixed value of the sample data, Y" (Raudenbush & Bryk, 2002, p. 52). Thus, the number of model parameters includes both fixed effects and variance/covariance components.

$$\mathbf{T} = \begin{bmatrix} \tau_{00} & & & & \\ \tau_{01} & \tau_{11} & & & \\ \tau_{02} & \tau_{12} & \tau_{22} & & \\ \vdots & \vdots & \vdots & \ddots & \\ \tau_{0q} & \tau_{1q} & \tau_{2q} & \cdots & \tau_{qq} \end{bmatrix} \tag{3.4}$$

Restricted Maximum Likelihood (REML)

In contrast, REML maximizes the joint likelihood of **T** and σ^2 given the observed sample data, **Y**. Therefore, when estimating variance components, REML takes the uncertainty due to loss of degrees of freedom from estimating fixed parameters into account, whereas FIML does not (Goldstein, 2011; Raudenbush & Bryk, 2002; Snijders & Bosker, 2012). Hence, under REML, the number of reported parameters includes only the variance and covariance components.

FIML vs. REML

When comparing models using REML, the fixed effects structure must be the same across the null and alternative models—competing models must differ only in their random effects. Conversely, under FIML, nested models can differ in both their fixed and random effects (Raudenbush & Bryk, 2002; Snijders & Bosker, 2012). Therefore, comparisons of models with differing fixed and random effects should utilize the deviance provided by FIML (Goldstein, 2011; McCoach & Black, 2008; McCoach et al., 2018; Snijders & Bosker, 2012).

The ability to compare the deviances of models that differ in terms of their fixed and/or random effects is a major advantage of FIML over REML. Although most statistical programs use REML as the default method of estimation, use FIML estimation when comparing two nested models with differing fixed effects (McCoach & Black, 2008).

When the number of clusters (level-2 units) is large, REML and FIML produce similar estimates of the variance components. However, for small numbers of clusters, FIML tends to underestimate variance components, and REML results may be more realistic (Raudenbush & Bryk, 2002). A simple formula, $(J-F)/J$, where J is the number of clusters and F is the number of fixed effects in the model, provides a rough approximation of the degree of downward bias in the FIML estimates (Raudenbush & Bryk, 2002). For example, when fitting a model with two fixed effects in a sample containing observations from 10 clusters, the level-2 variance components are likely to be 80% $(((10-2)/10)=.80)$ as large in FIML as they are in REML, which represents substantial underestimation of the variance components using FIML (McCoach & Cintron, 2021).

MODEL SELECTION

It seems unlikely that our models can ever fully capture the truth. Instead, they represent mere approximations of reality (Burnham & Anderson, 2004).[4] If we accept this basic premise, then how should we select among a competing set of models?

Model selection is a crucial part of the MLM process. Guided by theory and informed by data, balancing parsimony and model complexity, the ideal model should adequately describe the observed data to a satisfactory extent, but exclude unnecessary complications (Snijders & Bosker, 2012). Hence, the best model provides an adequate account of the data using the fewest parameters possible (Wagenmakers & Farrell, 2004).

In pursuit of this overarching goal, three general principles should guide the model selection process (Burnham & Anderson, 2004). First, parsimony is paramount. Including additional parameters may improve fit and cannot lead to worse model fit (Forster, 2000). Thus the critical question is whether the improvement in model fit justifies the inclusion of additional parameters. Second, using data to compare several plausible competing hypotheses generally provides more useful information than comparing a given model to an often-implausible null hypothesis (Burnham & Anderson, 2004). In reality, how many researchers legitimately hypothesize that most substantive parameters of interest are exactly zero in the population? Third, it is more desirable to judge the strength of evidence supporting a given theoretical model (Burnham & Anderson, 2004), than to assess the evidence against an atheoretical null model.

Criteria for Evaluating Model Fit

There are a variety of options for examining and comparing the model/data fit for a set of competing MLMs. We can compare *nested* models using the likelihood ratio test (i.e., deviance difference test). Note that two models are considered hierarchically nested when "the more complex model includes all of the parameters of the simpler model plus one or more additional parameters" (Raudenbush et al., 2000, pp. 80–81), within the same sample. In contrast, information criteria (ICs), such as the AIC and BIC, permit model fit assessment for a set of *nested* or *non-nested* models.

Likelihood Ratio Test (LRT)

The LRT compares the change in the deviances (i.e., –2*Log-Likelihood or –2LL) from two competing nested models. The null hypothesis is that the restricted/constrained model is the correct model; the alternative hypothesis posits that the unrestricted/unconstrained, more-complex model is correct (Dominicus et al., 2006). The null model (M_0) is more parsimonious (i.e., it contains fewer parameters) and the alternative model (M_1) is more parameterized.

In sufficiently large samples, under standard normal theory assumptions, and using the same set of observations, the difference between the deviances of two hierarchically nested models follows an approximate χ^2

distribution with degrees of freedom equal to the difference in the number of parameters being estimated between the two models (de Leeuw, 2004; Raudenbush & Bryk, 2002; Singer & Willett, 2003).

The deviance of the simpler model (D_0) is based on p_0 estimated parameters, whereas the deviance of the more-complex model (D_1) is associated with p_1 estimated parameters. Because the simpler model has fewer parameters ($p_0 < p_1$), the deviance of the simpler model must be at least as large as the deviance of the more-parameterized model ($D_0 \geq D_1$).

The LRT compares the difference in model deviances ($\Delta D = D_0 - D_1$) to the critical value of χ^2, with degrees of freedom equal to the difference in the number of estimated parameters ($\Delta p = p_1 - p_0$). If the model with the larger number of parameters fails to reduce the deviance by a substantial amount, we retain the more-parsimonious model (M_0). However, when the change in deviance (ΔD) exceeds the critical value of χ^2 with $p_1 - p_0$ degrees of freedom, the additional parameters result in statistically significantly improved model fit, which favors the more-parameterized model (M_1). Therefore, a statistically significant decrease in the deviance favors the more-complex model (M_1), whereas a non-statistically significant decrease favors the less-complex model (M_0; McCoach & Black, 2008; Raudenbush & Bryk, 2002).

LRTs for Random Effects. Evaluating variance components is less straightforward. LRTs are less appropriate for testing random effects/variance components because variance estimates cannot be less than zero (Berkhof & Snijders, 2001; Raudenbush & Bryk, 2002). The value for a variance component under the null hypothesis (0) lies on the boundary of the parameter space for this effect (Fitzmaurice et al., 2007; Jennrich & Schluchter, 1986; Lee & Braun, 2012; Snijders & Bosker, 2012; Verbeke & Molenberghs, 2003). In other words, the variance component is a boundary parameter: It cannot be normally distributed around a mean of zero if the null hypothesis is true (Dominicus et al., 2006; Stoel et al., 2006). Therefore, the distribution of a variance under the null hypothesis does not exist on the interior of the full parameter space (Gutierrez et al., 2001; Raudenbush & Bryk, 2002; Snijders & Bosker, 2012).

The distributions under the null hypotheses are necessarily different for fixed effects and variance components. Using two-tailed tests to evaluate whether variances are statistically different from zero is inappropriate (Raudenbush & Bryk, 2002). Furthermore, the computation and use of standard errors around these variance estimates becomes especially problematic as variance components approach zero (Baayen et al., 2008). The utility/appropriateness of confidence intervals and other statistical tests based on typical standard errors is also questionable, and the unadjusted LRT is too conservative for testing MLMs with random effects (Self & Liang, 1987; Stoel et al., 2006; Stram & Lee, 1994). In other words, the test tends to produce *p*-values that are larger than they should be (Baayen et al., 2008;

Dominicus et al., 2006; Stoel et al., 2006), which results in elevated Type II error rates (i.e., researchers failing to reject the null hypothesis when it is false; Dominicus et al., 2006; Stoel et al., 2006). Decreased power (Berkhof & Snijders, 2001; Stoel et al., 2006) may increase the likelihood of constraining variances of slopes that should be allowed to randomly vary.

A modified χ^2 sampling distribution (typically, a mixture of multiple χ^2 distributions) is used to test parameters that lie at the boundary of the parameter space (0) under the null hypothesis. For any parameter with a true value of zero in the population, approximately half of the estimates of that effect should be positive, and half should be negative (Snijders & Bosker, 2012). Because variance components cannot be negative, their sampling distributions cannot include negative values (i.e., their sampling distributions may only contain positive values and zero). Thus, these distributions must be bounded at zero. This resembles a mixture of zero and the positive χ^2 distributions. In other words, when investigating the inclusion of a single boundary parameter, 50% of the variance component's sampling distribution indicates the finding of zero variance, and 50% of the distribution represents positive variance values (Berkhof & Snijders, 2001; Miller, 1977; Self & Liang, 1987; Snijders & Bosker, 2012; Stoel et al., 2006).

To test random effects, the correct critical value comes from the $\bar{\chi}^2$ distribution, which represents this mixture of χ^2 distributions (Snijders & Bosker, 2012), with degrees of freedom equal to the number of random slopes being tested (e.g., $df = rs$) and the number of random slopes being tested plus one (e.g., $df = rs + 1$). In the simplest case, as described previously, each distribution receives equal weight (i.e., .5 for each distribution) because it is approximately equally probable that the boundary parameter could be zero or a positive, non-zero value (Stoel at al., 2006). For example, when testing a single variance parameter, if exactly 50% of the resulting parameter estimates are negative and exactly 50% of the estimates are positive, the mixture proportions for the $\chi^2_{(0)}$ and $\chi^2_{(1)}$ distributions would each be .50, representing a 50/50 mixture of the χ^2 distributions with zero and one degree of freedom, respectively.

However, the complexity of this deviance testing procedure can escalate quickly, and there is an added complication. When considering whether to include or exclude a random slope variance, we must also consider the covariance(s) between that random slope and any/all other random slopes/intercepts in the model because if a variance equals zero, then all of the covariances between it and other parameters must also be zero (Snijders & Bosker, 2012). However, covariances are not boundary parameters. They can be positive or negative, and there are no outer bounds on their magnitudes, unlike the ±1.00 bounds that exist for correlation coefficients (Stoel et al., 2006).

Generally, we prefer to estimate an unstructured residual variance/covariance (tau or **T**) matrix. However, this is not the only option. Although HLM and R[5] estimate the **T** matrix as unstructured by default, SAS, SPSS, and Stata do not. To estimate the unstructured **T** matrix in other statistical software packages, SAS users specify *TYPE=UN* when implementing PROC MIXED (SAS Institute Inc., 2022), Stata users specify *cov(unstructured)* or *cov(un)* (StataCorp, 2017), and SPSS users specify *covtype(un)* or *cov(un)* on either the *random* or *repeated* subcommand lines, as applicable (University of California, Los Angeles Institute for Digital Research & Education, 2020). See Chapter 11, "Within-Subject Residual Variance/Covariance Structures in Longitudinal Data Analysis," in this volume for additional information about variance/covariance structures.

Recommendations for Testing Random Effects

Typically, when comparing two models that differ by a single fixed effect (but no random effects), the χ^2 critical value is 3.841 at $\alpha = .05$. In contrast, if the models differ by a single random effect, and therefore only one τ, it is generally acceptable to conduct a one-tailed hypothesis test using the χ^2 critical value corresponding to $p = .10$ to test for statistical significance at $\alpha = .05$ (Snijders & Bosker, 2012). The critical value of χ^2 with one degree of freedom is 2.706 when $p < .10$. So if two models differ by one parameter and it is a variance component, we use the χ^2 critical value of 2.706. Furthermore, the critical values for the adjusted LRT are 5.14 for a variance and a covariance, 7.05 for a variance and two covariances, and 8.76 for a variance and three covariances (Snijders & Bosker, 2012). Snijders and Bosker (2012) present a more-detailed discussion of this issue and display a table with the correct critical values for the comparison of models that differ by one or more randomly varying slopes. For competing models that differ in both their fixed- and random-effect structures, determination of the proper LRT distribution depends on the number of boundary parameters of interest (k, with $k + 1$ distributions combined to formulate the appropriate mixture distribution) and the number of unconstrained parameters of interest (u, with u degrees of freedom for the first distribution in the mixture; Stoel et al., 2006).

For these reasons, Baayen et al. (2008) suggested that calculating standard errors around these variance estimates makes little sense. Some statistical packages (like HLM, M*plus*, and Stata) produce and report standard errors for variance components, some (like R) do not, and some offer user-specified options to obtain such estimates. For example, SAS includes the *covtest* option in Proc Mixed, which produces standard errors and statistical tests for covariance estimates (SAS Institute Inc., 2022). In addition, HLM provides univariate χ^2 tests for the randomly varying slopes and intercept. This χ^2 tests the null hypothesis that the level-2 (intercept and slope) variances are 0. Therefore, a statistically significant χ^2 suggests the necessity of the random effect and a non-statistically significant χ^2 suggests that the random effect could be removed. However, Raudenbush and Bryk (2002) caution that these χ^2 tests provide approximate, not exact, probability values because they are simple univariate tests and are estimated only for clusters that have sufficient data to compute a separate OLS regression.

Other Issues With Modeling Random Slopes

In addition to the boundary issues associated with testing variance components, there are several additional issues to consider when modeling random slopes. First, when estimating near-zero or unnecessary random effects (i.e., when the random slope variance is not significantly different from 0), the model may iterate repeatedly, decreasing computational speed. Even worse, the model may fail to converge (Palardy, 2011; McCoach et al., 2018). Such "failure to converge is not due to the defects of the estimation algorithm" (Bates et al., 2015, p. 19), but is instead "a straightforward consequence of attempting to fit a model that is too complex to be supported by the data" (Bates et al., 2015, p. 19). One solution to this problem is to fit an MLM with a more-parsimonious variance/covariance structure (Bates et al., 2015). Interestingly though, an MLM that fails to converge in one statistical package may converge when estimated in another package (McCoach et al., 2018).

Furthermore, when estimating an unstructured **T** matrix, each additional random slope increases the complexity of the model, more so than adding fixed effects. Note that the variance component structure in an estimated MLM represents a partitioning of the residual variance (i.e., unexplained variance in the outcome). Estimating additional variance components increases the computational load of model estimation as the unstructured **T** matrix includes the random slope *variances* as well as the *covariances* between each new random slope and each existing random intercept/slope in the model. Therefore, the number of estimated parameters increases by the number of random slopes we add to our model, as well as the number of covariances between those new random slopes and all existing random parameters. This added model complexity, in turn, increases computational burden and model estimation time. Therefore, it is important to model variance components judiciously. Incorporate theoretically defensible random effects into MLMs but resist the temptation to include all possible random effects by default, without carefully considering their purpose and necessity. Just because it is possible to specify a random slope parameter does not mean that it is advisable.

Information Criteria (ICs)

Fit criteria provide one tool for evaluating and comparing sets of competing models (Konishi & Kitagawa, 2008). As explained by Burnham and colleagues (2011), information criteria provide a formal method to quantify the strength of the data-based evidence for, and ranking of, each competing hypothesis and represent one aspect of multimodel inference.

Additionally, there are several advantages to consulting ICs during model selection, rather than relying solely upon deviance statistics and χ^2 difference tests to evaluate the goodness of fit of MLMs. First, ICs such as the AIC (Akaike, 1973) and BIC (Schwarz, 1978) allow for comparison of non-nested models, assuming the sample remains constant across models. Furthermore, ICs quantify the degree to which the given model represents an improvement over comparison models.

Criteria like the AIC and BIC generally feature two components: the deviance (which conveys the fit of the estimated model), as well as some penalty on the deviance for each additional estimated parameter (Dominicus et al., 2006; Lee & Ghosh, 2009; Zucchini et al., 2011). Although the number of estimated model parameters changes across models, the per-parameter penalties of the AIC and BIC remain constant for a given sample size. When using ICs to compare two models, we favor the model with the lower IC.

The Akaike Information Criterion (AIC). To compute the AIC, we simply multiply the number of parameters by two and add this product to the deviance statistic. The formula for the AIC is:

$$\text{AIC} = D + 2p \tag{3.5}$$

where D is the deviance and p represents the number of estimated parameters. The second term, $2p$, imposes a two-point penalty for each additional estimated parameter.

The model with the lowest AIC is considered the best fitting model.[6] Notably, the AIC tends to favor more-complex models (Bozdogan, 1987; Whittaker et al., 2012, 2013; Whittaker & Stapleton, 2006) than other available model fit indices (such as the BIC or LRT) because of its small per-parameter penalty. For example, the critical value of χ^2 with one degree of freedom at $\alpha = .05$ is 3.841 (or 2.706 for a 1-*df* change involving a variance). When comparing two models that differ by a single degree of freedom, the LRT imposes a more-stringent penalty for rejecting the simpler model. In fact, this is true for comparisons of models that differ by up to seven parameters:[7] Using the LRT results in an equivalent or more-parsimonious model than the AIC. Conversely, when comparing models that differ by more than seven parameters, the AIC favors more-parsimonious models than the LRT.

The Bayesian Information Criterion (BIC). The BIC is equal to the sum of the deviance and the product of the natural log of the sample size and the number of parameters. As with the AIC, we prefer the model with the lowest BIC. The formula for the BIC is:

$$\text{BIC} = D + \ln(n) * p \tag{3.6}$$

where D is the deviance ($-2LL$), p is the number of parameters estimated in the model, and n is the sample size. Therefore, the BIC's per-parameter penalty equals the natural log of the sample size. Comparatively, the per-parameter penalty for the BIC is typically larger than that of the AIC; the sample size must be less than eight for the BIC's per-parameter penalty to drop below AIC's two-point per-parameter penalty (Claeskens & Hjort, 2008; Schwarz, 1978). Consequently, when the AIC favors the simpler (less-parameterized model), the BIC also favors the simpler model. Likewise, when the BIC favors the more-complex (more-parameterized model), the AIC also favors this model.

The BIC incorporates the Bayesian perspective of examining the probability of a specified model, given the observed data in hand (Sclove, 1987). The BIC approximates the log of a Bayes factor (assuming a unit information prior—for more information, see Kass & Raftery, 1995; Kass & Wasserman, 1995; Raftery, 1999). Following the Bayesian paradigm, Bayes factors represent the ratio of the marginal likelihoods of observing the data given each competing model (Lee & Ghosh, 2009; Weakliem, 2016). Therefore, Bayes factors larger than 1.00 indicate stronger support for Model A (listed in the numerator), Bayes factors of exactly 1.00 demonstrate equal support for each model, and Bayes factors less than 1.00 suggest stronger support for Model B (appearing in the denominator):

$$\text{Bayes factor} = \frac{p(data|Model_A)}{p(data|Model_B)}. \tag{3.7}$$

Additionally, the difference between BIC values for two competing models forms a Bayes factor, according to the following formula:

$$\text{BIC}_B - \text{BIC}_A = 2\log(K) \tag{3.8}$$

such that calculating $e^{((\text{BIC}_B - \text{BIC}_A)/2)}$ provides a comparison of the likelihoods of the data given each competing model (Weakliem, 2016).

Raftery (1995) introduced guidelines regarding the magnitude of such BIC differences to guide interpretations of the strength of evidence for one model over another: BIC values differing by 0 to <2 points offer "Weak" evidence for the preferred model; 2 to <6 points provide "Positive" evidence; 6 to <10 points yield "Strong" evidence; and 10+ points represent "Very Strong" evidence (Raftery, 1995, p.140).

Determining the Correct Sample Size for Computing the BIC in MLM. What sample size should be used to compute the BIC: the total number of units at the lowest level or the number of units at the highest level? In single-level models, the sample size assumes complete independence of observations under simple random sampling. In clustered designs, the effective sample size (e.g., $N_{\text{effective}}$) adjusts the total number of level-1 units for the two-stage sampling procedure used to collect multilevel data (Snijders & Bosker, 2012). The effective sample size (n_{eff}) for a given sample is a function of the degree of non-independence in the sample and the number of observations per cluster. To estimate how much the clustered nature of the data impacts the standard errors requires accounting for both the homogeneity within clusters (ICC) and the average cluster size ($\bar{n}_{.j}$). The formula for n_{eff} is simply (Snijders & Bosker, 2012):

$$n_{\text{eff}} = \frac{N}{DEFF} = \frac{N}{1+\rho(\bar{n}_j - 1)} \tag{3.9}$$

where $\bar{n}_j$ is the average cluster size, ρ is the ICC, and N is the total sample size.

Given these multiple conceptualizations of sample size within a multilevel framework, there is no definitive consensus as to which sample size (the total sample size, the number of clusters, or the effective sample size [n_{eff}]) is most appropriate for computing the BIC (Skrondal & Rabe-Hesketh, 2004). Furthermore, different software packages compute the BIC differently—for example, SAS PROC MIXED uses the number of clusters (Beretvas & Murphy, 2013; Vallejo et al., 2014), whereas SPSS's MIXED procedure uses the total number of observations (IBM, n.d.). By default, STATA (StataCorp LP, n.d.) and R (Comets et al., 2017; Nathoo & Masson, 2016) also utilize the number of observations to calculate the BIC; however, STATA offers an option for the analyst to re-specify the sample size value used to compute the BIC [*n*( )] (StataCorp LP, n.d.). In contrast, M*plus* uses the total number of people as the sample size, which translates to using a sample of size *n* when calculating the BIC for cross-sectional MLMs (i.e., BIC-L1) and a sample of size *j* when computing the BIC for longitudinal models (i.e., BIC-L2).[8] Therefore, even though SPSS and SAS produce identical −2*LL* and AIC values, the BIC value differs across these programs. Because the BIC imposes a steeper per-parameter penalty as the sample size increases, the BIC value produced by SPSS is larger than the BIC value produced by SAS, and it tends to favor more-parsimonious models. Hence, the sample size used to compute the BIC could change the outcome(s) of the model selection process, leading to distinct model preferences.

Furthermore, some software programs do not provide AIC and BIC values in their generated model output. Neither HLM nor MLwiN (two software packages explicitly designed for MLM) provides users with IC values to aid in the model comparison/selection process (Vallejo et al., 2011).

However, both the AIC and the BIC can be computed easily from the deviance statistic, with one caveat—FIML is generally considered to be the most-appropriate estimation method for computing ICs (Verbeke & Molenberghs, 2000) when comparing models that differ in their fixed effects.

Comparing the AIC and BIC

The AIC and BIC do not just differ in the per-parameter penalties they assess. The purposes of the AIC and BIC are also fundamentally different, a point that is often ignored in practice. Whereas the AIC "is designed to find the best approximating model to the unknown true data-generating model, the BIC is designed to find the most probable model given the data" (Vallejo et al., 2011, p. 22). Because the AIC and BIC take different approaches to model selection, using these ICs may result in different conclusions about which model to favor.

In addition, a researcher's modeling objectives may also influence the choice of IC. For example, predictive models often include as many explanatory variables as possible, no matter how little each variable contributes. In such a situation, using the AIC[9] might be preferable. In contrast, when building explanatory models, researchers may seek the simplest model to explain the phenomenon of interest and have limited interest in less-influential predictors. The BIC focuses on parsimony by allowing fewer predictors to remain in the model and generates evidence for or against individual models.

Researchers also regularly use a variety of other ICs across disciplines and statistical packages. Table 3.1 describes a few additional common ICs.

TABLE 3.1 Other Common Information Criteria

Information Criterion	Definition
Sample size-adjusted Bayesian Information Criterion (SABIC; Sclove, 1987)	$SABIC = -2LL + \ln((n+2)/24) * p$, where n is the sample size and p is the number of estimated model parameters
Hannan and Quinn Information Criterion (HQIC, Hannan & Quinn, 1979)	$HQIC = -2LL + 2 * p * \ln(\ln(n))$, where n represents sample size and p represents the number of estimated model parameters
Consistent AIC (CAIC; Bozdogan, 1987)	$CAIC = -2LL + [\ln(n) + 1] * p$, where n = sample size and p = the number of estimated model parameters
Finite sample corrected AIC (AICC, Hurvich & Tsai, 1989)	$AICC = -2LL + (2 * p * n)/(n - p - 1)$, where n is the sample size and p is the number of estimated model parameters

TABLE 3.2 Model Fit Criteria Penalties for Models That Differ by 1, 2, 3, and 4 Parameters

	1 parameter	2 parameters	3 parameters	4 parameters
AIC	2.00	4.00	6.00	8.00
BIC (n = 10)	2.30	4.61	6.91	9.21
LRT (var)	2.71	5.14	7.05	8.76
LRT (trad)	3.84	5.99	7.82	9.49
BIC (n = 50)	3.91	7.82	11.73	15.64
BIC (n = 100)	4.61	9.22	13.83	18.44
BIC (n = 1000)	6.91	13.82	20.73	27.64
BIC (n = 10,000)	9.21	18.42	27.63	36.84

Note: This table has been reproduced. It originally appeared in the book, *Introduction to Modern Modelling Methods* (McCoach & Cintron, 2021).

What Happens When Model Fit Criteria Disagree?

The AIC, BIC, and LRT may favor different models. Table 3.2 (from McCoach & Cintron, 2021) displays the total penalty imposed by each criterion for models that differ by one, two, three, and four parameters. For example, imagine that we want to compare two models that differ by one fixed effect. Our total sample size is 1,000 people, nested within 50 clusters. In such scenarios, how much must the deviance decrease to favor the model that includes the fixed effect? Using the AIC, the deviance must be 2.00 points lower; using the BIC-L2 (BIC calculated with the total number of level-2 units), the deviance must be 3.91 points lower; and using the BIC-L1 (BIC calculated with the total number of level-1 units), the deviance must be 6.91 points lower to favor the model that includes the fixed effect. Therefore, differences between 2.00 and 6.91 lead to ambiguous conclusions (McCoach & Cintron, 2021).

Hence, in addition to computing appropriate model fit criteria (as shown previously), we recommend examining measures of model adequacy, such as those presented in Rights and Sterba's (2019) MLM variance decomposition framework (see next section). Computation of model adequacy measures can bolster a theory-driven argument for retaining or eliminating specific parameters and/or models from consideration and provide insight into the necessity of a given parameter by addressing the following questions:

1. What proportion of within, between, and/or total variance does a specific parameter of interest explain?
2. Does removing the parameter substantially reduce the predictive ability of the model?

Ultimately though, researchers must decide which error would be more serious: omitting a potentially necessary parameter or including an unnecessary parameter. Theoretically speaking, if excluding a potentially important parameter is more grievous, then we recommend favoring the more-parameterized model. Conversely, if including an unnecessary parameter is more problematic, then we suggest selecting the simpler model. In general, we tend to favor retaining potentially important fixed effects and eliminating unnecessary variance components (i.e., random effects). Such choices should be informed by the field of study and the specific research context, problem, and questions. Model selection decisions should always consider both the fit and the predictive ability of the MLM, especially when different fit criteria lead to different modeling decisions. Therefore, the next section introduces methods for quantifying explained variance in MLM.

Model Adequacy

In MLM, we generally evaluate model fit relative to other competing models. We can also evaluate model adequacy (i.e., explanatory power of the model—does the model do a good job of explaining scores on the outcome variable?). We can assess model adequacy both relative to competing models and in an absolute sense.

In single-level regression models, an important determinant of the utility of a model is the proportion of variance explained by the model, or R^2. In MLM, the outcome variance is decomposed into multiple pieces called variance components, which exist at each level of the MLM. In addition, in random-coefficients models, the relationships between the dependent variable and level-1 independent variables vary as a function of the level-2 unit or cluster variable. Therefore, quantification of the variance explained by a given set of predictors is much more complicated in MLM than in the single-level case.

So, deciding how to compute and report variance-explained measures in MLM requires explicit consideration of the context and goals of the research. Conceptually, we may be interested in measuring variance explained within clusters, variance explained between clusters, and/or total variance explained (both within and between clusters). A cluster-level (level-2) variable cannot possibly explain within-cluster variance. However, a cluster-level variable can explain between-cluster variance; and because it can explain between-cluster variance, it can also explain some of the total variance for a given model. Similarly, group-mean centered level-1 (within-cluster) variables cannot explain between-cluster variance. Imagine a situation in which 5% of the variance in the outcome variable lies between clusters and 95% of the variance lies within clusters. A variable that explains 80% of the between-cluster variance only explains 4% (80% * 5%) of the total variance. In contrast, a variable that explains only 5% of the within-cluster variance explains 4.75% of the total variance for the specified model.

Proportional Reduction in Variance Statistics

The most-commonly reported multilevel pseudo-R^2 statistics (i.e., measures of the proportion of outcome variance explained by a model) are Raudenbush and Bryk's (2002) *proportional reduction in variance statistics.* In models that do not include randomly varying slopes, we can estimate the proportional reduction in level-1 residual variance and the proportional reduction in variance for the level-2 intercept (residual) variance. We can also compute the proportional reduction in variance for the level-2 slope (residual) variance for slopes that randomly vary across clusters.

The proportional reduction in variance compares the residual variance from the full (more-parameterized) model to the residual variance from a simpler "base" model. If the full model explains additional variance, then the residual variance should decrease. In such a scenario, the full residual variance for the baseline model should be greater than the residual variance for the full model.

To compute the proportional reduction in level-1 variance, we subtract the residual (unexplained) variance of the more-parameterized, "fitted" level-1 model from the variance of the simpler, unconditional "base" model. Then, we divide this difference by the variance from the base model. The formula for this computation is:

$$\frac{\hat{\sigma}_b^2 - \hat{\sigma}_f^2}{\hat{\sigma}_b^2} \tag{3.10}$$

where $\hat{\sigma}_b^2$ is the estimated level-1 variance for the unconditional/base model and $\hat{\sigma}_f^2$ is the estimated level-1 variance for the more-parameterized, fitted model (Raudenbush & Bryk, 2002). After establishing a base model, we can compare subsequent fitted MLMs by determining the amount of residual variance each explains over and above that base model.

At level 2, $\hat{\tau}_{qq}$ are the estimated variance components for the intercept (β_{0j}) and each slope ($\beta_{1j}, \beta_{2j}, \ldots, \beta_{qj}$) that randomly varies across clusters. To calculate the proportional reduction in variance for a specific variance component, use the following formula:

$$\frac{\hat{\tau}_{qq_b} - \hat{\tau}_{qq_f}}{\hat{\tau}_{qq_b}} \tag{3.11}$$

where $\hat{\tau}_{qq_b}$ is the estimated variance of parameter q in the base model and $\hat{\tau}_{qq_f}$ is the estimated variance of q in the fitted model (e.g., $\tau_{11}, \tau_{22}, \ldots, \tau_{qq}$). Level-2 proportion reduction in variance statistics should only be used to compare models with identical level-1 models (Raudenbush & Bryk, 2002, p. 150).

However, note that Raudenbush and Bryk's (2002) proportion reduction in variance statistic does not behave like the familiar R^2 from single-level regression. It is simply a comparison of the residual variances from two models, so we cannot: (a) interpret it as the absolute amount of variance explained in the dependent variable; (b) compute it without estimating two separate models; or (c) calculate it for models that differ in their randomly varying slope structure. Furthermore, the proportion reduction in variance can be negative. This frequently occurs when comparing the level-2 intercept variance from a null model (i.e., an unconditional, random-effects ANOVA with no predictors at either level) to the level-2 intercept variance from a fitted model with a group-mean centered predictor at level 1.

Variance Decomposition Framework for MLM

More recently, Rights and Sterba (2019) developed "an integrative framework of R^2 measures for MLMs with random intercepts and/or slopes based on a completely full decomposition of variance" (p. 309). This framework allows for easy computation of existing variance-explained measures and alleviates the need to estimate multiple models.

However, partitioning total outcome variance into within- and between-cluster variance requires group-mean centering all level-1 predictors. To retain the cluster-level variance in the group-mean centered predictor, we re-introduce the aggregate of the group-mean centered level-1 variables into the level-2 model.[10] If we decompose all predictors into within- and between-cluster variables, then the model-implied total outcome variance contains five specific sources of variation: variance attributable to level-1 predictors via fixed slopes (f_1); level-2 predictors via fixed slopes (f_2); level-1 predictors via random slope variation and covariation (v); cluster-specific outcome means via random intercept variation (m);[11] and level-1 residuals (σ^2). Decomposing the model-implied total variance into these five sources enables the computation of a variety of variance-explained measures,[12] each of which provides potential insights into the model's predictive capability (Rights & Sterba, 2019).

Careful examination of this framework reveals a couple of important observations. First, assuming group-mean centered level-1 variables, three sources of variation contain only within-cluster variance: level-1 predictors via fixed slopes (f_1); level-1 predictors via random slope variation and covariation (v); and level-1 residuals (σ^2). Therefore, we can evaluate the proportion of within-cluster variance explained by the specified set of level-1 predictors (f_1) as well as the variances and covariances of their associated randomly varying slopes (v), and we can also distinguish these sources of explained variance from the remaining unexplained level-1 residual variance.

Second, the remaining sources of variation (i.e., level-2 predictors via fixed slopes (f_2) and cluster-specific outcome means via random intercept variation (m)) contain only between-cluster variance. In other words, between-cluster variance is either explained by our modeled set of level-2 variables (f_2) or captured by allowing intercepts to randomly vary (m).

Although a cluster-level variable cannot explain within-cluster variance, a cross-level interaction involving a between-cluster (level-2) variable and a within-cluster (level-1) variable can. Multiplying the cluster-level variable by the within-cluster (level-1) variable removes any relationship between the cluster-level variable and cluster-specific outcome means, but it does not necessarily remove any existing relationship between the within-cluster variable and the group-mean centered outcome values. Therefore, the cross-level interaction behaves like any other level-1 predictor included in the model: it can contribute to the variance attributable to level-1 predictors via fixed slopes (f_1), and to the variance attributable to level-1 predictors via random slope variation and covariation (v). Also, any within-cluster interactions (a within-cluster variable interacts with another within-cluster variable) function as level-1 predictors and any cluster-level interactions (a cluster-level variable interacts with a cluster-level variable) function as level-2 predictors.

Understanding proportion of variance-explained measures. To calculate the R^2 measures described in Rights and Sterba (2019) requires the model parameter estimates (i.e., the level-1 residual variance, fixed-effect estimates for all predictors, variance components for all random intercept and slope parameters, and all covariances among these variance components) as well as the sample variance-covariance matrix for the specific set of predictors included in the MLM. Given this information, and under the assumption that all level-1 variables are group-mean centered (including categorical variables), the following formulae produce Rights and Sterba's proposed R^2-measure components. For example, f_1 represents variance explained by fixed level-1 predictors:

$$f_1 = \gamma_W \Phi_W \gamma_W, \tag{3.12}$$

where γ_W is the vector of level-1 fixed effect estimates (ignoring the intercept term), and Φ_W is the variance-covariance matrix of the level-1 predictors included in the model. This set of level-1 predictors must include any within-cluster or cross-level interaction product variables as well.

The f_2 component (i.e., variance explained by the fixed level-2 predictors) is

$$f_2 = \gamma_B \Phi_B \gamma_B, \tag{3.13}$$

where γ_B is the vector of level-2 fixed effect estimates, and Φ_B is the variance-covariance matrix of the level-2 predictors included in the model. This set of level-2 predictors must include any cluster-level interaction product variables.

The v component (i.e., variance explained by random intercept/slope variation and covariation) is

$$v = tr(\mathbf{T\Sigma}), \tag{3.14}$$

where $tr(\,)$ is the trace function (the sum of the diagonal elements of a matrix), $\mathbf{T}$ is the variance-covariance matrix of the random effects (the "Tau matrix"), and $\mathbf{\Sigma}$ is the variance-covariance matrix of the level-1 predictors with randomly varying slopes in the MLM (this includes a variance of 0 for the intercept and covariances of 0 between the intercept and all of the predictors because the intercept is a constant).

The final two components do not require any additional computation–the m component is the estimate of random intercept variance (i.e., τ_{00}), and σ^2 is the level-1 residual variance estimate.

The model-implied level-1 outcome variance is the sum of the f_1, v, and σ^2 components; the model-implied level-2 outcome variance is the sum of the f_2 and m components; and the model-implied total outcome variance is the sum of all five components.

Notably, group-mean centering every level-1 variable allows the contribution of fixed effects to be partitioned into two components, f_1 and f_2, which in turn allows for computation of the within, between, and total R^2 measures. When modeling non-group-mean-centered level-1 variables (as researchers commonly model categorical predictors), it is still possible to compute total R^2 measures. However, it is not possible to decompose f into f_1 and f_2; therefore, it is not possible to separately compute the within- and between-level R^2 measures. The overall f component represents variance attributable to all predictors via fixed slopes:

$$f = \gamma\Phi\gamma, \tag{3.15}$$

where γ is the vector of fixed effect estimates for all predictors included in the model (ignoring the intercept term) and $\mathbf{\Phi}$ is the variance-covariance matrix of all predictors included in the model. Additionally, if there are any non-group-mean centered level-1 variables, the m component is:

$$m = \mu'\mathrm{T}\mu, \tag{3.16}$$

where μ is a column vector containing the means of the predictors with random slopes (including a 1 as its first element to represent the random intercept).

Using these five main sources of variation (i.e., f_1, f_2, v, m, σ^2), we can compute several variance-explained measures. For example, the proportion of within-cluster outcome variance explained by level-1 predictors via fixed slopes ($R_w^{2(f_1)}$) indicates how well the *fixed effects* for the set of level-1 predictors (slopes) explain within-cluster variance in the outcome variable:

$$R_w^{2(f_1)} = \frac{f_1}{f_1 + v + \sigma^2} \tag{3.17}$$

If any level-1 slopes randomly vary across clusters, we compute the proportion of within-cluster variance explained by level-1 predictors via random-slope variation/covariation:

$$R_w^{2(v)} = \frac{v}{f_1 + v + \sigma^2} \tag{3.18}$$

This indicates the proportion of within-cluster variance explained by allowing slopes to randomly vary across clusters. Adding cross-level interactions to explain between-cluster variability in the slopes decreases v and increases f_2. Therefore, $R_w^{2(v)}$ is the slope variance that is not explained by level-2 (between-cluster) variables.

Inspecting $R_w^{2(v)}$ also provides insight into the pervasive question: should all, some, or none of the slopes in the MLM randomly vary? To address this issue, we can examine the $R_w^{2(v)}$ separately for each randomly varying slope to determine how much within-cluster variance each slope explains. Specifically, we fit a model in which only one slope randomly varies. The $R_w^{2(v)}$ for the model with a single randomly varying slope indicates the proportion of within-cluster variance attributable to that random effect.

The total proportion of within-cluster outcome variance explained by level-1 predictors via fixed slopes and random slope variation/covariation ($R_w^{2(f_1 v)}$) is the sum of the variance attributable to level-1 predictors via fixed slopes (f_1) and via random-slope variation and covariation (v), divided by the total level-1 variance:

$$R_w^{2(f_1 v)} = \frac{f_1 + v}{f_1 + v + \sigma^2} = 1 - \frac{\sigma^2}{f_1 + v + \sigma^2} \tag{3.19}$$

This is analogous to computing Raudenbush and Bryk's (2002) proportion reduction in level-1 residual variance. One minus $R_w^{2(f_1 v)}$ is the proportion of within-cluster variance not explained by either the fixed effects or the randomly varying slopes for the specified set of level-1 predictors.

$$1-R_{w}^{2(f_1v)}=\frac{\sigma^2}{f_1+v+\sigma^2} \tag{3.20}$$

In addition, at level 2, the proportion of between-cluster outcome variance explained by level-2 predictors via fixed slopes ($R_b^{2(f_2)}$) captures the degree to which the set of level-2 predictors explains between-cluster variance in the outcome variable.

$$R_b^{2(f_2)}=\frac{f_2}{f_2+m}=\frac{f_2}{f_2+\tau_{00(f)}} \tag{3.21}$$

This is equivalent to Raudenbush and Bryk's proportion reduction in level-2 variance.

Note that one minus the proportion of between-cluster outcome variance explained by level-2 predictors via fixed slopes is the proportion of the between-cluster variance that remains unexplained by the model. This is the proportion of level-2 residual variance in the intercept:

$$R_b^{2(m)}=1-\frac{f_2}{f_2+m}=\frac{m}{f_2+m}=\frac{\tau_{00(f)}}{f_2+\tau_{00(f)}} \tag{3.22}$$

Furthermore, the proportion of total (outcome) variance explained by cluster-specific outcome means via random-intercept variation ($R_t^{2(m)}$) is the conditional ICC. The conditional ICC indicates the proportion of (residual) between-cluster variance in the intercept after accounting for all the variables in the model.[13] It indicates how much variance is explained by allowing the intercept to randomly vary across clusters (Rights & Sterba, 2019) $R_t^{2(m)}$ is

$$R_t^{2(m)}=\frac{m}{m+f_2+f_1+v+\sigma^2}=\frac{\tau_{00(f)}}{\tau_{00(f)}+f_2+f_1+v+\sigma^2} \tag{3.23}$$

Notably, the proportion of total variance explained by level-1 and level-2 fixed effects ($R_t^{2(f)}$) is somewhat analogous to an *R*-squared measure in single-level regression—it captures the proportion of outcome variance explained by fixed effects across levels of the MLM:

$$R_t^{2(f)}=\frac{f_1+f_2}{f_1+f_2+v+m+\sigma^2} \tag{3.24}$$

$R_t^{2(f)}$ can be decomposed into its two constituent pieces: $R_t^{2(f_1)}$ and $R_t^{2(f_2)}$. $R_t^{2(f_1)}$ is the proportion of total variance explained by level-1 slopes and cross-level interactions (given that all level-1 predictors are group-mean centered):

$$R_t^{2(f_1)} = \frac{f_1}{f_1 + f_2 + v + m + \sigma^2} \tag{3.25}$$

($R_t^{2(f_2)}$) is the proportion of total variance explained by level-2 predictors:

$$R_t^{2(f_2)} = \frac{f_2}{f_1 + f_2 + v + m + \sigma^2} \tag{3.26}$$

Finally, we may wish to report the proportion of outcome variance explained by the variances and covariances of our model's randomly varying slopes ($R_t^{2(v)}$):

$$R_t^{2(v)} = \frac{v}{f_1 + f_2 + v + m + \sigma^2} \tag{3.27}$$

Computing Proportion of Variance-Explained Measures. We can easily calculate Rights and Sterba's (2019) R^2 measures and their constituent components using any software that performs matrix calculations.[14] We present the general strategy here, followed by an applied example. Imagine we estimate an MLM with a random intercept and a random slope and save the results. (Let's assume we group-mean centered all level-1 predictors.) In R, we could use the following code to estimate the f_1 component:

```
f1 <- t(______)%*%var(______)%*%______
```

Here, the "`t()`" function transposes a vector/matrix, the "`%*%`" operator represents matrix multiplication, and the "`var()`" function produces the variance-covariance matrix for the variables identified within the parentheses. We replace the first and last blanks with a column vector of our level-1 fixed-effect estimates (excluding the intercept estimate), for example, "`c(0.123,-0.456)`" (the numbers inserted in this section's example code are all fabricated). Next, we replace the second blank with the level-1 predictor variables corresponding to those fixed-effect estimates, in the same order as the effects appear in the previous vector, for example, "`cbind(x1,x2)`".[15] If there are no level-1 predictors, then we set this component equal to zero ("`f1 <- 0`").

We use similar code to estimate the f_2 component:

```
f2 <- t(______)%*%var(______)%*%______
```

However, we now replace the first and last blanks with a column vector of our level-2 fixed-effect estimates, for example, "`c(1.234,0.567)`." Then, we replace the second blank with the level-2 predictor variables corresponding to those fixed-effect estimates (in the same order as their effects appear in the previous vector), for example, "`cbind(x3,x4)`." If there are no level-2 predictors, then we set this component equal to zero ("`f2 <- 0`").

Next, we estimate the v component with the following code:

```
v <- sum(diag(______%*%var(______)))
```

Here, the "`diag()`" function returns a vector of the diagonal elements of the specified matrix; the "`sum()`" function calculates the sum of these diagonal elements. Therefore, we replace the first blank with the variance-covariance matrix for our random slopes. For two randomly varying level-1 slopes, this would be a 2×2 matrix containing random slope variances on the diagonal and their covariance on the off-diagonal, for example, "`matrix(c(0.123,-0.456,-0.456,0.789),2,2)`." This matrix is basically the tau matrix with no random intercept elements. Next, we replace the second blank with the level-1 predictor variables that feature random slopes (in the same order as their variances appear in the previous matrix). If all our level-1 predictors have random slopes, this would be the same code as in the "`var()`" function for f_1: `cbind(x1,x2)`. However, if our model has no random slopes, we set v equal to zero ("`v <- 0`").

In contrast, the m and σ^2 components require no additional computation. The m component is equal to the model's random intercept variance estimate (e.g., "`m <- 0.579`") and the σ^2 component is equal to the level-1 residual variance estimate (e.g., "`s2 <- 0.421`"). Thus, the code to specify each of the components for this example is

```
f1 <- t(c(0.123,-0.456))%*%var(cbind(x1,x2))%*
   %c(0.123,-0.456)
f2 <- t(c(1.234,0.567))%*%var(cbind(x3,x4))%*%c(1.234,0.567)
v <- sum(diag(matrix(c(0.123,-0.456,-0.456,0.789),2,2)%*%var
   (cbind(x1,x2))))
m <- 0.579
s2 <- 0.421
```

Once we specify the f_1, f_2, v, m, and σ^2 components, we can easily estimate the model-implied total outcome variance, level-1 outcome variance, and level-2 outcome variance:

```
t_var<- f1 + f2 + v + m + s2
w_var<- f1 + v + s2
b_var<- f2 + m
```

Then, we can estimate each of Rights and Sterba's (2019) R^2 measures with the code in Table 3.3.

TABLE 3.3 R Code for Computing Select Rights and Sterba's (2019) R^2 Measures

R Code	Estimated R^2 Measure
`R2_f1_t <- f1/t_var`	$R_t^{2(f_1)}$
`R2_f2_t <- f2/t_var`	$R_t^{2(f_2)}$
`R2_v_t <- v/t_var`	$R_t^{2(v)}$
`R2_m_t <- m/t_var`	$R_t^{2(m)}$
`R2_f1_w <- f1/w_var`	$R_w^{2(f_1)}$
`R2_v_w <- v/w_var`	$R_w^{2(v)}$
`R2_f2_b <- f2/b_var`	$R_b^{2(f_2)}$
`R2_m_b <- m/b_var`	$R_b^{2(m)}$

If we do not group-mean center all the level-1 predictors, then we cannot separately estimate the within and between R^2 measures. We *can* use the same code to estimate the total R^2 measures, but we must replace the f_1 and f_2 components with the more-general f component:

```
f <- t(______)%*%var(______)%*%______
```

In this scenario, we replace the first and last blanks with a column vector of our fixed-effect estimates, regardless of level, for example, "`c(0.123,-0.456,1.234,0.567)`." Then, we replace the second blank with the predictor variables corresponding to those fixed-effect estimates (in the same order as the effects appear in the previous vector, for example, "`cbind(x1,x2,x3,x4)`." In a model with no predictors at any level, set f equal to zero ("`f <- 0`"). Then, we calculate the $R_t^{2(f)}$ measure by dividing f by the total outcome variance, yielding an estimate of the proportion of total variance explained by the set of the predictors.

Additionally, if we do not transform all the level-1 predictors with random slopes to have means of zero (e.g., group-mean centering, grand-mean centering, or standardizing), then we need to use the following code for the m component:

```
m <- t(c(colMeans(cbind(1,_____))))%*%______%*%c
    (colMeans(cbind(1,_____)))
```

Here, we replace the first and last blanks with the predictor variables that have random effects in the same order in which they appear in the tau matrix, for example, "`x1,x2`." Then, we replace the second blank with a matrix containing the variance-covariance matrix for our random intercept and slopes, for example, "`matrix(c(0.579,0.246,0.357,0.246,0.123,-0.456,0.357,-0.456,0.789),3,3)`."

Remarkably, all the R^2 measures discussed thus far represent "single-source" R^2 measures because they describe the proportion of total outcome variance attributable to a single model component. However, we can also compute R^2 measures for multiple components (i.e., "combination-source" R^2 measures) by summing the "single-source" R^2 measures for the components of interest. For example, to estimate the proportion of total outcome variance attributable to everything other than level-1 residuals, we can calculate $R_t^{2(fvm)}$. This quantity represents the sum of $R_t^{2(f_1)}$, $R_t^{2(f_2)}$, $R_t^{2(v)}$, and $R_t^{2(m)}$ for group-mean centered level-1 predictors; it denotes the sum of $R_t^{2(f)}$, $R_t^{2(v)}$, and $R_t^{2(m)}$ for non-group-mean centered level-1 predictors.

EXAMPLE[16]

This example demonstrates how both model fit and model adequacy measures can inform model selection decisions. For consistency, we used the data presented in Chapter 2 which contained 14,176 students clustered within 1,029 schools. The outcome variable, externalizing behavior, was measured on a 4-point Likert rating scale ($M = 1.63$, $SD = 0.64$). Our main predictor of interest was socioeconomic status (SES; $M = -0.04$, $SD = 0.81$).

For pedagogical purposes, we modeled the SES predictor two ways. First, we group-mean centered SES at the student level and reintroduced its grand-mean centered contextual effect at the school level. Next, we modeled SES, grand-mean centered, at the student level. We specifically estimated these distinct models for several reasons: (a) to demonstrate a non-nested model comparison (our group-mean centered FITTED and grand-mean centered ALTERNATIVE models are not nested); (b) to illustrate the within and between variance-explained measures available under group-mean centering; (c) to compare the multiple decompositions of variance for group-mean centered variables to the compacted, total outcome variance measures available for grand-mean centered predictors (see explanation of R^2 results); and (d) to account for the ubiquity of grand-mean centering in extant literature. The Alternative Model is not a model that would normally be included. Instead, we recommend selecting a centering technique prior to conducting analyses.

As such, we estimated[17] the following four MLMs:

1. *Unconditional Model:* This model included a random effect for the intercept, but no predictors.

	Multilevel Equations		Combined/Mixed Equation	
Level 1:	$CEXTERN_{ij} = \beta_{0j} + r_{ij}$	(3.28)	$CEXTERN_{ij} = \gamma_{00} + u_{0j} + r_{ij}$	(3.29)
Level 2:	$\beta_{0j} = \gamma_{00} + u_{0j}$			

2. *Fitted Model #1:* The first fitted model included SES group-mean centered at the student level and SES grand-mean centered at the school level. It featured a random effect for the intercept and fixed effects for both SES slopes.

	Multilevel Equations		Combined/Mixed Equation	
Level 1:	$CEXTERN_{ij} = \beta_{0j} + \beta_{1j}{*}(SES_{ij}) + r_{ij}$	(3.30)	$CEXTERN_{ij} = \gamma_{00} + \gamma_{01}{*}SCHL\_SES_j$	(3.31)
Level 2:	$\beta_{0j} = \gamma_{00} + \gamma_{01}{*}(SCHL\_SES_j) + u_0$		$+ \gamma_{10}{*}SES_{ij} + u_{0j} + r_{ij}$	
	$\beta_{1j} = \gamma_{10}$			

3. *Fitted Model #2:* The second fitted model also included SES group-mean centered at the student level and SES grand-mean centered at the school level. This model included a randomly varying student-level SES slope.

	Multilevel Equations		Combined/Mixed Equation	
Level 1:	$CEXTERN_{ij} = \beta_{0j} + \beta_{1j}{*}(SES_{ij}) + r_{ij}$	(3.32)	$CEXTERN_{ij} = \gamma_{00} + \gamma_{01}{*}SCHL\_SES_j$	(3.33)
Level 2:	$\beta_{0j} = \gamma_{00} + \gamma_{01}{*}(SCHL\_SES_j) + u_{0j}$		$+ \gamma_{10}{*}SES_{ij} + u_{0j} + u_{1j}{*}SES_{ij} + r_{ij}$	
	$\beta_{1j} = \gamma_{10} + u_{1j}$			

4. *Alternative Model:* The Alternative Model included SES grand-mean centered at the student level. (Note, that the contextual effect for SES at the school level was not included in this model). This model featured random effects for the intercept and the student-level SES slope.

	Multilevel Equations		Combined/Mixed Equation	
Level 1:	$CEXTERN_{ij} = \beta_{0j} + \beta_{1j}{*}(SES_{ij}) + r_{ij}$	(3.34)	$CEXTERN_{ij} = \gamma_{00} + \gamma_{10}{*}SES_{ij} + u_{0j}$	(3.35)
Level 2:	$\beta_{0j} = \gamma_{00} + u_{0j}$		$+ u_{1j}{*}SES_{ij} + r_{ij}$	
	$\beta_{1j} = \gamma_{10} + u_{1j}$			

See Table 3.4 for parameter estimates and interpretations for each of these four estimated models.

As indicated previously in this chapter, we can compare competing models with the Likelihood Ratio Test if (and only if) the tested models are nested. This approach would be appropriate for comparing each of our two fitted models (in which we group-mean centered SES) to the unconditional model or for comparing our Alternative Model, with SES grand-mean centered, to the unconditional model. However, the fitted and alternative models were not nested. Therefore, we turned to the AIC and BIC to help inform our selection from a model fit perspective.

As shown in Table 3.5, comparing Fitted Models 1 and 2, the AIC and BIC did not provide consistent conclusions—the AIC preferred Fitted Model 2, whereas the BIC-L1 and BIC-L2 favored Fitted Model 1. (Again, the

TABLE 3.4 Results and Interpretations for Four Competing Multilevel Models

Unconditional Model (3 parameters)						
Fixed	Est.	*SE*	*t* ratio	*df*	*p*	Interpretation
Intercept, γ_{00}	1.635	0.008	208.862	1028	< .001	The ICC (.076) indicated that approximately 7.6% of the variance in Externalizing Behavior can be explained by which school the student attends. Absent any additional student- or school-level information, the expected student-level Externalizing Behavior score is 1.64. There is statistically significant variability in these behavior scores across schools.
Random	*SD*	Var.	*df*	χ^2	*p*	
(u_0) τ_{00}	(0.175)	0.031	1028	2,228.375	< .001	
(r) σ^2	(0.611)	0.373	—	—	—	
Fitted Model #1: SES group-mean centered, No random SES slope (5 parameters)						
Fixed	Est.	*SE*	*t* ratio	*df*	*p*	Interpretation
Intercept, γ_{00}	1.639	0.008	213.022	1027	< .001	We expect a student at his/her school's mean on SES, in a school of average SES, to have an Externalizing Behavior score of 1.64. For each unit above his/her school's mean a student is on SES, expected Externalizing Behavior decreases by .08 points, on average. For each unit above the mean a school is on SES, expected Externalizing Behavior decreases by .10 points. Statistically significant variation in Externalizing Behavior across schools remains unexplained.
SCHL_SES, γ_{01}	–0.096	0.015	–6.519	1027	< .001	
SES, γ_{10}	–0.081	0.009	–9.376	13146	< .001	
Random	*SD*	Var.	*df*	χ^2	*p*	
(u_0) τ_{00}	(0.168)	0.028	1027	2145.578	< .001	
(r) σ^2	(0.609)	0.370	—	—	—	

(continued)

TABLE 3.4 Results and Interpretations for Four Competing Multilevel Models (continued)

Fitted Model #2: SES group-mean centered, Random SES slope (7 parameters)						
Fixed	Est.	*SE*	*t* ratio	*df*	*p*	Interpretation
Intercept, γ_{00}	1.639	0.008	213.134	1027	< .001	A student at his/her school's mean on SES, in a school of average SES, is expected to have an Externalizing Behavior score of 1.64. For each unit above the school's mean a student is on SES, expected Externalizing Behavior decreases by .08 points, on average. For each unit above the grand mean a school is on SES, expected Externalizing Behavior decreases by .10 points. Statistically significant variation in Externalizing Behavior across schools remains unexplained. In addition, there is statistically significant variation in the impact of SES on Externalizing Behavior scores.
SCHL_SES, γ_{01}	–0.098	0.015	–6.692	1027	< .001	
SES, γ_{10}	–0.081	0.009	–9.367	1028	< .001	
Random	*SD*	Var.	*df*	χ^2	*p*	
(u_0) τ_{00}	(0.169)	0.028	901	1979.682	< .001	
(u_1) τ_{11}	(0.067)	0.005	902	980.331	.035	
(r) σ^2	(0.607)	0.369	—	—	—	
τ_{10} [correlation]	$r = -0.393$					
Alternative Model: SES grand-mean centered, Random SES slope (6 parameters)						
Fixed	Est.	*SE*	*t* ratio	*df*	*p*	Interpretation
Intercept, γ_{00}	1.635	0.008	215.088	1028	< .001	A student at the overall mean on SES has an expected Externalizing Behavior score of 1.64. For each unit a student is above the overall mean on SES, expected Externalizing Behavior decreases by .09 points. Statistically significant residual variation exists in both Externalizing Behavior scores and the impact of SES on Externalizing Behavior scores.
SES, γ_{10}	–0.087	0.008	–11.373	1028	< .001	
Random	*SD*	Var.	*df*	χ^2	*p*	
(u_0) τ_{00}	(0.163)	0.027	902	1505.641	< .001	
(u_1) τ_{11}	(0.078)	0.006	902	982.260	.032	
(r) σ^2	(0.607)	0.368	—	—	—	
τ_{10} [correlation]	$r = -0.295$					

TABLE 3.5 Parameter Estimates From Four Competing Multilevel Models

	Unconditional Model			Fitted Model 1			Fitted Model 2			Alternative Model		
Fixed	**Est.**	***SE***	***p***	**Est.**	***SE***	***p***	**Est.**	***SE***	***p***	**Est.**	***SE***	***p***
Intercept, γ_{00}	1.635	0.008	< .001	1.639	0.008	< .001	1.639	0.008	< .001	1.635	0.008	< .001
SCHL_SES, γ_{01}	—	—	—	–0.096	0.015	< .001	–0.098	0.015	< .001	—	—	—
SES, γ_{10}	—	—	—	–0.081	0.009	< .001	–0.081	0.009	< .001	–0.087	0.008	< .001
Random	**Var.**	χ^2	***df***	**Var.**	χ^2	***df***	**Var.**	χ^2	***df***	**Var.**	χ^2	***df***
$(u_0)\ \tau_{00}$	0.031	2,228.375	1028	0.028	2145.578	1027	0.028	1979.682	901	0.027	1505.641	902
$(u_1)\ \tau_{11}$	—	—	—	—	—	—	0.005	980.331	902	0.006	982.26	902
$(r)\ \sigma^2$	0.373	—	—	0.370	—	—	0.369	—	—	0.368	—	—
τ_{10} [correlation]							–.393			–.295		
Est. Pars	3			5			7			6		
Deviance	26,995.955			26,856.367			26,848.734			26,844.508		
AIC	27,001.955			26,866.367			26,862.734			26,856.508		
BIC-L1	27,024.632			26,904.163			26,915.649			26,901.863		
BIC-L2	27,016.764			26,891.049			26,897.288			26,886.126		

Note: ICC = .076. Unconditional Model = Multilevel model with no predictors. Fitted Model 1 = Multilevel model with SES group-mean centered and no random SES slope. Fitted Model 2: Multilevel model with SES group-mean centered and a random SES slope. Alternative Model = Multilevel model with SES grand-mean centered and a random SES slope.

magnitudes of the differences in the AIC and BIC values were fairly small, relative to the values of these criteria.) Our Alternative Model featured the smallest AIC, BIC-L1 (BIC calculated with the total number of level-1 units), and BIC-L2 (BIC calculated with the total number of level-2 units) values, indicating preference for this model over the three competing models (though the magnitudes of the differences in AIC and BIC values were not large). (Brief, conceptual model interpretations appear in Table 3.4, followed by tabulated parameter estimates and model fit information in Table 3.5.)

This situation underscores the point that modelers cannot, and should not, be mindless enforcers of conventional rules of thumb or arbitrary statistical standards. Although model/data fit is certainly an important aspect of model selection, it should not be the only consideration. Therefore, we turned our attention back to the Rights and Sterba (2019) variance decomposition framework to supplement the IC approach and further contribute to a thoughtful evaluation of model adequacy for each of these models.

To calculate Rights and Sterba's (2019) R^2 measures for these models using the R code provided earlier, first, we import the data file into R. Let's call the data frame "dat" for convenience. Then we use the following code to add grand-mean centered versions of Student and School SES to the dataframe:[18]

```
dat[12]<-dat$z_fSES-mean(dat$z_fSES)
dat[13]<-dat$school_ses-mean(dat$school_ses)
```

The grand-mean centered version of Student SES is required for the Alternative Model, and grand-mean centered version of School SES is required for Fitted Models 1 and 2. Table 3.3 provides the exact code needed to calculate the measures for the four example models after these variables are created.

Next, let's take a closer look at the variance decomposition code for Fitted Model 2. First, we calculated the f_1 component: "`f1 <- t(_____)%*%var(_____)%*%_____`." The first and last blanks of this code need to be filled with a column vector of our level-1 predictor fixed effects estimates. In this case, we only had one level-1 predictor, Student SES, so our column vector contained only the fixed effect estimate for Student SES from Table 3.6: "`c(-0.081068)`." (Note: it's best to use as many digits as possible for these calculations, so these numbers contain more digits than those presented in Table 3.4). The middle blank indicated the position of group-mean centered Student SES in our data frame: "`dat[,11]`." Next, we computed the f_2 component: "`f2 <- t(_____)%*%var(_____)%*%_____`." For this line of code we referenced information about the level-2 predictors. The only level-2 predictor in Fitted Model 2 was School SES, so we filled the first and last blanks with a column vector containing the fixed effect estimate of School SES from Table 3.6: "`c(-0.098388)`." The middle blank indicated the position

TABLE 3.6 R Code to Produce Rights and Sterba's R2 Measures for the Example Models

Model	Declaring Model-Implied Variance Components	Calculating Model-Implied Total Variances	Calculating Total R^2 Measures	Calculating Within and Between R^2 Measures
Unconditional Model	f1 <- 0 f2 <- 0 v <- 0 m <- 0.03078 s2 <- 0.37315	t_var<- f1 + f2 + v + m + s2 w_var<- f1 + v + s2 b_var<- f2 + m	R2_f1_t <- f1/t_var R2_f2_t <- f2/t_var R2_v_t <- v/t_var R2_m_t <- m/t_var	R2_f1_w <- f1/w_var R2_v_w <- v/w_var R2_f2_b <- f2/b_var R2_m_b <- m/b_var
Fitted Model #1	f1 <- t(c(-0.081075))%*%var(dat[,11]) %*%c(-0.081075) f2 <- t(c(-0.095938))%*%var(dat[,13]) %*%c(-0.095938) v <- 0 m <- 0.02835 s2 <- 0.37046	t_var<- f1 + f2 + v + m + s2 w_var<- f1 + v + s2 b_var<- f2 + m	R2_f1_t <- f1/t_var R2_f2_t <- f2/t_var R2_v_t <- v/t_var R2_m_t <- m/t_var	R2_f1_w <- f1/w_var R2_v_w <- v/w_var R2_f2_b <- f2/b_var R2_m_b <- m/b_var
Fitted Model #2	f1 <- t(c(-0.081068))%*%var(dat[,11]) %*%c(-0.081068) f2 <- t(c(-0.098388))%*%var(dat[,13]) %*%c(-0.098388) v <- sum(diag(matrix(c(0.00455),1,1) %*%var(dat[,11]))) m <- 0.02844 s2 <- 0.36864	t_var<- f1 + f2 + v + m + s2 w_var<- f1 + v + s2 b_var<- f2 + m	R2_f1_t <- f1/t_var R2_f2_t <- f2/t_var R2_v_t <- v/t_var R2_m_t <- m/t_var	R2_f1_w <- f1/w_var R2_v_w <- v/w_var R2_f2_b <- f2/b_var R2_m_b <- m/b_var
Alternative Model	f <- t(c(-0.086595))%*%var(dat[,12]) %*%c(-0.086595) v <- sum(diag(matrix(c(0.00607),1,1) %*%var(dat[,12]))) m <- 0.02655 s2 <- 0.36809	t_var<- f + v + m + s2	R2_f_t <- f/t_var R2_v_t <- v/t_var R2_m_t <- m/t_var	–

Note: Because the Alternative Model did not use group-mean centered level-1 predictors, it was not possible to separate the *f* component into f_1 and f_2 or to calculate the within and between variance-explained measures.

of grand-mean centered School SES in our data frame: "`dat[,13]`." Finally, we calculated the v component: "`v <- sum(diag(_____%*%var(_____)))`." We filled the first blank with our tau matrix, excluding all elements involving the intercept. In this example, that included the random slope variance for Student SES: "`matrix(c(0.00455),1,1)`." The second blank indicated the position of group-mean centered Student SES in our data frame: "`dat[,11]`." The final two components were the simplest because they required no calculation. The m component was the random intercept variance estimate: "`m <- 0.02844`" and the σ^2 component was the level-1 residual variance estimate: "`s2 <- 0.36864`." After defining all five components, we ran the code in Table 3.6 to calculate the model-implied total variances and R^2 measures depicted in Table 3.7.

The Rights and Sterba R^2 measures elucidated how model adequacy varied across different models of the same outcome. For instance, it was obvious that Fitted Model #1 did a better job predicting the outcome than the unconditional model. However, these R^2 measures allowed us to determine the degree to which Fitted Model 1 outperformed the unconditional model (in terms of model adequacy), as well as the source of the improvement (see Table 3.7). For example, adding group-mean centered SES as a predictor at level 1 explained about 0.6% of the total outcome variance. Adding School SES as a predictor at level 2 similarly explained about 0.6% of the total outcome variance. Thus, the two predictors together explained about 1.2% of the total outcome variance in Fitted Model 1. Although School SES only explained 0.6% of the *total* outcome variance, it explained about 8.4% of the *between-cluster* outcome variance, leaving about 91.6% of the between-cluster outcome variance unexplained.[19] Of course, this model's adequacy

TABLE 3.7 R2 Measures for the Four Competing Multilevel Models

R^2 Measure	Unconditional Model	Fitted Model #1	Fitted Model #2	Alternative Model
$R_t^{2(f_1)} =$	0	0.006	0.006	—
$R_t^{2(f_2)} =$	0	0.006	0.007	—
$R_t^{2(f)} =$	0	0.013	0.013	0.012
$R_t^{2(v)} =$	0	0	0.004	0.010
$R_t^{2(m)} =$	0.076	0.070	0.070	0.066
$1-R_t^{2(fvm)} =$	0.924	0.917	0.912	0.912
$R_w^{2(f_1)} =$	0	0.007	0.007	—
$R_w^{2(v)} =$	0	0	0.005	—
$1-R_w^{2(f_2m)} =$	1.000	0.993	0.989	—
$R_b^{2(f_2)} =$	0	0.084	0.088	—
$R_b^{2(m)} =$	1.000	0.916	0.912	—

could surely be improved to explain some of the 91.7% of remaining total outcome variance.

Fitted Model 2 allowed the group-mean centered SES slope to randomly vary. This random slope variation accounted for about 0.4% of the total outcome variance, and about 0.5% of the within-cluster outcome variance.

The Alternative Model featured grand-mean centered SES at level 1 and excluded the School SES variable at level 2. The overall results were similar: The Alternative Model still left about 91% of the total outcome variance unexplained. However, choosing not to group-mean center the level-1 predictors prevented us from obtaining the more fine-grained within- and between-cluster R^2 measures. For example, in Fitted Model 2, the fixed components explained about 1.3% ($R_t^{2(f)} = .013$) of the total outcome variance: 0.6% by fixed effects of level-1 predictors ($R_t^{2(f_1)} = .006$) and 0.7% explained by fixed effects of level-2 predictors ($R_t^{2(f_2)} = .007$). In the Alternative Model, it was not possible to discern each level's contribution to the $R_t^{2(f)}$. However, grand-mean centered SES explained 1.2% of the total variance in the outcome variable ($R_t^{2(f)}$), which is virtually identical to the proportion of variance explained by the combination of individual- and cluster-level SES variables featured in Fitted Model #2. In contrast, Fitted Model 2's $R_t^{2(f)}$ was slightly larger than the Alternative Model's $R_t^{2(f)}$ because the former utilized information from both the group-mean centered Student SES and the School SES variables, whereas the Alternative Model only incorporated information from the Student SES variable (which contained all of the information about group-mean centered Student SES, but none of the information about School SES). If the correlation between uncentered Student SES and School SES was large, the $R_t^{2(f)}$ values for these models would be similar; if the correlation was small, then the $R_t^{2(f)}$ values would differ more. In this example, the correlation between Student SES and School SES was moderately high ($r = .65$), which explains why the models had similar $R_t^{2(f)}$ values.

Overall, these R^2 measures demonstrated the weak predictive ability of SES when explaining externalizing behavior. Most of the original between-cluster and within-cluster variance remained unexplained. From a more practical perspective, the predicted values produced by these models would do a poor job of predicting observed outcome values. Put differently, the $R_t^{2(f)}$ measure is analogous to the traditional R^2 measure used in linear regression. In standard regression, the predicted values only utilize information from the fixed components of the model: the $R_t^{2(f)}$ is the proportion of model-implied total outcome variance explained by all the fixed components included in the model. Therefore, an $R_t^{2(f)}$ of .013 would indicate that SES explained only 1.3% of the variance in the outcome variable and imply that the expected correlation between predicted values and observed values would be about .11 (.11 is the square root of the $R_t^{2(f)}$ value).

CHAPTER SUMMARY

Measures of model fit and model adequacy provide useful tools for comparing, selecting, and interpreting MLMs in applied research contexts. When used together, they offer distinct, but complementary, perspectives on the utility of theoretical models, allowing researchers to make informed and justifiable modeling decisions. In addition, techniques for evaluating both model fit and model adequacy continue to provide fruitful areas for methodological research. However, these tools are best viewed as heuristic guides, rather than statutes. As such, it is important to use these measures thoughtfully and selectively, with attention to their limitations.

RECOMMENDED READINGS

Technical Readings

Hamaker, E. L., van Hattum, P., Kuiper, R. M., & Hoijink, H. (2011). Model selection based on information criteria in multilevel modeling. In J. J. Hox & K. L. Roberts (Eds.), *Handbook of Advanced Multilevel Analysis* (pp. 231–256). Taylor & Francis.

Konishi, S., & Kitagawa, G. (2008). *Information criteria and statistical modeling.* Springer Science+Business Media.

Rights, J. D., & Sterba, S. K. (2019). Quantifying explained variance in multilevel models: An integrative framework for defining R-squared measures. *Psychological Methods, 24*(3), 309–338. https://doi.org/10.1037/met0000184

Applied Readings

Burnham, K. P., & Anderson, D. R. (2004). Multimodel inference: Understanding AIC and BIC in model selection. *Sociological Methods & Research, 33*(2), 261–304. https://doi.org/10.1177/0049124104268644

Vrieze, S. I. (2012). Model selection and psychological theory: A discussion of the differences between the Akaike information criterion (AIC) and the Bayesian information criterion (BIC). *Psychological Methods, 17*(2), 228–243. https://doi.org/10.1037/a0027127

Weakliem, D. L. (2016). *Hypothesis testing and model selection in the social sciences.* Guilford Press.

TRY THIS!

Using the same data set described in this chapter (available at the online appendix, http://modeling.uconn.edu) and previously discussed in

Chapter 2, consider that we are now interested in the following research questions: (a) "How do Externalizing Behavior scores differ for male and female students?" and (b) "Is the effect of gender on externalizing behavior constant across schools, or does it vary?" The data set includes 14,176 students clustered within 1,029 schools. The outcome variable is externalizing behavior, measured on a 4-point Likert rating scale ($M = 1.63$, $SD = 0.64$). Our new level-1 predictor of interest is the dichotomous dummy variable, MALE (coded 0/1, such that "1" indicates the student is male); the sample includes about as many females as males (51.31% male, 48.69% female). Note that MALE should be modeled as an uncentered predictor for each of the following models. How would centering MALE affect your interpretation of the results?

With this information and using the example from this chapter as a reference, estimate each of the following three MLMs. Report and interpret the results for each model.

- Which model do you prefer, in terms of model/data fit?
- Which model do you prefer, in terms of model adequacy?
- Compare/contrast your decisions from these two perspectives (model fit vs. model adequacy).
- Then, considering BOTH model fit and model adequacy, determine which model you would favor overall. Defend your selection.

1. *Unconditional Model:* A model with a random effect for the intercept, but no predictors.
2. *Try This Model 1*: This model should include MALE, uncentered at the student level. It should contain a random effect for the intercept and a fixed effect for the MALE slope.

	Multilevel Equations	Combined/Mixed Equation
Level 1:	$CEXTERN_{ij} = \beta_{0j} + \beta_{1j} * (\text{MALE}_{ij}) + r_{ij}$	
Level 2:	$\beta_{0j} = \gamma_{00} + u_{0j}$	
	$\beta_{1j} = \gamma_{10}$	

3. *Try This Model 2*: This model should include MALE, uncentered at the student level. It should contain random effects for the intercept and the MALE slope.

	Multilevel Equations	Combined/Mixed Equation
Level 1:		$CEXTERN_{ij} = \gamma_{00} + \gamma_{10} * MALE_j + u_{0j} + u_{1j}$
Level 2:		$* MALE_{ij} + r_{ij}$

NOTES

1. In this chapter, we focus exclusively on maximum likelihood estimation, but it is also possible to use fully Bayesian methods to estimate MLMs.
2. In addition, the school mean that is computed from a school with 1,000 students is likely to be a much better estimate of the school's performance than a school mean that is computed from only two students, a point to which we return when we discuss Empirical Bayes estimates.
3. Note that, implicitly, the deviance is both a function of model fit, as well as sample size.
4. Burnham and Anderson express this sentiment more eloquently than we do: "Are models ever true, in the sense that full reality is represented exactly by a model we can conceive and fit to the data, or are models merely approximations?" (Burnham & Anderson, 2004, p. 262).
5. To fit a model with a random intercept and slope where the intercept and slope are uncorrelated, use the formula $(1 \mid g) + (0 + x \mid g)$, where x is the predictor and g is the grouping factor; the 0 in the second term suppresses the intercept.
6. The competing model with the smallest AIC value should also feature the smallest Kullback-Leibler distance and represent the best-fitting model among a set of comparison models (Vrieze, 2012). This is precisely why interpreting IC values only makes sense with respect to other competing models and why attempting to interpret ICs in isolation is a wasted effort.
7. The number seven assumes that we are using the standard critical values for chi-square with alpha = .05, not critical values that have been adjusted for boundary issues in the variances. With a difference of seven parameters, the traditional, unadjusted LRT imposes a penalty of 14.07 points on the deviance, whereas the AIC features a 14.00-point penalty. However, with a difference of eight parameters, the traditional LRT's critical value is 15.51, whereas the AIC's penalty is 16.00 points. Therefore, for a difference of up to seven parameters, the traditional LRT imposes a larger penalty on the deviance than does the AIC; in contrast, for eight or more additional estimated parameters, the AIC's penalty exceeds that of the unadjusted LRT.
8. This is because growth models in M*plus* typically analyze data in wide or multivariate format, with repeated observations nested within individuals. Hence, the sample size for computing the BIC in M*plus* is actually the total number of level-2 units for longitudinal models and the total number of level-1 units for organizational models. (This appears to be true as of M*plus*, Version 8.)
9. Akaike (1973) developed the AIC with a focus on prediction—he saw the ability to forecast future outcomes as the function and utility of statistical models.
10. There is one exception: if the level-1 variable has only within-cluster variance, and has no between-cluster variance, then this is not necessary. For example, if each of the clusters contained an even number of people who were randomly assigned to either the treatment or control group, then regardless of coding/centering ([0, 1] dummy coding or [–1, 1] effect coding), the cluster mean for every cluster is identical. In such a scenario, it is unnecessary to add the aggregate back in at level 2. In fact, in such a situation, doing so would be problematic, given the complete lack of variability in the cluster means (The

cluster means would be constant, not variable across all clusters). Using an effect code ([+1, −1] or [+1/2, −1/2]) produces the same result as group-mean centering in this scenario.

11. m is technically $m = \tau_{00}$ as long as the level-1 predictors all have means of 0.
12. The authors showed correspondence between their integrative framework and other R^2 measures commonly used in MLM. See Rights and Sterba (2019) for more details.
13. Note: If both v and m were 0, there would be no need for an MLM—we would require only one residual component, σ^2.
14. There are also software packages that exist to aid in computing these measures. See the website associated with this book for more information.
15. The "`cbind()`" function in R combines column vectors together (i.e., it binds c's). If you consider each variable in your data set as a column vector of values, then this function simply combines the variables given in its parentheses into a single object containing those column vectors attached to each other side-by-side in the order they appeared in the parentheses.
16. To obtain copies of the data and code presented in this chapter, visit: www.modeling.uconn.edu
17. We utilized FIML estimation in HLM 8 to estimate the example models presented in this chapter. The tabulated results include fixed effects with robust standard errors.
18. The student-level SES variable in this dataset is on a z-score metric, so its mean is already close to zero, but the code is still provided here for illustrative purposes. Also, it is good to always be sure that the mean of the variable is zero so that your model interpretations are accurate.
19. This can also be thought of as what's leftover at level 2 that is only explained by school membership, so it can be thought of as explained or unexplained variance depending on how you talk about it.

REFERENCES

Akaike, H. (1973). Information theory and an extension of the maximum likelihood principle. In B. N. Petrov & B. F. Csaki (Eds.), *Second International Symposium on Information Theory* (pp. 267–281). Academiai Kiado.

Baayen, R. H., Davidson, D. J., & Bates, D. M. (2008). Mixed-effects modeling with crossed random effects for subjects and items. *Journal of Memory and Language, 59,* 390–412. https://doi.org/10.1016/j.jml.2007.12.005

Bates, D., Mächler, M., Bolker, B., & Walker, S. (2015). Fitting linear mixed-effects models using lme4. *Journal of Statistical Software, 67*(1), 1–48. http://dx.doi.org/10.18637/jss.v067.i01

Beretvas, S. N., & Murphy, D. L. (2013). An evaluation of information criteria use for correct cross-classified random effects model selection. *The Journal of Experimental Education, 81*(4), 429–463. https://doi.org/10.1080/00220973.2012.745467

Berkhof, J., & Snijders, T. A. B. (2001). Variance component testing in multilevel models. *Journal of Educational and Behavioral Statistics, 26*(2), 133–152. https://doi.org/10.3102/10769986026002133

Bozdogan, H. (1987). Model selection and Akaike's information criterion (AIC): The general theory and its analytical extensions. *Psychometrika, 52*(3), 345–270. https://doi.org/10.1007/BF02294361

Burnham, K. P., & Anderson, D. R. (2004). Multimodel inference: Understanding AIC and BIC in model selection. *Sociological Methods & Research, 33*(2), 261–304. https://doi.org/10.1177/0049124104268644

Burnham, K. P., Anderson, D. R., & Huyvaert, K. P. (2011). AIC model selection and multimodel inference in behavioral ecology: Some background, observations, and comparisons. *Behavioral Ecological Sociobiology, 65*, 23–35. https://doi.org/10.1007/s00265-010-1029-6

Claeskens, G., & Hjort, N. L. (2008). *Model selection and model averaging*. Cambridge University Press.

Comets, E., Lavenu, A., & Lavielle, M. (2017). Parameter estimation in nonlinear mixed effect models using Saemix, an R implementation of the SAEM algorithm. *Journal of Statistical Software, 80*(3), 1–41. http://dx.doi.org/10.18637/jss.v080.i03de

Dominicus, A., Skrondal, A., Gjessing, H. K., Pedersen, N. L., & Palmgren, J. (2006). Likelihood ratio tests in behavioral genetics: Problems and solutions. *Behavior Genetics, 36*(2), 331–340. https://doi.org/10.1007/s10519-005-9034-7

Fitzmaurice, G. M., Lipsitz, S. R., & Ibrahim, J. G. (2007). A note on permutation tests for variance components in multilevel generalized linear mixed models. *Biometrics, 63*(3), 942–946. https://doi.org/10.1111/j.1541-0420.2007.00775.x

Forster, M. R. (2000). Key concepts in model selection: Performance and generalizability. *Journal of Mathematical Psychology, 44*(1), 205–231. https://doi.org/10.1006/jmps.1999.1284

Goldstein, H. (2011). *Multilevel statistical models (Kendall's library of statistics 3)* (4th ed.). Edward Arnold.

Gutierrez, R. G., Carter, S., & Drukker, D. M. (2001). On boundary-value likelihood-ratio tests [sg160]. *Stata Technical Bulletin, 60*, 15–18. https://EconPapers.repec.org/RePEc:tsj:stbull:y:2001:v:10:i:60:sg160

IBM (n.d.). *Information Criteria*. https://www.ibm.com/support/knowledgecenter/de/SSLVMB_23.0.0/spss/tutorials/mixed_diet_info_02.html?view=embed

Jennrich, R. I., & Schluchter, M. D. (1986). Unbalanced repeated-measures models with structured covariance matrices. *Biometrics, 42*(4), 805–820. https://doi.org/10.2307/2530695

Kass, R. E., & Raftery, A. E. (1995). Bayes factors. *Journal of the American Statistical Association, 90*(430), 773–795. https://doi.org/10.2307/2291091

Kass, R. E., & Wasserman, L. (1995). A reference Bayesian test for nested hypotheses and its relationship to the Schwarz criterion. *Journal of the American Statistical Association, 90*(431), 928–934. https://doi.org/10 2307/2291327

Konishi, S., & Kitagawa, G. (2008). *Information criteria and statistical modeling*. Springer Science+Business Media.

Lee, O. E., & Braun, T. M. (2012). Permutation tests for random effects in linear mixed models. *Biometrics, 68*(2), 486–493. https://www.jstor.org/stable/23270450

Lee, H., & Ghosh, S. K. (2009). Performance of information criteria for spatial models. *Journal of Statistical Computation and Simulation, 79*(1), 93–106. https://doi.org/10.1080/00949650701611143

de Leeuw, J. (2004). Multilevel analysis: Techniques and applications (book review). *Journal of Educational Measurement, 41*(1), 73–77. http://www.jstor.org/stable/1435073

McCoach, D. B., & Black, A. C. (2008). Evaluation of model fit and adequacy. In A. A. O'Connell & D. B. McCoach (Eds.), *Multilevel modeling of educational data* (pp. 245–272). Information Age Publishing.

McCoach, D. B., & Cintron, D. W. (2021). *Introduction to modern modelling methods.* SAGE Publications.

McCoach, D. B., Rifenbark, G. G., Newton, S. D., Li, X., Kooken, J., Yomtov, D., Gambino, A., & Bellara, A. (2018). Does the package matter? A comparison of five common multilevel modeling software packages. *Journal of Educational and Behavioral Statistics, 43*(5), 594–627. https://doi.org/10.3102/1076998618776348

Miller, J. J. (1977). Asymptotic properties of maximum likelihood estimates in the mixed model of the analysis of variance. *The Annals of Statistics, 5*(4), 746–762. https://www.jstor.org/stable/2958794

Nathoo, F. S., & Masson, M. E. J. (2016). Bayesian alternatives to null-hypothesis significance testing for repeated-measures designs. *Journal of Mathematical Psychology, 72,* 144–157. https://doi.org/10.1016/j.jmp.2015.03.003

Palardy, G. J. (2011). Review of HLM 7. *Social Science Computer Review, 29*(4), 515–520. https://doi.org/10.1177/0894439311413437

Raftery, A. (1995). Bayesian model selection in social research. *Sociological Methodology, 25,* 111–163. https://doi.org/10.2307/271063

Raftery, A. E. (1999). Bayes factors and BIC: Comment on "A critique of the Bayesian information criterion for model selection." *Sociological Methods & Research, 27*(3), 411–427. https://doi.org/10.1177/0049124199027003005

Raudenbush, S. W., & Bryk, A. S. (2002). *Hierarchical linear models: Applications and data analysis methods* (2nd ed.). SAGE Publications.

Raudenbush S., Bryk, A., Cheong, Y., & Congdon, R. (2000). *HLM manual.* SSI International.

Rights, J. D., & Sterba, S. K. (2019). Quantifying explained variance in multilevel models: An integrative framework for defining R-squared measures. *Psychological Methods, 24*(3), 309–338. https://doi.org/10.1037/met0000184

SAS Institute Inc. (2022, February 2). *SAS help center: The MIXED procedure.* https://documentation.sas.com/doc/en/pgmsascdc/9.4_3.5/statug/statug_mixed_toc.htm

Schwarz, G. (1978). Estimating the dimension of a model. *The Annals of Statistics, 6*(2), 461–464. https://www.jstor.org/stable/2958889

Sclove, S. L. (1987). Application of model-selection criteria to some problems in multivariate analysis. *Psychometrika, 52*(3), 333–343. https://doi.org/10.1007/BF02294360

Self, S. G., & Liang, K. (1987). Asymptotic properties of maximum likelihood estimators and likelihood ratio tests under nonstandard conditions. *Journal of the*

American Statistical Association, 82(398), 605–610. https://doi.org/10.2307/2289471

Singer, J. D., & Willett, J. B. (2003). *Applied longitudinal data analysis: Modeling change and event occurrence.* Oxford University Press.

Skrondal, A., & Rabe-Hesketh, S. (2004). *Generalized latent variable modeling: Multilevel, longitudinal, and structural equation models.* Chapman & Hall/CRC Press.

Snijders, T. A. B., & Bosker, R. J. (2012). *Multilevel analysis: An introduction to basic and advanced multilevel modeling* (2nd ed.). SAGE Publications.

StataCorp LP. (n.d.). *BIC note—Calculating and interpreting BIC.* http://www.stata.com/manuals14/rbicnote.pdf

StataCorp. (2017). *Stata multilevel mixed-effects reference manual release 15.* StataCorp LLC.

Stoel, R. D., Garre, F. G., Dolan, C., & van den Wittenboer, G. (2006). On the likelihood ratio test in structural equation modeling when parameters are subject to boundary constraints. *Psychological Methods, 11*(4), 439–455. https://doi.org/10.1037/1082-989X.11.4.439

Stram, D. O., & Lee, J. W. (1994). Variance components testing in the longitudinal mixed effects model. *Biometrics, 50*(4), 1171–1177. https://www.jstor.org/stable/2533455

University of California, Los Angeles Institute for Digital Research & Education Statistical Consulting Group. (2020, December 23). *SPSS mixed command.* https://stats.idre.ucla.edu/spss/seminars/spss-mixed-command/

Vallejo, G., Fernández, M. P., Livacic-Rojas, P. E., & Tuero-Herrerro, E. (2011). Selecting the best unbalanced repeated measures model. *Behavior Research Methods, 43*(1), 18–36. https://doi.org/10.3758/s13428-010-0040-1

Vallejo, G., Tuero-Herrero, E., Núñez, J. C., & Rosário, P. (2014). Performance evaluation of recent information criteria for selecting multilevel models in behavioral and social sciences. *International Journal of Clinical and Health Psychology, 14*(1), 48–57. https://doi.org/10.1016/S1697-2600(14)70036-5

Verbeke, G., & Molenberghs, G. (2000). *Linear mixed models for longitudinal data.* Springer.

Verbeke, G., & Molenberghs, G. (2003). The use of score tests for inference on variance components. *Biometrics, 59*(2), 254–262. https://doi.org/10.1111/1541-0420.00032

Vrieze, S. I. (2012). Model selection and psychological theory: A discussion of the differences between the Akaike information criterion (AIC) and the Bayesian information criterion (BIC). *Psychological Methods, 17*(2), 228–243. https://doi.org/10.1037/a0027127

Wagenmakers, E. J., & Farrell, S. (2004). AIC model selection using Akaike weights. *Psychonomic Bulletin & Review 11,* 192–196. https://doi.org/10.3758/BF03206482

Weakliem, D. L. (2016). *Hypothesis testing and model selection in the social sciences.* Guilford Press.

Whittaker, T. A., Chang, W., & Dodd, B. G. (2012). The performance of IRT model selection methods with mixed-format tests. *Applied Psychological Measurement, 36*(3), 159–180. https://doi.org/10.1177/0146621612440305

Whittaker, T. A., Chang, W., & Dodd, B. G. (2013). The impact of varied discrimination parameters on mixed-format item response theory model selection. *Educational Psychological Measurement, 73*(3), 471–490. https://doi.org/10.1177/0013164412472188

Whittaker, T. A., & Stapleton, L. M. (2006). The performance of cross-validation indices used to select among competing covariance structure models under multivariate nonnormality conditions. *Multivariate Behavioral Research, 41*(3), 295–335. https://doi.org/10.1207/s15327906mbr4103_3

Zucchini, W., Claeskens, G., & Nguefack-Tsague, G. (2011). Model selection. In M. Lovric (Ed.), *International Encyclopedia of Statistical Science* (pp. 830–833). Springer. https://www.encyclopediaofmath.org/images/b/b4/Model_selection.pdf

CHAPTER 4

CAUSAL INFERENCE IN MULTILEVEL SETTINGS

Chris Rhoads
University of Connecticut

Eva Yujia Li
Yale University

Not many statistical phrases make it into the common vernacular. "Correlation is not causation" is one of them. This phrase is no less true in a multilevel context. Multilevel regression models can be quite complex relative to their standard regression counterparts. Nonetheless, just like standard regression models, the building blocks of multilevel regression models are correlations and partial correlations. Therefore, causal claims from data cannot be justified solely on the basis of a multilevel regression analysis of a data set. Instead, such claims can be justified only by some combination of features of the research design and untestable assumptions. This chapter explores the ways in which multilevel models can (and cannot) help to facilitate appropriate causal inferences from data. In particular, multilevel models may allow a researcher to account for aspects of the research design that would not arise in non-causal settings (e.g., random assignment).

Multilevel Modeling Methods with Introductory and Advanced Applications, pages 95–126

Additionally, the assumptions necessary to obtain unbiased (or close to unbiased) estimates of average causal effects may be more credible when multilevel models are utilized than assumptions associated with other analytic approaches. Finally, using multilevel models in settings where causal inference is desired confers many of the same advantages that accrue when using these models in purely descriptive settings.

This chapter explores the reasons listed above for utilizing a multilevel model to draw appropriate causal inferences from data. As is always the case when considering multilevel models, the discussion presupposes the existence of a population with a hierarchical (or clustered) structure. We utilize a running example whereby this population consists of students nested within classrooms nested within schools. However, the ideas presented below apply to any sort of population with a hierarchical structure.

This chapter considers two distinct research designs that can be used to estimate causal effects: randomized experiments and observational studies. For simplicity, we consider research studies that compare units exposed to a single *treatment* condition with units exposed to a *control* condition. Randomized experiments are characterized by the ability of the researcher to intervene in the system being studied to randomly assign units in the population to receive either the treatment or the control condition. Randomization may occur at any level of the hierarchy (i.e., the design may randomize schools, classrooms or students). Randomization is typically achieved through the use of readily available computer based random number generators. On the other hand, in observational studies units are exposed to either the treatment or the control condition through a process that is not controlled by the researcher and is almost certainly not fully known to the researcher.

DEFINING CAUSAL EFFECTS IN MULTILEVEL SETTINGS

Any discussion of causality with data must begin with a definition of a measurable causal effect. This chapter does so by using the potential outcomes notation typically referred to as Rubin's causal model (Holland, 1986; Rubin, 1974).

Following the usual notational conventions, define $Y_i(1)$ as the value of the outcome variable, Y, that would be obtained if student i were exposed to the treatment condition and $Y_i(0)$ as the value of the outcome variable that would be obtained if student i were exposed to the control condition. $Y_i(1)$ and $Y_i(0)$ are the *potential outcomes* that are the essential elements of Rubin's causal model. The causal effect for individual i is then defined as

$$Y_i(1) - Y_i(0) \tag{4.1}$$

The fundamental problem of causal inference (Holland, 1986) is that $Y_i(1)$ and $Y_i(0)$ can never be observed for the same individual at the same time. The consequence of the fundamental problem is that, except under very specialized circumstances where strong identifying assumptions are plausible, it is impossible to estimate the causal effect of an intervention for a particular individual. Instead, the so-called *statistical solution* to the fundamental problem seeks to estimate an average causal effect for a particular population of students (Holland, 1986).

For both experiments and observational studies, the necessary condition for unbiased estimation of an average causal effect of interest is the independence of treatment assignment from potential outcomes. The analysis of data from randomized experiments is typically much simpler than the analysis of data from observational studies because the necessary independence assumption is guaranteed by the randomization process. However, in an observational study this assumption is unlikely to hold unless the researcher has made efforts to adjust statistically for pre-existing differences between groups with respect to observable pre-treatment variables (covariates).

Because of the increased complexity of observational studies, most ideas below are presented in the context of randomized experiments. Unless otherwise indicated, the same considerations also hold for observational studies. The final part of the chapter considers the additional complexities introduced by observational studies with hierarchically structured populations.

The need to estimate *average* causal effects leads immediately to a complication when causal effects need to be defined and estimated in hierarchically structured populations. Namely, different average causal effects may be of interest, and these different causal effects can be estimated in different ways, even when using the same dataset. To gain insight, it is necessary to elaborate further the basic potential outcomes notation defined above. Assume that the outcome variable is measured at the student level. Define $Y_{ijk}(1)$ as the potential outcome under the treatment condition for student i within classroom j within school k, and define $Y_{ijk}(0)$ in an analogous fashion for the control condition. Define $Y_{jk}(1)$ as the average value of the outcome variable measured in classroom j and school k when students in that classroom are exposed to the treatment condition, and define $Y_{jk}(0)$ in an analogous fashion. $Y_{jk}(1)$ and $Y_{jk}(0)$ are classroom level potential outcomes. Potential outcomes at the school level, $Y_k(1)$ and $Y_k(0)$, can be defined in an analogous manner.

Unless the number of students per classroom and the number of classrooms per school are constant, the average student level causal effect does not equal the average classroom level causal effect which does not equal the average school level causal effect. Typically, causal effects at the student level are of primary interest, and most educational studies choose the average student level causal effect as the target of estimation (What Works Clearinghouse, 2020).

While the average student level causal effect is typically of most interest, there are situations where a different average causal effect may be of interest. For instance, a superintendent considering adopting a new math curriculum may be more interested in the average causal effect on schools in their district than the average causal effect on students. While these average causal effects are likely to be similar, they will not be the same, particularly if the adoption of the new math curriculum induces certain students to leave the district (see the example in the next two paragraphs).

It may also be the case that credible estimates of causal effects can be obtained only for, for example, a school level average causal effect but not for a student level average causal effect. As an example, consider an educational experiment that randomizes schools to either implement a "whole school" reform or to a control condition. Suppose the study measures student level outcomes for students attending the randomized schools 2 years after the initial randomization.

Randomization to the treatment condition in this study can affect the observed average outcome at the school level through at least two different mechanisms. The first is by effecting a change in the outcome experienced by individual students in these schools (i.e., the intervention has non-zero treatment effects on individuals). The second is by inducing certain students to change schools (i.e., the intervention may have a non-zero average treatment effect on schools, even if it has a zero treatment effect on all individuals being studied). Under this scenario, the school level randomization would not justify making a causal inference about student level average causal effects. Indeed, the outcome measurement may not even be available for students who have moved out of the schools participating in the study during the 2 years between randomization and measurement of outcomes. However, unbiased estimation of the average impact on schools is still possible, provided that the researcher is careful to clarify that the observed treatment effect may be the result of both mechanisms listed above (What Works Clearinghouse, 2020).

THE INFERENTIAL POPULATION TAKE TWO: TREATING UNITS IN THE STUDY AS SAMPLED (OR NOT)

Another issue to consider when estimating causal effects in multilevel settings is whether units being studied are considered fixed or, instead, sampled from a population of units. The identical issue arises in multilevel descriptive studies, where the data analyst must decide whether to treat a certain level in the hierarchy as fixed or random (Hedges & Vevea, 1998; Schochet, 2008; Townsend et al., 2013). For instance, in a descriptive study of standardized math test scores for 500 students nested within 20 schools, a data analyst might take one of two approaches.

The multilevel modeling approach would fit a 2-level model with students nested within schools. This approach is appropriate if it is reasonable to treat the schools in the study as a random sample from a very large (technically infinite) population of schools, and students within schools as a random sample from a very large (technically infinite) population of students. In this situation statistical quantities estimated by the 2-level model appropriately account for the sampling error that results from sampling schools and students.

A different approach would be to fit a standard OLS regression model with dummy variables for each of the 20 schools (while either omitting the intercept from the model or treating one of the schools as the reference category). The inclusion of the dummy variables accounts for the correlation between observations within the same school because, conditional on the dummy variables, it is reasonable to assume that outcomes are independent. This approach is appropriate if statistical inferences are meant to apply only to students within the 20 schools in the data set, but the students within schools are regarded as sampled from a population of students (or potential students) within a given school. Indeed, if neither schools nor students are regarded as sampled, there would be no point in reporting standard errors of estimates, since only descriptive statistics are necessary to summarize the information in the data.

Strictly speaking, the decision about which of the two approaches to use should be determined by the nature of the data collection. If data were collected by first sampling schools and then students within schools, the multilevel model would be the correct approach. If data were collected by taking a simple random sample of students within each of 20 predetermined schools, then the OLS approach would be appropriate. However, in applied work, data analysts often act as if clusters such as schools are sampled even when a formal sampling process did not occur. Usually the reason for this approach is the desire to generalize results to a larger population of schools than those represented in the data set and the consequent need to account for the error involved in making such generalizations.

In randomized experiments it is even less likely that formal sampling of schools in a study has occurred. Almost universally, schools participating in randomized experiments have been recruited for the study and participate voluntarily. Thus, these schools are not randomly sampled and are not likely to be representative of a larger population of schools. Despite this fact, it is common to treat schools in randomized experiments as though they were randomly sampled. Recent work has attempted to improve on this practice by summarizing the extent to which results from a randomized experiment might be generalizable (Tipton, 2014), suggesting recruiting practices that might improve the generalizability of randomized experiments (Tipton, 2013a; Tipton et al., 2014) and proposing statistical adjustments to experimental results in order to improve generalizability (Stuart et al., 2015; Tipton, 2013b). It should be noted that existing work in this area has not explicitly made use of multilevel models, as adjustments are typically based on only school level covariate data.

STABLE UNIT TREATMENT VALUE ASSUMPTION (SUTVA)

Notice that the formulation of an individual level causal effect articulated in Equation 4.1 has implicitly made two assumptions. Collectively, these two assumptions are referred to as the stable *unit treatment value assumption* (SUTVA; Rubin, 1986). The first part of the SUTVA assumption is that treatments are "stable," in other words, there are not different versions of treatment. Without this assumption it would not be possible to represent the potential outcome of individual *i* under, for example, the control condition, with the single number, $Y_i(0)$, since there would be different outcomes for different versions of treatment. To some extent this issue can be resolved by imagining averaging across different versions of treatment (Rubin, 1990).

The second part of the SUTVA assumption is that there is not "interference" with respect to treatment effects. This "no interference" assumption stipulates that the potential outcome experienced by individual *i* cannot depend on the treatment experienced by other individuals in the population of interest. If this were not so it would be impossible to represent the potential outcome of an individual with a single number, since the potential outcome would depend on the treatment assignments of other individuals in the population.

This "no interference" aspect of the SUTVA assumption is of particular interest when considering causal inference in hierarchically structured populations. It may be that the SUTVA assumption is violated at a certain level of the hierarchy but is reasonable at a different level of the hierarchy. For instance, consider an experiment that seeks to determine the causal effect on student literacy of providing remedial tutoring to struggling readers. At the student level it is likely that the SUTVA assumption is violated. In other words, it is likely that the literacy level of students who share the same classroom will depend on whether or not other students in the classroom are also receiving tutoring. However, it may be reasonable to assume that the potential outcomes of students in a particular classroom are independent of the treatment received by students in different classrooms. In this case SUTVA is satisfied at the classroom level (Hill, 2013).

The plausibility of SUTVA at different levels of the hierarchy has implications for how the average causal effect of interest is defined and for the research design that is chosen. For instance, in the example above the researcher might choose to focus on the average causal effect at the classroom level. Even if it is instead desirous to estimate an individual level average causal effect, estimation may proceed without complications provided randomization is done at the classroom level (or at a higher level in the hierarchy, such as the school level). All other things equal, individual level randomization would not be recommended when the no interference assumption does not hold at the individual level (Hill, 2013; Rubin et al., 2004).

MULTILEVEL MODELS FOR RANDOMIZED EXPERIMENTS

This section makes explicit how multilevel models can be utilized to account for features of the research design and for the inferential goals of the study when analyzing data from randomized experiments. A continuous dependent variable is assumed, however straightforward modifications can be made in order to accommodate the use of multilevel models for logistic, ordinal or count data (see Chapter 7, "Multilevel Logistic and Ordinal Models," in this volume). The below also assumes that primary interest is in estimating average treatment effects at the student level. It is, however, straightforward to modify the models described below to estimate different average causal effects of interest. For instance, the model (H3) described below is appropriate when randomizing schools and estimating individual level average treatment effects. However, if school level average treatment effects are desired, the data can first be reduced to school averages and then a standard OLS regression model can be used with the school level data.

There are two main ways that higher level units in a hierarchically structured population can be utilized in experiments: (a) as blocks and (b) as clusters. A higher level unit is considered a block if randomization of lower level units occurs in such a way that each block is guaranteed to contain both treatment and control units. Randomized experiments with blocking are referred to as "randomized block" or "multi-site" designs (Spybrook et al., 2014). A higher level unit is considered a cluster if randomization occurs in such a way that all lower level units within a higher level unit are guaranteed to receive the same treatment. These experiments are called "cluster randomized" designs. A single study can contain both blocking and clustering. For instance, consider a study with 10 schools with four classrooms per school and 20 students per classroom. If individual students within classrooms are randomized to a treatment or control condition this is a randomized block design. If entire schools are randomized to a treatment or control condition this is a cluster randomized design. If individual classrooms within schools are randomized to treatment or control then this is a blocked, cluster randomized design.

The multilevel approach described in this section is appropriate if units at a particular level in the hierarchy are viewed as sampled from a much larger population. This assumption is typically invoked in educational studies because even if schools, classrooms and/or students are not formally sampled, researchers usually desire to generalize results to units beyond those explicitly studied. Multilevel models provide a method of accounting for additional error in the estimated average causal effects when generalizing to a larger population. An alternative, finite-population based, approach to the analysis of experiments in multi-level settings is described at the end of this section.

Strictly Hierarchical, Cluster Randomized Designs

A hierarchical experimental design is when randomization occurs at the highest level of the hierarchy (for instance, at level 3 in a 3-level model or level 2 in a 2-level model). When using multilevel models to analyze data from experiments the level at which randomization occurred is always considered a level in the model. Furthermore, in multilevel models, when a higher level is treated as random it is typical to also treat lower levels in the hierarchy as random. For instance, in a design that randomizes schools, if information about students' membership in classrooms is also available, the study would typically be analyzed using a 3 level model with students at level 1, classrooms at level 2 and schools at level 3. It would not make sense to include fixed effects of classrooms (dummy variables for classrooms) at level 2 when schools at level 3 are treated as sampled because a different sample of schools at level 3 would result in different classrooms at level 2.

The following describes a typical multilevel model used to analyze data from a three level, hierarchical, cluster randomized design, with randomization at the highest level. Let i index students, j index classrooms and k index schools. The model allows for the possibility of modeling additional covariates (measured prior to exposure to treatment) at each level of the model. The relevant 3-level, hierarchical linear model (Model H3) is expressed by the equations below.

Level 1:

$$Y_{ijk} = \pi_{0jk} + \Sigma\pi_{Pjk}(\mathrm{IX_P})_{ijk} + e_{ijk}$$

Level 2:

$$\pi_{0jk} = \beta_{00k} + \Sigma\beta_{0Qk}(CX_Q)_{jk} + u_{0jk}$$
$$\pi_{Pjk} = \beta_{P0k} \tag{4.2}$$

Level 3:

$$\beta_{00k} = \gamma_{000} + \gamma_{001}\mathrm{TREAT}_k + \Sigma\gamma_{00R}(\mathrm{SX_R})_k + r_{00k}$$
$$\beta_{P0k} = \gamma_{P00}$$
$$\beta_{0Qk} = \gamma_{0Q0}$$

Y_{ijk} is the outcome measurement score for student i within classroom j and school k. TREAT is a dummy variable coded 0 for schools not exposed to treatment and 1 for schools that are exposed. $(\mathrm{IX_P})_{ijk}$ are student level covariates (measured prior to treatment exposure), $(\mathrm{CX_Q})_{jk}$ are classroom level covariates (measured prior to treatment exposure) and $\mathrm{SX_R}$ are school level covariates (measured prior to treatment exposure). The estimate of γ_{001} is the estimated average treatment effect and the standard error of γ_{001} is the standard error of this estimate. The terms e_{ijk}, u_{0jk}, r_{00k} and are, respectively, level 1, level 2 and level 3 random effect terms which are assumed to have independent, mean zero normal distributions with different variances.

In principle one could also allow level 1 and level 2 slopes to vary, either across classrooms only or across both classrooms and schools (in the case of student level covariates) or across schools (in the case of classroom level covariates). However, in randomized experiments where interest is mainly in the estimation of treatment effects there is no reason to introduce this additional complexity to the model.

On the other hand, if one objective of the researcher is to study the moderating effects of observed student and/or classroom level variables, then there is reason to elaborate the above models by including the treatment dummy variable in the equations for τ_{P0k} and/or τ_{0Qk}. The associated regression coefficients would allow the researcher to examine the extent to which treatment effects vary across these observed student and/or classroom level variables.

Multi-Site or Randomized Block Designs

This subsection presents multilevel models that can be used to analyze data from randomized experiments where higher level units are used as blocks. This implies a *restricted randomization* procedure whereby, within each block, units are randomly assigned to conditions. Typically, randomization attempts to ensure an equal number of units in the treatment and control conditions, as this approach maximizes statistical power (see Chapter 5, "Statistical Power for Linear Multilevel Models," in this volume). However, there may be other reasons (such as costs or the preferences of research partners) to deviate from equal allocation (Bloom, 1995).

While continuing to focus attention on a population with a 3-level hierarchical structure, there are now two designs to consider. In the first, classrooms are used as blocks and randomization occurs at level 1 (the individual level). In the second, schools are used as blocks and randomization occurs at level 2 (the classroom level).

Possible Multilevel Models for a Randomized Block Design With Treatment at Level 1

Because randomization is at the lowest level of the hierarchy, it is reasonable to fit one of multiple different models, depending on whether classrooms and/or schools are treated as fixed or random. If both classrooms and schools are treated as fixed then a multilevel model is unnecessary. A single level regression model could be used with dummy variables for each classroom included in the model. As noted above, if schools are treated as random it is typical to also treat classrooms as random. In this case the relevant model (model MS1a), is

$$
\begin{aligned}
&\text{Level 1:}\\
&\quad Y_{ijk} = \pi_{0jk} + \pi_{1jk}\text{TREAT}_{ijk} + \Sigma\pi_{\text{P}jk}(\text{IX}_{\text{P}})_{ijk} + e_{ijk}\\
&\text{Level 2:}\\
&\quad \pi_{0jk} = \beta_{00k} + \Sigma\beta_{0\text{Q}k}(\text{CX}_{\text{Q}})_{jk} + u_{0jk}\\
&\quad \pi_{1jk} = \beta_{10k} + u_{1jk}\\
&\quad \pi_{\text{P}jk} = \beta_{\text{P}0k}\\
&\text{Level 3:}\\
&\quad \beta_{00k} = \gamma_{000} + \Sigma\gamma_{00\text{R}}(\text{SX}_{\text{R}})_k + r_{00k}\\
&\quad \beta_{10k} = \gamma_{100} + r_{10k}\\
&\quad \beta_{\text{P}0k} = \gamma_{\text{P}00}\\
&\quad \beta_{0\text{Q}k} = \gamma_{0\text{Q}0}
\end{aligned}
\tag{4.3}
$$

The notation is defined as it was for the hierarchical design (with obvious modifications to account for treatment entering the model at level 1). The estimate of γ_{100} is the estimated average treatment effect and the standard error of γ_{100} is the standard error of this estimate. To understand how the treatment effect varies as a function of measured classroom and/or school level variables (i.e., a moderation analysis), additional terms can be added to the equations for β_{10k} and/or τ_{1jk}.

Notice that there are now two random effects at level 2 and two random effects at level 3. The additional random effects, u_{1jk} and r_{10k}, are random slope terms that, respectively, model treatment effect variation across classrooms and schools. Failure to include these terms in the statistical model amounts to the assumption that treatment effects are constant for all classrooms and all schools in the study. If this assumption is violated the standard error of the treatment effect estimate will be downwardly biased. The reason for this downward bias is as follows. When classrooms and schools are each sampled from a larger population of schools then the true sampling variance of the treatment effect estimate depends on the residual error within each classroom and treatment group, the variation in treatment effects across classrooms, and the variation in treatment effects across schools. However, when the u_{1jk} and r_{10k} terms are omitted, the estimated standard error of γ_{100} from the multi-level model depends only on the level-1 variance component estimate, ignoring the variation in treatment effects that might exist at higher levels.

A different multilevel model is used in order to treat classrooms as sampled (random) but schools as fixed. In this case, only a 2-level model is necessary, with dummy variables for schools included in the equation for the random intercept at level 2. The model (Model MS1b) can be written as follows. In this case level 1 represents students and level 2 represents classrooms.

$$\begin{aligned}
&\text{Level 1:}\\
&\quad Y_{ij} = \beta_{0k} + \beta_{1j}\text{TREAT}_{ij} + \Sigma\beta_{Pj}(\text{IX}_{\text{P}})_{ij} + e_{ij}\\
&\text{Level 2:}\\
&\quad \beta_{0j} = \gamma_{00} + \Sigma\gamma_{0k}(\text{CX}_{\text{Q}})_j + \Sigma\gamma_{0\text{R}}(\text{SCHOOL}_{\text{R}})_j + u_{0j}\\
&\quad \beta_{1j} = \gamma_{10} + u_{1j}\\
&\quad \beta_{Pj} = \gamma_{P0}
\end{aligned} \tag{4.4}$$

It is again necessary to include the random slope term at level 2 to ensure accurate estimation of the standard error of the average treatment effect. If desired, school dummy variables may be included in the equation for β_{1j} to obtain school specific estimates of the average treatment effect.

Possible Multilevel Models for a Blocked, Cluster Randomized Design With Treatment at Level 2

If blocking is at the school level and randomization is at the classroom level, then there are again two possible models, depending on whether schools are treated as fixed or random. If schools are treated as fixed, then the 2-level model (Model MS2a) is specified as follows. Level 1 units are students and level 2 units are classrooms. Because treatment is at the classroom level, the dummy variable for treatment enters the model as a level 2 predictor. Unless there is a desire to look at cross-level interactions, the dummy variable for treatment only enters the equation for the intercept at level 2.

$$\begin{aligned}
&\text{Level 1:}\\
&\quad Y_{ij} = \beta_{0j} + \Sigma\beta_{\mathrm{P}j}(\mathrm{IX_P})_{ij} + e_{ij}\\
&\text{Level 2:}\\
&\quad \beta_{0j} = \gamma_{00} + \gamma_{01}\mathrm{TREAT}_j + \Sigma\gamma_{0\mathrm{Q}}(\mathrm{CX_Q})_j + \Sigma\gamma_{0\mathrm{R}}(\mathrm{SCHOOL_R})_j + u_{0j}\\
&\quad \beta_{\mathrm{P}j} = \gamma_{\mathrm{P}0}
\end{aligned} \tag{4.5}$$

The school dummy variables may be interacted with the treatment variable at level 2 to obtain school specific estimates of the average treatment effect.

If schools are treated as random, the model (Model MS2b) is three levels, with students as level 1 units, classrooms as level 2 units, and schools as level 3 units. Since schools are viewed as sampled, the coefficient for treatment (which appears at level 2) is allowed to randomly vary across schools at level 3. As noted above, this is necessary to obtain a correct standard error for the treatment effect when schools are viewed as sampled.

$$\begin{aligned}
&\text{Level 1:}\\
&\quad Y_{ijk} = \pi_{0jk} + \Sigma\pi_{\mathrm{P}jk}(\mathrm{IX_P})_{ijk} + e_{ijk}\\
&\text{Level 2:}\\
&\quad \pi_{0jk} = \beta_{00k} + \beta_{01k}\mathrm{TREAT}_{jk} + \Sigma\beta_{0\mathrm{Q}k}(\mathrm{CX_Q})_{jk} + u_{0jk}\\
&\quad \pi_{\mathrm{P}jk} = \beta_{\mathrm{P}0k}\\
&\text{Level 3:}\\
&\quad \beta_{00k} = \gamma_{000} + \Sigma\gamma_{00\mathrm{R}}(\mathrm{SX_R})_k + r_{00k}\\
&\quad \beta_{01k} = \gamma_{010} + r_{01k}\\
&\quad \beta_{\mathrm{P}0k} = \gamma_{\mathrm{P}00}\\
&\quad \beta_{0\mathrm{Q}k} = \gamma_{0\mathrm{Q}0}
\end{aligned} \tag{4.6}$$

Again, the random slope at level 3 is necessary to ensure an accurate estimate of the standard error of the treatment effect estimate.

The Finite-Population, Design-Based Alternative

This chapter focuses on the multilevel modeling approach to analyzing data from randomized experiments and observational studies conducted in hierarchically structured settings. This approach is most appropriate when all levels in the hierarchy (e.g., schools, classrooms, and students) are viewed as samples from a very large super-population, even if formal sampling did not occur and even if the population is never explicitly articulated.

Some authors have argued against adopting this perspective in randomized experiments and instead argue for treating all units in the hierarchy as fixed. That is, potential outcomes are regarded as fixed, but unknown, quantities. The only part of the data generating process that is assumed to be stochastic (random) is the randomization process that determines which potential outcome (or set of outcomes, in the case of cluster randomization) is observed. For example, in a design that randomized schools, the observed data, Y_{ijk}, would be assumed to be generated from the model $Y_{ijk} = T_j * Y_{ijk}(1) + (1 - T_j) * Y_{ijk}(0)$, where T_j is a Bernoulli random variable. Estimation and statistical inference about the average causal effect is then based strictly on the distribution of possible average treatment effect estimates that is induced by the variation in these Bernoulli random variables.

This "finite-population, randomization based" approach to causal inference in experiments was first discussed by Neyman (1923/1990) in the context of completely randomized designs and has since been adapted to more complex designs (e.g., Hansen & Bowers, 2009; Imai, 2008; Middleton & Aronow, 2015; Schochet, 2013). Authors writing from this perspective point out that commonly used estimators of average treatment effects (including estimators based on multilevel regression) can be biased depending on the average treatment effect of interest, the clustering structure and the estimation method that is used (Imai et al., 2009; Middleton, 2008; Middleton & Aronow, 2015; Schochet, 2013).

For instance, Middleton and Aronow (2015) note that, when within-cluster sample sizes vary, the sample size in each treatment group is not fixed, but rather is a random variable that depends on the result of the randomization process. Estimators that treat these sample sizes as fixed will therefore be biased. Similarly, Schochet (2013) shows that Neyman's finite population based model implies a different weighting scheme for schools than what is used in the super-population model that undergirds standard multilevel model based estimates.

In addition to the potential for bias discussed above, even when the estimates themselves are unbiased, regression based standard error estimates (including multilevel regression) can be biased (Freedman, 2008; Schochet, 2013). Finally, adjusting for baseline covariates via regression (including multilevel regression) can introduce bias even when the treatment effect estimate from the unadjusted model is unbiased (Freedman, 2008; Rubin, 1974; Schochet, 2010).

Of course, in a hierarchically structured population, each level of the hierarchy can be regarded as either fixed or random (sampled). As such, Imai et al. (2009) note that in the case of a 2-level hierarchically structured population, four different average treatment effects can be defined, each requiring different estimators with different statistical properties. See the original paper for further explanation.

The randomization based approach to inference is different from the multilevel modeling approach that is the main focus of this chapter. Readers interested in the randomization based approach should consult the above references. Additionally, software for analyzing experiments using the randomization based approach is now freely available at www.rct-yes.com.

MULTILEVEL OBSERVATIONAL STUDIES

Randomized experiments are not always feasible for many reasons. First, individuals or organizations may not agree to be randomized. For example, parents or schools may not agree to be randomized to study a new curriculum, because they may believe such curriculum is beneficial and want to be in the treatment condition, or they may believe such curriculum is harmful and want to opt out of the study. Secondly, there may be ethical reasons that prevent randomization. For studying the effect of a possibly life-saving medicine, it may be unethical to give patients a placebo. Third, researchers may be interested in retrospectively studying the effect of a treatment that was experienced by certain individuals and was not randomly assigned. Lastly, it may be too costly to plan and implement a RCT. It is often still desirable to learn about causal effects even when assignment to treatment has occurred outside of the control of the researcher. These studies are called *observational studies*, and this section describes analytic techniques for causal inference when these observational studies occur in hierarchically structured populations.

Due to the additional complexities involved in observational studies, discussion is limited to situations with a single level of clustering (i.e., students nested in schools). This approach helps illustrate the basic ideas involved, which can then be extended to more complex multilevel structures. Limiting discussion to 2 levels of clustering narrows the focus to two basic situations: (a) *multi-site* observational studies where treatment is at level 1 (e.g., some students in a given school are exposed to each of the two conditions); and (b) *hierarchical* observational studies where treatment is at level 2 (e.g., all students in a given school are exposed to the same condition).

As noted previously, unbiased estimation of a population average causal effect is possible when treatment assignment is independent of potential outcomes. Random assignment can ensure that this condition is met. In observational studies, this condition will only be met if statistical procedures can be utilized to remove the influence of *confounding variables*, which are defined as variables that correlate with both selection into treatment and the outcome of interest. Of course, statistical procedures can only account for the effect of observed confounders. Thus the hope is that any unobserved confounders are sufficiently correlated with observed confounders that techniques for removing or minimizing the influence of observed confounders will suffice to remove most bias. Multilevel models can also help to account for unobserved level 2 confounding variables. The assumption that units under study experience treatment essentially at random after statistical adjustment for observed confounders (i.e., there are no unobserved confounders) is known as the *unconfounded treatment assignment* assumption.

Defining Average Treatment Effects in Observational Studies

The distinction between different average causal effects discussed previously applies to both experimental and observational studies. The average student level causal effect may be different than the average classroom level causal effect, which may be different than the average school level causal effect. Additionally, it is possible that SUTVA is violated at a lower level of the hierarchy but plausible at a higher level. This section assumes primary interest is in student level causal effects and that SUTVA holds at the student level. However, when conceptualizing an observational study of causal effects, the researcher should keep in mind the issues around SUTVA and the competing definitions of average causal effects that were introduced previously.

In observational studies there is yet another distinction between different average treatment effects that must be acknowledged. If treatment effects are different for different units (whether those units be schools or students) then average treatment effects will differ for different collections of units. Since units self-select the condition to which they will be exposed in observational studies, the units that experience the treatment condition may be quite different from the units that experience the control condition. It is therefore necessary to distinguish between two different causal effects that might be of interest, the average treatment effect on the treated (ATT) and the average treatment effect on all individuals in the study population (ATE). This distinction does not arise in a randomized study because the randomization procedure ensures that the units that receive treatment are a random subset of the study population (making the two average treatment effects identical in a randomized study).

The fact that causal inference depends crucially on adjusting for potential confounders in observational studies can also lead to different decisions about how to define the average causal effect of interest. Consider the example of a school randomized experiment in which assignment to the treatment condition may cause compositional changes in the student population within a given school. In this case, randomization does not identify average causal effects at the individual level and so researchers may have to be satisfied with estimating an average causal effect on schools, even if understanding impacts on students would be preferable (USDOE, 2017). In observational studies researchers may be more willing to adjust the composition of naturally occurring clusters in an attempt to remove bias from average treatment effect estimates. If the statistical adjustment is successful in ensuring that the unconfounded treatment assignment condition is met, then individual level average causal effects would be identified, even if treatment is applied at the cluster level. However, it may be somewhat difficult to characterize the population for whom the average causal effect is estimated. See the Zubizarreta and Keele (2017) paper discussed later in this chapter for an example.

Methods for Removing the Influence of Observed Confounding Variables

Regression

One technique for removing the influence of confounding variables is to include these variables in a regression model along with an indicator variable for the treatment of interest. If the data is from a hierarchically structured population and either (a) higher level units are treated "as-if" randomly sampled or (b) treatment is received by all individuals within some group in the hierarchy (e.g., all students in a classroom receive the treatment) then analysis using a multilevel model is appropriate.

The regression approach to handling confounding results in the analyst using the same sorts of multilevel models described previously for experiments. For instance, in a population with students nested within classrooms nested within schools, where entire schools receive treatment, Model H3 would be appropriate. For the same population where some individuals within each classroom receive the treatment and some do not, and where classrooms are treated as sampled (random) but schools are treated as fixed effects, model MS1b would be appropriate. Regression and multilevel regression models are implicitly estimating the ATE.

Feller and Gelman (2015) discuss two important ways in which multilevel regression models can be useful for causal inference. First, multilevel models provide an efficient way to adjust for confounding. For instance, suppose that school membership is viewed as an important confounding variable in a study with a large number of schools. Rather than including a large number of dummy variables for schools in a propensity score model or an outcome model, it may be preferable to instead use schools as a level in a multi-level model.

Second, Feller and Gelman (2015) note that multilevel models provide for *partial pooling* of cluster level parameters. This feature of multilevel models is not unique to causal settings or observational studies, but partial pooling has a particularly useful application to understanding variation in treatment effects in causal studies. Consider a multi-site observational study. There may be interest in understanding how treatment effects vary across the sites. Typically, the site-specific treatment effects will be quite imprecise and overly dispersed relative to the true variation in treatment effects. The partial pooling property of multi-level models pulls these site-specific estimates towards the population average treatment effect, producing more efficient estimates of the site-specific treatment effects and a better estimate of the true variation in treatment effects across sites.

The crucial difference between observational studies and experiments is that the covariate set at each level must include all important confounding variables for the model to successfully produce an unbiased estimate of the average treatment effect. Equally important, the functional relationship between the entire covariate set and the outcome variable must be correctly specified in the model. This will usually mean that the model should include non-linear and interaction terms. As a result of the need to model the effects of multiple confounders and to avoid the negative effects of an underspecified model, multilevel models for observational studies are generally much more complex than multilevel models for experiments.

Previous research has shown the negative consequences of mis-specified models for causal inference in observational studies (e.g., Cochran & Rubin, 1973; Rubin, 2001). As a result, recent research has focused on computationally intensive regression techniques that are able to fit regression functions with an extremely flexible form. By effectively removing the problem of model dependence from the causal inference problem, these methods have the potential to provide estimates of average causal effects that are much more trustworthy than estimates obtained from traditional regression modeling. Bayesian additive regression trees (BART) are a particularly promising example of this approach (Hill, 2011). Multilevel adaptations of BART are just now starting to be developed. At the time this chapter is being written there does not seem to be a statistical or methodological publication devoted solely to multilevel BART. However, supplementary material from a recent publication in *Nature* implements a multilevel BART model (Yeager et al., 2019), and a recent review article suggests that more work on multilevel BART may be on the way (Hill et al., 2020).

Methods Based on Propensity Scores and Balancing Confounders

Regression based approaches attempt to remove the influence of confounding variables by using statistical adjustment to properly account for the relationship between those confounders and the outcome variable. A different set of techniques instead adjusts the data in order to achieve balance across the treatment and control group with respect to potential confounding variables. These techniques, which seek to break the link between the confounders and treatment assignment, are sometimes referred to as *matching methods* (Rubin, 2006; Stuart, 2010), although this name is misleading since not all methods grouped under this heading involve matching. In this section we consider matching, sub-classification and re-weighting as examples of matching methods. We first review the application of these techniques in simple settings without clustering. We then discuss extensions to a (2-level) multilevel setting.

Propensity scores and matching methods in single level contexts. Most applications of matching methods utilize the *propensity score*, as such it is a central tool in using matching methods to achieve balance between treatment and control cases in an observational study. If $\mathbf{X}$ represents the vector of confounding variables and Z represents the indicator variable for treatment assignment, the propensity score (PS) is defined as $e(\mathbf{X}) = P(Z = 1 \mid \mathbf{X})$. Rosenbaum and Rubin (1983) showed that the propensity score is a *balancing score*, meaning that conditional on $e(\mathbf{X})$, the distribution of X is the same in the two groups. The practical implication is that, to equate the distribution of $\mathbf{X}$ across the treatment and control groups, it is sufficient to equate the distribution of $e(\mathbf{X})$ across the groups.

Propensity Score Estimation. Propensity scores are usually estimated with a logistic regression model (Rosenbaum & Rubin, 1983). Correct specification of this logistic regression model is important: If the relationship between $\mathbf{X}$ and $e(\mathbf{X})$ is mis-specified the propensity score will no longer have the balancing score property. Other methods for estimating the propensity score have been proposed, such as neural networks (Keller et al., 2015) and boosted regression (McCaffrey et al., 2004). These non-parametric methods avoid the need to pre-specify a functional form for the propensity score estimation model. It is not clear how they might be extended to multilevel settings, so they are not considered further in this chapter.

Using Propensity Scores to Achieve Balance. Once propensity scores have been estimated there are multiple ways to use these scores to achieve balance with respect to confounding variables. The most common methods are matching, weighting, and sub-classification. We review each of these in turn.

Matching. If there is interest in estimating the ATT, and if there are fewer treated units available for the study than control students, matching may be used to reduce bias. Certain matching techniques, such as optimal full matching (Hansen, 2004), do not require fewer treated units than control units, but we do not discuss those techniques here. A very simple (but still effective) algorithm is *low-to-high one-to-one nearest neighbor matching with respect to the propensity score.* In this algorithm, starting with the case with the lowest propensity score, each treatment case is matched to the control case with the nearest propensity score. Any leftover control cases are discarded. The matched dataset will be better balanced with respect to **X** than the original dataset.

Weighting. Weighting methods may be used to estimate either the ATT or the ATE. If an estimate of the ATT is desired, then treatment cases are given a weight of 1, and control cases are given a weight of $e(\mathbf{X})/(1-e(\mathbf{X}))$. If an estimate of the ATE is desired, then treatment cases are given a weight of $1/e(\mathbf{X})$, and control cases are given a weight of $1/(1-e(\mathbf{X}))$. Assuming correct estimation of the propensity score, the re-weighted data set will be balanced with respect to the elements of **X**. Weighting methods can be used without discarding any cases.

Sub-Classification. One drawback of weighting methods is that extreme weights can inflate the variance of the estimated average treatment effect. Sub-classification, a way to smooth the propensity score weights, is a procedure where units with similar propensity scores are grouped together. For units in the same subclass, the average mean difference in treatment and control outcomes is then computed. These estimates may be combined to estimate the overall average treatment effect. If estimates are combined with weights proportional to the total number of treatment cases in the subclass, then an estimate of the ATT is obtained. If estimates are combined with weights proportional to the total number of cases in the subclass overall, then an estimate of the ATE is obtained.

Using Adjusted Dataset to Estimate Treatment Rffects. Once an adjusted dataset has been created using matching, weighting, or sub-classification, average treatment effects can be computed using the adjusted data. Although a simple comparison of means is possible, it is preferable to utilize a regression model to make additional adjustments for confounders. This approach results in causal inferences that are *doubly robust* (Ho et al., 2007; Schafer & Kang, 2008). In this context *doubly robust* means that the average treatment effect is correctly estimated if either the propensity score model or the regression model is correctly specified, but it is not necessary for both to be correctly specified.

EXTENDING PROPENSITY SCORES AND MATCHING METHODS TO MULTILEVEL SETTINGS

Multi-Site Observational Studies

The study of propensity scores and matching methods in multi-level contexts is a new and developing area of research. Most work thus far has focused on the multi-site case where there are individuals in both the treatment and control conditions within each level 2 unit. Existing research has focused on the use of multilevel logistic models to estimate propensity scores and multilevel models for continuous outcomes to estimate average treatment effects. Often, the research examines ways in which the propensity score model might compensate for misspecification in the outcome model and vice-versa. This work has examined both weighting and matching approaches in 2 level settings. We discuss matching approaches first.

Matching

In observational studies there may be unobserved features of clusters that contribute to confounding. One way to control for unobserved cluster level confounders is to *utilize within cluster matching* (WM). This approach involves estimation of a separate PS model within each cluster using individual level covariates and only matching treatment units to control units within the same cluster (Rosenbaum, 1986). A difficulty with WM is that small cluster sizes and/or imbalance in the proportion of treatment and control students within a cluster will often mean it is impossible to obtain accurate PS estimates within clusters and also very difficult to obtain good matches within clusters.

When the data will not allow for WM techniques it is possible to use *pooled matching* (PM) techniques. These techniques do not insist that matches come from the same cluster but rather match any treatment case to any available control case. Arpino and Mealli (2011) conducted simulations to look at pooled matching with an omitted cluster level confounder under three different propensity score models: (a) ignoring clustering; (b) a logistic model with dummy variables (fixed effects) for clusters, and (c) a 2 level logistic model. The second two approaches each removed most of the bias in estimates of the ATT, with the fixed effects model having slightly better performance. Kim and Seltzer (2007) discussed three types of 2-level logistic models for estimating the propensity score: random intercepts, simple random slopes, and random slope models that also model the interaction between individual level and cluster level covariates. After performing matching and checking covariate balance, Kim and Seltzer found that either type of model with random slopes was sufficient to achieve good covariate balance, but models with only random intercepts were not sufficient.

Kim and Steiner (2015) suggested a compromise approach that combines features WM and PM. Kim and Steiner first identified homogeneous classes of clusters using multilevel latent class logistic regression—an extension of standard latent class models that was adapted to multilevel settings. Once the latent class model was built, each cluster was classified into a latent class. Second, they built separate PS models for each class. These models included individual and cluster level characteristics and their interactions. Matches were then formed across clusters but separately for each latent class.

Rickles and Seltzer (2014) considered a two-stage matching (2SM) technique that extends WM in the case where an insufficient number of within cluster matches are available. In the first stage, the procedure attempts to find matches from within the same cluster. In the second stage, groups of homogeneous sites are defined (not unlike the approach described in the previous paragraph). Treatment cases that are unmatched in the first stage are matched to control cases from a cluster within the same grouping. Finally, an additional adjustment is made when the treatment case is matched to a control case from a different cluster. Empirical Bayes estimates of the site-specific intercepts and slopes for control group cases are used to adjust the outcome value of the control case to make it closer to what it likely would have been had it come from the same cluster.

Rickles and Seltzer (2014) compared the performance of WM, PM, and 2SM in a simulation study with a variety of treatment assignment mechanisms and three types of propensity score models: single level (SL PS), 2 level random intercept (RI PS) and 2 level random intercept and slope (RIS PS). Matching was performed within a caliper so that treatment cases for whom a close match could not be found were discarded. Rickles and Seltzer found that 2SM was able to balance level 1 observed and unobserved covariates better than PM and almost as well as WM while retaining more treatment cases for analysis than either of the other two approaches. However, neither PM nor 2SM performed nearly as well as WM with respect to balancing site level unobservables (although WM matched many fewer cases). Interestingly, with respect to balancing site level unobservables, 2SM outperformed PM when a SL PS model was used but the reverse was true when either an RI or RIS PS model was used.

Weighting

Li et al. (2013) and Leite et al. (2015) both considered weighting techniques applied to multi-site observational studies. Li et al. (2013) investigated the same three propensity score models explored by Arpino and Mealli (2011); namely, ignoring clustering, a multilevel model and a single level model with fixed effects of clusters. They used the three sets of weights generated by these models in conjunction with three, analogous, regression models which were used to estimate average causal effects: one that ignored clustering, a multilevel model and a single level model with fixed effects of clusters. The study was an extension of Arpino and Mealli (2011) in the sense that multilevel regression models were used in conjunction with weights to produce doubly robust inference. In contrast, Arpino and Mealli did not use doubly robust models and used matching instead of weighting. Li et al. (2013) found it was much more important to account for the multilevel structure in the outcome model than in the propensity score model. Bias was smallest when using a multilevel outcome model and the impact of different methods of generating propensity score weights was minimal.

Similar to the Li et al. (2013) study, Leite et al. (2015) conducted a simulation study to consider the influence of factors like cluster size, type of propensity score model, and type of matching technique on the bias of ATT estimates. Importantly, this study considered an additional issue that arises when using weights in multilevel settings. There are multiple reasonable ways to scale the level 1 weights and different ways of scaling may be preferable in different situations (Pfefferman et al., 1998; Stapleton, 2002). Weights can be scaled: (a) to sum to the total sample size (normalized), (b) such that the within cluster weights sum to the cluster size, or (c) such that the within cluster weights sum to the effective sample size. Leite et al. found that the scaling of weights had no influence on the bias of the ATT, but had a large effect on the estimation of the between cluster variance of the ATT. None of the weighting methods did a particularly good job of recovering the between cluster ATT variance, but the method that formed weights using the effective sample size performed substantially better than the other two weighting methods.

Hierarchical Observational Studies

Most of the research on multilevel observational studies has involved multi-site studies. However, two recent papers by Keele and colleagues (Pimentel et al., 2018; Zubizarreta & Keele, 2017) considered matching in observational studies where treatment is experienced at the cluster level. Zubizarreta and Keele noted that common practice in multilevel matching is to first match clusters and then match units within clusters. Unfortunately, this will usually lead to suboptimal balance. Instead, they introduced a *dynamic cardinality matching* algorithm that relies on integer programming in order to match pairs of units within a particular treatment and a particular control cluster. The algorithm repeats this process across all possible combinations of treatment and control clusters. Once within cluster matches are found across all possible cluster pairs the algorithm matches clusters in a way to ensure optimal within cluster balance. The algorithm finds the largest set of within cluster units that can be matched while still meeting a set of balance constraints. This approach, of course, changes the composition of the clusters in the study, resulting in an estimated average causal effect that applies to a synthetic population that can be hard to characterize.

Pimentel et al. (2018) presented a multilevel matching algorithm based on network flows that achieves many of the same objectives as the integer programming algorithm of Zubizarreta and Keele (2017), but with some important differences. The Pimentel et al. algorithm is much faster than the integer programming algorithm and so can be used on larger datasets.

Unlike the Zubizarreta and Keele algorithm, it cannot guarantee a pre-specified level of balance for specific covariates. However, the algorithm can be modified to ensure that all treated units are retained in the analysis, ensuring that the method estimates the ATT, rather than an average treatment effect for a synthetic population.

CHAPTER SUMMARY

This chapter has illustrated a number of the particularities involved when performing causal inference from data in multilevel settings. The hierarchical structure of a population/dataset can be exploited to define different average causal effects of interest and to render more plausible the assumptions necessary to define and estimate certain causal effects. Additionally, multilevel models can help facilitate appropriate statistical inferences by ensuring standard errors are appropriately matched to the inferential population of interest, accounting for the way in which units are sampled into the study and accurately modeling assignment to treatment. In observational studies the hierarchical structure of the population may be exploited by, for example, performing matching within clearly defined clusters or groups of clusters, using weights generated from multilevel logistic models or by using multilevel models to help protect against the omission of cluster level confounders from the regression model predicting outcomes. In each of these cases the clustered nature of the data is utilized to provide additional robustness against unmeasured cluster level confounding variables in observational studies. Finally, in hierarchical observational studies, new, computationally intensive algorithms may be used to ensure balance on both individual and cluster level covariates. In sum, using multilevel models to exploit the clustered nature of the data can help improve inference in both observational and experimental studies.

RECOMMENDED READINGS

Technical Readings

Draper, D. (1995). Inference and hierarchical modeling in the social sciences. *Journal of Educational and Behavioral Statistics*, 20, 115–147.

Schochet, P. (2013). Estimators for clustered education RCTs using the Neyman model for causal inference. *Journal of Educational and Behavioral Statistics, 38*(3), 219–238.

Stuart, E. (2010). Matching methods for causal inference: A review and a look forward. *Statistical Science, 25*(1), 1–21.

Applied Readings

Hong, G., & Raudenbush, S. W. (2006). Evaluating kindergarten retention policy: A case study of causal inference for multilevel observational data. *Journal of the American Statistical Association, 101*, 901–910.

Rickles, J., & Seltzer, M. (2014). A two-stage propensity score matching strategy for treatment effect estimation in a multisite observational study. *Journal of Educational and Behavioral Statistics, 39*(6), 612–636.

Tipton, E., Fellers, L., Caverly, S., Vaden-Kiernan, M., Borman, G., Sullivan, K., & Ruiz de Castilla, V. (2016) Site selection in experiments: A follow-up evaluation of site recruitment in two scale-up studies. *Journal of Research on Educational Effectiveness, 9*(sup1), 209–228.

TRY THIS!

For this exercise, we will use the sesame_trunc.csv file (provided on the online appendix http://modeling.uconn.edu), which is a modified version of the sesame.dta file that is available as a companion to Gelman and Hill (2006). The data come from a randomized experiment where children were encouraged to watch *Sesame Street.* Further details of the study can be found in Bogatz and Ball (1971) and Murphy (1991).

Randomization occurred within different sites. We do not utilize the indicator of child randomization status. Instead, we approximate an observational study by utilizing a different variable in the dataset, *regular,* which indicates whether a particular child regularly watched *Sesame Street* or not. We desire to understand the causal effect of regularly watching *Sesame Street* on a posttest measure of letter recognition (*postlet*). Since regular was not randomized (indeed, it could not be randomized), there is the need to adjust for potential confounding variables. We explore different methods to adjust for confounding, including regression adjustment and propensity score weighting. We fit two-level random intercept models with children nested within sites, as well as models with fixed effects of sites and models that ignore sites. The variables included in the dataset are described below. Analyses use the R package *lme4* (Bates et al., 2011).

id: subject identification number

site: 1 = Three to five year old disadvantaged children from inner city areas in various parts of the country.
2 = Four year old advantaged suburban children.
3 = Advantaged rural children.
4 = Disadvantaged rural children.
5 = Disadvantaged Spanish speaking children.

sex: male = 1, female = 0

age: age in months

setting: setting in which *Sesame Street* was viewed, 1 = home, 0 = school

prebody: pretest on knowledge of body parts (scores range from 0–32)

prelet: pretest on letters (scores range from 0–58)

preform: pretest on forms (scores range from 0–20)

prenumb: pretest on numbers (scores range from 0–54)

prerelat: pretest on relational terms (scores range from 0–17)

preclasf: pretest on classification skills

postlet: posttest for letter recognition (0–58). Outcome variable.

peabody: mental age score obtained from administration of the Peabody Picture Vocabulary test as a pretest measure of vocabulary maturity

siteh: (for $h = 2, \ldots, 5$): Dummy variables for site membership to use in fixed effects models.

regular: frequency of viewing: 0 = rarely watched the show, 1 = watched once/week or greater.

A. First, we will fit "naïve" models that ignore potential confounding.

1. Fit a 2-level model with *regular* as the only fixed effect in the model.
2. Fit the same model as in 1, only now use fixed, rather than random, site effects.
3. Fit the same model as in 2, only now ignore clustering by removing the dummy variables for sites.

Describe the differences between the three "naïve" models. What is the difference between the three models in terms of the estimate of the impact of regular viewing on letter recognition? What is the difference between the three models in terms of the standard error of the estimate of the impact of regular viewing on letter recognition?

B. Next, we adjust for confounding by adding potential confounder variables to the models. In most observational studies all pre-treatment variables that are possibly correlated with the outcome are added to the model. However, variables that are measured after treatment has become active, and so are potentially impacted by treatment, are usually left out of statistical models. Including post-treatment variables in the regression equations could cause bias by "adjusting away" some of the true impact of treatment. In the current example this means that we will exclude the *setting* variable, but we will include *sex*, *age*, and all of the pretest variables (including *peabody*).

How has the estimated treatment effect changed for each model as a result of including the potential confounding variables?

C. Now we estimate propensity scores and use them to obtain weights. We fit three propensity score models specified in Li et al. (2013) which are analogous to the three outcome models we have fit so far. Namely,

4. Fit a 2-level logistic regression model including *sex*, *age*, and all of the pretest variables as predictors.

   - Assuming that the data has been named "mlmdat", the model can be fit with the code:
     ```
     mlmprop <-glmer(regular~sex+age+prebody+prelet+
     preform+prenumb+prerelat+ preclasf+ peabody+
     (1 | site), family = binomial, control=glmer
     Control(optimizer="bobyga"), data = mlmdat)
     ```
   - Obtain predicted values with the code:
     `mlm.prop.pred <- predict(mlmprop,re.form=NA).`
     This corresponds to marginal predicted values averaging across the random effects (see chapter 7, "Multilevel Logistic and Ordinal Models," in this volume).
   - Convert to predicted probabilities with the code (requires the R package "boot" to be loaded):
     ```
     mlm.prob.pred <- inv.logit(mlm.logit.pred)
     ```
   - Assign weights to obtain an estimate of the ATE with the code:
     ```
     leng <- length(mlmdat$regular)
     mlm.weight <- vector(length=leng)
     for (i in 1:leng) {
       ifelse(mlmdat2$regular[i] != 0,mlm.weight[i]
       <- 1/mlm.prob.pred[i],mlm.weight[i] <- 1/(1-
       mlm.prob.pred[i]))
     }
     ```

5. Fit a single-level logistic regression model with the same predictors as 4, but also add dummy variables for sites.

   - The model can be fit with the code:
     ```
     fixprop <- glm(regular~sex+age+prebody+prelet
     +preform+prenumb+prerelat+ preclasf+ peabody
     +site2+site3+site4_site5), family = binomial,
     data = mlmdat)
     ```

- Obtain predicted probabilities directly with the code:
```
fix.pred <- predict(fixprop,type=c("response"))
```
- Assign weights to obtain an estimate of the ATE with the code:
```
fix.weight <- vector(length=leng)
for (i in 1:leng)
  {ifelse(mlmdat$regular[i] != 0,fix.weight[i]
  <- 1/fix.pred[i],fix.weight[i] <- 1/(1-fix
  pred[i]))
}
```

6. Perform the same steps as in 5, but remove the site dummies from the initial model. Save the weights from this step to a vector named "ig.weight".

   Once the three sets of weights are obtained, we add them to the data frame with the code: `mlmdat <- cbind(mlmdat,mlm.weight,ig.weight,fix.weight)`

D. Following Li et al. (2013) we use each of these three sets of weights to compute two different non-parametric estimators of the ATE. The first, which Li et. al. (2013) term a *marginal estimator* compares weighted means for the treatment and control group across the entire sample. The second, which Li et al. (2013) term a *clustered estimator,* first obtains separate (weighted) ATE estimates within each cluster and then obtains the overall average ATE weighted by the total weights in each cluster. Code to obtain the estimates is provided below.

Code to get marginal estimates:

7. Using the weights based on 2-level propensity score models, we run the following code:

```
margmlm<- aggregate(id~regular,mlmdat,
   function(i) weighted.mean(mlmdat$postlet[i],
   mlmdat$mlm.weight[i]))
margmlm$id[2]-margmlm$id[1]
```

Analogous steps can be followed to obtain estimates for the other two sets of weights.

Code to get clustered estimates:

8. To get the necessary weighted estimate we proceed as follows. First, add indicator of site by treatment status to the dataset with the following code.
```
sitebytreat <- mlmdat$regular*10+mlmdat$site
mlmdat2 <- cbind(mlmdat,sitebytreat)
```

Next, get weighted mean differences per cluster using mlm weights as follows.

```
cmlm<- aggregate(id~sitebytreat,mlmdat2,function
    (i) weighted.mean(mlmdat2$postlet[i],mlmdat2$
    mlm.weight[i]))
cmlm_md1 <- cmlm$id[6]-cmlm$id[1]
cmlm_md2 <- cmlm$id[7]-cmlm$id[2]
cmlm_md3 <- cmlm$id[8]-cmlm$id[3]
cmlm_md4 <- cmlm$id[9]-cmlm$id[4]
cmlm_md5 <- cmlm$id[10]-cmlm$id[5]
cmlm_md <- c(cmlm_md1,cmlm_md2,cmlm_md3,cmlm_
    md4,cmlm_md5)
```

Then, get the sum of the weights per cluster as follows.

```
cmlmw<- aggregate(mlm.
    weight~sitebytreat,mlmdat2,sum)
cmlm_w1 <- cmlmw$mlm.weight[6]+cmlmw$mlm.
    weight[1]
cmlm_w2 <- cmlmw$mlm.weight[7]+cmlmw$mlm.
    weight[2]
cmlm_w3 <- cmlmw$mlm.weight[8]+cmlmw$mlm.
    weight[3]
cmlm_w4 <- cmlmw$mlm.weight[9]+cmlmw$mlm.
    weight[4]
cmlm_w5 <- cmlmw$mlm.weight[10]+cmlmw$mlm.
    weight[5]
cmlm_w <- c(cmlm_w1,cmlm_w2,cmlm_w3,cmlm_
    w4,cmlm_w5)
```

Finally, compute the desired estimate as follows.

```
weighted.mean(cmlm_md,cmlm_w)
```

Analogous steps can be followed to obtain estimates for the other two sets of weights.

How do these six estimates compare to each other and to the regression adjusted estimates obtained previously? Do the estimates differ more or less than you expected? Provide a range of values that represents your best guess about what the true average causal effect of watching *Sesame Street* regularly on letter recognition might be?

REFERENCES

Arpino, B., & Mealli, F. (2011). The specification of the propensity score in multilevel observational studies. *Computational Statistics and Data Analysis, 55,* 1770–1780.

Bates, D., Maechler, M., & Bolker, B. (2011). *lme4: Linear mixed-effects models using S4 classes.* http://CRAN.R-project.org/package=lme4

Bloom, H. S. (1995). Minimum detectable effects: A simple way to report statistical power of experimental designs. *Evaluation Review, 19,* 547–556.

Bogatz, G. A., & Ball, S. (1971). *The second year of Sesame Street: A continuing evaluation* (two volumes). Educational Testing Service.

Cochran, W. G., & Rubin, D. B. (1973). Controlling bias in observational studies: A review. *Sankhya, 35*(4), 417–446.

Feller, A., & Gelman, A. (2015). Hierarchical models for causal effects. In R. Scott & S. Kosslyn (Eds.), *Emerging trends in the social and behavioral sciences.* SAGE Publications.

Freedman, D. (2008). On regression adjustments to experimental data. *Advances in Applied Mathematics, 40,* 180–193.

Gelman, A., & Hill, J. (2006). *Data analysis using regression and multilevel/hierarchical models.* Cambridge University Press.

Hansen, B. B. (2004). Full matching in an observational study of coaching for the SAT. *Journal of the American Statistical Association, 99*(467), 609–618.

Hansen, B., & Bowers, J. (2009) Attributing effects to a cluster-randomized get-out-the-vote campaign. *Journal of the American Statistical Association, 104,* 873–885.

Hedges, L. V., & Vevea, J. (1998). Fixed and random effects models in meta-analysis, *Psychological Methods, 3*(4), 486–504.

Hill, J. (2011). Bayesian nonparametric modeling for causal inference. *Journal of Computational and Graphical Statistics, 20*(1), 217–240.

Hill, J. (2013). Multilevel models and causal inference. In M. Scott, J. Simonoff, & B. Marx (Eds.), *The Sage handbook of multilevel modeling.* SAGE.

Hill, J., Linero, A., & Murray, J. (2020). Bayesian additive regression trees: A review and look forward. Annual Review of Statistics and Its Application, 7(1), 251–278.

Ho, D., Imai, K., King, G., & Stuart, E. (2007). Matching as nonparametric preprocessing for reducing model dependence in parametric causal inference. *Political Analysis,* 15, 199–236.

Holland, P. (1986). Statistics and causal inference. *Journal of the American Statistical Association, 81*(396), *945–960.*

Imai, K. (2008). Variance identification and efficiency analysis in randomized experiments under the matched-pair design. *Statistics in Medicine, 27,* 4857–4873.

Imai, K., King, G., & Nall, C. (2009). The essential role of pair matching in cluster-randomized experiments, with application to the Mexican Universal Health Insurance Evaluation. *Statistical Science, 24*(1), 29–53.

Keller, B., Kim, J.-S., & Steiner, P. M. (2015). Neural networks for propensity score estimation: Simulation results and recommendations. In L. A. van der Ark, D. M. Bolt, S.-M. Chow, J. A. Douglas, & W.-C. Wang (Eds.), *Quantitative psychology research* (pp. 279–291). Springer.

Kim, J., & Seltzer, M. (2007). *Causal inference in multilevel settings in which selection processes vary across schools* (CSE Technical Report 708). National Center for Research on Evaluation, Standards, and Student Testing.

Kim, J. S., & Steiner, P. M. (2015). Multilevel propensity score methods for estimating causal effects: A latent class modeling strategy. In L. A. van der Ark, D. M. Bolt, S.-M. Chow, J. A. Douglas, & W.-C. Wang (Eds.), *Quantitative psychology research* (pp. 293–306). Springer.

Leite, W. L., Jimenez, F., Kaya, Y., Stapleton, L. M., MacInnes, J. W., & Sandbach, R. (2015). An evaluation of weighting methods based on propensity scores to reduce selection bias in multilevel observational studies. *Multivariate Behavioral Research, 50*(3), 265–284.

Li, F. , Zaslavsky, A., & Landrum, M. B. (2013). Propensity score weighting with multilevel data. *Statistics in Medicine, 32*(19), 3373–3387.

McCaffrey, D., Ridgeway, G., & Morral, A. (2004). Propensity score estimation with boosted regression for evaluating causal effects in observational studies. *Psychological Methods, 9*(4), 403–425.

Middleton, J. (2008). Bias of the regression estimator for experiments using clustered random assignment. *Statistics and Probability Letters, 78,* 2654–2659.

Middleton, J. A., & Aronow, P. M. (2015). Unbiased estimation of the average treatment effect in cluster-randomized experiments. *Statistics, Politics, and Policy, 6*(1–2), 39–75.

Murphy, R. T. (1991). *Educational effectivenes of Sesame Street: A review of the first twenty years of research, 1969–1989* (Report RR-91-55). Educational Testing Service.

Neyman, J. (1990). On the application of probability theory to agricultural experiments: Essay on principles, Section 9 (D. M. Dabrowska & T. P. Speed, Trans./ Eds.). *Statistical Science, 5*(4), 465–480. (Originally published in 1923).

Pimentel, S. D., Page, L., Lenard, M., & Keele, L. (2018). Optimal multilevel matching using network flows: an application to a summer reading intervention. *Annals of Applied Statistics, 12*(3), 1479–1505.

Pfeffermann, D., Skinner, C. J., Holmes, D. J., Goldstein, H., & Rasbash,J. (1998). Weighting for unequal selection probabilities in multi-level models. *Journal of the Royal Statistical Society, Series B, 60,* 23–56.

Rickles, J. H., & Seltzer, M. (2014). A two-stage propensity score matching strategy for treatment effect estimation in a multisite observational study. *Journal of Educational and Behavioral Statistics, 39*(6), 612–636.

Rosenbaum, P. R. (1986). Dropping out of high school in the United States: An observational study. *Journal of Educational Statistics, 11*(3), 207–224.

Rosenbaum, P. R., & Rubin, D. B. (1983). The central role of the propensity score in observational studies for causal effects. *Biometrika, 70*(1), 41–55.

Rubin, D. B. (1974). Estimating causal effects of treatments in randomized and nonranomdized studies. *Journal of Educational Psychology, 66*(5), 688–701.

Rubin, D. B. (1986). Statistics and causal inference: Comment: Which ifs have causal answers. *Journal of the American Statistical Association, 81*(396), 961–962.

Rubin, D. B. (1990). Comment: Neyman (1923) and causal inference in experiments and observational studies. *Statistical Science, 5*(4), 472–480.

Rubin, D. B. (2001). Using propensity scores to help design observational studies: Application to the tobacco litigation. *Health Services & Outcomes Research Methodology, 2,* 169–188.

Rubin, D. B. (2006). *Matched sampling for causal inference.* Cambridge University Press.

Rubin, D. B., Stuart, E. A., & Zanetto, E. L. (2004). A potential outcomes view of value-added assessment in education. *Journal of Educational and Behavioral Statistics, 29*(1), 103–116.

Schafer, J. L., & Kang, J. (2008). Average causal effects from nonrandomized studies: A practical guide and simulated example. *Psychological Methods, 13,* 279–313.

Schochet, P. (2008). Statistical power for random assignment evaluations of education programs. *Journal of Educational and Behavioral Statistics, 33*(1), 62–87.

Schochet, P. (2010). Is regression adjustment supported by the Neyman model for causal inference? *Journal of Statistical Planning and Inference, 140,* 246–259.

Schochet, P. (2013). Estimators for clustered education RCTs using the Neyman model for causal inference. *Journal of Educational and Behavioral Statistics, 38*(3), 219–238.

Spybrook, J., Hedges, L.V., & Borenstein, M. (2014). Understanding statistical power in cluster randomized trials: Challenges posed by differences in notation and terminology. *Journal of Research in Educational Effectiveness, 7,* 384–406.

Stapleton, L. M. (2002). The incorporation of sample weights into multilevel structural equation models. *Structural Equation Modeling, 9,* 475–503.

Stuart, E. (2010). Matching methods for causal inference: A review and a look forward. *Statistical Science, 25*(1), 1–21.

Stuart, E., Bradshaw, A., & Leaf, C. (2015). Assessing the generalizability of randomized trial results to target populations. *Prevention Science, 16*(3), 475–485.

Tipton, E. (2013a). Stratified sampling using cluster analysis: A sample selection strategy for improved generalizations from experiments. *Education Review, 37*(2), 109–139.

Tipton, E. (2013b). Improving generalizations from experiments using propensity score subclassification: Assumptions, properties, and contexts. *Journal of Educational and Behavioral Statistics, 38*(3), 239–266.

Tipton, E. (2014). How generalizable is your experiment? An index for comparing experimental samples and populations. *Journal of Educational and Behavioral Statistics, 39*(6), 478–501.

Tipton, E., Hedges, L., Vaden-Kiernan, M., Borman, G., Sullivan, K., & Caverly, S. (2014). Sample Selection in Randomized Experiments: A New Method Using Propensity Score Stratified Sampling. *Journal of Research on Educational Effectiveness, 7*(1), 114–135.

Townsend, Z., Buckley, J., Harada, M., & Scott, M. (2013). The choice between fixed and random effects. In M. Scott, J. Simonoff, & B. Marx (Eds.), *The SAGE handbook of multilevel modeling* (pp. 73–87). SAGE.

What Works Clearinghouse. (2020). *What Works Clearinghouse Standards Handbook, version 4.1.* U.S. Department of Education, Institute of Education Sciences, National Center for Education Evaluation and Regional Assistance.

Yeager, D. S., Hanselman, P., Walton, G. M., Murray, J., Crosnoe, R., Muller, C., Tipton, E., Schneider, B., Hulleman, C. S., Hinojosa, C. P., Paunesku, D., Romero, C., Flint, K., Roberts, A., Trott, J., Iachan, R., Buontempo, J., Yang, S. M.,

Carvalho, C. M., . . . Dweck, C. S. (2019). A national experiment reveals where a growth mindset improves adolescent achievement. *Nature, 573,* 364–369.

Zubizarreta, J. R., & Keele, L. (2017), Optimal multilevel matching in clustered observational studies: A case study of the school voucher system in Chile. *Journal of the American Statistical Association, 112,* 547–560.

CHAPTER 5

STATISTICAL POWER FOR LINEAR MULTILEVEL MODELS

Jessaca Spybrook
Western Michigan University

Benjamin M. Kelcey
University of Cincinnati

Nianbo Dong
University of North Carolina at Chapel Hill

In educational contexts, cluster randomized trials (CRT) are becoming increasingly common in studies seeking to assess the efficacy of an intervention (Spybrook, Shi, et al., 2016). In CRTs, intact clusters (e.g., schools) are assigned to treatment conditions rather than individuals (e.g., students). CRTs are advantageous in that they permit researchers to accommodate existing school structures and interventions that are designed to operate at the school level (Bloom, 2005; Boruch & Foley, 2000; Cook, 2005; Spybrook & Raudenbush, 2009). To yield rigorous evidence of whether a program

Multilevel Modeling Methods with Introductory and Advanced Applications, pages 127–164

works, however, such studies must be carefully designed. A principal consideration in the design of CRTs is the power or probability with which a study can detect an effect if it exists.

In the past 20 years, the field of education has made great strides in the design of CRTs and estimation of statistical power for the average treatment effect, or the main effect. Raudenbush (1997) presented seminal power calculations for detecting the main effect in 2-level CRTs. Since then, numerous others have contributed to this body of literature by extending the power calculations to designs such as 3-level CRTs and blocked designs (Hedges & Rhoads, 2010; Konstantopoulos, 2008a; Konstantopoulos, 2008b; Raudenbush & Liu, 2000; Raudenbush & Liu, 2001; Raudenbush et al., 2007; Schochet, 2008).

This body of literature, however, has largely been developed for detecting only main effects. Although powering the study for the main effect of treatment addresses the *what works* question, there is a growing recognition that there are critical explanatory questions that need to be addressed if we are to fully understand the validity and value of substantive theories and interventions in education. One critical line of inquiry that is largely missing in conventional study designs concerns treatment effect moderation—or questions examining *for whom* and *under what circumstances* an intervention works. For example, it may be that an intervention is more effective in urban schools compared to rural schools such that school context moderates the treatment effect.

Similarly, questions delineating the mechanisms through which an intervention operates are critical to assessing the validity of, and advancing scientific theories underlying, an intervention (i.e., mediation questions). Black box approaches that only estimate the main or average effect of, for example, an experimental curriculum on student achievement, are limited because the results can inform us if the experimental curriculum was causally related to increases in student achievement but cannot explain why or provide valid assessments of the theory of action that underlies the curriculum. In this way, mediation analyses provide a platform to delineate the *mechanisms* by which interventions or variables of interest impact valued outcomes.

The purpose of this chapter is to present an overview and introduction to power calculations for main effects, moderation effects, and mediation effects for CRTs. One chapter could not possibly cover the breadth of multilevel designs and model complexities that exist. Hence, for pedagogical purposes, we limit the scope of this chapter to cross-sectional designs and specifically to 2-level CRTs in which individuals are nested within clusters and clusters are randomly assigned to condition. Further, we only consider one binary moderator at either the individual or cluster level and one mediator at the individual or cluster level. We encourage readers to consult the recommended readings for extensions of power analyses for longitudinal models, 3-level CRTs, and other more complex model specifications.

We begin by describing an example used throughout the chapter. We then present models, power calculations, and examples for main effects, moderation effects, and mediation effects, respectively. All power calculations employ PowerUp! (Dong & Maynard, 2013), PowerUp!-Moderator (Dong et al., 2016a), and PowerUp!-Mediator (Dong et al., 2016b) software (www.causalevaluation.org).

GUIDING EXAMPLE

The example presented here is used to illustrate the power calculations for the main effect, moderation effects, and mediation effects in the remainder of the chapter. Suppose that a group of researchers developed a new whole school reform (WSR) model focused on improving achievement. The WSR model was tested in a small pilot study and the evidence suggested it has potential. As such, the researchers are planning an efficacy study to test the model. In the efficacy trial, the WSR model will be implemented at the school level. They hypothesize that the WSR model will improve the academic climate at the school, which will lead to greater student achievement. Additionally, the researchers believe students will be more engaged, which will also lead to higher achievement. A 2-level CRT is planned with students nested in schools and in which entire schools are randomly assigned to receive either the new WSR model or continue with current activities. The outcome of interest is student academic achievement, and for the purposes of this example we focus on math achievement. The efficacy sample includes small schools, defined as those with less than 300 students, and large schools, those with more than 300 students. There are three primary questions guiding the study:

1. Is the new WSR model more effective than the current activities?
2. Is the new WSR model more effective for
   a. small schools versus large schools?
   b. boys versus girls?
3. To what extent do the following mediate the relationship between the WSR model and student math achievement?
   a. academic climate
   b. student engagement

These three research questions correspond to three types of effects: main effects, moderation effects, and mediation effects, respectively. Research Question 1 is associated with the main effect of treatment. In practical terms, it seeks to answer the *what works* question and is the most common question in education evaluation studies. Research Question 2 examines whether there is a differential effect, or moderation effect. In more concrete language, it focuses on identifying *under what conditions* and/or *for whom* the treatment is effective. Given the multilevel nature of the study, there may be school-level moderators such as school size (in Part A), or student-level moderators such as gender (in Part B). Research Question 3 explores mediation. In real-world terms, it seeks to understand *how* the program works, or through what *mechanisms* the program operates. Similar to the questions around moderation, the potential mediators may be at the school level, such as the academic climate of the school, or at the student level, such as student engagement. We discuss the statistical models and the power calculations for each of the three types of effects in the next sections.

Main Effects

The main effect of treatment represents the average effect of the WSR model on student math achievement.

The Model

Based on notation from Raudenbush and Bryk (2002), the model for a 2-level CRT in hierarchical form is given in Equations 5.1 and 5.2. The level-1, or student-level model is:

$$Y_{ij} = \beta_{0j} + e_{ij}$$
$$e_{ij} \sim N(0, \sigma^2) \tag{5.1}$$

where Y_{ij} is the math achievement for individual $i = \{1, \ldots, n\}$ in school $j = \{1, \ldots, J\}$; β_{0j} is the mean math achievement for school j; and e_{ij} is the residual error associated with students, with variance σ^2.

The level-2 model, or school-level model, is:

$$\beta_{0j} = \gamma_{00} + \gamma_{01} T_j + u_{0j}$$
$$u_{0j} \sim N(0, \tau_{00}^2) \tag{5.2}$$

where γ_{00} is the grand mean math achievement; γ_{01} is the mean difference between the treatment and control groups or the main effect of treatment; T_j is a school-level treatment indicator, coded as –½ for control and ½ for treatment; and u_{0j} is the residual error associated with schools, with variance τ_{00}^2.

For simplicity, we assume equal allocation of schools to treatment and control, or $J/2$ schools per condition, though this assumption may be relaxed (see Bloom et al., 2007). The treatment effect is estimated by:

$$\hat{\gamma}_{01} = \bar{Y}_E - \bar{Y}_C \tag{5.3}$$

where $\bar{Y}_E$ is the mean for the treatment group and $\bar{Y}_C$ is the mean for the control group. The variance of the estimated treatment effect is (Raudenbush, 1997):

$$Var(\hat{\gamma}_{01}) = 4(\tau_{00}^2 + \sigma^2 / n) / J, \tag{5.4}$$

where τ_{00}^2 is the between-cluster variance, σ^2 is the within-cluster variance, n is the within-cluster sample size, and J is the total number of clusters. Note that the value 4 in the expression for the variance is a result of assuming a balanced design, or $J/2$ clusters per condition. Equation 5.4 can also be written for the more general case in which P is the proportion of clusters assigned to the treatment condition, such that

$$Var(\hat{\gamma}_{01}) = \left(\tau_{00}^2 + \frac{\sigma^2}{n}\right) / JP(1-P).$$

For a balanced design, or $P = 0.5$, the two expressions are equivalent.

Testing the Main Effect of Treatment

We can test γ_{01} using a t-test. Under the null hypothesis, the test statistic follows a t distribution. However, under the alternative hypothesis, the test statistic follows a noncentral t distribution, T'. Since power calculations assume there is a difference between groups, the noncentral t distribution, T', is critical. The mean of the noncentral t distribution is defined by the noncentrality parameter, which is a ratio of the true difference between the groups to the standard error of the estimated difference between the two groups, as shown in Equation 5.5:

$$\lambda = \sqrt{\frac{\gamma_{01}^2}{4[\tau_{00}^2 + \sigma^2 / n] / J}} \tag{5.5}$$

It is common to standardize the parameters and re-express λ as:

$$\lambda = \sqrt{\frac{\delta^2}{4[\rho + (1-\rho)/n]J}}, \tag{5.6}$$

where

$$\rho = \frac{\tau_{00}^2}{\tau_{00}^2 + \sigma^2}$$

is the intraclass correlation (ICC), or proportion of variance between clusters, and

$$\delta = \frac{\gamma_{01}}{\sqrt{\tau_{00}^2 + \sigma^2}}$$

is the standardized effect size.

The statistical power for a two-sided test is:

$$1-\beta = 1 - P\left[T'(J-2,\lambda) < t_{1-\alpha/2,J-2}\right] + P\left[T'(J-2,\lambda) \leq t_{\alpha/2,J-2}\right]. \quad (5.7)$$

The noncentrality parameter is strongly related to the power of the test. As the noncentrality parameter increases, the power increases. Hence it is important to consider how the sample sizes, the ICC, and the effect size are related to the noncentrality parameter and the power to detect the main effect of treatment.

In terms of sample sizes, from Equation 5.6 we can see that increasing the total number of clusters, J, has a greater effect on increasing the noncentrality parameter, and hence the power, than increasing the total number of individuals per cluster, n, holding everything else constant. This point is illustrated in Figures 5.1 and 5.2, which display power curves. In both cases,

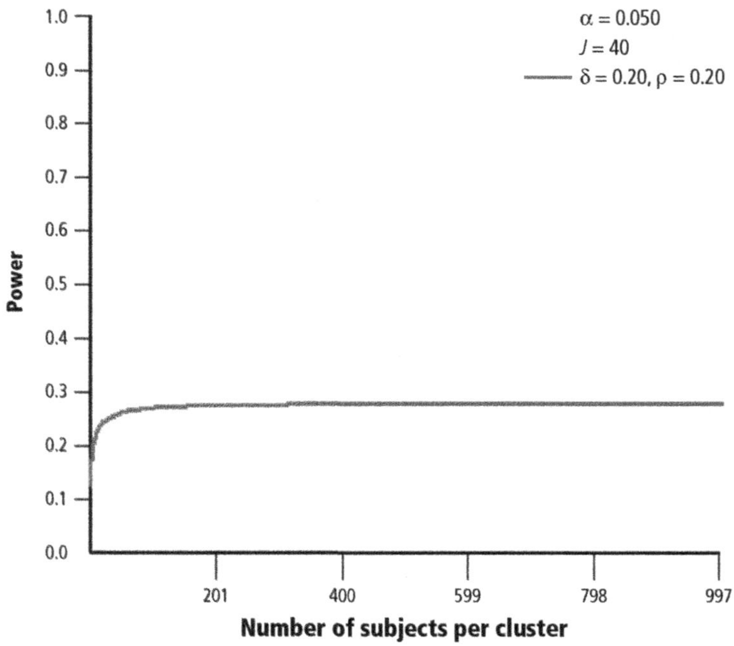

Figure 5.1 Power versus the number of subjects per cluster.

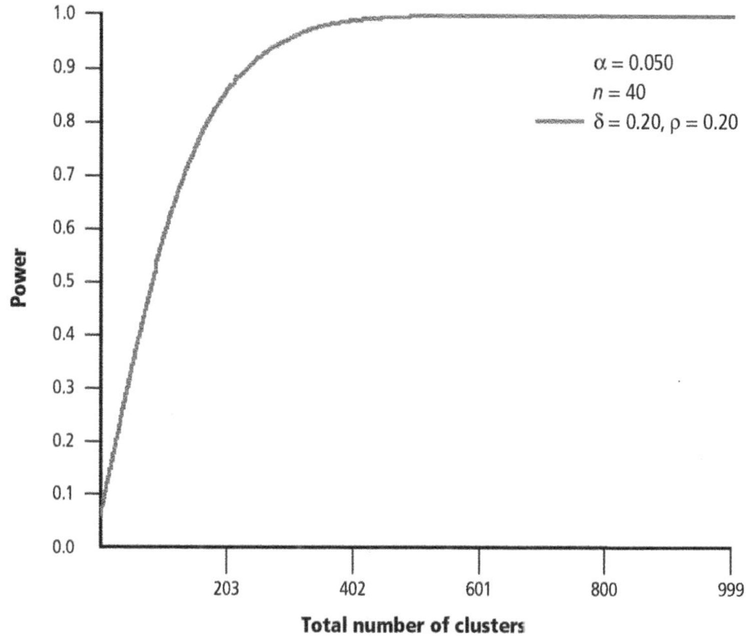

Figure 5.2 Power versus the total number of clusters.

the effect size of interest and the ICC are set to 0.20. The total sample size (i.e., number of clusters * number of individuals per cluster) is the same in both figures. However, in Figure 5.1, the number of individuals per cluster varies along the *x*-axis and the total number of clusters is set to 40. In Figure 5.2, the total number of clusters varies along the *x*-axis and the total number of individuals is set to 40. The two power curves look very different. In Figure 5.1, even as the number of individuals per cluster increases to nearly 1,000 for a total overall sample size of 1,000 * 40 = 40,000, the power to detect the effect is only approximately 0.26. In other words, increasing the number of individuals per cluster has limited impact on the power. In Figure 5.2, the total number of clusters varies along the *x*-axis. Power approaches 1.00 as the total number of clusters increases, hence the total number of clusters has a much greater impact on increasing the power to detect the main effect.

The ICC also impacts the noncentrality parameter and consequently the power. Consider the extreme case, with an ICC = 0. In this case, there is no variation between clusters and the study functions like an individually randomized trial in that the power is driven by the total sample size, $J * n$. As the ICC increases, or when there is more variation between clusters, more clusters will be necessary to achieve adequate power to detect a given effect size. Figure 5.3 illustrates this point. The two power curves show power assuming an ICC of 0.15 and an ICC of 0.05 and holding all other parameters

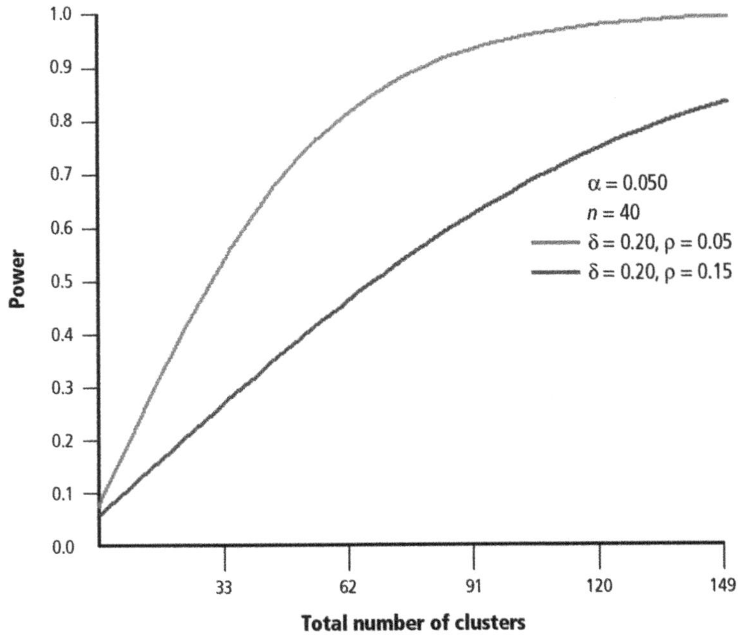

Figure 5.3 Power versus the total number of clusters for two different ICC values.

constant across the two cases. A total of 46 clusters are needed to achieve power = 0.80 assuming an ICC = 0.05, whereas 124 clusters are needed to achieve the same level of power for an ICC = 0.15.

The effect size also impacts the power. Holding all else constant, there will be greater power to detect a larger effect than a smaller effect. This can be seen in Figure 5.4. The two power curves demonstrate the power, assuming true effect sizes of 0.50 and 0.20. For example, 22 clusters are needed to detect the effect size of 0.50 with power = 0.80 whereas 126 are needed to detect the effect size of 0.20 for the same level of power.

Example: Main Effect of Treatment

There are a variety of programs that can be used for power calculations including G*POWER (Erfelder et al., 1996), PINT (Snijders et al., 2007), and Optimal Design (Raudenbush et al., 2011). We use PowerUp! (Dong & Maynard, 2013; www.causalevaluation.org) in this chapter because PowerUp! also has parallel programs for calculating power for moderation and mediation in 2-level CRTs. We demonstrate power calculations for the main effect of treatment using our earlier guiding example. Recall that students are nested within schools, and schools are assigned to condition. Suppose that the researchers are able to recruit 40 schools, 20 to each condition, with 100 students per school.

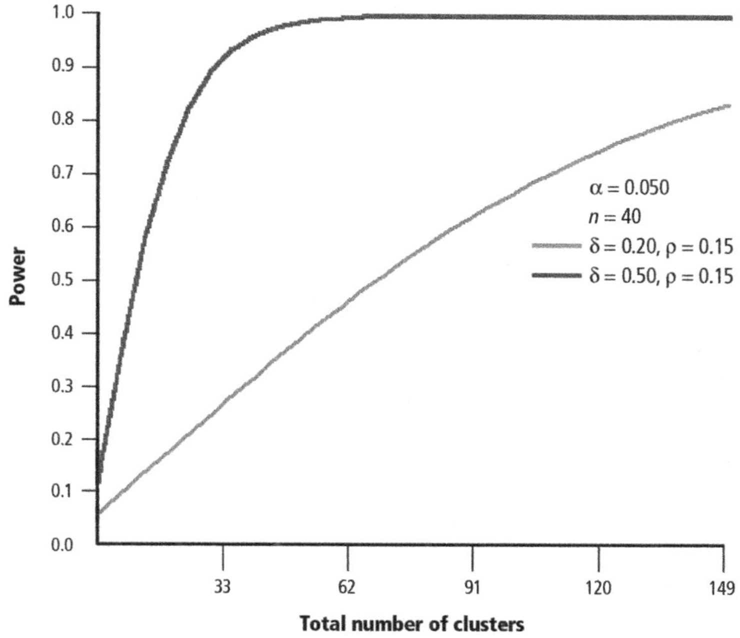

Figure 5.4 Power versus the total number of clusters for two different effect size values.

To conduct the power analysis, the researchers must also estimate the ICC. In this case, suppose that the researchers estimate an ICC = 0.15. ICC's are context specific and can heavily influence the power calculations; hence, care should be taken in estimating these values. The most-common method for estimating ICCs is to consult databases of empirical estimates of ICCs and find ICCs for similar outcomes and contexts to that of the study. Results from empirical analyses of ICCs can be found in published literature (for an overview of key papers, see Spybrook, 2013; Spybrook & Kelcey, 2016) and for some outcomes, on websites (e.g., http://stateva .ci.northwestern.edu/). ICCs can also be estimated via a pilot study, or researchers may consult the published literature in a particular domain.

Suppose that for this study, the researchers are interested in detecting a treatment effect of 0.25 standard deviations. Like the ICC, the magnitude of the treatment effect is also context specific. In general, cluster randomized trials of educational interventions are often powered to detect Cohen's *d* effect sizes in the range of 0.20 to 0.30 standard deviation units (Spybrook, Shi, et al., 2016). This range is consistent with empirical evidence around benchmarks for effect sizes in education intervention studies conducted by Hill et al. (2008), which examined effect sizes from completed studies, gaps that exist between different subgroups, and typical growth in each year of schooling.

We used PowerUp! for the calculations. Step-by-step instructions for getting started in PowerUp! are available at www.causalevaluation.org and can be used to walk through the available options. Figure 5.5 displays the output from PowerUp! after inputting the design parameters noted above: 40 schools (20 per condition), 100 students per school, ICC = 0.15, and the desired effect size of 0.25. The PowerUp! worksheet is fully annotated, and users should only enter the design parameters that are relevant for their particular study, setting other parameters to 0. For this study, the power associated with the effect size of 0.25 is 0.49, which is below the value that is typically acceptable in the social sciences, 0.80. We can use PowerUp! to determine the smallest effect size that can be detected with power = 0.80 under these assumptions, known as the minimum detectable effect size (MDES; Bloom, 1995). Figure 5.6 shows these results and suggests an MDES = 0.362.

Extensions

To increase the power to detect an effect of a given magnitude, it is common to include a covariate at level 1, level 2, or both levels. The stronger the correlation between the covariate and the outcome, or the larger the explanatory power of the covariate, the greater the increase in power holding all other parameters constant. For pedagogical purposes, we present the models with a level-1 covariate, such as student-level pretest, and the aggregate level-2 covariate. These models easily extend to the case of only a level-1 covariate or only a level-2 covariate.

The level-1 model with student-level covariates can be written as:

$$\begin{aligned} Y_{ij} &= \beta_{0j} + \sum_q \beta_{qj} X_{qij} + e_{ij} \\ e_{ij} &\sim N(0, \sigma^2_{|X_Q}) \end{aligned} \tag{5.8}$$

for $i \in \{1,2,\ldots,n\}$ persons per cluster, $j \in \{1,2,\ldots,J\}$ clusters, where Y_{ij} is the outcome for person i in cluster j; β_{0j} is the adjusted mean for cluster j; X_{qij} is the value of the qth student-level covariate for student i in school j; β_{qj} is the coefficient associated with the qth covariate for school j; e_{ij} is the residual error associated with each person, conditional on the Q covariates; and $\sigma^2_{|X_Q}$ is the residual within-cluster variance.

The new level-2, or cluster-level, model is:

$$\begin{aligned} \beta_{0j} &= \gamma_{00} + \gamma_{01} T_j + \sum_j \gamma_{0s} W_{sj} + u_{0j} \\ \beta_{qj} &= \gamma_{q0} \\ u_{0j} &\sim N(0, \tau^2_{|W}) \end{aligned} \tag{5.9}$$

where γ_{00} is the adjusted grand mean; γ_{01} is the mean difference between the

Model 3.1: MDES Calculator for Two-Level Cluster Random Assignment Design (CRA2_2)— Treatment at Level 2		
Assumptions		**Comments**
Alpha Level (α)	0.05	Probability of a Type I error
Two-tailed or One-tailed Test?	2	
Power (1-β)	0.49	Statistical power (1-probability of a Type II error)
Rho (ICC)	0.15	Proportion of variance in outcome that is between clusters
P	0.50	Proportion of Level 2 units randomized to treatment: $J_T / (J_T + J_C)$
R_1^2	0.00	Proportion of variance in Level 1 outcomes explained by Level 1 covariates
R_2^2	0.00	Proportion of variance in Level 2 outcome explained by Level 2 covariates
g*	0	Number of Level 2 covariates
n (Average Cluster Size)	100	Mean number of Level 1 units per Level 2 cluster (harmonic mean recommended)
J (Sample Size [# of Clusters])	40	Number of Level 2 units
M (Multiplier)	2.00	Computed from T_1 and T_2
T_1 (Precision)	2.02	Determined from alpha level, given two-tailed or one-tailed test
T_2 (Power)	0.03	Determined from given power level
MDES	**0.252**	Minimum Detectable Effect Size

Figure 5.5 Power for the main effect assuming an effect size of 0.25. *Note:* "Model 3.1" refers to model naming convention in PowerUp! software.

Model 3.1: MDES Calculator for Two-Level Cluster Random Assignment Design (CRA2_2)— Treatment at Level 2		
Assumptions		**Comments**
Alpha Level (α)	0.05	Probability of a Type I error
Two-tailed or One-tailed Test?	2	
Power (1-β)	0.80	Statistical power (1-probability of a Type II error)
Rho (ICC)	0.15	Proportion of variance in outcome that is between clusters
P	0.50	Proportion of Level 2 units randomized to treatment: $J_T / (J_T + J_C)$
R_1^2	0.00	Proportion of variance in Level 1 outcomes explained by Level 1 covariates
R_2^2	0.00	Proportion of variance in Level 2 outcome explained by Level 2 covariates
g*	0	Number of Level 2 covariates
n (Average Cluster Size)	100	Mean number of Level 1 units per Level 2 cluster (harmonic mean recommended)
J (Sample Size [# of Clusters])	40	Number of Level 2 units
M (Multiplier)	2.88	Computed from T_1 and T_2
T_1 (Precision)	2.02	Determined from alpha level, given two-tailed or one-tailed test
T_2 (Power)	0.85	Determined from given power level
MDES	**0.362**	Minimum Detectable Effect Size

Figure 5.6 The minimum detectable effect size for the main effect. *Note:* "Model 3.1" refers to model naming convention in PowerUp! software.

treatment and control group or the main effect of treatment; T_j is a treatment indicator, coded −½ for control and ½ for treatment; W_{sj} is the value of the sth school-level covariate for school j; γ_{0s} is the coefficient associated with the sth school-level covariate; u_{0j} is the residual error associated with each cluster, conditional on the S covariates; and $\tau^2_{|W}$ is the residual variance between clusters. Note that we assume the level-1 covariates are fixed and γ_{q0} represents the mean coefficient associated with the qth covariate.

The proportion of variance explained at level 1,

$$R^2_{L1} = \frac{\sigma^2 - \sigma^2_{X_Q}}{\sigma^2},$$

and at level 2,

$$R^2_{L2} = \frac{\tau^2_{00} - \tau^2_{W_s}}{\tau^2_{00}},$$

reduces the variance in the outcome, and thereby increases the power. The impact on power can be seen through the conditional noncentrality parameter:

$$\lambda_{|Q,W} = \sqrt{\frac{\delta^2}{4\left[(1-R^2_{L2})\rho+(1-R^2_{L1})(1-\rho)/n\right]/J}} \tag{5.10}$$

By reducing the variance, the noncentrality parameter increases and hence the power increases. The power is defined as in Equation 5.7 but with the new noncentrality parameter, $\lambda_{|Q,W}$, and J-2-S degrees of freedom. Note that the proportion of variance explained at level 2 is more critical for increasing power than the proportion of variance explained at level 1. As noted earlier, this is because the larger the ICC, or the more variance between clusters, the more clusters are needed, holding all else constant. From Equation 5.10, we can see that the proportion of variance explained at level 2 helps decrease the ICC, whereas the proportion of variance explained at level 1 only decreases the within-cluster variance.

The importance of including covariates to increase the power can be seen through the guiding example. Suppose that in addition to the design parameters described in our "Example" section, the researchers have access to pretest scores for all students. They plan to include the student-level pretest at level 1 and the aggregate at level 2. Recall that study data are not available during the planning stage, so the researcher needs to estimate the percent of variance explained by the pretest at both levels. To do so, we employ the same strategies used to estimate ICCs, by consulting empirical literature or published databases, results from a pilot study, or reported R^2 values from other studies. In this case, suppose the researchers estimate $R^2_{L1} = 0.40$ and $R^2_{L2} = 0.53$. Figure 5.7 displays the new results for the power

Model 3.1: MDES Calculator for Two-Level Cluster Random Assignment Design (CRA2_2)— Treatment at Level 2		
Assumptions		**Comments**
Alpha Level (α)	0.05	Probability of a Type I error
Two-tailed or One-tailed Test?	2	
Power (1-β)	0.80	Statistical power (1-probability of a Type II error)
Rho (ICC)	0.15	Proportion of variance in outcome that is between clusters
P	0.50	Proportion of Level 2 units randomized to treatment: $J_T / (J_T + J_C)$
R_1^2	0.40	Proportion of variance in Level 1 outcomes explained by Level 1 covariates
R_2^2	0.53	Proportion of variance in Level 2 outcome explained by Level 2 covariates
g*	1	Number of Level 2 covariates
n (Average Cluster Size)	100	Mean number of Level 1 units per Level 2 cluster (harmonic mean recommended)
J (Sample Size [# of Clusters])	40	Number of Level 2 units
M (Multiplier)	2.88	Computed from T_1 and T_2
T_1 (Precision)	2.03	Determined from alpha level, given two-tailed or one-tailed test
T_2 (Power)	0.85	Determined from given power level
MDES	**0.250**	Minimum Detectable Effect Size

Figure 5.7 Power for the main effect of treatment with covariates. *Note:* "Model 3.1" refers to model naming convention in PowerUp! software

analyses, including information for these covariates. The power to detect a main effect of 0.25 is now 0.80 compared to 0.49 in the previous case with no covariates. Given the sample size, the researchers are able to achieve an acceptable level of power by including covariates in the design.

MODERATION EFFECTS

For the 2-level CRT, we consider two different types of moderators, a school-level moderator (school size) and a student-level moderator (gender). Considering power for moderation effects is critical in the design phase of a study so that (a) it is clear from the outset whether a study is powered to address questions related to *for whom* and *under what circumstances*; and (b) assuming that moderation effects are critical, design modifications can be made during the planning phase to increase the power to detect moderation effects to an acceptable level. We begin with a discussion of power for cluster-level moderators followed by individual-level moderators.

Cluster-Level Moderator Model

In a 2-level CRT, a cluster-level moderator is at the same level as the treatment, hence it can be thought of as part of a same-level interaction. In a case without additional covariates, the level-1 model is identical to Equation 5.1. The level-2 model, or cluster-level model is:

$$\begin{gathered}\beta_{0j} = \gamma_{00} + \gamma_{01}T_j + \gamma_{02}S_j + \gamma_{03}(T_jS_j) + u_{0j} \\ u_{oj} \sim N(0, \tau^2_{00|S})\end{gathered} \tag{5.11}$$

where γ_{00} is the grand mean; γ_{01} is the mean difference between the treatment and control group; T_j is a treatment indicator, coded −½ for control and ½ for treatment; S_j is a school size indicator with −½ for small and ½ for large schools; γ_{02} is the school size effect; γ_{03} is the treatment by school size interaction; and u_{0j} is the residual error associated with clusters, with variance $\tau^2_{00|S}$. Note that we used effect coding rather than dummy coding to facilitate the interpretation of the interactions and main effects and to reduce multicollinearity between predictors. The proportion of level-2 variance explained by the moderator S and the interaction of S and T is

$$R^2_{|S} = 1 - \frac{\tau^2_{00|S}}{\tau^2_{00}}.$$

Comparing small and large schools across treatment groups, the cluster-level moderator effect is estimated by:

$$\hat{\gamma}_{03} = \left[\bar{Y}_E^{large} - \bar{Y}_E^{small}\right] - \left[\bar{Y}_C^{large} - \bar{Y}_C^{small}\right] \tag{5.12}$$

The variance of the estimated moderator effect is:

$$Var(\hat{\gamma}_{03}) = 16\left[(1 - R_{|S}^2)\tau_{00}^2 + \sigma^2 / n\right] / J \tag{5.13}$$

Note that the multiplier in the variance for the estimated moderator effect is 16, which is 4 times larger than for the main effect of treatment. This is for the balanced case and assumes $J/4$ clusters per condition. For unbalanced designs, see Spybrook, Kelcey, et al. (2016) and Dong et al. (2021).

Testing the Cluster-Level Moderator Effect

As in the case of the main effect, we can test γ_{03} using a t-test. Under the alternative hypothesis, the test statistic follows a noncentral t distribution, T', and the noncentrality parameter is:

$$\lambda_{|S} = \sqrt{\frac{\gamma_{03}^2}{16\left[(1 - R_{|S}^2)\tau_{00}^2 + \sigma^2 / n\right] / J}} \tag{5.14}$$

For consistency with the main effect calculations, we standardize by setting $\tau_{00}^2 + \sigma^2 = 1$. Hence, the noncentrality parameter using standardized notation is:

$$\lambda_{|S} = \sqrt{\frac{\delta_{CL\text{mod}}^2}{16\left[(1 - R_{|S}^2)\rho + (1-\rho) / n\right] / J}} \tag{5.15}$$

where

$$\delta_{CL\text{mod}} = \frac{\gamma_{03}}{\tau_{00}^2 + \sigma^2}.$$

The statistical power for a two-sided test is

$$1 - \beta = 1 - P\left[T'(J-4, \lambda_{|S}) < t_{1-\alpha/s, J-4}\right] + P\left[T'(J-4, \lambda_{|S}) \leq t_{\alpha/2, J-4}\right]. \tag{5.16}$$

As in the case of main effects, individual- and cluster-level covariates can also be included to increase the power to detect cluster-level moderation. Assuming we include an individual-level covariate (X) and a cluster-level covariate (W) in the models, the new noncentrality parameter is:

$$\lambda_{|SWX} = \sqrt{\frac{\delta^2_{CL\text{mod}}}{16\left[(1-R^2_{|WS})\rho+(1-R^2_{|X})(1-\rho)/n\right]/J}}. \tag{5.17}$$

Note that, just like in the case of the main effects, the covariates explain both level-1 and level-2 variance. Also similar to the main effects case, explaining the level-2 variance yields greater gains in power than explaining the level-1 variance (see Spybrook, Kelcey, et al. [2016], Dong et al. [2021], and Dong et al. [2018] for more details).

Example: Cluster-Level Moderator Effect

We demonstrate the power calculations for a cluster-level moderator using PowerUp!-Moderator. As we introduced earlier, suppose there is a total of 40 schools, 20 per condition, with 10 small and 10 large schools in each condition. We assume an ICC = 0.15. We assume the student-level pretest explains 40% of the variance in the outcome at level 1 and the aggregate pretest and school size variables explain 60% of the variance in the outcome at level 2. The expected magnitude for a moderator effect in a 2-level CRT is less clear. In general, because a moderator effect is a difference in the treatment effect for two groups, theory suggests it will be smaller than the main effect of treatment (Aguinis et al., 2005). However, there is little empirical evidence to guide the expected magnitude for the moderator effect. Given that lack of empirical evidence, we assume a cluster-level moderator effect of 0.125, which is half that of the main effect.

Figure 5.8 presents the findings. The power to detect a cluster-level moderator effect = 0.125 is very low, 0.11. One reason for such low power is the magnitude of the moderator effect of interest, which is half that of the main effect. The second reason is clear in a comparison of the variance for the main effect of treatment (see the denominator in Equation 5.6) and the variance for the cluster-level moderator effect (see the denominator in Equation 5.10). Holding the R^2 values equal across the two equations and assuming a fully balanced design, the variance for the cluster-level moderator effect has a multiplier of 16, which is 4 times that of the main effect. The larger variance leads to a smaller noncentrality parameter, and lower power. For the researchers designing this study, many more schools would need to be added to have adequate power to detect a cluster-level moderator effect of 0.125.

Individual-Level Moderator Model

In a 2-level CRT, an individual-level moderator is at level 1 whereas the treatment is at level 2; thus, this effect can be conceived of as a cross-level interaction. The level-1 model with the individual-level moderator is:

Model CRA2-2: Power Calculator for Two-Level Cluster Random Assignment Design — Treatment at Level 2 and Binary Moderator at Level 2		
Assumptions		**Comments**
Alpha Level (α)	0.05	Probability of a Type I error
Two-tailed or One-tailed Test?	2	
Effect Size Difference	0.125	Effect Size Difference regarding Cohen's d.
Rho (ICC)	0.15	Proportion of variance in outcome that is between clusters
P	0.50	Proportion of Level 2 units randomized to treatment: $J_T / (J_T + J_C)$
Q	0.50	Proportion of Level 2 units in Moderator subgroup: $J_1 / (J_1 + J_0)$
R_1^2	0.40	Proportion of variance in Level 1 outcomes explained by Level 1 covariates
R_2^2	0.60	Proportion of variance in Level 2 outcome explained by Level 2 covariates
g*	1	Number of Level 2 covariates excluding the moderator and moderator*Treatment
n (Average Cluster Size)	100	Mean number of Level 1 units per Level 2 cluster (harmonic mean recommended)
J (Sample Size [# of Clusters])	40	Number of Level 2 units
Noncentrality Parameter	0.72	Automatically computed from the above assumptions
Power (1-β)	**0.11**	Statistical power (1-probability of a Type II error)

Figure 5.8 Power for cluster-level moderator effect. *Note:* "Model CRA2-2" refers to model naming convention in PowerUp! software.

$$\begin{aligned} Y_{ij} &= \beta_{0j} + \beta_{1j} X_{ij} + e_{ij} \\ e_{ij} &\sim N(0, \sigma^2_{|x}) \end{aligned} \tag{5.18}$$

where Y_{ij} is the math achievement for individual $i = \{1, \ldots, n\}$ in school $j = \{1, \ldots, J\}$; β_{0j} is the mean achievement in school j; X_{ij} is an indicator for gender, with −½ for boys and ½ for girls; β_{1j} is the gender gap in school j; and e_{ij} is the residual error associated with students. For simplicity, we hold the gender effect constant within clusters and do not include the aggregated version of gender in the level-2 model. The school-level model is:

$$\begin{aligned} \beta_{0j} &= \gamma_{00} + \gamma_{01} T_j + u_{0j} \\ \beta_{1j} &= \gamma_{10} + \gamma_{11} T_j \\ u_{0j} &\sim N(0, \tau^2_{00}) \end{aligned} \tag{5.19}$$

where γ_{00} is the mean achievement across schools; T_j is a treatment indicator, coded −½ for control and ½ for treatment; γ_{01} is the average treatment effect; γ_{10} is the gender gap; γ_{11} is the treatment by gender interaction; and u_{0j} is the error associated with mean achievement across schools with variance τ^2_{00}.

Comparing boys to girls across treatment groups, the individual-level moderator effect is estimated by:

$$\hat{\gamma}_{11} = \left[\bar{Y}_E^G - \bar{Y}_E^B\right] - \left[\bar{Y}_C^G - \bar{Y}_C^B\right] \tag{5.20}$$

The variance of the estimated moderator effect is:

$$Var(\hat{\gamma}_{11}) = 16\left[(1 - R^2_{|x})\sigma^2\right] / nJ \tag{5.21}$$

We assume an equal number of boys and girls per cluster. For unbalanced cases, see Spybrook, Kelcey, et al. (2016) and Dong et al. (2021). The variance for the individual-level moderator effect looks different than the main effect and the cluster-level moderator effect, since the between-school variance, τ_{00}, does not contribute to the variance of the estimated moderator effect. This is because within each school there are boys and girls, and hence the school effect is removed.

Testing the Individual-Level Moderator Effect

We can test γ_{11} using a *t*-test. Under the alternative test, the test statistic follows a non-central *t* distribution, T', and the noncentrality parameter is:

$$\lambda_{|x} = \frac{\gamma_{11}^2}{\left[16(1-R_{|x}^2)\sigma^2\right]nJ} \tag{5.22}$$

We standardize the same way as we did for the main effects and cluster-level moderator such that:

$$\lambda_x = \sqrt{\frac{\delta_{IND\mathrm{mod}}^2}{\left[16(1-R_{|x}^2)(1-\rho)\right]/nJ}}, \tag{5.23}$$

where

$$\delta_{IND\mathrm{mod}} = \frac{\gamma_{11}}{\sqrt{\tau_{00}^2+\sigma^2}}$$

The statistical power for a two-sided test is:

$$1-\beta = 1-P\left[T'(nxJ-J-2,\lambda_{|X}) < t_{1-\frac{\alpha}{2},nxJ-J-2}\right] + P\left[T'(nxJ-J-2,\lambda_{|X}) \le t_{\alpha/2,nxJ-J-2}\right] \tag{5.24}$$

We can also include additional level-1 covariates to increase the power to detect an effect of a given magnitude.

Example: Individual-Level Moderator Effect

We continue with our guiding example and examine the power to detect individual-level moderation effects. As in the case of the cluster-level moderator, the expected magnitude of an individual-level moderator lacks empirical research. Following the same logic for the cluster-level moderator, we examine the power to detect an effect size difference, or an individual-level moderator effect, of 0.125. Figure 5.9 presents the findings for 40 total clusters, 100 students per cluster, with an ICC = 0.15. We assume that pretest is also included in this model such that together, pretest and gender explain 50% of the variance in the outcome. The power to detect an effect size difference of 0.125 for boys and girls is 0.86. The power for the individual-level moderator is much higher than that of the cluster-level moderator for the same-sized effect. This is for two reasons. First, the between-cluster variance is not a part of the variance of the estimated moderator effect. Second, the number of individuals per cluster is as influential as the total number of clusters in the power calculations.

Model CRA2-1N: Power Calculator for Two-Level Cluster Random Assignment Design — Treatment at Level 2 and Binary Moderator at Level 1 (Nonrandomly varying moderator slope model)		
Assumptions		**Comments**
Alpha Level (α)	0.05	Probability of a Type I error
Two-tailed or One-tailed Test?	2	
Effect Size Difference	0.125	Effect Size Difference regarding Cohen's d.
Rho (ICC)	0.15	Proportion of variance in outcome that is between clusters
P	0.50	Proportion of Level 2 units randomized to treatment: $J_T / (J_T + J_C)$
Q	0.50	Proportion of Level 1 units in Moderator subgroup: $n_1 / (n_1 + n_0)$
$R_1^{\ 2}$	0.50	Proportion of variance in Level 1 outcomes explained by Level 1 covariates
g*	2	Number of Level 1 covariates excluding the moderator
n (Average Cluster Size)	100	Mean number of Level 1 units per Level 2 cluster (harmonic mean recommended)
J (Sample Size [# of Clusters])	40	Number of Level 2 units
Noncentrality Parameter	3.03	Automatically computed from the above assumptions
Power (1-β)	**0.86**	Statistical power (1-probability of a Type II error)

Figure 5.9 Power for individual-level moderator effect. *Note:* "Model CRA2-1N" refers to model naming convention in PowerUp! software.

MEDIATION EFFECTS

Similar to moderation, we next consider mediators at two different levels: the school level (e.g., academic climate) and the student level (e.g., student engagement). We denote the first case as 2-2-1 mediation (treatment at level 2, mediator at level 2, outcome at level 1) and the second case as 2-1-1 mediation (treatment at level 2, mediator at level 1, outcome at level 1). Just like in the case for moderator effects, it is important to conduct an a priori power analysis to identify total sample sizes (school and student) that provide a sufficient level of power to detect a mediation effect of a given magnitude under the particular design constraints.

The basis for detecting mediation differs slightly from main effects or moderator effects. Hence, before we discuss the two mediation cases, we provide a brief overview using a simple mediation model as shown in Figure 5.10. The total effect, denoted c, corresponds to the main effect of treatment. In simple mediation contexts, this effect can be decomposed into two parts, the indirect effect and the direct effect. The indirect effect, which is typically captured as the product of the effect of the treatment on the mediator and the effect of the mediator on the outcome, is most-commonly used to describe a mediation effect. The direct effect is the portion of the total effect that does not flow through the mediator. In order to estimate these effects, we need to specify an outcome model and a mediator model. The outcome model allows us to estimate the b path, or the effect of the mediator on the outcome, and the c' path, or the direct effect. The mediator model allows us to estimate the a path, or the effect of the treatment on the mediator. In the remainder of this section, we extend the concepts from this simple model to the 2-2-1and 2-1-1 cases.

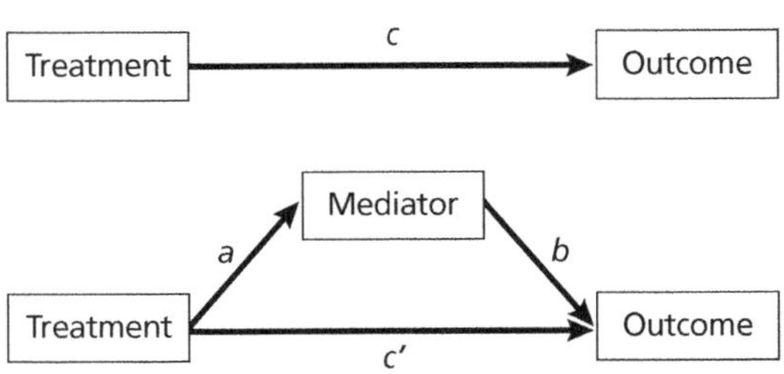

total effect = indirect effect + direct effect

$$c = ab + c'$$

Figure 5.10 Simple mediation model.

2-2-1 Mediation Model

The mediator model for the 2-2-1 case is:

$$M_j = \pi_0 + aT_j + \pi_1 W_j + \pi_2 \bar{X}_j + \varepsilon_j^M \qquad (5.25)$$
$$\varepsilon_j^M \sim N(0, \sigma_{M|}^2)$$

where M_j is the mediator for school j; W_j is a school-level covariate with π_1 as its coefficient; X_j is the aggregate of a student-level covariate with π_2 as its coefficient; T_j is a treatment indicator coded as –½ for control and ½ for treatment; a is the treatment's impact on the mediator; and ε_j^M is the error term with conditional normal distribution $\varepsilon_j^M \sim N(0, \sigma_{M|}^2)$.

The outcome model is:

$$Y_{ij} = \beta_{0j} + \beta_1(X_{ij} - \bar{X}_j) + \beta_2 V_{ij} + \varepsilon_{ij}^Y \quad \text{with} \quad \varepsilon_{ij}^Y \sim N(0, \sigma_{Y|}^2) \qquad (5.26)$$

and

$$\beta_{0j} = \gamma_{00} + bM_j + c'T_j + \gamma_{01} W_j + \gamma_{02} \bar{X}_j + u_{0j} \quad \text{with} \quad u_{0j} \sim N(0, \tau_{Y|}^2) \qquad (5.27)$$

where Y_{ij} is the outcome for student i in school j; X_{ij} is a student-level covariate that potentially varies across students and schools with coefficient β_1; V_{ij} is a student-level covariate that only varies among students (no school-level variance); and ε_{ij}^Y is the level-1 error term with conditional normal distribution $\varepsilon_{ij}^Y \sim N(0, \sigma_{Y|}^2)$. At the school-level, γ_{01} and γ_{02} are the path coefficients for the school-level covariate, W, and aggregate of the student-level covariate, $\bar{X}_j$; T_j is the treatment indicator coded as –½ for control and ½ for treatment; b is the conditional relationship between the mediator and the outcome; c' is the direct effect of the treatment; and u_{0j} is the organization-level random intercept (with conditional normal distribution, $u_{0j} \sim N(0, \tau_{Y|}^2)$. The models for both the mediator and the outcome are more complex because they both include covariates. This is necessary in order to satisfy the assumptions for causal interpretations in multilevel mediation. A full discussion of these assumptions is beyond the scope of this chapter, but we encourage readers to familiarize themselves with this topic (Kelcey, Dong, Spybrook, & Cox, 2017; VanderWeele, 2008; VanderWeele, 2010).

As previously noted, the typical way to estimate the mediation effect in the 2-2-1 case is by the product of the treatment-mediator path, or the a path, and the mediator-outcome path, or the b path. We denote this mediation effect as $ME_{221} = ab$.

Testing the Cluster-Level Mediator Effect

There are several different tests we could employ for the mediation effects including the Sobel test, the joint test, the predictive partial posterior test, and the Monte Carlo interval test. In this chapter we introduce the joint test because of its balance between performance and simplicity. Details on other tests can be found in Kelcey, Dong, Spybrook, and Shen (2017). The joint test is based on two simultaneous *t*-tests, one for the *a* path and one for the *b* path (Raudenbush & Bryk, 2002; Kenny & Judd, 2014). Under the alternative hypothesis, the test statistics follow a noncentral *t* distribution, T', and the noncentrality parameters for the test of each path are:

$$\lambda_a = \sqrt{\frac{a^2}{\sigma_a^2}} \quad \text{and} \quad \lambda_b = \sqrt{\frac{b^2}{\sigma_b^2}}. \tag{5.28}$$

The expressions for the variance for the *a* path and *b* path can be found in Kelcey, Dong, Spybrook, and Shen (2017) and Kelcey et al. (2018). For simplicity, we do not provide the full expressions here. However, note that we can standardize the expressions much like we did in the case of the main effects, such that:

$$\rho = \frac{\tau_Y^2}{\tau_Y^2 + \sigma_Y^2};$$

$\sigma_M^2 = 1$; the *a* and c' paths are now standardized mean group differences; and the *b* path is a standardized regression coefficient for the school-level variable. The variance of the *a* and *b* paths can be expressed in terms of ρ, R_{YL1}^2, R_{YL2}^2, and R_M^2 where ρ is defined above, R_{YL1}^2 is the proportion of outcome variance at the individual-level explained by individual-level covariates, R_{YL2}^2 is the proportion of outcome variance at the cluster-level explained by cluster-level covariates, and R_M^2 is the proportion of the mediator variance explained by cluster-level covariates. Much like in the case of power for main effects, the variance decreases and power increases as the ICC decreases and the percentage of outcome variance explained at each level increases, holding all other parameters constant. However, the variance explained in the mediator by covariates has a more complex role in that its contribution to power depends on the values of other parameters (for details, see Kelcey & Shen, 2020).

The statistical power for a two-sided test is the product of the powers to detect the *a* and *b* paths:

$$1-\beta = \begin{pmatrix} 1-P\left[T'(J-q-1,\lambda_a) < t_{1-\frac{\alpha}{2},J-q-1}\right] \\ +P\left[T'(J-q-1,\lambda_a) \leq t_{1-\frac{\alpha}{2},J-q-1}\right] \end{pmatrix} * \begin{pmatrix} 1-P\left[T'(J-q-1,\lambda_b) < t_{1-\frac{\alpha}{2},J-q-1}\right] \\ +P\left[T'(J-q-1,\lambda_b) \leq t_{1-\frac{\alpha}{2},J-q-1}\right] \end{pmatrix} \tag{5.29}$$

where the degrees of freedom are often approximated by $J - q - 1$, where q is the number of cluster-level predictors in the model (e.g., Kenny & Judd, 2014). Note that the power is highly dependent on the decomposition of the mediation effect.

Example: Cluster-Level Mediator Effect

We use PowerUp!-Mediator to demonstrate the power calculations. Recall that in our guiding example, we assume there are 40 schools, 20 per condition, and 100 students per school. The ICC = 0.15. We assume a student-level pretest is available and explains 40% and 52% of the variance in the outcome at the individual and cluster levels, respectively. Now, we assume that the covariates explain 50% of the variance in the mediator. The researchers are interested in detecting a total effect, that is, main effect, of 0.25. They hypothesize that the standardized treatment-mediator path is $a = 0.50$, the mediator-outcome path is $b = 0.30$, and the direct effect is $c' = 0.10$. Hence, the standardized overall mediator effect is $0.50 * 0.30 = 0.15$. Figure 5.11 shows the results from PowerUp!-Mediator. Under the joint test, the power to detect the mediation effect is 0.643. To detect the mediator effect under these assumptions, a total of 58 schools would be necessary (see Figure 5.12). However, under a different decomposition of the mediator effect—that is, a larger standardized treatment-mediator path of $a = 0.60$ and a smaller mediator-outcome path of $b = 0.25$—while maintaining earlier assumptions regarding an overall mediator effect of 0.15 and direct effect of $c' = 0.10$, and with 40 schools, the design yields power of 0.822 to detect the mediation effect under the joint test. This example illustrates the dependency of power on the decomposition of the mediation effect.

Model CRA2-2-1: Power Calculator for Two-Level Cluster Random Assignment Design — Treatment at Level 2 and Mediator at Level 2		
Parameter Values		**Comments**
Alpha Level (α)	0.05	Probability of a Type I error
Effect Size: *a*	0.50	Treatment-mediator path coefficient as standardized mean difference
Effect Size: *b*	0.30	Mediator-outcome path coefficient as standardized regression coefficient
Effect Size: *c′*	0.10	Direct treatment-outcome path coefficient as standardized mean difference
Intraclass Correlation (ρ)	0.15	Proportion of variance in outcome that is between clusters
R^2_{YL1}	0.40	Proportion of outcome variance at individual-level explained by individual-level covariates
R^2_{YL2}	0.52	Proportion of outcome variance at cluster-level explained by cluster-level covariates (excluding the treatment and mediator)
R^2_{M}	0.50	Proportion of mediator variance explained by cluster-level covariates (excluding the treatment)
n (Average Cluster Size)	100	Mean number of individuals per cluster (harmonic mean recommended)
J (Sample Size [# of Clusters])	40	Total number of clusters (total=treatment +contol)
P	0.50	Proportion of level 2 units randomized to treatment: $J_T / (J_T + J_C)$
Mediaiton Effect (ME)	**0.150**	ME = a*b
Power (1-β) for ME: Sobel Test	**0.633**	Statistical power to detect the Mediation Effect (ME) for two-tailed test using Sobel test
Power (1-β) for ME: Joint Test	**0.643**	Statistical power to detect the Mediation Effect (ME) for two-tailed test using joint test
Power (1-β) for ME: MC	**0.682**	Statistical power to detect the Mediation Effect (ME) for two-tailed test using Monte Carlo simulation (2000 replications)
	Run MC	

Figure 5.11 Power for 2-2-1 mediation with 40 schools. *Note:* "Model CRA2-2-1" refers to model naming convention in PowerUp! software.

Model CRA2-2-1: Power Calculator for Two-Level Cluster Random Assignment Design — Treatment at Level 2 and Mediator at Level 2		
Parameter Values		**Comments**
Alpha Level (α)	0.05	Probability of a Type I error
Effect Size: *a*	0.50	Treatment-mediator path coefficient as standardized mean difference
Effect Size: *b*	0.30	Mediator-outcome path coefficient as standardized regression coefficient
Effect Size: *c'*	0.10	Direct treatment-outcome path coefficient as standardized mean difference
Intraclass Correlation (ρ)	0.15	Proportion of variance in outcome that is between clusters
R^2_{YL1}	0.40	Proportion of outcome variance at individual-level explained by individual-level covariates
R^2_{YL2}	0.52	Proportion of outcome variance at cluster-level explained by cluster-level covariates (excluding the treatment and mediator)
R^2_M	0.50	Proportion of mediator variance explained by cluster-level covariates (excluding the treatment)
n (Average Cluster Size)	100	Mean number of individuals per cluster (harmonic mean recommended)
J (Sample Size [# of Clusters])	58	Total number of clusters (total=treatment +contol)
P	0.50	Proportion of level 2 units randomized to treatment: $J_T / (J_T + J_C)$
Mediaiton Effect (ME)	**0.150**	ME = a*b
Power (1-β) for ME: Sobel Test	**0.791**	Statistical power to detect the Mediation Effect (ME) for two-tailed test using Sobel test
Power (1-β) for ME: Joint Test	**0.807**	Statistical power to detect the Mediation Effect (ME) for two-tailed test using joint test
Power (1-β) for ME: MC	**0.838**	Statistical power to detect the Mediation Effect (ME) for two-tailed test using Monte Carlo simulation (2000 replications)
	Run MC	

Figure 5.12 Power for 2-2-1 mediation with 58 schools. *Note:* "Model CRA2-2-1" refers to model naming convention in PowerUp! software.

2-1-1 Mediation Model

The 2-1-1 mediation model follows a similar set-up to the 2-2-1 mediation model. The mediator model is:

$$M_{ij} = \pi_{0j} + \pi_1(X_{ij} - \bar{X}_j) + \pi_2 V_{ij} + \varepsilon_{ij}^M \quad \text{with} \quad \varepsilon_{ij}^M \sim N(0, \sigma_{M|}^2) \tag{5.30}$$

and

$$\pi_{0j} = \zeta_{00} + aT_j + \zeta_{01}W_j + \zeta_{02}\bar{X}_j + u_{0j}^M \quad \text{with} \quad u_{0j}^M \sim N(0, \tau_{M|}^2) \tag{5.31}$$

where M_{ij} represents the mediator value for student i in school j; X_{ij} is a student-level covariate (with π_1 as its path coefficient); $\bar{X}_j$ is the school-level variable or mean aggregate of the student-level variable (with ζ_{02} as its path coefficient); V_{ij} is a student-level covariate that only varies among students (no school-level variance); T_j is the treatment assignment variable (with path coefficient a); ε_{ij}^M is the error term; and u_{0j}^M is the school-specific random effect.

The outcome model is:

$$Y_{ij} = \beta_{0j} + b_1(M_{ij} - \bar{M}_j) + \beta_1(X_{ij} - \bar{X}_j) + \beta_2 V_{ij} + \varepsilon_{ij}^Y \quad \text{with} \quad \varepsilon_{ij}^Y \sim N(0, \sigma_{Y|}^2) \tag{5.32}$$

and

$$\beta_{0j} = \gamma_{00} + B\bar{M}_j + c'T_j + \gamma_{01}W_j + \gamma_{02}\bar{X}_j + u_{0j}^Y \quad \text{with} \quad u_{0j}^Y \sim N(0, \sigma_{Y|}^2) \tag{5.33}$$

where Y_{ij} is the outcome for student i in school j; X_{ij} is a student-level covariate (with β_1 as path coefficient); $M_{ij} - \bar{M}_j$ is the school-centered student-level mediator (with coefficient b_1); $\bar{M}_j$ is the mean of the mediator in school j (with path coefficient B); c' is the treatment-outcome conditional path coefficient; γ_{01} is the coefficient for W; γ_{02} is the coefficient for $\bar{X}_j$; and ε_{ij}^Y and u_{0j}^Y are the level-1 and 2 error terms, respectively.

In the 2-1-1 mediation model, we may be interested in three options: detecting the overall mediation effect, the lower-level mediation effect, or the upper-level mediation effect. In this chapter, we focus on the overall mediation effect, which is the extent to which the effect of the treatment, that is, the WSR model on math achievement flows through changes in student engagement at the individual level and the school level. Lower-level mediation tracks the extent to which the effect of the WSR model on math achievement flows through changes only in individual student engagement, whereas upper-level mediation tracks the extent to which the effect flows through changes only in the aggregate engagement of individuals at the school-level. We refer interested readers to Kelcey et al. (2018)

and Pituch and Stapleton (2012) for additional detailed discussion of these three options.

The overall mediation effect is denoted $ME_{211} = aB$, where a is the treatment-mediator path and B is the mediator-outcome path. Note that under a set of assumptions and centering within schools, B quantifies how changes in student engagement at both the individual and school levels are conditionally related to math achievement (see VanderWeele, 2010; Kelcey, Dong, Spybrook, & Cox, 2017). Hence B is the sum of the portion of the overall mediator effect attributed to student-level association, denoted b_1 in Equation 5.32, and the portion of the overall mediator effect attributed to the school-level association, which is not depicted in the model formulation above (see Kelcey, Dong, Spybrook, & Cox [2017] for more on this topic).

Testing the Individual-Level Mediator Effect

Just like the cluster-level mediator case, there are multiple tests that can be employed. We provide details for the joint test here and encourage readers to consult Kelcey, Dong, Spybrook, and Cox (2017) for other options. As in the 2-2-1 case, the joint test is based on simultaneous t-tests for the a and B paths. Under the alternative hypothesis, the test statistics follow a noncentral t distribution, T', and the noncentrality parameter for the tests of each path are:

$$\lambda_a = \sqrt{\frac{a^2}{\sigma_a^2}} \quad \text{and} \quad \lambda_B = \sqrt{\frac{B^2}{\sigma_B^2}} \tag{5.34}$$

For more details on these variance expressions, see Kelcey, Dong, Spybrook, and Cox (2017). We can standardize parameters much like we did in the case of the main effects, such that

$$\rho_Y = \frac{\tau_Y^2}{\tau_Y^2 + \sigma_Y^2}$$

and

$$\rho_M = \frac{\tau_M^2}{\tau_M^2 + \sigma_M^2}.$$

As such, the magnitude of the a and c' paths are now standardized mean group differences, and the B path is a standardized regression coefficient for the school-level variable. We can also express the variances in terms of ρ_Y, ρ_M, R^2_{ML1}, R^2_{ML2}, R^2_{YL1}, and R^2_{YL2}, where ρ_Y and ρ_M are defined as above, R^2_{ML1} is the proportion of mediator variance at the individual-level explained by individual-level covariates, R^2_{ML2} is the proportion of mediator variance at the school or cluster level that is explained by cluster-level covariates, R^2_{YL1} is the proportion of outcome variance at the individual-level that is explained by individual-level covariates, and R^2_{YL2} is the proportion of outcome variance at the cluster-level that is explained by cluster-level covariates.

The statistical power for a two-sided test is the product of the powers to detect the a and B paths:

$$1-\beta=\begin{pmatrix} 1-P\left[T'(J-q-1,\lambda_a)<t_{1-\frac{\alpha}{2},J-q-1}\right] \\ +P\left[T'(J-q-1,\lambda_a)\leq t_{1-\frac{\alpha}{2},J-q-1}\right] \end{pmatrix} * \begin{pmatrix} 1-P\left[T'(J-q-1,\lambda_B)<t_{1-\frac{\alpha}{2},J-q-1}\right] \\ +P\left[T'(J-q-1,\lambda_B)\leq t_{1-\frac{\alpha}{2},J-q-1}\right] \end{pmatrix} \tag{5.35}$$

where the degrees of freedom are often approximated by $J-q-1$, where q is the number of cluster-level predictors in the model (e.g., Kenny & Judd, 2014).

Example: Individual-Level Mediator Effect

We demonstrate how PowerUp!-Mediator can be used to conduct power calculations for 2-1-1 mediation. Recall that in the guiding example, researchers hypothesize that the treatment operates via changes in student engagement, which leads to increased math achievement. They have a total of 40 schools and plan to randomly assign half to each condition. There are 100 individuals per school. They estimate the ICC for the outcome as 0.15, and use the same ICC for the mediator. They anticipate that the inclusion of covariates in the models will explain 40% of the variance in the mediator at levels 1 and 2 and 70% of the variance in the outcome at levels 1 and 2. They hypothesize that the overall standardized treatment-mediator path is $a = 0.50$, the overall mediator-outcome path is $B = 0.30$, and the direct effect is $c' = 0.10$. Further, they hypothesize that $b_1 = 0.10$, the part of the overall mediator-outcome association that will be at the student-level. Note that the overall mediator effect is the same as in the 2-2-1 case, 0.50 * 0.30 = 0.15. Figure 5.13 shows the results. The power for the joint test for the overall mediation effects is 0.499. To achieve power of 0.80, an additional 40 schools would need to be included in the study for a total of 80 clusters, as shown in Figure 5.14.

Model CRA2-1-1: Power Calculator for Two-Level Cluster Random Assignment Design — Treatment at Level 2 and Mediator at Level 1		
Assumptions		**Comments**
Alpha Level (α)	0.05	Probability of a Type I error
Effect Size: a	0.50	Treatment-mediator path coefficient as standardized mean difference
Effect Size: b_1	0.10	Mediator-outcome path coefficient at the individual-level as a standardized regression coefficient
Effect Size: B	0.30	Overall mediator-outcome path coefficient as a standardized regression coefficient (overall coef=individual coef+cluster coef)
Effect Size: c'	0.10	Direct treatment-outcome path coefficient as standardized mean difference
Intraclass Correlation (ρ_M)	0.15	Proportion of variance in the mediator (M) that is between clusters
Intraclass Correlation (ρ_Y)	0.15	Proportion of variance in the outcome (Y) that is between clusters
$R^2_M{}^{L1}$	0.40	Proportion of mediator variance at individual-level explained by individual-level covariates
$R^2_M{}^{L2}$	0.40	Proportion of mediator variance at cluster-level explained by cluster-level covariates (excluding the treatment)
$R^2_Y{}^{L1}$	0.70	Proportion of outcome variance at individual-level explained by individual-level covariates (excluding mediator)
$R^2_Y{}^{L2}$	0.70	Proportion of outcome variance at cluster-level explained by cluster-level covariates (excluding the treatment and mediator)
n (Average Cluster Size)	100	Mean number of individuals per cluster (harmonic mean recommended)
J (Sample Size [# of Clusters])	40	Total number of clusters (total=treatment +contol)
P	0.50	Proportion of level 2 units randomized to treatment: $J_T / (J_T + J_C)$
Lower-Level Mediaiton Effect (LME)	0.05	LME = a * b1
Upper-Level Mediaiton Effect (UME)	0.10	UME = a * (B-b1)
Overall Mediaiton Effect (OME)	0.15	OME = a * B
Power (1-β) for LME: Sobel Test	**1.000**	Statistical power to detect the lower-level mediation effects for two-tailed test using Sobel test
Power (1-β) for LME: Joint Test	**1.000**	Statistical power to detect the lower-level mediation effects for two-tailed test using joint test
Power (1-β) for LME: MC	**1.000**	Statistical power to detect the lower-level mediation effects for two-tailed test using Monte Carlo simulation (2000 replications)
Power (1-β) for UME: Sobel Test	**0.263**	Statistical power to detect the upper-level mediation effects for two-tailed test using Sobel test
Power (1-β) for UME: Joint Test	**0.256**	Statistical power to detect the upper-level mediation effects for two-tailed test using joint test
Power (1-β) for UME: MC	**0.283**	Statistical power to detect the upper-level mediation effects for two-tailed test using Monte Carlo simulation (2000 replications)
Power (1-β) for OME: Sobel Test	**0.501**	Statistical power to detect the overall mediation effects for two-tailed test using Sobel test
Power (1-β) for OME: Joint Test	**0.499**	Statistical power to detect the overall mediation effects for two-tailed test using joint test
Power (1-β) for OME: MC	**0.508**	Statistical power to detect the overall mediation effects for two-tailed test using Monte Carlo simulation (2000 replications)
	Run MC	

Figure 5.13 Power for 2-1-1 mediation with 40 schools. *Note:* "Model CRA2-1-1" refers to model naming convention in PowerUp! software.

Model CRA2-1-1: Power Calculator for Two-Level Cluster Random Assignment Design — Treatment at Level 2 and Mediator at Level 1		
Assumptions		**Comments**
Alpha Level (α)	0.05	Probability of a Type I error
Effect Size: a	0.50	Treatment-mediator path coefficient as standardized mean difference
Effect Size: b_1	0.10	Mediator-outcome path coefficient at the individual-level as a standardized regression coefficient
Effect Size: B	0.30	Overall mediator-outcome path coefficient as a standardized regression coefficient (overall coef=individual coef+cluster coef)
Effect Size: c'	0.10	Direct treatment-outcome path coefficient as standardized mean difference
Intraclass Correlation (ρ_M)	0.15	Proportion of variance in the mediator (M) that is between clusters
Intraclass Correlation (ρ_Y)	0.15	Proportion of variance in the outcome (Y) that is between clusters
$R^{2}_{M}{}^{L1}$	0.40	Proportion of mediator variance at individual-level explained by individual-level covariates
$R^{2}_{M}{}^{L2}$	0.40	Proportion of mediator variance at cluster-level explained by cluster-level covariates (excluding the treatment)
$R^{2}_{Y}{}^{L1}$	0.70	Proportion of outcome variance at individual-level explained by individual-level covariates (excluding mediator)
$R^{2}_{Y}{}^{L2}$	0.70	Proportion of outcome variance at cluster-level explained by cluster-level covariates (excluding the treatment and mediator)
n (Average Cluster Size)	100	Mean number of individuals per cluster (harmonic mean recommended)
J (Sample Size [# of Clusters])	80	Total number of clusters (total=treatment +contol)
P	0.50	Proportion of level 2 units randomized to treatment: J_T / (J_T + J_C)
Lower-Level Mediaiton Effect (LME)	0.05	LME = a * b1
Upper-Level Mediaiton Effect (UME)	0.10	UME = a * (B-b1)
Overall Mediaiton Effect (OME)	0.15	OME = a * B
Power (1-β) for LME: Sobel Test	**1.000**	Statistical power to detect the lower-level mediation effects for two-tailed test using Sobel test
Power (1-β) for LME: Joint Test	**1.000**	Statistical power to detect the lower-level mediation effects for two-tailed test using joint test
Power (1-β) for LME: MC	**1.000**	Statistical power to detect the lower-level mediation effects for two-tailed test using Monte Carlo simulation (2000 replications)
Power (1-β) for UME: Sobel Test	**0.465**	Statistical power to detect the upper-level mediation effects for two-tailed test using Sobel test
Power (1-β) for UME: Joint Test	**0.464**	Statistical power to detect the upper-level mediation effects for two-tailed test using joint test
Power (1-β) for UME: MC	**0.461**	Statistical power to detect the upper-level mediation effects for two-tailed test using Monte Carlo simulation (2000 replications)
Power (1-β) for OME: Sobel Test	**0.793**	Statistical power to detect the overall mediation effects for two-tailed test using Sobel test
Power (1-β) for OME: Joint Test	**0.803**	Statistical power to detect the overall mediation effects for two-tailed test using joint test
Power (1-β) for OME: MC	**0.806**	Statistical power to detect the overall mediation effects for two-tailed test using Monte Carlo simulation (2000 replications)
	Run MC	

Figure 5.14 Power for 2-1-1 mediation with 80 schools. *Note:* "Model CRA2-1-1" refers to model naming convention in PowerUp! software.

CHAPTER SUMMARY

A priori power analyses are critical in the planning phase of a study. This chapter focused on power analyses for main effects and for cluster-level and individual-level moderator and mediator effects in 2-level CRTs. We demonstrated how calculations differ depending on the effect of interest. PowerUp!, PowerUp!-Moderator, and PowerUp!-Mediator are easy to use and freely available tools for conducting power calculations for the three types of effects discussed in this chapter, and we highly recommend these accessible tools to researchers planning CRTs. Two other widely used programs that we also recommend for calculating power for main effects in CRTs include Optimal Design (Raudenbush et al., 2011) and CRT Power (Borenstein et al., 2012). Below we suggest recommended readings on the models demonstrated in this chapter, some extensions for 3-level and longitudinal designs, and an applied example for readers.

RECOMMENDED READINGS

As set forth at the beginning of this chapter, we view this chapter as an introduction to power for main effects, moderation effects, and mediation effects for CRTs. There are many possible extensions to what we present in this chapter. For the methods demonstrated here we encourage readers to consult the following references for more technical details, and for extensions including more complex designs and alternative assumptions.

Power for Main Effects in CRTs

Liu, X. S. (2014). *Statistical power analysis for the social and behavioral sciences: Basic and advanced techniques.* Routledge.

Moerbeek, M., & Teerenstra, S. (2016). *Power analysis of trials with multilevel data.* Chapman & Hall.

Raudenbush, S. W. (1997). Statistical analysis and optimal design for cluster randomized trials. *Psychological Methods, 2*(2), 173–185. https://doi.org/10.1037/1082-989X.2.2.173

Schochet, P. Z. (2008). Statistical power for random assignment evaluations of education program. *Journal of Educational and Behavioral Statistics, 33*(1), 62–87. https://doi.org/10.3102/1076998607302714

Power for Moderation Effects in CRTs

Spybrook, J., Kelcey, B., & Dong, N. (2016). Power for detecting treatment by moderato effects in two and three-level cluster randomized trials. *Journal of*

Educational and Behavioral Statistics, 41(6), 605–627. https://doi.org/10.3102/1076998616655442

Dong, N., Kelcey, B., & Spybrook, J. (2018). Power analyses for moderator effects in three-level cluster randomized trials. *Journal of Experimental Education, 86*(3), 489–514. https://doi.org/10.1080/00220973.2017.1315714

Dong, N., Spybrook, J., Kelcey, B., & Bulus, M. (2021). Power analyses for moderator effects with (non)random slopes in cluster randomized trials. *Methodology, 17*(2), 92–110. doi: https://doi.org/10.5964/meth.4003

Power for Mediation Effects in CRTs

Kelcey, B., Dong, N., Spybrook, J., & Cox, K. (2017). Statistical power for causally-defined indirect effects in group-randomized trials with individual-level mediators. *Journal of Educational and Behavioral Statistics, 42*(5), 499–530. https://doi.org/10.3102/1076998617695506

Kelcey, B., Dong, N., Spybrook, J., & Shen, Z. (2017). Experimental power for indirect effect in group-randomized studies with group level mediators. *Multivariate Behavioral Research, 52*(6), 699–719. https://doi.org/10.1080/00273171.2017.1356212

Kelcey, B., Spybrook, J., & Dong, N. (2018). Sample size planning for cluster-randomized interventions probing multilevel mediation. *Prevention Science, 20*(3), 407–418. https://doi.org/10.1007/s11121-018-0921-6

Power for Longitudinal Designs

Brandmaier, A. M., von Oertzen, T., Ghisletta, P., Hertzog, C., & Lindenberger, U. (2015, March 24). LIFESPAN: A tool for the computer-aided design of longitudinal studies. *Frontiers in Psychology, 6*(272). https://doi.org/10.3389/fpsyg.2015.00272

GLIMMPSE software and documentation: http://samplesizeshop.org/

Kreidler, S. M., Muller, K. E., Grunwald, G. K., Ringham, B. M., Coker-Dukowitz, Z. T., Sakhadeo, U. R., Barón, A. E., & Glueck, D. H. (2013). GLIMMPSE: Online power computation for linear models with and without a baseline covariate. *Journal of Statistical Software, 54*(10).

TRY THIS!

For this exercise, please use PowerUp!, PowerUp!-Moderator, and PowerUp!-Mediator.

Suppose a group of researchers are designing a study to examine the efficacy of a new reading curriculum designed to engage and motivate students around the curricular content. They plan to randomly assign 24 schools to the treatment condition and 24 schools to the comparison condition. In each school, they plan to test 80 fifth graders, approximately 40 boys and 40 girls. The outcome of interest is reading achievement, as measured by a state test.

Question 1: What is the power to detect a main effect of treatment of 0.20 standard deviations? 0.30 standard deviations?

Question 2: Suppose that the researchers were interested in whether the treatment had a differential effect by gender. What is the power to detect a moderator effect of 0.10 standard deviations? 0.15 standard deviations?

Question 3: Suppose that researchers intended to further unpack the treatment effects by probing the extent to which the curriculum impacted student achievement through improvements in student motivation. What is the power to detect an overall mediation effect of 0.30, assuming the standardized treatment-mediator path coefficient is 0.60 and the standardized mediator-outcome path coefficient is 0.50?

After completing each power analysis, prepare a written summary of the analyses and results. Be sure to include enough detail (e.g., the values of the design parameters; the sample size at each level) so that the analysis could be replicated.

Here are some guidelines to follow and/or consider as you complete this exercise.

- Assume the typical conventions: two-sided test, $\alpha = 0.05$.
- Make your own assumptions about which covariates may be available.
- Use empirical evidence to estimate the necessary design parameters, e.g., the intraclass correlation coefficient(s) (ρ) and percent of variance explained (R^2). One key resource for locating empirical estimates is the Online Intraclass Correlation Database, http://stateva.ci.northwestern.edu/ (see Spybrook & Kelcey [2016] and Spybrook [2013] for additional references).

REFERENCES

Aguinis, H., Beaty, J. C., Boik, R. J., & Pierce, C. A. (2005). Effect size and power in assessing moderating effects of categorical variables using multiple regression: A 30-year review. *Journal of Applied Psychology, 90*(1), 94–107.

Bloom, H. S. (1995). Minimum detectable effects: A simple way to report the statistical power of experimental designs. *Evaluation Review, 19*(5), 547–556.

Bloom, H. S. (2005). Randomizing groups to evaluate place-based programs. In H. S. Bloom (Ed.), *Learning more from social experiments: Evolving analytic approaches* (pp. 115–172). Russell Sage Foundation.

Bloom, H. S., Hill, C. J., Black, A. R., & Lipsey, M. (2008). Performance trajectories and performance gaps as achievement effect-size benchmarks for educational interventions. *Journal of Research on Educational Effectiveness, 1*(4), 289–328.

Bloom, H. S., Richburg-Hayes, L., & Black, A. R. (2007). Using covariates to improve precision for studies that randomize schools to evaluate educational interventions. *Educational Evaluation and Policy Analysis, 29*(1), 30–59.

Borenstein, M., Hedges, L. V., & Rothstein, H. (2012). *CRT power.* Biostat.

Boruch, R. F., & Foley, E. (2000). The honestly experimental society. In L. Bickman (Ed.), *Validity and social experiments: Donald Campbell's legacy* (pp. 199–239). SAGE Publications.

Cook, T. D. (2005). Emergent principles for the design, implementation, and analysis of cluster-based experiments in social science. *The ANNALS of the American Academy of Political and Social Science, 599*(1), 176–198.

Dong, N., Kelcey, B., Spybrook, J., & Maynard, R. (2016a). PowerUp!-Moderator: A tool for calculating statistical power and minimum detectable effect size of the moderator effects in cluster randomized trials (Version 1.06) [Software]. Available from https://www.causalevaluation.org/power-analysis.html

Dong, N., Kelcey, B., Spybrook, J., & Maynard, R. (2016b). PowerUp!-Mediator: A tool for calculating statistical power and minimum detectable effect size of the mediator effects in cluster randomized trials (Version 1.06) [Software]. Available from https://www.causalevaluation.org/power-analysis.html

Dong, N., Kelcey, B., & Spybrook, J. (2018). Power analyses for moderator effects in three-level cluster randomized trials. *Journal of Experimental Education, 86,* 489–514.

Dong, N., & Maynard, R. (2013). PowerUp!: A tool for calculating minimum detectable effect sizes and minimum required sample sizes for experimental and quasi-experimental design studies. *Journal of Research on Educational Effectiveness, 6*(1), 24–67.

Dong, N., Spybrook, J., Kelcey, B., & Bulus, M. (2021). Power analyses for moderator effects with (non)random slopes in cluster randomized trials. *Methodology, 17*(2), 92–110. doi: https://doi.org/10.5964/meth.4003

Erdfelder, E., Faul, F., & Buchner, A. (1996). GPOWER: A general power analysis program. *Behavior Research Methods, Instruments, & Computers, 28,* 1–11.

Hedges, L. V., & Rhoads, C. (2010). *Statistical power analysis in education research* (NCSER 2010–3006). National Center for Special Education Research, U.S. Department of Education.

Hill, C. J., Bloom, H. S., Black, A. R., & Lipsey, M. W. (2008). Empirical benchmarks for interpreting effect sizes in research. *Child Development Perspectives, 2*(3), 172–177.

Kelcey, B., Dong, N., Spybrook, J., & Cox, K. (2017). Statistical power for causally-defined indirect effects in group-randomized trials with individual-level mediators. *Journal of Educational and Behavioral Statistics, 42*(5), 499–530.

Kelcey, B., Dong, N., Spybrook, J., & Shen, Z. (2017). Experimental power for indirect effect in group-randomized studies with group level mediators. *Multivariate Behavioral Research, 52,* 699–719.

Kelcey, B., & Shen, Z. (2020). Strategies for efficient experimental design in studies proving 2-1-1 mediation. *Journal of Experimental Education, 88*(2), 311–334.

Kelcey, B., Spybrook, J., & Dong, N. (2018). Sample size planning for cluster-randomized interventions probing multilevel mediation. *Prevention Science, 20*(3), 407–418. https://doi.org/10.1007/s11121-018-0921-6

Kenny, D. A., & Judd, C. M. (2014). Power anomalies in testing mediation. *Psychological Science, 25*, 334–339.

Konstantopoulos, S. (2008a). The power of the test for treatment effects in three-level block randomized designs. *Journal of Research on Educational Effectiveness, 1*(4), 265–288.

Konstantopoulos, S. (2008b). The power of the test for treatment Effects in three-level cluster randomized designs. *Journal of Research on Educational Effectiveness, 1*(1), 66–88.

Pituch, K. A., & Stapleton, L. M. (2012). Distinguishing between cross- and cluster-level mediation processes in the cluster randomized trial. *Sociological Methods & Research, 41*, 630–670.

Raudenbush, S. W. (1997). Statistical analysis and optimal design for cluster randomized trials. *Psychological Methods, 2*(2), 173–185.

Raudenbush, S. W., & Bryk, A. S. (2002). *Hierarchical linear models: Applications and data analysis methods.* SAGE.

Raudenbush, S. W., & Liu, X. (2000). Statistical power and optimal design for multisite randomized trials. *Psychological Methods, 5*(2), 199.

Raudenbush, S. W., & Liu, X. (2001). Effects of study duration, frequency of observation, and sample size on power in studies of group differences in polynomial change. *Psychological Methods, 6*(4), 387.

Raudenbush, S. W., Martinez, A., & Spybrook, J. (2007). Strategies for improving precision in group-randomized experiments. *Educational Evaluation and Policy Analysis, 29*(1), 5–29.

Raudenbush, S. W., Spybrook, J., Congdon, R., Liu, X, Martinez, A., Bloom, H. S., & Hill, C. J. (2011). Optimal Design Software Plus Empirical Evidence (Version 3.0) [Software]. http://wtgrantfoundation.org/resource/optimal-design-with-empirical-information-od

Schochet, P. Z. (2008). Statistical power for random assignment evaluations of education programs. *Journal of Educational and Behavioral Statistics, 33*(1), 62–87.

Snijders, T. A. B., Bosker, R., & Guldemond, H. (2007) Power analysis IN Two-level designs (Version 2.12) [Software]. https://www.stats.ox.ac.uk/~snijders/multilevel.htm#progPINT

Spybrook, J. (2013). Introduction to a special issue on design parameters for cluster randomized trials in education. *Evaluation Review, 37*(6), 435–444.

Spybrook, J., & Kelcey, B. (2016). Introduction to three special issues on design parameter values for planning cluster randomized trials in the social sciences. *Evaluation Review, 40*(6), 491–499.

Spybrook, J., Kelcey, B., & Dong, N. (2016). Power for detecting treatment by moderator effects in two and three-level cluster randomized trials. *Journal of Educational and Behavioral Statistics, 41*(6), 605–627.

Spybrook, J., & Raudenbush, S. W. (2009). An examination of the precision and technical accuracy of the first wave of group randomized trials funded by the Institute of Education Sciences. *Educational Evaluation and Policy Analysis, 31*(3), 298–318.

Spybrook, J., Shi, R., & Kelcey, B. (2016). Progress in the past decade: An examination of the precision of cluster randomized trials funded by the U.S. Institute of Education Sciences. *International Journal of Research and Method in Education, 39*(3), 255–267.

VanderWeele, T. J. (2008). Ignorability and stability assumptions in neighborhood effects research. *Statistics in Medicine, 27*, 1934–1943.

VanderWeele, T. J. (2010). Direct and indirect effects for neighborhood-based clustered and longitudinal data. *Sociological Methods & Research, 38*, 515–544.

CHAPTER 6

CROSS-CLASSIFIED RANDOM-EFFECTS MODELS

Audrey J. Leroux
Georgia State University

S. Natasha Beretvas
The University of Texas at Austin

Researchers commonly use multilevel modeling techniques to model the clustering inherent in social science settings (such as students within classrooms grouped in schools and patients tended by nurses grouped in hospitals). In addition, the range of possible models has expanded quickly. One of the more complicated extensions involves the modeling of cross-classified random effects. Although the multilevel models presented so far in this text are examples of purely nested data structures, data do not always qualify as fitting a pure hierarchy. In educational research, an example of a cross-classified data structure is students' cross-classification by middle and high school. High schools are not purely clustered by middle school, nor are middle schools purely clustered within high schools. In other words, students who attend a certain set, or cluster, of high schools do not typically all come from a single middle school. Similarly, students from a cluster of

Multilevel Modeling Methods with Introductory and Advanced Applications, pages 165–208

middle schools do not all attend the same high school. Rather, this structure entails students cross-classified by middle and high school. Using a cross-classified random-effects model (CCREM) appropriately models this cross-classification.

Most multilevel modeling textbooks mention this form of modeling (e.g., Hox, 2010; Raudenbush & Bryk, 2002; Snijders & Bosker, 2012) and it has gained increasing use, especially in the last 10 years, in fields such as medicine (Cafri & Fan, 2018), public health (De Clercq et al., 2014; Dunn et al., 2015), survey research (Brunton-Smith et al., 2017), veterinary medicine (Aunsmo et al., 2009; Elghafghuf et al., 2014), criminology (Brunton-Smith et al., 2018; Johnson, 2012), and educational research (Kim et al., 2010; Leckie, 2009; Lei et al., 2018) among other fields. Although the technical sophistication of the model seems to discourage it from being used more often, cross-classified (CC) data structures, however, provide a commonly encountered reality that one needs to model appropriately.

The first part of this chapter distinguishes examples of purely nested versus CC data structures. This chapter also presents formulation of the model and assumptions made, starting with the simplest 2-level model (with two CC factors at level 2) in Part I with extensions to 3-level examples in Part II. Network graphs (Rasbash & Browne, 2001) help present the distinction between pure hierarchical and CC data structures. There are already several very useful textbook chapters on CCREMs, and the reader should reference each of them (Goldstein, 2011; Hox, 2010; Rasbash & Browne, 2001; Raudenbush & Bryk, 2002; Snijders & Bosker, 2012). The current chapter contributes to these pedagogical expositions by offering two worked examples that detail how to use software to estimate the relevant CCREM parameters as well as how to interpret the relevant output.

The first example provided includes a 2-level CC structure with students (level 1) cross-classified by middle and high schools (level-2 CC factors). In the 3-level example, the data structure contains measurement occasions (level 1) for each examinee (level 2), with examinees cross-classified by rater and classroom (the two level-3 CC factors). The chapter explores explanatory variables at the measurement occasion, individual student, middle school, and/or high school level as sources of the possible variability. The examples demonstrate estimation of the relevant parameters and interpretation of results using simulated datasets. Specifically, this chapter shows how to use HLM 8 (student edition) software and R (version 4.1.2) software package lme4 (version 1.1-28; Bates et al., 2015) for estimation of CCREM parameters with 2- and 3-level models.

PART I: BACKGROUND AND 2-LEVEL CCREMS

Data Structures

Pure Hierarchy, 2-Level Data

Researchers commonly assume that their data structures involve *pure* clustering of units within groups. Using a 2-level model is appropriate to study some research questions. For example, a researcher might be interested in investigating the impact of school effects on students' performance on a high school exam. Use of the 2-level model involves the assumption that each student attended a single high school. Using Raudenbush and Bryk's (2002) formulation of the 2-level unconditional model, the level-1 outcome, Y_{ij}, representing the score of student i from school j, would be

$$Y_{ij} = \beta_{0j} + r_{ij}, \tag{6.1}$$

and at level 2

$$\beta_{0j} = \gamma_{00} + u_{0j}, \tag{6.2}$$

or together as

$$Y_{ij} = \gamma_{00} + u_{0j} + r_{ij}. \tag{6.3}$$

Figure 6.1 is a network graph that provides an example of a small, purely hierarchical dataset consisting of 12 students (A through L) attending four schools. As seen in Figure 6.1, the data's structure is a pure cluster with students clustered within *only one* high school.

Pure Hierarchy, 3-Level Data

A researcher might be interested in extending the model to consider the possible dependency associated with middle schools clustered in the same high school (as well as that of students within the same middle school). Clearly, a researcher could estimate a 3-level model in which level 1 represents

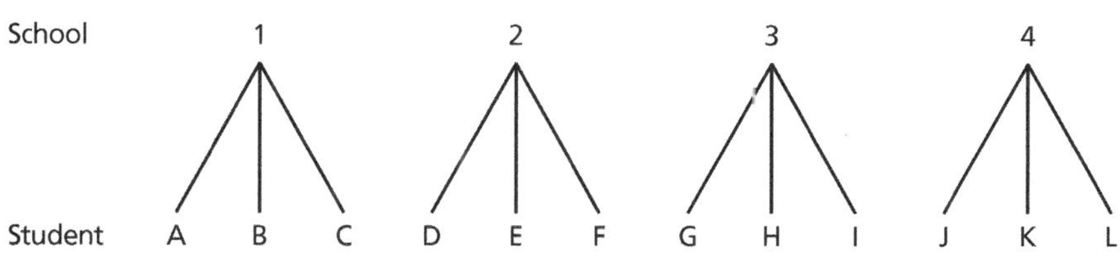

Figure 6.1 Network graph of pure hierarchy of students (level 1) clustered within schools (Level 2).

students, level 2 represents the middle schools, and level 3 models the high schools from which middle schools enroll their students. Inclusion of this additional level necessitates the addition of an extra index, k, to represent the third level. The 3-level unconditional model would be at level 1

$$Y_{ijk} = \pi_{0jk} + e_{ijk}, \tag{6.4}$$

at level 2

$$\pi_{0jk} = \beta_{00k} + r_{0jk}, \tag{6.5}$$

and at level 3

$$\beta_{00k} = \gamma_{000} + u_{00k}, \tag{6.6}$$

or as a combined equation

$$Y_{ijk} = \gamma_{000} + u_{00k} + r_{0jk} + e_{ijk}. \tag{6.7}$$

Figure 6.2 provides an example of a dataset that meets the assumption of the pure nesting of students within middle schools and of middle schools within high schools.

Cross-Classified Hierarchy

In Figure 6.2, the pure clustering within levels is again evident. However, the structure of real educational datasets frequently does not mimic this pure clustering. For the 3-level example given here, it is unlikely that middle schools send their students only to a single high school. In other words, students from high school i might have attended a middle school

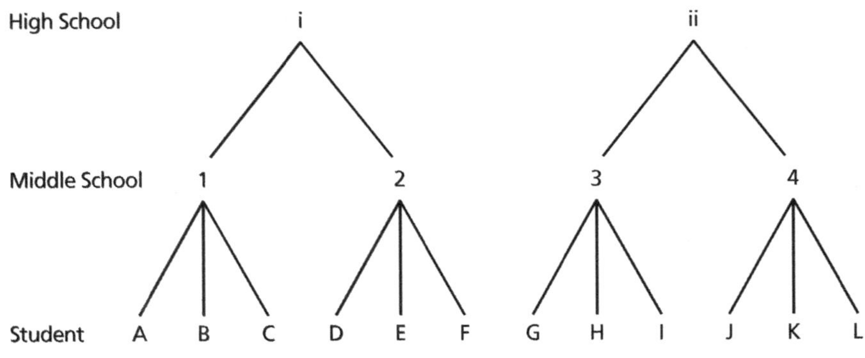

Figure 6.2 Network graph of pure hierarchy of students (level 1) grouped within middle schools (level 2) clustered within high schools (level 3).

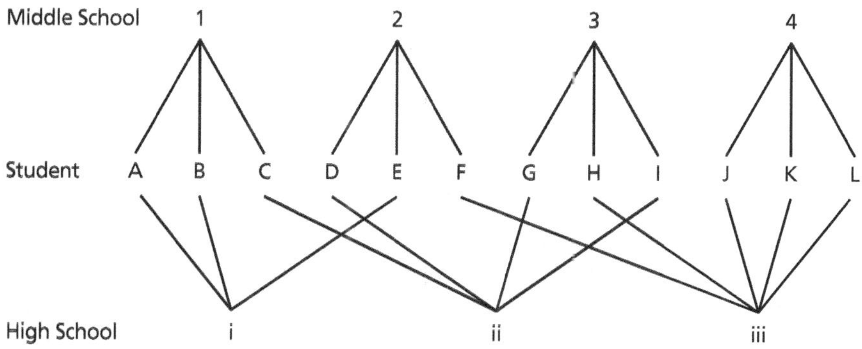

Figure 6.3 Network graph of cross-classified data structured with students grouped within cross-classifications of middle school and high school.

that students from high school *ii* also attended. Similarly, it is unlikely that high schools draw their students from only a single middle school.

Figure 6.3 more clearly depicts the lack of pure clustering of middle schools within high schools. As displayed in Figure 6.3, middle schools are not purely clustered within high schools, nor are high schools purely nested within middle schools. This means that there might be a cross-classification in the dataset; specifically, the dataset contains students *cross-classified* by middle schools and high schools. In this example, the cross-classification occurs at level 2.

Consequences of Ignoring Cross-Classified Data Structures

Before discussing how to model CC data structures, this section describes methods used by researchers to handle these more complex data structures. The first and most commonly encountered alternative involves the researcher's ignorance of one of the cross-classification factors. For example, the researcher might only model the high school attended and ignore the possible effects of middle schools, thereby mis-specifying the model and possibly leading to spurious conclusions.

As a second alternative, for a data structure with the cross-classification of students by middle and high school, a researcher may delete the smallest possible number of students to leave a set of middle schools purely clustered within each high school. This second alternative can lead to an unnecessarily reduced dataset with a concomitant lack of statistical power and a reduction in the generalization of the resulting inferences. Use of a CCREM avoids these analytic and inferential issues and provides an appropriate model to use with CC data structures. The next section provides the formulation of a 2-level CCREM, demonstrates the estimation of an actual model, and interprets the estimated parameters.

Models for Cross-Classified Data

This section uses the example involving middle and high schools as the cross-classified level-2 factors to demonstrate formulation and interpretation of a 2-level CCREM. To gauge the partitioning of variance into its various components, just like with purely hierarchical data, the first model typically investigated in the CCREM is the fully unconditional model. At level 1, the model is

$$Y_{i(j_1,j_2)} = \beta_{0(j_1,j_2)} + r_{i(j_1,j_2)}, \tag{6.8}$$

where $Y_{i(j_1,j_2)}$ represents the outcome (say, a math achievement test score) for student i in middle school j_1 and high school j_2. The intercept, $\beta_{0(j_1,j_2)}$, represents the predicted math achievement score for students from the specific combination of middle school j_1 and high school j_2 (i.e., the predicted cell means). The residual, $r_{i(j_1,j_2)}$, represents the deviation of a student's score from the student's middle and high schools' predicted intercept value (assumed normally distributed with mean zero and variance, σ_r^2).

In the current chapter, formulation of the CCREM employs a mixture of the notations of Raudenbush and Bryk (2002), Hox (2010), and Rasbash and Browne (2001). The CCREMs described here use Raudenbush and Bryk's (2002) levels' formulation, and indices appearing together within parentheses (such as j_1 and j_2, here) represent CC factors, per Hox (2010). Then, using Rasbash and Browne's (2001) notation, the same letter (j) identifies CC factors at the same level (here, level 2). The subscripted number associated with the index (thus, j_1 versus j_2) only distinguishes the factors themselves.

At level 2, the level-1 intercept is a random effect in the unconditional model

$$\beta_{0(j_1,j_2)} = \gamma_{000} + u_{0j_10} + u_{00j_2} + u_{0(j_1,j_2)}. \tag{6.9}$$

Note that the 2-level CCREM needs three subscripts. The first subscript represents the level-1 unit (here, student) and two more subscripts identify each of the cross-classified factors (here, middle and high school). The overall intercept, γ_{000}, represents the grand-mean math achievement score. The middle school residual, u_{0j_10}, represents the middle school effect for middle school j_1 (averaged across high schools) and is assumed normally distributed with a mean of zero and variance of $\sigma^2_{u0j_10}$ (Raudenbush & Bryk, 2002; Snijders & Bosker, 2012). Similarly, the high school residual,

u_{00j_2}, represents the high school effect for high school j_2 (averaged across middle schools) and is assumed $\sim N(0,\sigma^2_{u00j_2})$. The random interaction effect, $u_{0(j_1,j_2)}$, represents the residual beyond that predicted by the grand mean, γ_{000}, and the two random main effects, u_{0j_10} and u_{00j_2}, and is assumed $\sim N(0,\sigma^2_{u0(j_1,j_2)})$. Typically, the sample size per cell is insufficient to provide a good estimate of $\sigma^2_{u0(j_1,j_2)}$ as distinct from the student-level (within CC cells) variance, σ^2_r (Raudenbush & Bryk, 2002), although see Shi et al. (2010) for a study investigating this issue. Frequently, the CCREM does not include this random effect, $u_{0(j_1,j_2)}$, nor its variance, $\sigma^2_{u0(j_1,j_2)}$, (e.g., in the CCREM chapters of the Hox, 2010; Rasbash & Browne, 2001; and Raudenbush & Bryk, 2002 textbooks) and the remainder of this chapter omits the random effect as well (where omission implies setting its variance to zero).

One can combine the level-1 and level-2 equations (Equations 6.8 and 6.9) to provide the overall model of interest

$$Y_{i(j_1,j_2)} = \gamma_{000} + u_{0j_10} + u_{00j_2} + r_{i(j_1,j_2)}, \tag{6.10}$$

which clearly demonstrates the partitioning of the variability in students' math achievement scores into the component between middle schools, u_{0j_10}, the component between high schools, u_{00j_2}, and the remaining variability between students within cells, $r_{i(j_1,j_2)}$.

Similar to analyses of purely hierarchical data, one can use the variance components to describe the proportions of variability at each level (Raudenbush & Bryk, 2002). The correlation between test scores for students in the same middle school, j_1, although who have attended different high schools, j_2 and j_2' (where $j_2 \neq j_2'$), is

$$\rho_{Y_{i(j_1,j_2)},Y_{i'(j_1,j_2')}} = \frac{\sigma^2_{u0j_10}}{\sigma^2_{u0j_10} + \sigma^2_{u00j_2} + \sigma^2_r}. \tag{6.11}$$

The corresponding correlation for students who attended the same high school, j_2, although who are attending different middle schools, j_1 and j_1', is

$$\rho_{Y_{i(j_1,j_2)},Y_{i'(j_1',j_2)}} = \frac{\sigma^2_{u00j_2}}{\sigma^2_{u0j_10} + \sigma^2_{u00j_2} + \sigma^2_r}. \tag{6.12}$$

The correlation between scores for students attending the same middle and high schools (i.e., proportion of between-cells variability) is

$$\rho_{Y_{i(j_1,j_2)},Y_{i'(j_1,j_2)}} = \frac{\sigma^2_{u0j_10} + \sigma^2_{u00j_2}}{\sigma^2_{u0j_10} + \sigma^2_{u00j_2} + \sigma^2_r}. \tag{6.13}$$

Lastly, subtracting Equation 6.13 from one provides the proportion of the total variability that is within cells.

If there is substantial variability within cross-classification (level-1) variability, then a student-level descriptor could explain some of that variability (in Equation 6.8). For example, a researcher might want to investigate the relationship between students' math self-concept, X, and their performance on a math achievement test. Equation 6.8 would become

$$Y_{i(j_1,j_2)} = \beta_{0(j_1,j_2)} + \beta_{1(j_1,j_2)} X_{i(j_1,j_2)} + r_{i(j_1,j_2)}. \tag{6.14}$$

And at level 2, the coefficient, $\beta_{1(j_1,j_2)}$, describing the relationship between X and the outcome, could be a fixed effect

$$\begin{cases} \beta_{0(j_1,j_2)} = \gamma_{000} + u_{0j_10} + u_{00j_2} \\ \beta_{1(j_1,j_2)} = \gamma_{100} \end{cases}. \tag{6.15}$$

In Equation 6.15, the residuals represent deviations around the intercept, γ_{000}. And when X is grand-mean centered, γ_{000} now represents the predicted math achievement score for a student with the average self-concept score. Therefore, here, γ_{000} is an adjusted mean math achievement score. The reader should review the introductory chapters in this book that provide detailed expositions of the use of centering. Similar to modeling pure hierarchical data, there are several centering choices of a level-1 predictor in a CCREM. One can enter X into the CCREM as uncentered, centered at the grand-mean, or centered at the cell mean (although, methodological research has yet investigated cell-mean centering). The demonstrations in this chapter either grand-mean centered or left uncentered the predictors in the applied examples to facilitate interpretation of the relevant model's parameters.

Alternatively, the researcher might wish to model the slope for the covariate, X, to vary randomly across middle and high schools, in which case Equation 6.15 would become

$$\begin{cases} \beta_{0(j_1,j_2)} = \gamma_{000} + u_{0j_10} + u_{00j_2} \\ \beta_{1(j_1,j_2)} = \gamma_{100} + u_{1j_10} + u_{10j_2} \end{cases}, \tag{6.16}$$

leading to the inclusion of two random effects, u_{1j_10} and u_{10j_2}.

If substantial variability remains in the intercept between high schools ($\sigma^2_{u00j_2}$) and middle schools ($\sigma^2_{u0j_10}$), then the model can include descriptors of those CC factors. For example, a researcher might hypothesize that a measure of a middle school's resources for math classrooms, Z, could explain some of the variability in achievement test scores after controlling for self-concept. Thus, the level-2 model (Equation 6.16) becomes

$$\begin{cases} \beta_{0(j_1,j_2)} = \gamma_{000} + \gamma_{010}Z_{j_1} + u_{0j_10} + u_{00j_2} \\ \beta_{1(j_1,j_2)} = \gamma_{100} + u_{1j_10} + u_{10j_2} \end{cases}. \tag{6.17}$$

As expected, inclusion of the middle school predictor changes interpretation of the intercept, γ_{000}. If Z is grand-mean centered, γ_{000} now represents the adjusted (for student math self-concept) math achievement score for a student from a middle school with the average amount of math resources. There are several ways to model the relationship between the CC factor (middle school predictor, Z) and the intercept. One could model the slope as fixed across levels of the other CC factors (i.e., across high schools). Alternatively, the relationship between a middle school's math resources and a student's adjusted math achievement score (i.e., the coefficient, γ_{010}) could vary randomly across high schools (Snijders & Bosker, 2012). In this case, Equation 6.17 becomes

$$\begin{cases} \beta_{0(j_1,j_2)} = \gamma_{000} + (\gamma_{010} + u_{01j_2})Z_{j_1} + u_{0j_10} + u_{00j_2} \\ \beta_{1(j_1,j_2)} = \gamma_{100} + u_{1j_10} + u_{10j_2} \end{cases}, \tag{6.18}$$

where u_{01j_2} represents the variability across high schools in the coefficient for Z. Last, a high school descriptor such as average high school class size, W, could explain some of the variability in students' math achievement between high schools in the relationship between achievement (controlling for self-concept) and middle school math resources by adding W to Equation 6.18:

$$\begin{cases} \beta_{0(j_1,j_2)} = \gamma_{000} + \gamma_{001}W_{j_2} + (\gamma_{010} + \gamma_{011}W_{j_2} + u_{01j_2})Z_{j_1} + u_{0j_10} + u_{00j_2} \\ \beta_{1(j_1,j_2)} = \gamma_{100} + u_{1j_10} + u_{10j_2} \end{cases}. \tag{6.19}$$

Remember that grand mean-centered student math self-concept, X, is a level-1 predictor (see Equation 6.14), thereby changing interpretation of the intercept to represent the predicted math achievement for a student with average math self-concept. This intercept varies across both middle and high schools (Equation 6.15). Equation 6.17 includes the quantity of middle school math resources, Z, to explain some of the variability in this intercept across middle schools. The relationship between the middle school predictor, Z, and the intercept then varies across high schools in Equation 6.18. High school class size, W, explains some of the variability among high schools in the relationship between Z and the intercept (see Equation 6.19). The addition of the high school predictor, W, to explain variability between middle schools' Z slope is formally equivalent to adding an interaction between W and Z to the model (Raudenbush & Bryk, 2002).

In other words, the first line of Equation 6.19 can assess whether W moderates the relationship between Z and the intercept. Finally, the intercept in Equation 6.19 now represents the predicted math achievement score for a student with average math self-concept controlling for middle school math resources and high school class size.

The reader could also investigate the relationship between students' adjusted math achievement and a high school's average class size, W (grand-mean centered). The level-2 equation (Equation 6.17) for the intercept now includes the high school descriptor

$$\begin{cases} \beta_{0(j_1,j_2)} = \gamma_{000} + \gamma_{010}Z_{j_1} + \gamma_{001}W_{j_2} + u_{0j_10} + u_{00j_2} \\ \beta_{1(j_1,j_2)} = \gamma_{100} + u_{1j_10} + u_{10j_2} \end{cases}, \quad (6.20)$$

such that W's coefficient, γ_{001}, is fixed across middle schools. Alternatively, the relationship between a high school's average class size and achievement could vary randomly across middle schools

$$\begin{cases} \beta_{0(j_1,j_2)} = \gamma_{000} + \gamma_{010}Z_{j_1} + (\gamma_{001} + u_{0j_11})W_{j_2} + u_{0j_10} + u_{00j_2} \\ \beta_{1(j_1,j_2)} = \gamma_{100} + u_{1j_10} + u_{10j_2} \end{cases}. \quad (6.21)$$

Or, as in Equation 6.19, an interaction introduced into Equation 6.21 could explain variability among middle schools in the relationship between high school class size and math achievement. This changes the set of coefficients in front of W in Equation 6.21 from $(\gamma_{001} + u_{0j_11})$ to $(\gamma_{001} + \gamma_{011}Z_{j_1} + u_{0j_11})$ and further complicates estimation and interpretation of the model parameters. In Equation 6.21, the math achievement score predicted for a student with average math self-concept varies across middle and high schools. Middle school math resources, high school class size, and the interaction of the two predictors partly explains the variability in this intercept across the two CC factors.

This section expanded the original unconditional model (Equation 6.10) to include CC factors' descriptors designed to explain variability in the intercept (see Equation 6.21). It also includes a level-1 predictor, X, modeled in Equation 6.21 such that its relationship with Y varies randomly across middle and high schools. As demonstrated with the intercept, the reader could also add relevant descriptors to explain variability across middle and high schools in the relationship between student math self-concept (X) and math achievement (Y). While it might be tempting to take advantage of the sophisticated flexibility of CCREMs, the addition of multiple covariates and associated random effects make the models increasingly difficult to estimate and interpret (Raudenbush & Bryk, 2002). As with any kind of modeling, parsimony is strongly encouraged (Snijders & Bosker, 2012).

The next section demonstrates the use of HLM 8 and R software to estimate the parameters of a 2-level CCREM. The demonstration uses a simulated dataset with the content matching the example just provided. Note that HLM software currently only can estimate 2- and 3-level CCREMs with two CC factors at the higher level. R, SAS, Stata, and MLwiN can estimate CCREMs with additional levels and CC factors. However, HLM software is a common choice of multilevel modelers, and thus, this chapter demonstrates its use.

2-Level CCREM Example

An educational researcher might be interested in the relationship between students' performance on math achievement and descriptors of the student and of the middle and high school that the student attended. This simulated dataset contains information for 829 students consisting of student, middle school, and high school IDs, and scores on the following variables: the outcome variable (math achievement scores), a student-level predictor (math self-concept), a middle school measure (school math resources), and a high school measure (average class size). The CC nature of the dataset with students cross-classified by middle and high school attended is evident in Table 6.1. There are no students attending some of the combinations of middle and high school. For example, no students attend high school 1 and middle schools 11 through 17. The number of students per cell ranges from 8 to 22 with an average per-cell (including the empty cells) sample size of 14.8.

Use of HLM Software for 2-Level CCREM Example

CCREM Dataset Construction for Analysis Using HLM Software. As when using HLM for purely hierarchical datasets, the user must construct a dataset outside of HLM software. The demonstrations used SPSS to construct the datasets for the examples here. With units cross-classified at a higher level, the dataset should identify each case using identifiers for each of the relevant CC factors. For the current dataset (**CC_2lvl.sav**), this means that each student's information should be associated with the student's middle and high school identifiers. For example, there were 17 students identified as having attended middle school 1 and high school 1 (see Table 6.1). Aligned with each combination of identifiers are each case's values on the relevant level-1 and level-2 predictors. Lastly, the user must sort the dataset in ascending order by the middle school identifier (i.e., "row factor" identifier).

After constructing the dataset, the user must construct a multivariate data matrix (MDM) file for the analysis. In the HLM package, after clicking on **File**, requesting **Make new MDM file**, and identifying **Stat package input**, the user then sees the screen displayed in Figure 6.4. The default

TABLE 6.1 Cross-Classification of Students by Middle and High School Attended

Middle School	High School												
	1	2	3	4	5	6	7	8	9	10	11	12	Total
1	17	13	17	21									68
2	17		15			14		11					57
3	13		14					14					41
4	13							16					29
5	12	16		11	15		15						69
6		15	20	20				13					68
7			15		8		16						39
8		15	13			11	11						50
9				20	8	18							46
10					18	16	19	17					70
11									11	14	21	18	64
12									18		14	15	47
13									14	18	10		42
14									22	12	18	18	70
15										14	16		30
16										13	14		27
17											12		12
Total	72	59	94	72	49	59	61	71	65	71	105	51	829

Note: Empty cells indicate no students in the dataset were from the associated combination of middle and high school.

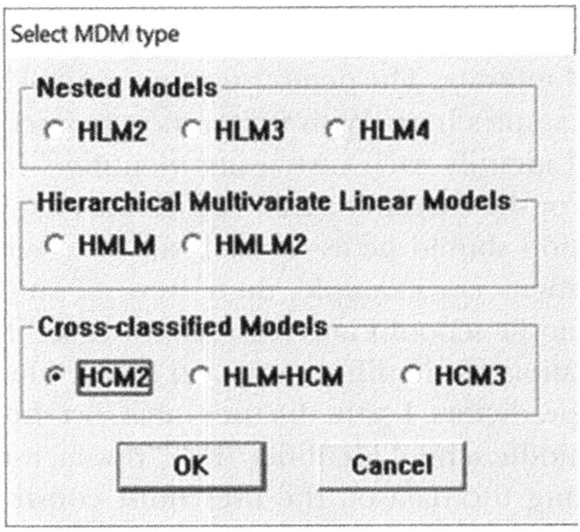

Figure 6.4 Window for selection of 2-level cross-classified data structure when building mdm file.

for this screen is **HLM2**; however, the user needs to click on **HCM2** to indicate that the MDM is for a 2-level CC dataset. Next, similar to modeling for purely hierarchical data, a screen appears (see Figure 6.5) from which the user must identify the relevant file. After naming the MDM file and selecting the data file, the user clicks on **Choose Variables** to identify the relevant ID variables and variables included for later modeling. However, the **Choose Variables** screen that appears for CC datasets is a little different (see Figure 6.6).

The HLM manual recommends using the CC factor with more units as the "row" factor and the other factor as the "column" factor (see Table 6.1 to understand what is meant by "row" and "column"), although it does not affect the results if the user makes the opposite choice. In the current dataset, there are 17 middle schools and 12 high schools. Thus, this demonstration includes middle school as the row factor and high school as the column factor when constructing the MDM. Therefore, when **Choosing Variables**, the user can choose **MS_ID** as the **rowid** (row identifier) and **HS_ID** as the **colid**. The remaining point-and-click commands are the same as for purely hierarchical datasets.

Estimating a 2-Level CCREM in HLM Software. The first model to estimate is the fully unconditional model (see Equations 6.8 and 6.9, summarized in Equation 6.10). First, the user identifies the outcome by clicking on **MATH_ACH** and selecting the only available option, thereby indicating that

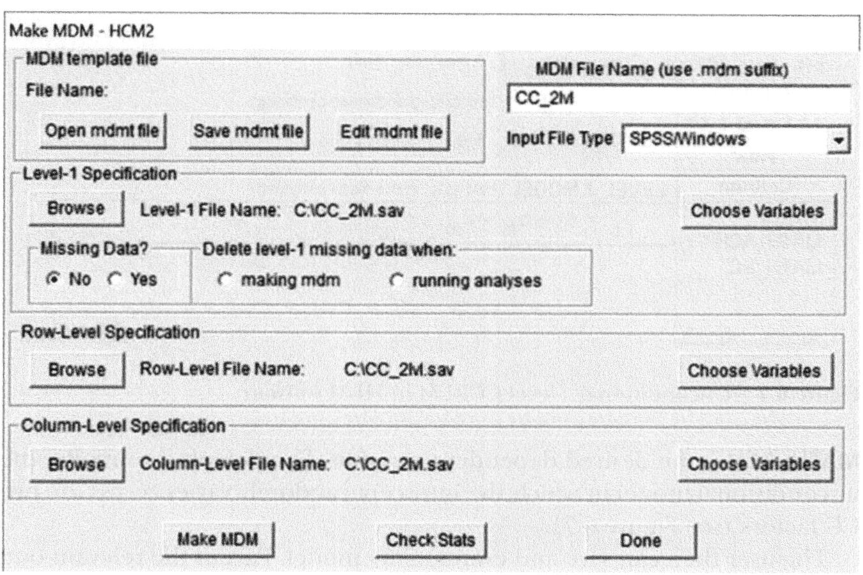

Figure 6.5 Window for identification of data file when constructing MDM file.

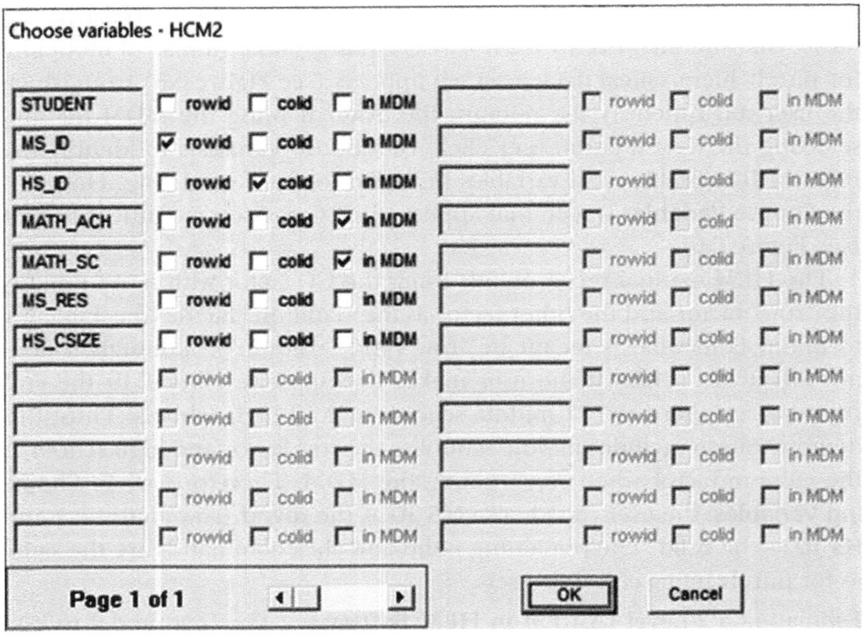

Figure 6.6 Choose variables screen for level-1 specification when constructing mdm for cross-classified dataset.

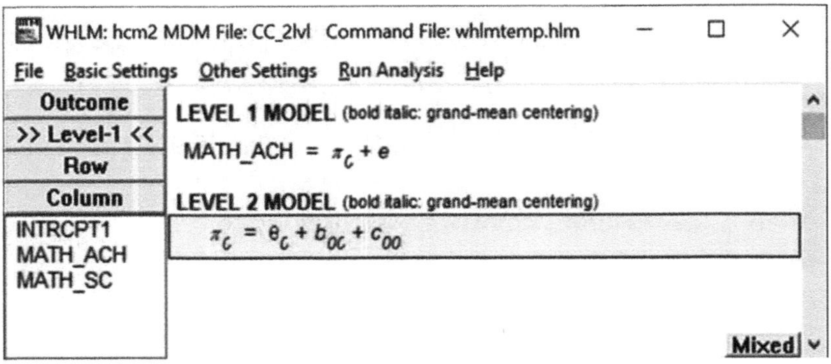

Figure 6.7 Unconditional 2-level CCREM in HLM software.

MATH_ACH is the desired dependent variable. The default is to model the unconditional model in which the intercept randomly varies across the two CC factors (see Figure 6.7).

The user then can save and estimate the model. Part of the relevant output follows. The reader should note that HLM output associates whichever CC factor is the "row" variable (here, middle school) with the *b* random effects and the other factor (here, high school) with the *c* random effects. Full maximum likelihood is the estimation procedure used. The HLM output presented in this chapter includes bolding, italicizing, and underling of important values.

Table 6.2 also summarizes the results of interest in the CCREM unconditional model column of data. The fixed effect estimates and tests of statistical significance appear as the first results in the output followed by the random effect results. The researcher can use the variance components (see Equations 6.11, 6.12, and 6.13) to obtain the partitioning of the variability. Summing together the three variance components $(\sigma^2_{u0j_10} + \sigma^2_{u00j_2} + \sigma^2_r)$ yields: 80.95 + 9.22 + 9.42 = 99.59, which is the total variance (the denominator in the equations). The proportion of variability within cells (i.e., within cross-classifications of middle and high schools once middle and high school effects have been partialled out) was 81.3% (80.95/99.59). There was a substantial amount of variability found between middle schools (9.5% from 9.42/99.59) as well as between high schools (9.3% from 9.22/99.59), supporting the importance of modeling the cross-classification. The fixed effect estimate of the coefficient for the intercept was 50.29, providing the predicted average math achievement score (found to differ from zero, $p < .001$).

Final estimation of fixed effects:

Fixed Effect	Coefficient	Standard Error	*t*-ratio	Approx. *d.f.*	*p*-value
For INTRCPT1, π_0					
INTERCEPT, θ_0	***50.289157***	1.205156	41.728	800	<0.001

Final estimation of row and level-1 variance components:

Random Effect	Standard Deviation	Variance Component	*d.f.*	χ^2	*p*-value
INTRCPT1/ICPTROW, b_{00j}	3.06899	***9.41869***	16	119.71679	<0.001
level-1, *e*	8.99704	***80.94675***			

Final estimation of column level variance components:

Random Effect	Standard Deviation	Variance Component	*d.f.*	χ^2	*p*-value
INTRCPT1/ICPTCOL, c_{00k}	3.03676	***9.22193***	11	114.39497	<0.001

Statistics for the current model

Deviance = ***6048.168892***
Number of estimated parameters = ***4***

TABLE 6.2 Parameters Estimates for the Models Estimated with the 2-Level Cross-Classified Dataset

		CCREM						2-Level Model	
		Uncond.	Model 1	Model 2	Model 3			Uncond.	Model 1
Fixed effects						Fixed effects			
Intercept	γ_{000}	50.29***	50.31***	50.43***	50.43***	Intercept	γ_{00}	50.06***	50.05***
MATH_SC	γ_{100}		0.23***	0.23***	0.23***	*MATH_SC*	γ_{10}		0.22***
MS_RES	γ_{010}			0.63*	0.63*				
HS_SIZE	γ_{001}				0.02				
Variance components						Variance components			
Student	σ^2_r	80.95	78.38	78.40	78.40	Student	σ^2_r	88.97	86.56
Middle school	$\sigma^2_{u0j_10}$	9.42	9.67	6.15	6.15	High school	σ^2_{u0j}	10.33	10.86
High school	$\sigma^2_{u00j_2}$	9.22	9.57	9.47	9.46				
Model deviance		6,048.17	6,022.92	6,017.32	6,017.32			6,096.80	6,076.20

Note: Fixed effects are point estimates of the specified coefficients. Variance components provide variance estimates of the specified residuals. This table only notes the statistical significance of fixed effects. *MATH_SC* = Math self-concept (centered). *MS_RES* = Middle school math resources (centered). *HS_SIZE* = High school class size (centered). Uncond. = Unconditional; Model 1 includes *MATH_SC* as a fixed level-1 predictor; Model 2 includes *MATH_SC* as a fixed level-1 predictor and *MS_RES* as a fixed (CC factor) predictor; Model 3 is the same as Model 2 except including *HS_SIZE* as a fixed (CC factor) predictor.

$^{*}p < .05$ $^{***}p < .001$

To show the results of mis-specifying the multilevel model by ignoring the clustering of students within middle schools, this demonstration estimated a purely hierarchical (i.e., 2-level) model. This model ignores the middle school enrollment by having students (level-1 units) clustered only within high schools (level-2 units). The results of the unconditional 2-level model appear in the right-hand portion of Table 6.2. The calculation of the intraclass correlation coefficient (Raudenbush & Bryk, 2002) revealed that when not modeling middle school cross-classification, the proportion of variability between high schools (10.4%, from 10.33/[88.97 + 10.33]) and between students (89.6%, from 88.97/[88.97 + 10.33]) were higher than the CCREM's estimates of 9.5% and 81.3%, respectively. The ignored variability between middle schools appears re-distributed into partly the variability between students (within cross-classifications) and partly between high schools. This clearly could impact the statistical significance testing and interpretation of results when modeling the CC structure (using the CCREM) versus when ignoring it (using the 2-level model).

Returning to the CCREM analysis, the researcher might add the student-level covariate (grand-mean centered math self-concept score) to the unconditional model (Equations 6.14 and 6.15) to explain some of the variability found at each level. To do this in HLM software, simply click on the relevant predictor (**MATH_SC**) and specify it as grand-mean centered. The result is the screen shown in Figure 6.8.

The default in HLM is for predictors added to a model as fixed effects (seen in Figure 6.8, in which the font of the slope's random effects [u_{1j_10} and u_{10j_2}] is only very faintly visible, and resulting in the model in Equation 6.15 rather than Equation 6.16). Clicking on each of these effects (u_{1j_10} and u_{10j_2}) converts the slope to vary randomly across middle and high schools (see Figure 6.9), which is necessary if theoretically warranted. Due to the

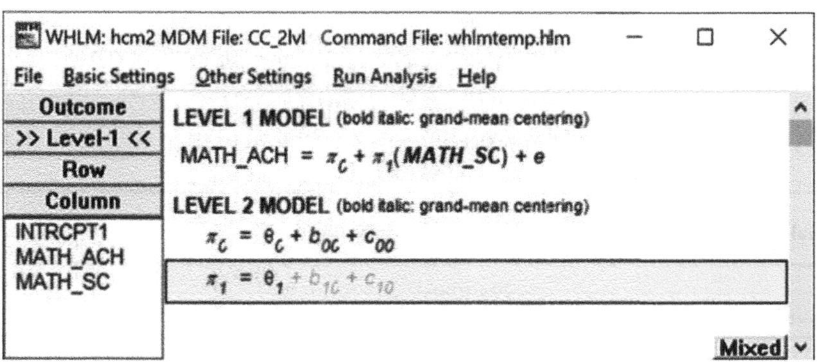

Figure 6.8 Conditional CCREM (with a fixed level-1 predictor) in HLM software.

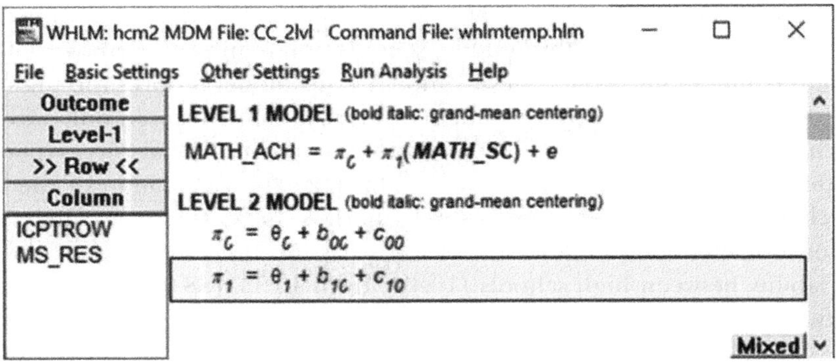

Figure 6.9 Conditional CCREM (with a randomly varying level-1 predictor) in HLM software.

increased number of random effects in the model, reaching convergence required additional iterations. (The unconditional model's estimation took a mere 6 iterations as compared with 249 for the model depicted in Figure 6.9.) The output appears below.

Final estimation of fixed effects:

Fixed Effect	Coefficient	Standard Error	*t*-ratio	Approx. *d.f.*	*p*-value
For INTRCPT1, π_0					
INTERCEPT, θ_0	***50.324645***	1.227605	40.994	771	<0.001
For MATH_SC, π_1					
INTERCEPT, θ_1	***0.227583***	0.049088	4.636	771	<0.001

Final estimation of row and level-1 variance components:

Random Effect	Standard Deviation	Variance Component	*d.f.*	χ^2	*p*-value
INTRCPT1/ICPTROW, b_{00j}	3.15307	***9.94184***	16	118.42762	<0.001
MATH_SC/ICPTROW, b_{10j}	0.01246	***0.00016***	16	8.84313	>0.500
level-1, *e*	8.83384	***78.03674***			

Final estimation of column level variance components:

Random Effect	Standard Deviation	Variance Component	*d.f.*	χ^2	*p*-value
INTRCPT1/ICPTCOL, c_{00k}	3.09320	***9.56786***	11	122.32000	<0.001
MATH_SC/ICPTCOL, c_{10k}	0.07180	***0.00515***	11	11.15316	0.431

Statistics for the current model

Deviance = *6020.224117*
Number of estimated parameters = *9*

Model comparison test

χ^2 statistic = 27.94478
Degrees of freedom = 5
p-value = <0.001

The relationship between *MATH_SC* and the outcome was positive and statistically significant ($\hat{\gamma}_{100} = 0.23$, $p < .001$) and reduced the unexplained level-1 variance, $\hat{\sigma}_r^2$, from 80.95 to 78.04 (by 3.6%). The value of this regression coefficient, $\hat{\gamma}_{100} = 0.23$, indicates that the higher students' reported math self-concept, the better their predicted math achievement score. In addition, given Student A has a math self-concept score that is 1 point higher than Student B's, then Student A's predicted math achievement test score is 0.23 points higher. The change in the deviance from the unconditional model (deviance = 6,048.17) to the model including the level-1 predictor, which added five parameters to those estimated in the unconditional model (deviance = 6,020.22), was statistically significant ($\chi^2(5) = 27.94$, $p < .001$), indicating that the model including the level-1 predictor resulted in a statistically significant improvement in model-fit. The small values found in the output for the slope's variability across middle and high schools (0.00016 and 0.00515, respectively) and their lack of statistical significance ($p > .500$ and $p = .431$, respectively) indicate that the effect might be better as fixed rather than random.

The next model estimated includes *MATH_SC* as a fixed effect (see Equations 6.14 and 6.15 and Figure 6.8), with results shown in Table 6.2 under Model 1. There was no difference found in the estimates of the resulting fixed effects (see Model 1 of CCREMs in Table 6.2). Only slight differences were found in the variance estimates between Model 1 ($\hat{\sigma}_r^2 = 78.38$, $\hat{\sigma}_{u0j_10}^2 = 9.67$, $\hat{\sigma}_{u00j_2}^2 = 9.57$) and the unconditional model ($\hat{\sigma}_r^2 = 80.95$, $\hat{\sigma}_{u0j_10}^2 = 9.42$, $\hat{\sigma}_{u00j_2}^2 = 9.22$). The deviance value for the model with the slope for *MATH_SC* fixed across middle and high schools was 6,022.92. The difference in the deviances was not statistically significant ($\chi^2(4) = 2.69$, $p > .500$) between the model in which the slope for *MATH_SC* was random versus fixed across middle and high schools, suggesting the fit of this more parsimonious (fixed slope) model was no worse than the fit of the more complex model.

A substantial amount of variability between middle and high schools remained unexplained in the intercept (predicted math achievement scores adjusted for math self-concept scores). To address this, the next model adds

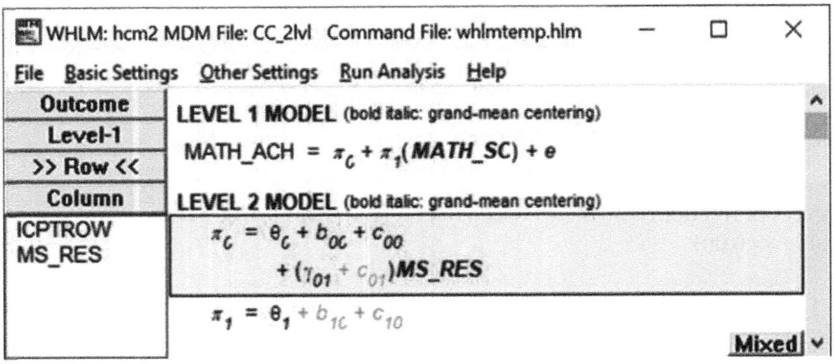

Figure 6.10 Conditional CCREM (with Fixed Level-1 and Cross-Classified Factor Predictors) in HLM Software

the measure of a middle school's math resources (*MS_RES*) as a grand-mean centered variable to the level-2 equation for the intercept. Initially, the coefficient for the relationship between *MS_RES* and the outcome was a fixed effect across high schools (see Equation 6.17). To accomplish this in HLM software, ensure that the list of "row" (i.e., middle school) variables appears on the left of the screen (click on **Row** so it becomes ">> *Row* <<"). Then, the user selects *MS_RES* as a grand-mean centered predictor. Again, the default is to include the predictor as a fixed effect (as demonstrated in Figure 6.10). The results for this model appear under Model 2 in Table 6.2. The fixed effect estimates from the output appear below.

Final estimation of fixed effects:

Fixed Effect	Coefficient	Standard Error	*t*-ratio	Approx. *d.f.*	*p*-value
For INTRCPT1, π_0					
INTERCEPT, θ_0	*50.428433*	1.129441	44.649	800	<0.001
MS_RES, γ_{01}	*0.625939*	0.240469	2.603	15	0.020
For MATH_SC, π_1					
INTERCEPT, θ_1	*0.227408*	0.044453	5.116	800	<0.001

As the results in Table 6.2 indicate, the addition of *MS_RES* did not substantially change the fixed effect estimate of the relationship between *MATH_SC* and *MATH_ACH* ($\hat{\gamma}_{100}$). The estimate of the intercept, ($\hat{\gamma}_{000} = 50.43$, $p < .001$), now represents the predicted score for a student, controlling for *MATH_SC* scores and for *MS_RES*. *MS_RES* positively predicted ($\hat{\gamma}_{010} = 0.63$, $p = .020$) *MATH_ACH* (controlling for *MATH_SC*), indicating that the more math resources a middle school offered, the better a student tended to perform on the math achievement test. More specifically, students with the same *MATH_SC* scores at middle schools that differ by 1 point on the *MS_RES* measure differ on their predicted math achievement test scores by 0.63 points. The middle school variance component (see Table 6.2) reduced from 9.42 (in the unconditional model) to 6.15 (in Model 2), indicating that *MS_RES* explained about 34.7% of the variability in math achievement scores (controlling for *MATH_SC*).

To assess how much the relationship between middle school math resources and the intercept varied across high schools, the demonstration involves estimation of Equation 6.18 (instead of Equation 6.17). To change the model in this way using HLM software, the user clicks on the faint u_{01j_2} so it becomes dark as appears in Figure 6.11. Table 6.2 does not show the resulting estimates. However, with the addition of the random effects, estimation of the model required 198 iterations to converge on a solution. The fixed effect estimates did not change. The deviance statistic of this model was 6,017.19 compared with 6,017.32 (for the fixed *MS_RES* model in Equation 6.17, i.e., Model 2). Addition of the two extra parameters (the variance of u_{01j_2} and its covariance with u_{00j_2}) did not improve the fit of the model.

This demonstration has not yet explored variability between high schools and, therefore, adds high school class size (*HS_SIZE*) to the level-2 equation as a grand-mean centered, fixed effect predictor of the intercept (see Equation 6.20). To accomplish this in HLM, the user must ensure that variables describing this CC factor identified as the "column" variable appear on the left-hand side of the HLM window (click on **Column** so it becomes ">> *Column* <<"). Next, select the *HS_SIZE* variable and add it as a fixed (across middle schools), grand-mean centered predictor of the intercept, resulting in the HLM window shown in Figure 6.12. The estimates resulting from this final model appear under Model 3 in Table 6.2.

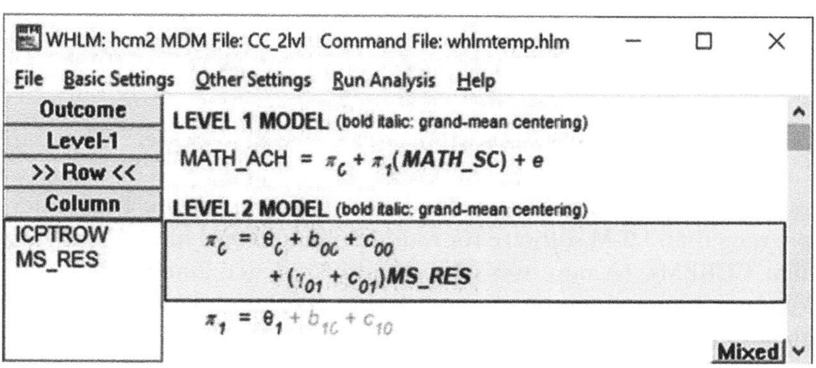

Figure 6.11 Conditional CCREM (with a Fixed Level-1 and a Random Cross-Classified Factor Predictor) in HLM Software

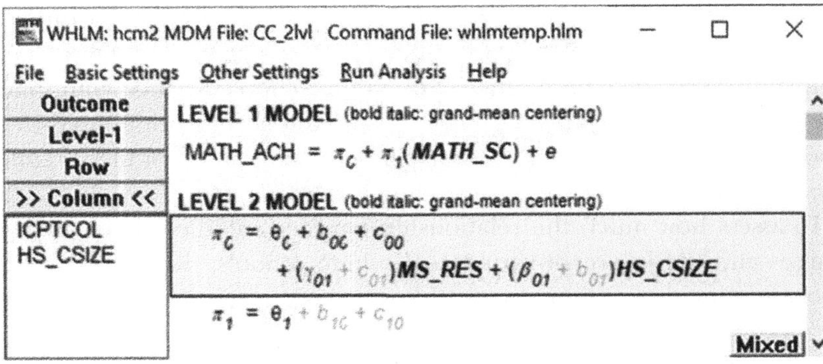

Figure 6.12 Conditional CCREM (with a Fixed Level-1 Predictor and Two Fixed Cross-Classified Factor Predictors) in HLM Software.

In this dataset, high school class size was not related to the intercept ($\hat{\gamma}_{001} = 0.02$, $p = .969$). This also is evident in the lack of change in the between-high school variance of the residuals, $\sigma^2_{u00j_2}$, which was 9.47 before and 9.46 after inclusion of *HS_SIZE* in the model. Note that there was no change in the deviance (a difference of 0.00161) between Models 2 and 3.

Overall, these results suggest the selection of Model 2. In Model 2, there was a positive relationship between math self-concept and math achievement, with the relationship modeled as fixed across middle and high schools. In addition, the amount of math resources in a middle school also had a positive relationship with students' math achievement scores (even after controlling for students' math self-concept). A substantial proportion (9.5%) of variability remained unexplained between middle schools and between high schools (6.2%), as well as between students within cross-classifications by middle and high school (78.7%). Additional sources of variability in mathematics achievement, particularly among individual students, should be investigated.

Use of R Software for 2-Level CCREM Example

CCREM Dataset Construction for Analysis Using R Software. The graphical user interface along with the levels' formulation in HLM software favors its use with new multilevel modelers. However, use of R is easier in some ways than HLM software for estimating multilevel models and in particular CCREMs. (A new user of R should reference Finch et al.'s [2019] book describing estimation of purely hierarchical multilevel models before embarking on the use of R for estimation of CCREMs). When using R to estimate multilevel models, the user again needs a single dataset with identifying variables distinguishing each unit and cross-classified factor, and, unlike for HLM, they do not need to sort the dataset. Though, the user must transform variables requiring centering prior to inclusion in the `lmer` function. Here, *MATH_SC_C* and *MS_RES_C* refer to grand-mean centered versions of these variables.

Estimating a 2-Level CCREM in R Software. Once the user reads the dataset into R, the code below estimates the unconditional CCREM in Equation 6.10.

```
uncon <- lmer(MATH_ACH ~ 1 + (1 | MS_ID) + (1 | HS_ID),
      data = CC_21v1, REML = FALSE)
```

Be sure to identify the dataset with the "`data =`" argument. As seen in the R code, there is no need to identify the data's structure as cross-classified. R recognizes the cross-classification from the dataset's structure and estimates a CCREM if a pure hierarchy of the groups (middle and high schools) does not exist. Note that including the "`REML = FALSE`" argument specifies maximum likelihood estimation rather than restricted maximum likelihood (REML) estimation. This demonstration made the change so that the R results correspond to the HLM results, but the user may want to choose the default REML estimation method.

For estimation of the unconditional CCREM, the user only needs the model formula "`MATH_ACH ~ 1 + (1 | MS_ID) + (1 | HS_ID)`" to specify the outcome variable (here, *MATH_ACH*) as varying randomly across the relevant CC factors. The "`(1 | MS_ID)`" and "`(1 | HS_ID)`" portions of the code correspond to the level-2 (CC factors') equations, one for middle schools and one for high schools, respectively. Using R to estimate the unconditional model, the following fixed and random effect results (obtained from using the "`summary(uncon)`" command) match those from HLM software:

```
Linear mixed model fit by maximum likelihood  ['lmerMod']
Formula: MATH_ACH ~ 1 + (1 | MS_ID) + (1 | HS_ID)
   Data: CC_21v1

     AIC      BIC   logLik deviance df.resid
  6056.2   6075.0  -3024.1   6048.2      825

Scaled residuals:
     Min       1Q   Median       3Q      Max
-2.75746 -0.64457 -0.01075  0.65916  3.07518

Random effects:
 Groups   Name        Variance Std.Dev.
 MS_ID    (Intercept)  9.418   3.069
 HS_ID    (Intercept)  9.222   3.037
 Residual             80.947   8.997
Number of obs: 829, groups:  MS_ID, 17; HS_ID, 12

Fixed effects:
            Estimate Std. Error t value
(Intercept)   50.289      1.205   41.73
```

Modification of the model formula as follows: "MATH_ACH ~ 1 + MATH_SC_C + (1 | MS_ID) + (1 | HS_ID)" can accomplish estimation of the model that includes student *MATH_SC_C* as a grand-mean centered, fixed predictor of the intercept (see Equations 6.12 and 6.14). The random effects' variance components and fixed effect estimates (shown below) correspond with those from HLM (see Model 1 in Table 6.2).

```
Linear mixed model fit by maximum likelihood  ['lmerMod']
Formula: MATH_ACH ~ 1 + MATH_SC_C + (1 | MS_ID) + (1 | HS_ID)
   Data: CC_2lvl

     AIC      BIC   logLik deviance df.resid
  6032.9   6056.5  -3011.5   6022.9      824

Scaled residuals:
     Min       1Q   Median       3Q      Max
-2.86157 -0.62949  0.00909  0.65224  3.06726

Random effects:
 Groups   Name        Variance Std.Dev.
 MS_ID    (Intercept)  9.674   3.110
 HS_ID    (Intercept)  9.573   3.094
 Residual             78.381   8.853
Number of obs: 829, groups:  MS_ID, 17; HS_ID, 12

Fixed effects:
            Estimate Std. Error t value
(Intercept) 50.30604    1.22207  41.164
MATH_SC_C    0.22527    0.04445   5.067

Correlation of Fixed Effects:
          (Intr)
MATH_SC_C 0.002
```

The R code used to estimate the model in which *MATH_SC_C* varies randomly across both middle and high schools (see Equation 6.16) is the same as for the fixed *MATH_SC_C* model except that the formula becomes: "MATH_ACH ~ 1 + MATH_SC_C + (1 + MATH_SC_C | MS_ID) + (1 + MATH_SC_C | HS_ID)". The results of the HLM analysis had indicated that the variances of the resulting slope residuals for middle schools, $\hat{\sigma}^2_{u1j_10}$, and for high schools, $\hat{\sigma}^2_{u10j_2}$, were very small. R reported that the random effects' variance-covariance matrices resulted in singular fits (i.e., were non-positive definite). This happens in scenarios such as the current one in which variance components' values are very close to zero. The R output appears below, although the estimates' values are questionable (and thus, not highlighted) due to the problem with the variance-covariance matrices.

```
Linear mixed model fit by maximum likelihood  ['lmerMod']
Formula: MATH_ACH ~ 1 + MATH_SC_C + (1 + MATH_SC_C | MS_ID) +
(1 + MATH_SC_C | HS_ID)
   Data: CC_2lvl

     AIC      BIC   logLik deviance df.resid
  6038.2   6080.7  -3010.1   6020.2      820
```

```
Scaled residuals:
     Min       1Q   Median       3Q      Max
-2.83449 -0.67695 -0.00251  0.64004  3.05513

Random effects:
 Groups   Name        Variance  Std.Dev. Corr
 MS_ID    (Intercept) 9.953e+00 3.154766
          MATH_SC_C   3.987e-05 0.006314 1.00
 HS_ID    (Intercept) 9.568e+00 3.093159
          MATH_SC_C   5.069e-03 0.071198 1.00
 Residual             7.804e+01 8.833836
Number of obs: 829, groups:  MS_ID, 17; HS_ID, 12

Fixed effects:
            Estimate Std. Error t value
(Intercept) 50.32455    1.22785  40.986
MATH_SC_C    0.22753    0.04893   4.651

Correlation of Fixed Effects:
          (Intr)
MATH_SC_C 0.332
optimizer (nloptwrap) convergence code: 0 (OK)
boundary (singular) fit: see help('isSingular')
```

To estimate a model in which a CC factor predictor (e.g., grand-mean centered middle school *MS_RES_C*) explains some of the variability in the intercept (as in Equation 6.17), the variable should be added to the model formula which becomes: "`MATH_ACH ~ 1 + MATH_SC_C + MS_RES_C + (1 | MS_ID) + (1 | HS_ID)`".

Including a Cross-Level Interaction. In the current model, variability was not substantial across middle or high schools in the slope for *MATH_SC_C*. If variability had been substantial, then a researcher might use a CC factor descriptor, *Z*, to explain some of that variability in the slope. In that case, Equation 6.17 would become

$$\begin{cases} \beta_{0(j_1,j_2)} = \gamma_{000} + \gamma_{010}Z_{j_1} + u_{0j_10} + u_{00j_2} \\ \beta_{1(j_1,j_2)} = \gamma_{100} + \gamma_{110}Z_{j_1} + u_{1j_10} + u_{10j_2} \end{cases}. \tag{22}$$

When using R, the user should be aware that the moderating variable translates into an interaction ($Z \times X$) between the level-2 variable, *Z*, and the level-1 variable, X (*MATH_SC_C*). The reader should review the sections of introductory chapters in this book and other sources (such as Snijders & Bosker, 2012) that discuss cross-level interactions. The cross-level interaction is easier to envision when combining the levels' equations (here, Equations 6.14 and 6.22) into a single equation

$$\begin{aligned} Y_{i(j_1,j_2)} = {} & \gamma_{000} + \gamma_{010}Z_{j_1} + u_{0j_10} + u_{00j_2} \\ & + (\gamma_{100} + \gamma_{110}Z_{j_1} + u_{1j_10} + u_{10j_2})X_{i(j_1,j_2)} + r_{i(j_1,j_2)} \end{aligned}. \tag{6.23}$$

The R code needed to estimate the model that includes the cross-level interaction (see Equation 6.23) follows. (The reader should note the term "MATH_SC_C * MS_RES_C" includes the variables *MATH_SC_C* and *MS_RES_C* along with their interaction.)

```
crosslevel <- lmer(MAT_ACH ~ 1 + MATH_SC_C * MS_RES_C
                   + (1 + MATH_SC_C | MS_ID)
                   + (1 + MATH_SC_C | HS_ID),
                   data = CC_21v1, REML = FALSE)
```

There are many other examples of 2-level CCREMs that one can estimate and additional educational research data structures that necessitate their use. To complement this introduction to the CCREM, Part II of this chapter presents an example of a 3-level CC data structure in which the cross-classification occurs at the highest level along with a demonstration of the use of HLM and R to estimate the 3-level CCREM's parameters.

PART II: 3-LEVEL CCREMS

A simple extension of the 2-level CC data structure described above in Part I could lead to the need to estimate a 3-level CCREM. For example, a longitudinal analysis of students' achievement across secondary schools could consider the impact of middle and high schools on that growth. Such a model contains measurement occasions (level 1) nested within students (level 2) cross-classified by middle and high schools attended (CC, level-3 factors).

Another example of a 3-level CCREM could include modeling change over time in students' scores on a writing test. Several raters, not assigned to specific classrooms, might grade each test, and thus, this example has students cross-classified by rater and classroom. Figure 6.13 depicts this example (for a tiny sample solely used for clarification purposes) where the cross-classification occurs at the third level. Specifically, the data structure is measurement occasions (level 1) nested within examinees (level 2) cross-classified by raters and classrooms (level 3, CC factors).

The next section provides the formulation of the 3-level CCREM depicted in Figure 6.13. It also demonstrates use of HLM 8 software and R software (lme4 package) for estimating fixed and random effect parameters of a 3-level CCREM in which the cross-classification (by two factors) occurs at level 3. This demonstration simulated the contents of the dataset to match the 3-level CCREM example just described.

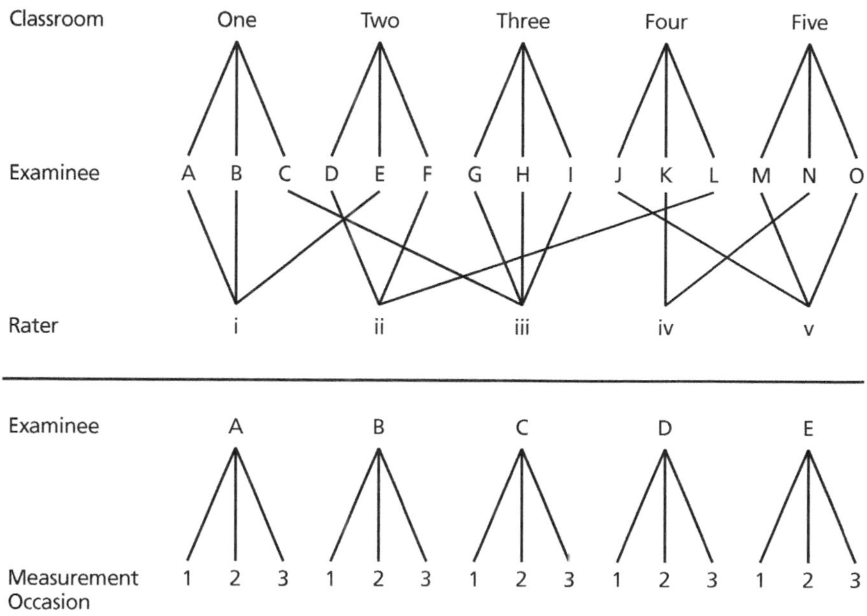

Figure 6.13 Network Graph of Cross-Classified Data Structured with Measurement Occasion (Level 1) Nested Within Examinee (Level 2) and Examinee Cross-Classified by Rater and Classroom (Level 3 CC Factors)

3-Level CCREM Example

The dataset used to demonstrate estimation of the model in Figure 6.13 consists of 1,000 examinees with scores at three time points spaced one month apart. A score of zero for the *TIME* variable represents the initial time point and it increases by a value of 1 (month) for each consecutive time point. The outcome variable, *WRITE*, is the student's score on a measure of writing ability. The dataset includes identification variables distinguishing raters, students, and classrooms. The dataset also includes the following variables: student's gender (coded such that *FEMALE* = 0 for boys and *FEMALE* = 1 for girls), a measure of the rater's experience as a rater (*EXPER*), and a measure of the socio-economic status of students in the classroom (*SES*). For the *SES* variable, a higher value indicates a higher *SES* for the classroom.

For simplification, and because only three time points of data are available, the example assumes a linear change over time in *WRITE* scores. The baseline linear growth model at level 1 would be

$$Y_{ti(j_1,j_2)} = \pi_{0i(j_1,j_2)} + \pi_{1i(j_1,j_2)} TIME_{ti(j_1,j_2)} + e_{ti(j_1,j_2)}. \tag{6.24}$$

This 3-level CCREM needs four subscripts, such that $Y_{ti(j_1,j_2)}$ is the writing score at time point t for student i assessed by rater j_1 in classroom j_2. At the student level (level 2) in the baseline model, the intercept and slope would vary across students as follows

$$\begin{cases} \pi_{0i(j_1,j_2)} = \beta_{00(j_1,j_2)} + r_{0i(j_1,j_2)} \\ \pi_{1i(j_1,j_2)} = \beta_{10(j_1,j_2)} + r_{1i(j_1,j_2)} \end{cases}. \tag{6.25}$$

Lastly, at level 3 (the cross-classification level), the intercept and slope would vary randomly across raters (u_{00j_10} and u_{10j_10}, respectively) and classrooms (u_{000j_2} and u_{100j_2}, respectively):

$$\begin{cases} \beta_{00(j_1,j_2)} = \gamma_{0000} + u_{00j_10} + u_{000j_2} \\ \beta_{10(j_1,j_2)} = \gamma_{1000} + u_{10j_10} + u_{100j_2} \end{cases}. \tag{6.26}$$

As with linear models for change, the intercept represents the predicted outcome (here, writing score) when $TIME = 0$. Thus, the intercept in the current model represents the predicted *WRITE* score at the first measurement occasion. The slope represents the linear change in writing scores over time (given a change in *TIME* of one unit).

A researcher might hypothesize that there are gender differences in the initial writing score (intercept) and in the linear change over time (slope). If that were the case, Equation 6.25 could include this level-2 (examinee) descriptor (uncentered):

$$\begin{cases} \pi_{0i(j_1,j_2)} = \beta_{00(j_1,j_2)} + \beta_{00(j_1,j_2)} FEMALE_{i(j_1,j_2)} + r_{0i(j_1,j_2)} \\ \pi_{1i(j_1,j_2)} = \beta_{10(j_1,j_2)} + \beta_{11(j_1,j_2)} FEMALE_{i(j_1,j_2)} + r_{1i(j_1,j_2)} \end{cases}, \tag{6.27}$$

thereby introducing additional parameters at the third level. The researcher might hypothesize a fixed effect of *FEMALE* across raters and across classrooms, and thus, would estimate the following level-3 model:

$$\begin{cases} \beta_{00(j_1,j_2)} = \gamma_{0000} + u_{00j_10} + u_{000j_2} \\ \beta_{01(j_1,j_2)} = \gamma_{0100} \\ \beta_{10(j_1,j_2)} = \gamma_{1000} + u_{10j_10} + u_{100j_2} \\ \beta_{11(j_1,j_2)} = \gamma_{1100} \end{cases}. \tag{6.28}$$

Lastly, the researcher might be interested in exploring the relationship between a rater's (grand-mean centered) degree of experience and the examinee's initial score while controlling for possible classroom (grand-mean centered) *SES* differences in the starting point and in students' rate of change over time in their writing scores. If this were the case, Equation 6.28 then becomes

$$\begin{cases} \beta_{00(j_1,j_2)} = \gamma_{0000} + \gamma_{0010}EXPER_{j_1} + \gamma_{0001}SES_{j_2} + u_{00j_10} + u_{000j_2} \\ \beta_{01(j_1,j_2)} = \gamma_{0100} \\ \beta_{10(j_1,j_2)} = \gamma_{1000} + \gamma_{1001}SES_{j_2} + u_{10j_10} + u_{100j_2} \\ \beta_{11(j_1,j_2)} = \gamma_{1100} \end{cases} . \quad (6.29)$$

Estimating a 3-Level CCREM in HLM Software

HLM 8 can estimate parameters of a 3-level CCREM with two cross-classification factors. For this example, the dataset (CC_3lvl.sav) identifies the measurement occasions for each student (level-2 unit) as well as includes the outcome variable (*WRITE*) and any level-1 variables (here, *TIME*). The data file also includes identifiers for each student, rater (first CC level-3 unit), and classroom (second CC level-3 unit) along with student, rater (row-factor), and classroom (column-factor) predictors (here, *FEMALE*, *EXPER*, and *SES*, respectively). The user must sort the dataset in ascending order by the level-2 (student) identifier. Once the user constructs the dataset, they can create the MDM file through similar steps previously described in Part I for the 2-level CCREM analysis (also see the HLM 8 manual for more information). One difference from the previous steps is that the user should select *HLM-HCM* (see Figure 6.4) for the MDM type.

Figure 6.14 provides the HLM software specification for the baseline individual growth model (see Equations 6.24, 6.25, and 6.26). Similar to estimating a 2-level CCREM in HLM software, note that the "row" factor (here, rater) is associated with the *b* random effects and the "column" factor (here, classroom) is associated with the *c* random effects. The bottom rows of the baseline model's estimates appearing in Table 6.3 record the random effect estimates. The HLM output for the variance estimates appears below.

Final estimation of level-1 and level-2 variance components:

Random Effect	Standard Deviation	Variance Component	*d.f.*	χ^2	*p*-value
INTRCPT1, e_{0jk}	4.98570	***24.85718***	931	24027.79765	<0.001
TIME, e_{1jk}	5.03657	***25.36705***	931	13993.30797	<0.001
σ^2, ε	2.41908	***5.85195***			

Final estimation of row level variance components:

Random Effect	Standard Deviation	Variance Component	*d.f.*	χ^2	*p*-value
INTRCPT1/INTRCPT2/ ICPTROW, b_{000}	5.20807	***27.12395***	34	980.96195	<0.001
TIME/INTRCPT2/ ICPTROW, b_{100}	1.97583	***3.90391***	34	153.35955	<0.001

Final estimation of column level variance components:

Random Effect	Standard Deviation	Variance Component	*d.f.*	χ^2	*p*-value
INTRCPT1/INTRCPT2/ ICPTCOL, c_{000}	4.50048	***20.25430***	33	224.12903	<0.001
TIME/INTRCPT2/ ICPTCOL, c_{100}	3.73527	***13.95224***	33	312.54928	<0.001

Statistics for the current model

Deviance = ***19106.470933***
Number of estimated parameters = ***12***

The results suggest that variability existed in the initial *WRITE* scores across students ($\hat{\sigma}^2_{r0} = 24.86$), raters ($\hat{\sigma}^2_{u00j_10} = 27.12$), and classrooms ($\hat{\sigma}^2_{u000j_2} = 20.25$). There also was variability in students' linear growth over time in *WRITE* scores across students ($\hat{\sigma}^2_{r1} = 25.37$), raters ($\hat{\sigma}^2_{u10j_10} = 3.90$), and classrooms ($\hat{\sigma}^2_{u100j_2} = 13.95$).

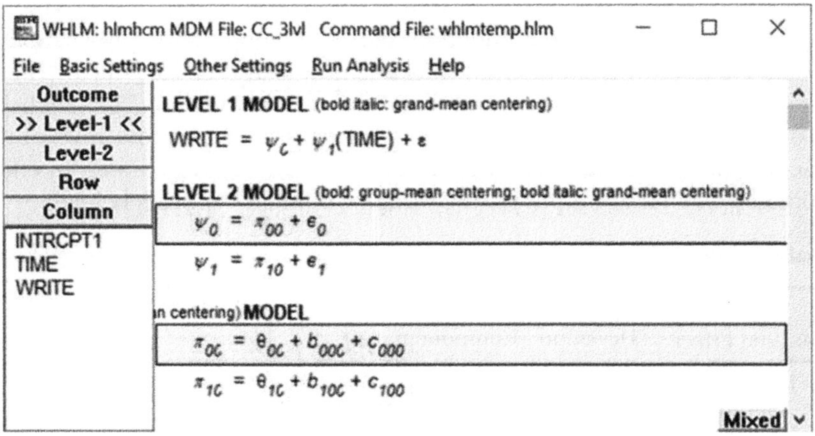

Figure 6.14 Baseline 3-Level CCREM in HLM Software

TABLE 6.3 Parameter Estimates for the 3-Level CCREMs

		Baseline	Model A	Model B	Model C
Fixed effects					
First timepoint					
Intercept	γ_{0000}	19.07***	17.64***	17.55***	17.55***
Student *FEMALE*	γ_{0100}		2.79***	2.81***	2.81***
Rater *EXPER*	γ_{0010}			1.20***	1.20***
Class *SES*	γ_{0001}			−0.58	
TIME slope					
Intercept	γ_{1000}	6.75***	6.18***	6.16***	6.16***
Student *FEMALE*	γ_{1010}		1.16**	1.17***	1.17***
Class *SES*	γ_{1001}			−2.29	
Variance components					
Intercept					
Student	σ^2_{r0}	24.86	22.91	22.92	22.92
Rater	$\sigma^2_{u00j_10}$	27.12	27.69	13.21	13.15
Class	$\sigma^2_{u000j_2}$	20.25	19.18	19.14	19.25
TIME slope					
Student	σ^2_{r1}	25.37	25.05	25.07	25.06
Rater	$\sigma^2_{u10j_10}$	3.90	3.79	3.83	3.90
Class	$\sigma^2_{u100j_2}$	13.95	13.74	12.27	13.58
Level-1 residual	σ^2_e	5.85	5.85	5.85	5.85
Model deviance		19,106.47	19,020.67	18,999.22	19,002.13

Note: Equations 6.24, 6.27, and 6.28 describe Model A; Equations 6.24, 6.27, and 6.29 describe Model B; Model C is the same as Model B but does not include *SES* as a predictor of the slope or the intercept.

** $p =< .01$, *** $p < .001$

The fixed effect estimates appear below.

Final estimation of fixed effects:

Fixed Effect	Coefficient	Standard Error	*t*-ratio	Approx. *d.f.*	*p*-value
For INTRCPT1, π_0					
INTRCPT2,					
INTERCEPT, θ_{00}	***19.073439***	1.204821	15.831	860	<0.001
For TIME, π_1					
INTRCPT2,					
INTERCEPT, θ_{10}	***6.754306***	0.755306	8.942	860	<0.001

Interpretation of the fixed effect estimates indicates that the average initial ($TIME = 0$) *WRITE* score (predicted to be 19.07) and the average monthly growth ($\hat{\gamma}_{1000} = 6.75$) differed from zero ($p < .001$). Table 6.3 includes the fixed and random effect estimates from the baseline 3-level CCREM.

The next model (Model A in Table 6.3) adds *FEMALE* as a predictor of the intercept and of the growth in *WRITE* over time to address possible gender differences in the intercept and slope (see Equation 6.27). The relationship between *FEMALE* and *WRITE* is fixed across raters and classrooms (see Equation 6.28). Estimates of the model appear in Table 6.3 and below.

Final estimation of fixed effects:

Fixed Effect	Coefficient	Standard Error	*t*-ratio	Approx. *d.f.*	*p*-value
For INTRCPT1, π_0					
INTRCPT2, INTERCEPT, θ_{00}	***17.644833***	1.209825	14.585	858	<0.001
FEMALE, INTERCEPT, θ_{01}	***2.787671***	0.341138	8.172	858	<0.001
For TIME, π_1					
INTRCPT2, INTERCEPT, θ_{10}	***6.177846***	0.767867	8.045	858	<0.001
FEMALE, INTERCEPT, θ_{11}	***1.155714***	0.341204	3.387	858	0.001

Final estimation of level-1 and level-2 variance components:

Random Effect	Standard Deviation	Variance Component	*d.f.*	χ^2	*p*-value
INTRCPT1, e_{0jk}	4.78639	***22.90952***	931	23032.19550	<0.001
TIME, e_{1jk}	5.00525	***25.05252***	931	13854.35962	<0.001
σ^2, ε	2.41909	***5.85198***			

Final estimation of row level variance components:

Random Effect	Standard Deviation	Variance Component	*d.f.*	χ^2	*p*-value
INTRCPT1/INTRCPT2/ ICPTROW, b_{000}	5.26195	***27.68812***	34	1077.28685	<0.001
TIME/INTRCPT2/ ICPTROW, b_{100}	1.94757	***3.79304***	34	150.33550	<0.001

Final estimation of column level variance components:

Random Effect	Standard Deviation	Variance Component	*d.f.*	χ^2	*p*-value
INTRCPT1/INTRCPT2/ ICPTCOL, c_{000}	4.37930	***19.17828***	33	219.96727	<0.001
TIME/INTRCPT2/ ICPTCOL, c_{100}	3.70654	***13.73847***	33	312.82697	<0.001

Statistics for the current model

Deviance = ***19020.670452***
Number of estimated parameters = ***14***

These results indicate that *FEMALE* was a statistically significant predictor of students' initial *WRITE* scores ($\hat{\gamma}_{0100} = 2.79, p < .001$) and their linear growth in *WRITE* scores ($\hat{\gamma}_{1100} = 1.16, p = .001$). Given the coding scheme for *FEMALE,* the positive effects indicate that girls' predicted initial *WRITE* scores are 2.79 points higher than boys, while the estimates of the growth rate are 7.33 for girls and 6.18 for boys. Adding *FEMALE* as a predictor reduced the student-level variance in initial scores ($\hat{\sigma}^2_{r0}$) from 24.86 to 22.91 (7.8%) and the student-level variance in growth rates ($\hat{\sigma}^2_{r1}$) from 25.37 to 25.05 (1.3%). For the sake of completeness, if a researcher wished to estimate a model in which the gender effect (on both slope and intercept) varied randomly across raters and classrooms, then the corresponding faint random effects for raters (u_{01j_10} and u_{11j_10}) and classrooms (u_{010j_2} and u_{110j_2}) should be clicked on so they become dark.

Given the gender differences detected, the next model retained *FEMALE* as a fixed predictor of the intercept and the slope (see Equation 6.28). In addition, the model adds a rater's level of experience (*EXPER*) and classroom *SES* as predictors of the intercept (both grand-mean centered). Last, the model also adds (grand-mean centered) *SES* as a predictor of students' growth in *WRITE* over time (see Equation 6.29 for the level-3 model). Figure 6.15 provides the HLM model specification used to estimate this model. The fixed- and random-effect estimates appear below in the HLM output format. Table 6.3 also reports these results under Model B.

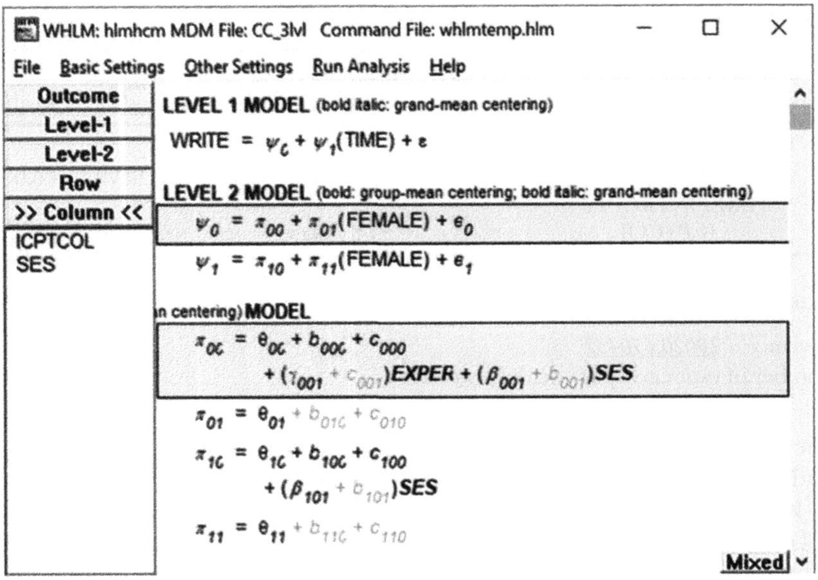

Figure 6.15 Conditional 3-Level CCREM (with a Fixed Level-2 Predictor and Two Fixed Cross-Classified Factor Predictors) in HLM Software.

Final estimation of fixed effects:

Fixed Effect	Coefficient	Standard Error	*t*-ratio	Approx. *d.f.*	*p*-value
For INTRCPT1, π_0					
INTRCPT2,					
INTERCEPT, θ_{00}	***17.554444***	1.017111	17.259	858	<0.001
EXPER, γ_{001}	***1.198792***	0.227148	5.278	32	<0.001
SES, β_{002}	***−0.582443***	1.683257	−0.346	32	0.732
FEMALE,					
INTERCEPT, θ_{01}	***2.809216***	0.341082	8.236	858	<0.001
For TIME, π_1					
INTRCPT2,					
INTERCEPT, θ_{10}	***6.157023***	0.740224	8.318	858	<0.001
SES, β_{101}	***−2.288426***	1.323184	−1.729	32	0.093
FEMALE,					
INTERCEPT, θ_{11}	***1.166356***	0.341329	3.417	858	<0.001

Final estimation of level-1 and level-2 variance components:

Random Effect	Standard Deviation	Variance Component	*d.f.*	χ^2	*p*-value
INTRCPT1, e_{0jk}	4.78775	***22.92257***	931	24754.34389	<0.001
TIME, e_{1jk}	5.00651	***25.06516***	931	13489.62474	<0.001
σ^2, ε	2.41908	***5.85193***			

Final estimation of row level variance components:

Random Effect	Standard Deviation	Variance Component	*d.f.*	χ^2	*p*-value
INTRCPT1/INTRCPT2/ ICPTROW, b_{000}	3.63427	***13.20794***	34	410.61958	<0.001
TIME/INTRCPT2/ ICPTROW, b_{100}	1.95671	***3.82872***	34	148.96769	<0.001

Final estimation of column level variance components:

Random Effect	Standard Deviation	Variance Component	*d.f.*	χ^2	*p*-value
INTRCPT1/INTRCPT2/ ICPTCOL, c_{000}	4.37481	***19.13893***	33	299.46123	<0.001
TIME/INTRCPT2/ ICPTCOL, c_{100}	3.50226	***12.26581***	33	285.61099	<0.001

Statistics for the current model

Deviance = ***18999.216729***
Number of estimated parameters = ***17***

The results indicate that *SES* was not a statistically significant predictor of the intercept ($\hat{\gamma}_{0001} = -0.58$, $p = .732$) nor of the slope ($\hat{\gamma}_{1001} = -2.29$, $p = .093$).

A final estimated model then removed *SES* from the level-3 equations for the intercept and slope:

$$\begin{cases} \beta_{00(j_1,j_2)} = \gamma_{0000} + \gamma_{0010} EXPER_{j_1} + u_{00j_10} + u_{000j_2} \\ \beta_{01(j_1,j_2)} = \gamma_{0100} \\ \beta_{10(j_1,j_2)} = \gamma_{1000} + u_{10j_10} + u_{100j_2} \\ \beta_{11(j_1,j_2)} = \gamma_{1100} \end{cases} . \quad (6.30)$$

The parameter estimates appear below and in Model C's column in Table 6.3.

Final estimation of fixed effects:

Fixed Effect	Coefficient	Standard Error	t-ratio	Approx. $d.f.$	p-value
For INTRCPT1, π_0					
INTRCPT2, INTERCEPT, θ_{00}	***17.551130***	1.017985	17.241	858	<0.001
EXPER, γ_{001}	***1.201855***	0.224165	5.361	32	<0.001
FEMALE, INTERCEPT, θ_{01}	***2.809908***	0.341077	8.238	858	<0.001
For TIME, π_1					
INTRCPT2, INTERCEPT, θ_{10}	***6.156050***	0.767371	8.022	858	<0.001
FEMALE, INTERCEPT, θ_{11}	***1.167226***	0.341314	3.420	858	<0.001

Final estimation of level-1 and level-2 variance components:

Random Effect	Standard Deviation	Variance Component	$d.f.$	χ^2	p-value
INTRCPT1, e_{0jk}	4.78786	***22.92360***	931	24642.25262	<0.001
TIME, e_{1jk}	5.00571	***25.05712***	931	13906.16536	<0.001
σ^2, ε	2.41907	***5.85190***			

Final estimation of row level variance components:

Random Effect	Standard Deviation	Variance Component	$d.f.$	χ^2	p-value
INTRCPT1/INTRCPT2/ ICPTROW, b_{000}	3.62684	***13.15395***	34	410.08090	<0.001
TIME/INTRCPT2/ ICPTROW, b_{100}	1.97421	***3.89750***	34	149.25545	<0.001

Final estimation of column level variance components:

Random Effect	Standard Deviation	Variance Component	$d.f.$	χ^2	p-value
INTRCPT1/INTRCPT2/ ICPTCOL, c_{000}	4.38800	***19.25458***	33	296.53108	<0.001
TIME/INTRCPT2/ ICPTCOL, c_{100}	3.68546	***13.58262***	33	305.29006	<0.001

Statistics for the current model

Deviance = ***19002.133619***
Number of estimated parameters = ***15***

SES was not a statistically significant predictor of the slope nor of the intercept; therefore, it is not surprising that the results do not change much between Models B and C. Model C still identifies a statistically significant gender difference in initial *WRITE* scores that favors girls over boys ($\hat{\gamma}_{0100} = 2.81$, $p < .001$) as well as a in monthly growth in *WRITE* scores ($\hat{\gamma}_{1100} = 1.17$, $p < .001$). The experience of the rater was statistically significantly related to initial writing scores ($\hat{\gamma}_{0010} = 1.20$, $p < .001$).

If a researcher chooses Model C as the final model, they likely would investigate additional rater and classroom descriptors to help explain the variability across raters and classrooms in students' first writing score and in their monthly change in writing scores. Over and above Model B, this final model only explained some of the variability across raters (using the measure of a rater's experience). The remaining variance component estimates did not change from the baseline model after the addition of the relevant predictors.

Estimating a 3-Level CCREM in R Software

The reader can use R (or SAS, Stata, or MLwiN) to estimate parameters of a CCREM that include additional levels. Use of lme4 package in R to estimate a 3-level CCREM is relatively simple and involves only the addition of the relevant additional "grouping factors" (i.e., levels). The R code needed to estimate the baseline individual growth model (see Equations 6.24, 6.25, and 6.26) is

```
base <- lmer(WRITE ~ 1 + TIME + (1 + TIME | STUDENT)
        + (1 + TIME | CLASS) + (1 + TIME | RATER),
        data = CC_31v1, REML = FALSE)
```

The reader should notice that, again, the formula statement identifies the ID distinguishing each of the levels. In the current model, level 2 is the student level, and "level 3" contains the two cross-classified factors (classroom and rater). The R output (obtained using the command "`summary(base)`") for the baseline model appears below and corresponds to that of the HLM software.

```
Linear mixed model fit by maximum likelihood  ['lmerMod']
Formula: WRITE ~ 1 + TIME + (1 + TIME | STUDENT) + (1 + TIME |
         CLASS) + (1 + TIME | RATER)
   Data: CC_3lvl

     AIC       BIC    logLik deviance df.resid
 19130.5   19202.5   -9553.2  19106.5     2988

Scaled residuals:
    Min      1Q  Median      3Q     Max
-2.5325 -0.4001  0.0172  0.4186  2.4491

Random effects:
 Groups   Name        Variance Std.Dev. Corr
 STUDENT  (Intercept) 24.860   4.986
          TIME        25.369   5.037    -0.03
 CLASS    (Intercept) 20.254   4.500
          TIME        13.952   3.735     0.11
 RATER    (Intercept) 27.123   5.208
          TIME         3.904   1.976    -0.38
 Residual              5.850   2.419
Number of obs: 3000, groups:  STUDENT, 1000; CLASS, 34; RATER, 34

Fixed effects:
            Estimate Std. Error t value
(Intercept)  19.0734     1.2049  15.830
TIME          6.7543     0.7554   8.941

Correlation of Fixed Effects:
     (Intr)
TIME -0.078
```

The following syntax accomplishes the estimation of Model C (the final model in Table 6.3), which includes *FEMALE* to address gender differences in the intercept and growth over time (see Equation 6.27) as well as a rater's grand-mean-centered level of experience (*EXPER_C*) as a predictor of the intercept (see Equation 6.30):

```
modelC <- lmer(WRITE ~ 1 + TIME * FEMALE + EXPER_C +
               (1 + TIME | STUDENT) + (1 + TIME | RATER),
               data = CC_31v1, REML = FALSE)
```

This model (Model C in Table 6.3) adds *EXPER_C* as a predictor of the intercept (by including *EXPER_C* in the formula statement). Addition of the interaction term "TIME * FEMALE" to the formula specifies *FEMALE* as a predictor of the intercept and of the growth in *WRITE* over time. Estimates of the model follow and correspond to the HLM output.

```
Linear mixed model fit by maximum likelihood  ['lmerMod']
Formula: WRITE ~ 1 + TIME * FEMALE + EXPER_C + (1 + TIME |
         STUDENT) + (1 + TIME | CLASS) + (1 + TIME | RATER)
   Data: CC_3lvl

     AIC       BIC    logLik deviance df.resid
 19032.1   19122.2   -9501.1  19002.1     2985
```

```
Scaled residuals:
     Min       1Q   Median       3Q      Max
-2.52504 -0.39726  0.01355  0.41564  2.48103

Random effects:
 Groups   Name        Variance Std.Dev. Corr
 STUDENT  (Intercept) 22.926   4.788
          TIME        25.059   5.006    -0.07
 CLASS    (Intercept) 19.254   4.388
          TIME        13.582   3.685     0.05
 RATER    (Intercept) 13.154   3.627
          TIME         3.897   1.974    -0.28
 Residual              5.850   2.419
Number of obs: 3000, groups:  STUDENT, 1000; CLASS, 34; RATER, 34

Fixed effects:
            Estimate Std. Error t value
(Intercept)  17.5511     1.0180  17.240
TIME          6.1560     0.7674   8.022
FEMALE        2.8099     0.3411   8.237
EXPER_C       1.2018     0.2242   5.361
TIME:FEMALE   1.1672     0.3413   3.420

Correlation of Fixed Effects:
            (Intr) TIME   FEMALE EXPER_
TIME        -0.053
FEMALE      -0.172  0.037
EXPER_C     -0.024 -0.006  0.020
TIME:FEMALE  0.028 -0.223 -0.163  0.006
```

SUMMARY

This chapter has described general scenarios in which CC data structures might exist and the corresponding basic formulation of the 2- and 3-level CCREM. The examples described use of HLM software and R package lme4 for estimation of parameters in a 2- and 3-level CCREM (in which the cross-classification occurred at the highest level). The reader should explore these same analyses using SAS, Stata, or MLwiN software, which are other commonly used multilevel modeling programs.

The CCREMs discussed here are of the simplest form. They only involved two cross-classified factors at a single level and only one level of CC factors. The reader should read additional texts that describe modeling of these more complex models that commonly exist in educational research (such as, Lei et al., 2018; Rasbash et al., 2010). Another simplification encountered in this chapter is the modeling only of interval-scaled outcomes. The CCREM also is useful for modeling dichotomous and ordinal outcomes for variables (e.g., Johnson, 2010), such as graduation or recidivism, or as another extension of the multilevel measurement model for use with multiple-choice item scores (Beretvas et al., 2005).

Researchers may encounter another form of CC data structures and source of impure clustering in which some lower-level units (e.g., individuals) are members of multiple higher-level units *of the same kind.* For instance, a dataset might have students clustered within high schools, but some of the students attended more than one high school. If a researcher is interested in modeling the effect of the higher-level clustering unit on an individual's outcome, then it seems important to model the contribution of all associated clustering units. This would be possible using a multiple membership random-effects model (MMREM). An additional complexity encountered when using the MMREM involves selection and specification of weights assigned to the set of clustering units associated with each lower-level unit. For example, in an educational scenario, weights might reflect an approximation of the proportion of time spent in each of the schools (level-2 units) by the student (level-1 unit). Readers should refer to Beretvas (2011), Chen and Leroux (2018), Chung and Beretvas (2012), Goldstein (2003), and Rasbash and Browne (2001) to see more demonstrations of the utility and estimation of multiple membership models.

The sophistication and flexibility of CCREMs are useful in modeling some of the data structures commonly encountered in educational research. However, there always are tradeoffs. Estimation of these models is computationally very complex and, thus, can be time-consuming. Although easy implementation with various multilevel software packages greatly facilitates the ability to estimate these models, the user still must ensure full understanding of the models' parameters and assumptions. In addition, due to the complexity, these models need much larger overall and per-cell sample sizes for precise and accurate estimation of the parameters (Luo & Kwok, 2009; Meyers & Beretvas, 2006). Despite these caveats and considerations, CCREMs allow researchers to model data that is not purely hierarchical because cross-classified structures occur quite frequently within educational data. Therefore, CCREMs provide an important analytic addition to the multilevel modeling arsenal of modeling tools. See this volume's website for access to all simulated data used for the examples presented here.

RECOMMENDED READINGS

Technical Readings

Luo, W., & Kwok, O. (2009). The impacts of ignoring a crossed factor in analyzing cross-classified data. *Multivariate Behavioral Research, 44*(2), 182–212. https://doi.org/10.1080/00273170902794214

Meyers, J. L., & Beretvas, S. N. (2006). The impact of inappropriate modeling of cross-classified data structures. *Multivariate Behavioral Research, 41*(4), 473–497. https://doi.org/10.1207/s15327906mbr4104_3

Shi, Y., Leite, W., & Algina, J. (2010). The impact of omitting the interaction between crossed factors in cross-classified random effects modelling.

British Journal of Mathematical and Statistical Psychology, 63(1), 1–15. https://doi.org/10.1348/000711008X398968

Applied Readings

Johnson, B. D. (2012). Cross-classified multilevel models: An application to the criminal case processing of indicted terrorists. *Journal of Quantitative Criminology, 28*(1), 163–189. https://doi.org/10.1007/s10940-011-9157-3

Leckie, G. (2009). The complexity of school and neighbourhood effects and movements of pupils on school differences in models of educational achievement. *Journal of the Royal Statistical Society: Series A (Statistics in Society), 172*(3), 537–554. https://doi.org/10.1111/j.1467-985X.2008.00577.x

Lei, X., Li, H., & Leroux, A. J. (2018). Does a teacher's classroom observation ratings vary across multiple classrooms? *Educational Assessment, Evaluation and Accountability, 30*(1), 27–46. https://doi.org/10.1007/s11092-017-9269-x

TRY THIS!

Suppose a researcher is interested in the effect of middle school urbanicity *and* elementary school percent of students receiving free/reduced lunch (FRL) on middle school science achievement, while controlling for student gender. Uh oh! This scenario is a bit of a conundrum for a typical multilevel model because the dataset is not purely clustered. (Note: the data come from the Tennessee's Student Teacher Achievement Ratio project.) There are 3,128 students cross-classified by 131 middle schools and 79 elementary schools in the data file, which is available at the online appendix, http://modeling.uconn.edu. Answer the following questions to practice and understand appropriately modeling cross-classified data structures.

1. What is the 2-level formulation of the *unconditional CCREM* for this scenario using appropriate variable names?
2. Estimate the unconditional CCREM and report the results of fixed effect and variance component estimates.
3. Calculate and interpret the proportion of variance attributable to variability among each cross-classified factor?
4. Estimate a conditional CCREM with the *uncentered* level-1 predictor, and test whether the slope for this variable randomly varies across middle schools. Report the results from the likelihood ratio test. What is your interpretation of the results for this test?
5. Next, test whether the gender differences in science achievement vary across elementary schools. Report the results from the likelihood ratio test. What is your interpretation of the results for this test?

6. Building upon the previous *best-fitting* model, let's add the cross-classified predictors to the level-2 model for the intercept. Be sure to *grand-mean center* percentage of students receiving FRL. Use "suburban" as the referent category for middle school urbanicity. In addition, include FRL percentage in the model for the gender variable's slope. Report the final estimates of your fixed effects and random effects' variance components.
7. What is the 2-level formulation using appropriate variable names of the final conditional CCREM you just estimated?
8. Write a paragraph or two describing the results of your final CCREM. Within your write-up of the results, be sure to interpret the fixed effects and variability explained in model comparisons.

REFERENCES

Aunsmo, A., Øvretveit, S., Breck, O., Valle, P. S., Larssen, R. B., & Sandberg, M. (2009). Modelling sources of variation and risk factors for spinal deformity in farmed Atlantic salmon using hierarchical-and cross-classified multilevel models. *Preventive Veterinary Medicine, 90*(1–2), 137–145. https://doi.org/10.1016/j.prevetmed.2009.03.015

Bates, D., Maechler, M., Bolker, B., & Walker, S. (2015). Fitting linear mixed-effects models using lme4. *Journal of Statistical Software, 67*(1), 1–48. https://doi.org/10.18637/jss.v067.i01

Beretvas, S. N. (2011). Cross-classified and multiple membership models. In J. J. Hox & J. K. Roberts (Eds.), *Handbook of advanced multilevel analysis* (pp. 313–334). Routledge.

Beretvas, S. N., Meyers, J. L., & Rodriguez, R. A. (2005). The cross-classified multilevel measurement model: An explanation and demonstration. *Journal of Applied Measurement, 6*(3), 322–341. https://pubmed.ncbi.nlm.nih.gov/15942074/

Brunton-Smith, I., Sturgis, P., & Leckie, G. (2017). Detecting and understanding interviewer effects on survey data by using a cross-classified mixed effects location-scale model. *Journal of the Royal Statistical Society: Series A (Statistics in Society), 180*(2), 551–568. https://doi.org/10.1111/rssa.12205

Brunton-Smith, I., Sturgis, P., & Leckie, G. (2018). How collective is collective efficacy? The importance of consensus in judgments about community cohesion and willingness to intervene. *Criminology, 56*(3), 608–637. https://doi.org/10.1111/1745-9125.12180

Cafri, G., & Fan, J. (2018). Between-within effects in survival models with cross-classified clustering: Application to the evaluation of the effectiveness of medical devices. *Statistical Methods in Medical Research, 27*(1), 312–319. https://doi.org/10.1177/0962280216628561

Chen, J., & Leroux, A. J. (2018). Residual normality assumption and the estimation of multiple membership random effects models. *Multivariate Behavioral Research, 53*(6), 898–913. https://doi.org/10.1080/00273171.2018.1533445

Chung, H., & Beretvas, S. N. (2012). The impact of ignoring multiple membership data structures in multilevel models. *British Journal of Mathematical and Statistical Psychology, 65*(2), 185–200. https://doi.org/10.1111/j.2044-8317.2011.02023.x

De Clercq, B., Pfoertner, T.-K., Elgar, F. J., Hublet, A., & Maes, L. (2014). Social capital and adolescent smoking in schools and communities: A cross-classified multilevel analysis. *Social Science & Medicine, 119*, 81–87. https://doi.org/10.1016/j.socscimed.2014.08.018

Dunn, E. C., Richmond, T. K., Milliren, C. E., & Subramanian, S. V. (2015). Using cross-classified multilevel models to disentangle school and neighborhood effects: An example focusing on smoking behaviors among adolescents in the United States. *Health and Place, 31*, 224–232. https://doi.org/10.1016/j.healthplace.2014.12.001

Elghafghuf, A., Stryhn, H., & Waldner, C. (2014). A cross-classified and multiple membership Cox model applied to calf mortality data. *Preventive Veterinary Medicine, 115*(1–2), 29–38. https://doi.org/10.1016/j.prevetmed.2014.03.012

Finch, W. H., Bolin, J. E., & Kelley, K. (2019). *Multilevel modeling using R* (2nd ed.). Taylor & Francis Group.

Goldstein, H. (2003). Multilevel modelling of educational data. In D. Courgeau (Ed.), *Methodology and epistemology of multilevel analysis* (pp. 25–42). Springer.

Goldstein, H. (2011). *Multilevel statistical models* (4th ed.). John Wiley & Sons.

Hox, J. J. (2010). *Multilevel analysis: Techniques and applications* (2nd ed.). Routledge.

Johnson, B. D. (2012). Cross-classified multilevel models: An application to the criminal case processing of indicted terrorists. *Journal of Quantitative Criminology, 28*(1), 163–189. https://doi.org/10.1007/s10940-011-9157-3

Johnson, I. Y. (2010). Class size and student performance at a public research university: A cross-classified model. *Research in Higher Education, 51*(8), 701–723. https://doi.org/10.1007/s11162-010-9179-y

Kim, Y. S., Petscher, Y., Foorman, B. R., & Zhou, C. (2010). The contributions of phonological awareness and letter-name knowledge to letter-sound acquisition—A cross-classified multilevel model approach. *Journal of Educational Psychology, 102*(2), 313–326. https://doi.org/10.1037/a0018449

Leckie, G. (2009). The complexity of school and neighbourhood effects and movements of pupils on school differences in models of educational achievement. *Journal of the Royal Statistical Society: Series A (Statistics in Society), 172*(3), 537–554. https://doi.org/10.1111/j.1467-985X.2008.00577.x

Lei, X., Li, H., & Leroux, A. J. (2018). Does a teacher's classroom observation ratings vary across multiple classrooms? *Educational Assessment, Evaluation and Accountability, 30*(1), 27–46. https://doi.org/10.1007/s11092-017-9269-x

Luo, W., & Kwok, O. (2009). The impacts of ignoring a crossed factor in analyzing cross-classified data. *Multivariate Behavioral Research, 44*(2), 182–212. https://doi.org/10.1080/00273170902794214

Meyers, J. L., & Beretvas, S. N. (2006). The impact of inappropriate modeling of cross-classified data structures. *Multivariate Behavioral Research, 41*(4), 473–497. https://doi.org/10.1207/s15327906mbr4104_3

Rasbash, J., & Browne, W. J. (2001). Modeling non-hierarchical structures. In A. H. Leyland and H. Goldstein (Eds.), *Multilevel modeling of health statistics* (pp. 93–105). John Wiley & Sons.

Rasbash, J., Leckie, G., Pillinger, R., & Jenkins, J. (2010). Children's educational progress: partitioning family, school and area effects. *Journal of the Royal Statistical Society: Series A (Statistics in Society), 173*(3), 657–682. https://doi.org/10.1111/j.1467-985X.2010.00642.x

Raudenbush, S. W., & Bryk, A. S. (2002). *Hierarchical linear models: Applications and data analysis methods* (2nd ed.). SAGE.

Shi, Y., Leite, W., & Algina, J. (2010). The impact of omitting the interaction between crossed factors in cross-classified random effects modelling. *British Journal of Mathematical and Statistical Psychology, 63*(1), 1–15. https://doi.org/10.1348/000711008X398968

Snijders, T. A. B., & Bosker, R. J. (2012). *Multilevel analysis: An introduction to basic and advanced multilevel modeling* (2nd ed.). SAGE.

CHAPTER 7

MULTILEVEL LOGISTIC AND ORDINAL MODELS

Ann A. O'Connell
The Ohio State University

Meng-Ting Lo
National Chiao Tung University

Jessica Goldstein
University of Connecticut

H. Jane Rogers
University of Connecticut

C.-Y. Joanne Peng
Indiana University and National Taiwan University

So far, the models covered in this book have assumed a normal or approximately normal distribution for the outcome of interest, and thus for the errors across the multiple levels of the model. However, many response variables of interest in education and the social sciences do not fit the normal distribution framework; most often, these non-normal outcomes involve

Multilevel Modeling Methods with Introductory and Advanced Applications, pages 209–254

dependent variables with discrete or limited response categories. Common statistical approaches for analyzing these kinds of data are drawn from a class of linear models where the outcomes follow a distribution from the exponential family (McCullagh & Nelder, 1989). These models are referred to as *generalized linear models* (GLMs) and can be applied to binary outcomes, count data, and ordinal or categorical data. In fact, models for normally distributed outcomes—often referred to as general linear models—are simply a special case of the generalized linear model where the normal distribution is used to describe the distribution of model errors. In the generalized case, binary outcomes typically are analyzed using the logistic distribution; count data by the Poisson or negative binomial distribution; and ordinal/ categorical data by the cumulative logistic or multinomial logistic models. This chapter presents logistic models for two types of categorical responses: (a) binary response variables and (b) ordinal response variables.

As an extension of single-level response models, multilevel analyses for categorical, count, or other types of non-normal data often are referred to as *generalized linear mixed models* (GLMMs; Fielding, 2003; McCulloch et al., 2008), although they have also been referred to as *hierarchical generalized linear models* (HGLMs; Raudenbush & Bryk, 2002). The application of GLMs in a multilevel framework parallels their applications in single-level research designs. For example, studies on factors related to dropping out of school, attaining proficiency for specific academic tasks, or enrolling in advanced placement courses in high school involve dichotomous outcomes that are analyzed best through logistic regression procedures.[1] When data are collected from individuals sampled across multiple classrooms, schools, regions, or other contexts, we expect the dichotomous responses to contain a within-context dependency, a situation that requires statistical adjustment for the resulting cluster effects through multilevel research designs and analyses. In some research situations, group- or context-level variables such as teacher preparation, school funding, or school climate may be used to help explain differences across contexts in the relationships between individual-level variables and the outcome of interest. In short, the logistic regression coefficients are presumed to vary across contexts, and GLMM provides the methodology with which to model this variation. Differences at the cluster- or context-level represent random effects and are typically assumed to be normally distributed (O'Connell et al., 2016; Stroup, 2013), similar to clustered models for continuous response variables. Overall, the use of a multilevel approach to analyze clustered dichotomous and other forms of non-normal data is a direct extension of the application of these kinds of models for single-level data, with the inclusion of random cluster-level effects.

Our demonstrations of GLMM analyses use HLM, Version 8 (Raudenbush et al., 2019) for mixed-effects logistic regression and R (R Core Team, 2020) with package mixor (Archer et al., 2018) for mixed-effects ordinal regression analysis. Our multilevel perspective was adapted from Dobson's (2002) single-level approach for describing GLMs. Accordingly, we emphasize four elements of multilevel model building throughout our examples:

1. expressing the outcome of interest in terms of a collection of explanatory or predictor variables,
2. parameter estimation and interpretation,
3. assessing quality of fit of the multilevel model to the data, and
4. statistical inference and hypothesis testing.

We begin with some background on the data set used for our examples and then provide a brief review of single-level GLMs for categorical data. Next, we provide the conceptual background for multilevel logistic models for dichotomous outcomes and the multilevel proportional odds model for ordinal outcomes. After setting the stage conceptually, we present specific data examples in-depth for each of these two multilevel models, highlighting critical aspects of each approach. Extensions to these models in other educational contexts are discussed in the final section, along with a brief review of some software considerations.

Background: The Early Childhood Longitudinal Study—Kindergarten Cohort

The data for the examples contained in this chapter were drawn from the Early Childhood Longitudinal Study-Kindergarten Class of 1998–1999 (ECLS-K),[2] which tracks the arithmetic and reading progress of a nationally representative sample of kindergarten children through the completion of eighth grade (eighth-grade data were released in 2008). In the ECLS-K, random samples of children were drawn from selected schools. The examples in this chapter analyze proficiency data for a subsample of these children as they neared the end of first grade.

The ECLS-K is a program of the U.S. National Center for Education Statistics (NCES, 2002) that, in part, assesses student proficiency in early literacy, mathematics, and general knowledge as a series of "stepping-stones," which reflect an ordinal set of skills that form the foundation for further learning (West et al., 2000). In the kindergarten and first-grade releases of the ECLS-K, proficiency for early numeracy is defined using a series of five hierarchical skills outlined in Table 7.1. Proficiency for these skills is assumed to follow the Guttman model so that mastering one skill requires the mastery of the previous skills (NCES, 2002).[3] The proficiency assessment contains clusters of items representing each hierarchical skill, and proficiency criteria were met if a child successfully answered three out of

TABLE 7.1 Proficiency Categories for the ECLS-K Measures for Early Numeracy

Proficiency Category	PROFMATH	ORDPROF[a]
0	Did not pass level 1	—
1	Can identify numbers and shapes	At most can identify numbers and shapes
2	Can understand relative size and recognize patterns	Can understand relative size and recognize patterns
3	Can understand ordinality and sequencing	Can understand ordinality and sequencing
4	Can solve simple addition and subtraction problems	Can solve simple addition and subtraction problems
5	Can solve simple multiplication and division problems	Can solve simple multiplication and division problems

[a] Category 0 was combined with Category 1 for the ordinal analysis and renamed ORDPROF.

the four items within each cluster. The highest proficiency level attained by each child was recorded in the database, with level 0 indicating that the first level had not been reached successfully.[4] This ordinal proficiency variable is referred to here as PROFMATH and has six levels (0 through 5). Because our analytic sample (described later) had only nine observations in the lowest category, we combined responses for the two lowest categories into category 1; the resulting distribution of PROFMATH—renamed as ORDPROF—had five levels (1 through 5). We also used this ordinal progression to create a dichotomous version of proficiency, PROF45, with levels 4 and 5 representing "proficient" versus all lower levels as "not proficient." Accordingly, our multilevel logistic examples use PROF45 as the outcome, and our multilevel proportional odds examples use the collapsed range of the ordinal scale, ORDPROF.

GENERALIZED LINEAR MODELS

Although this chapter assumes that the reader has had some exposure to logistic regression methods for single-level data, a brief introduction to logistic models is included here. More comprehensive treatments of logistic regression may be found in Agresti (2013, 2019), Hosmer et al. (2013), and McCullagh and Nelder (1989). In addition, O'Connell (2000, 2006) and Peng and Nichols (2003) provide instructive applications of logistic and logit-type models to ordinal and multinomial single-level data, respectively. In this chapter, our discussion centers around multilevel extensions of single-level logistic models for binary and ordinal response variables with applications to education data. For historical background on multilevel models for dichotomous responses, readers may want to review Wong and Mason (1985); Hedeker and Gibbons (1994) and Hedeker and Mermelstein (1998) do the same for multilevel ordinal models.

The terminology and estimation strategies for fitting GLMMs are fairly straightforward extensions of those used for multilevel models for continuous outcomes, coupled with modifications for the nature of the response variable. We will begin, however, with a single-level characterization of a GLM, identified by three related model components:

- A *random component*, for which the dependent variable Y follows one of the distributions from the exponential family, such as the normal, binomial, or inverse Gaussian;
- A *linear component*, which describes how a function, η, of the dependent variable Y depends on a collection of explanatory variables; and
- A *link function*, which describes the transformation of the expected value of the dependent variable Y to η (Fox, 2017).

The distribution of the random component defines the kind of generalized linear model that is used to represent the data (Liao, 1994). When the random component is specified by the normal distribution, the standard linear regression model for the response variable Y is obtained through the *identity* link function:

$$E(Y_i) = \eta_i = \beta_0 + \beta_1 X_{i1} + \beta_2 X_{i2} + \ldots + \beta_q X_{iq} \tag{7.1}$$

The *identity* link does not transform the dependent variable; the expected values for Y are summarized over the collection of explanatory variables, and there is a linear relationship between the expected values and the regression coefficients. Further, the errors from this model, $Y_i - E(Y_i)$, are assumed to be $N(0, \sigma^2)$ as well as homoscedastic.

For categorical dependent variables, such as dichotomous or ordinal outcomes, these familiar assumptions of the standard linear regression model no longer hold. However, if a suitable transformation of the dependent variable is applied, many of the useful model building strategies and interpretive properties of standard linear regression are still applicable. For GLMs, these transformations are from the exponential family (Fox, 2017; Liao, 1994; McCullagh & Nelder, 1989; Nelder & Wedderburn, 1972). Preferences for specific distributions and kinds of link functions exist, but one of the most commonly selected link functions is the logit link; other options are described in McCullagh and Nelder (1989). The logit link offers a connection between odds and probability, which are natural and familiar quantities for describing the distribution of a categorical variable. By definition, the logit is the natural log of the odds. These terms are explained below.

We focus the remainder of this section on dichotomous variables before describing extensions for ordinal outcomes. The usual approach to analyzing dichotomous outcomes is to begin by identifying and coding the two possible outcomes as either success ($Y = 1$) or failure ($Y = 0$). Thus, the mean value of Y becomes a *conditional* probability, $P(Y = 1|\underline{x})$; that is, the expected value represents the estimated proportion of successes given a particular set of values, $\underline{x}$, for the explanatory variables (or a local average if a predictor is continuous rather than discrete). However, since there are only two possible outcomes for the dependent variable, the errors from a model that seeks to predict probability of success are not normally distributed. In fact, the errors are heteroscedastic; the variance itself depends on the predicted values.

The probability distribution often associated with a dichotomous outcome is the *Bernoulli* distribution, which is a special case of the *binomial* distribution. The binomial distribution arises from counts of the number of successes in m attempts or trials. The Bernoulli distribution refers to a situation in which there is only one trial per person, and thus this single observation can be classified as either "success" or "failure." This is typically the case for most dichotomous data; that is, we don't count the number of times a student reaches proficiency for a specific arithmetic skill. Rather, we observe and record whether or not a student has attained proficiency at the time of data collection. In a Bernoulli distribution, the mean or expected value of Y is the proportion of observations possessing the characteristic of interest, such as $Y = 1$; the proportion or mean is written as $\pi(\underline{x})$ for the proportion of successes given a specific set of values for the collection of explanatory variables, $\underline{x}$. For a Bernoulli random variable, the variance is a function of the mean, $\pi(\underline{x}) * (1 - \pi(\underline{x}))$; this is the probability of success times the probability of failure.

Predicting the probability of success seems like a reasonable goal for the analysis of a dichotomous variable, but the relationship between probability of success and the collection of explanatory variables is inherently non-linear, requiring data transformations or other adjustments to aid in the process of deriving a good prediction model. If standard linear regression were applied to the dichotomous outcome data (i.e., the linear probability model), predictions could fall below 0 or exceed 1; these would be implausible values for predictions of probability. Generally, a non-linear relationship between predicted probability and the predictors is expected, where probability predictions begin to taper off somewhat and approach 0 or 1 more gradually than the predictions provided by a straight-line model (Agresti, 2013; Hosmer et al., 2013). The resulting pattern of this gradual trend is referred to as an S-shaped curve. Figure 7.1 illustrates this pattern

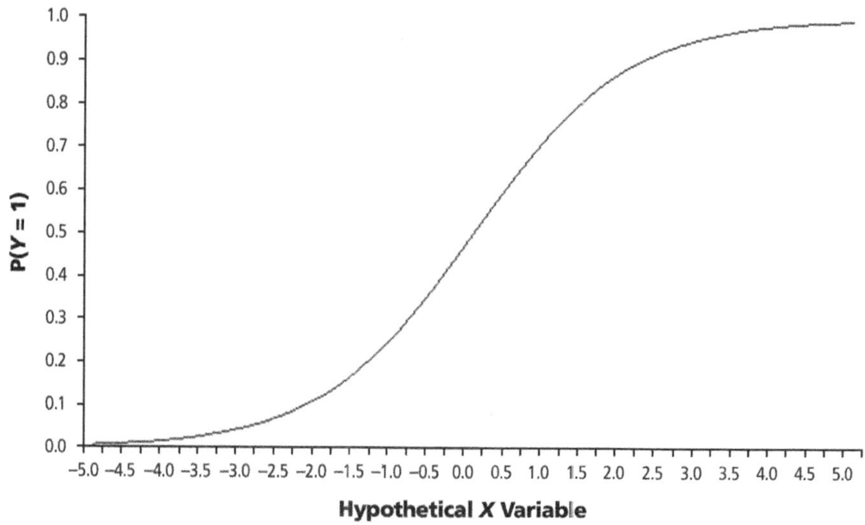

Figure 7.1 Example relationship between continuous predictor and probability of "success."

for a hypothetical explanatory variable and probability of "success" for a dichotomous variable *Y*.

These three issues—(a) probability predictions needing to be constrained between 0 and 1, (b) heteroscedastic errors, and (c) an S-shaped relationship of interest—can be addressed by applying models that allow a curvilinear relationship between $\pi(\underline{x})$ and $\underline{x}$. The logistic distribution is one of the most useful distributions for modeling this non-linear relationship between probability and a collection of predictors, primarily because it is easy to use and results in a fairly straightforward interpretation of predictor effects.

Logistic analyses for binary outcomes attempt to model the odds of an event's occurrence and to estimate the effects of explanatory variables on these odds. The odds for an event is a quotient that conveniently compares the probability that an event occurs (referred to as "success") to the probability that it does not occur (referred to as "failure," or the complement of success). When the probability of success is greater than the probability of failure, the odds are greater than 1.0; if the two outcomes are equally likely, the odds are 1.0; and if the probability of success is less than the probability of failure, the odds are less than 1.0.

For the ECLS-K data described above, suppose we are interested in studying the attainment of mathematics proficiency category 4 (addition and subtraction) or 5 (multiplication and division) among children at the end of first grade. The outcome can be described as binary: a child attains proficiency in category 4 or 5 (success) or the child does not (failure). The odds of proficiency are computed from the sample data by dividing the probability of reaching categories 4 or 5 (scored as $Y = 1$) by the probability of not reaching categories 4 or 5 (scored as $Y = 0$):

$$Odds = \frac{P(\text{"}success\text{"})}{P(\text{"}failure\text{"})} = \frac{P(Y=1)}{P(Y=0)} = \frac{P(Y=1)}{1-P(Y=1)} \tag{7.2}$$

To examine the impact on the odds of an independent variable, such as gender or the number of family risk characteristics, we construct the odds-ratio (*OR*), which compares the odds for different values of the explanatory variable. For example, if we want to compare the odds of reaching proficiency between males (coded $x = 1$) and females (coded $x = 0$), we would compute the following ratio:

$$OR = \frac{Odds(success|male)}{Odds(success|female)} = \frac{\dfrac{P(Y=1|x=1)}{1-P(Y=1|x=1)}}{\dfrac{P(Y=1|x=0)}{1-P(Y=1|x=0)}} \tag{7.3}$$

Odds ratios are bounded below by 0 but have no upper bound; thus, they can range from 0 to infinity. An *OR* of 1.0 indicates that an explanatory variable has no effect on the odds of success; that is, the odds of success for males are the same as the odds of success for females. Small values of the *OR* (< 1.0) indicate that the odds of success for the persons with the value of x used in the denominator (0 = females) are greater than the odds of success for the persons with the higher value of x used in the numerator (1 = males). The opposite is true for values of the *OR* that exceed 1.0; in that case, the odds for males of being proficient are greater than the odds of proficiency for females. The nature and type of coding used for the explanatory variables becomes important in interpretation; we use simple dummy or referent coding in the examples in this chapter. Other approaches to coding categorical predictor variables can change the interpretation of that variable's effect in the model; discussions of alternative approaches to categorizing qualitative data in logistic regression models can be found in Hosmer et al. (2013).

The *OR* is a measure of association between the binary outcome and an explanatory variable that provides an indication of how odds for the outcome ($Y = 1$) change as the explanatory variable increases or decreases. Although we seek a model for the probability of success, we use logistic regression to model the odds or, more specifically, to model the natural (base e) log of the odds, referred to as the *logits* of a distribution. These logits then can be used to determine variable effects and predicted probabilities given the collection of explanatory variables.

The logit transformation has many desirable properties. First, it eliminates the skewness inherent in estimates of the *OR* (Agresti, 2013), which can range from zero to infinity with a value of one indicating the *null* case of *no change in the odds*. The logit ranges from negative infinity to positive infinity, which eliminates the boundary problems associated with both the *OR* and probability. The logit model is linear in the parameters, which means that the effects of explanatory variables on the log of the odds are additive. The null case of no change in the odds is characterized by a parameter estimate of zero in the logit metric because $\log(1) = 0$, so that statistical tests around a null hypothesis of zero are conceptually familiar. Thus, the model is easy to work with and allows for application of model-building strategies that mirror those of ordinary linear regression.

This process can be extended to include more than one explanatory variable. If we let $\pi(Y = 1 | X_1, X_2, \ldots, X_q) = \pi(\underline{x})$ represent the probability of "success," or the outcome of interest (e.g., a child being in proficiency category 4 or 5), for a given set of q explanatory variables, then the logistic model can be written as

$$\log(Y') = \text{logit}\left[\pi(\underline{x})\right] = \log\left(\frac{\pi(\underline{x})}{1-\pi(\underline{x})}\right) = \beta_0 + \beta_1 X_1 + \beta_2 X_2 + \ldots + \beta_q X_q. \quad (7.4)$$

In this expression, Y' is simply a convenient way to emphasize the odds as a transformed outcome variable: Rather than predicting Y directly, we are predicting the natural log of the odds of $Y = 1$. The link function describes the process of "linking" the original Y to the transformed outcome: $\eta = \log(Y') = \log\left[\pi(\underline{x}) / \left(1 - \pi(\underline{x})\right)\right]$, which is referred to as the *logit link*. Solving for $\pi(\underline{x})$ gives us the familiar expression for the logistic regression model for the probability of success:

$$\begin{aligned} \pi(\underline{x}) &= \frac{\exp(\beta_0 + \beta_1 X_1 + \beta_2 X_2 + \ldots - B_q X_q)}{1 + \exp(\beta_0 + \beta_1 X_1 + \beta_2 X_2 + \ldots + B_q X_q)} \\ &= \frac{1}{1 + \exp\left[-(\beta_0 + \beta_1 X_1 + \beta_2 X_2 + \ldots + B_q X_q)\right]} \end{aligned} \quad (7.5)$$

For the ith person in a sample, the predicted logit can be written as

$$\text{logit}_i\left(\pi(\underline{x})\right) = \eta_i = b_0 + b_1 X_{i1} + b_2 X_{i2} + \ldots + b_q X_{iq}. \quad (7.6)$$

The logit, η_i, represents the log of the odds of "success" for the ith person, conditional on the set of predictors, and it provides a convenient metric in which to work: Logits are continuous, and the logit model is linear in the parameters. A shorthand expression for estimating probability based on the predicted logit for the *i*th person is

$$\pi(\underline{x}) = \frac{\exp(\eta_i)}{1 + \exp(\eta_i)} \tag{7.7}$$

Note that Equation 7.6 contains no specific error term. In the binomial distribution, the variance is a function of the conditional mean, $\pi(\underline{x})$, and thus the errors are described already by the choice of the binomial distribution rather than being estimated separately (Hosmer et al., 2013).

Parameter estimation in logistic regression is based on principles and methods of maximum likelihood and thus involves special iterative methods. This approach yields parameter estimates that "maximize the probability of obtaining the observed set of data" (Hosmer et al., 2013, p. 8). A value of −2 times the log-likelihood (−2LL) for a particular fitted model can be used to assess quality of fit through a general likelihood ratio test for comparison of competing models. The "deviance" of a model often is referred to as −2LL and, in a sense, represents how poorly a model fits the data. If deviance is reduced by a competing nested model, the competing model is preferred. A discussion of different approaches to assessing model fit for single-level logistic regression models is outside the scope of this chapter (see Menard, 2000, for work in this area), but we return to some of these ideas as we examine the models and applications of GLMMs.

THE MULTILEVEL LOGISTIC MODEL FOR DICHOTOMOUS OUTCOMES

So far, we have reviewed logit models for the analysis of single-level dichotomous data. In many research situations, however, observations are clustered within groups; educational data provide the most obvious context. Typically, children are nested within classrooms or schools, and schools are nested within districts or larger metropolitan units. It is likely that group-level variables might have an effect on an individual's response outcome, thus requiring a multilevel perspective of the data. Continuing our earlier example, if we let Y_{ij} represent the proficiency response (0 or 1) for the *i*th child in the *j*th school, where 1 indicates the child having reached Proficiency Level 4 or 5, then the expected value of Y is likely to vary based on child characteristics as well as by school characteristics. The sampling distribution for this collection of dichotomous data is Bernoulli, with $E(Y_{ij}|\pi_{ij}) = \pi_{ij}$ and $\text{Var}(Y_{ij}|\pi_{ij}) = \pi_{ij}(1-\pi_{ij})$, where π_{ij} is the probability of reaching proficiency for the *i*th child in the *j*th school. The proportion of children reaching proficiency for a given school, P_j, varies across schools depending on the proficiency outcomes for the randomly selected children sampled from within that school. The commonly used model for the link function is the logit link, and the resulting multilevel model resembles the single-level model, although now the regression coefficients from level 1 are allowed to vary across the different schools in the sample.

$$\begin{aligned}
&\text{Level 1:} \quad \eta_{ij} = \beta_{0j} + \beta_{1j}X_{1ij} + \beta_{2j}X_{2ij} + \beta_{qj}X_{qij} \\
&\text{Level 2:} \quad \beta_{qj} = \gamma_{q0} + \sum_{s=1}^{S_q} \gamma_{qs}W_{sj} + u_{qj}
\end{aligned} \tag{7.8}$$

This is the most general expression for the multilevel logistic model. A model in combined form can be written but is too bulky to include here; however, we provide the combined expression for each of our specific examples below. The distribution of regression coefficients at the school level is assumed to be normal, with the collection of residuals at level 2—referred to as random effects—described through covariance matrix T in Equation 7.9:

$$u_{qj} \sim N(\underline{0}, T) \tag{7.9}$$

THE MULTILEVEL PROPORTIONAL ODDS MODEL FOR ORDINAL OUTCOMES

Analysis strategies for multilevel ordinal data are extensions of those for single-level ordinal data, mirroring the process of adapting logistic regression procedures for multilevel dichotomous data. Reviews of different methods for analyzing ordinal data in general can be found in Fullerton (2009), O'Connell (2000, 2006), Agresti (2013), Azen and Walker (2011), Bender and Benner (2000), Clogg and Shihadeh (1994), Long (1997), and McCullagh (1980). Several multilevel and longitudinal texts include brief sections pertaining to the analysis of clustered ordinal data, such as Hedeker and Gibbons (2006), Skrondal and Rabe-Hesketh (2004), Hox et al. (2013), Heck et al. (2012), Snijders and Bosker (2012), and Raudenbush and Bryk (2002). For the most part, multilevel methods for analyzing continuous and dichotomous outcomes are far more developed than multilevel methods for ordinal data. However, substantive aims and the needs of more complex ordinal data structures are being more adequately met as estimation methods and software availability for multilevel ordinal analyses continue to improve (Fielding, 2003; Hedeker et al., 2018; Pinto & Sooriyarachchi, 2019).

The examples in this chapter focus on the application and interpretation of the multilevel proportional odds (PO) model, a particular type of analysis for ordinal data that sometimes is referred to as the cumulative odds model. The PO approach is based on the most commonly used representation of ordinal categories for analyzing single-level ordinal data. It characterizes a sequence of cumulative outcomes and uses a *cumulative logit link*, which is an extension of the logit link typically applied to dichotomous data (Agresti, 2013, 2019; Armstrong & Sloan, 1989; Long, 1997; McCullagh, 1980).

The PO model assumes proportional odds across successive cumulative categories. The assumption of proportionality is useful primarily because of its parsimony, and the PO model corresponds naturally to models in which the interest may be in ascertaining the overall likelihood of a response being at or below a specific outcome category. The proportional odds assumption is also called the cumulative odds or the equal slopes assumption. As an example, Table 7.1 shows the six possible proficiency outcomes for early numeracy for the ECLS-K *PROFMATH* ordinal scale: *PROFMATH* = 0, 1, 2, 3, 4, and 5. Cumulatively, the data can be partitioned into five "splits" as follows: $Y \leq 0$, $Y \leq 1$, $Y \leq 2$, $Y \leq 3$, $Y \leq 4$ (where Y represents each of the sequential response category possibilities). Since all observations are contained in what could be called the final split, $Y \leq 5$, this last cumulative representation is not necessary. Now, if we imagine a consecutive series of binary logistic regressions across all of these cumulative splits—each split being used to estimate the probability of a child's proficiency level being at or below that specific response category—the assumption of proportionality implies that the effect of any explanatory variable remains constant regardless of the response value identifying each split. For example, this means that the effect of gender would be assumed to be the same whether we are referring to the probability of a response being less than or equal to category 0 or to a response being less than or equal to category 3.

Proportionality is restrictive in the sense that equal log-odds ratios for the explanatory variables across all of the successive cumulative comparisons are assumed, which means that the log-odds of proficiency do not depend on response category. In many research situations involving single-level data, this assumption is a reasonable one. Hosmer et al. (2013) state that "inferences from fitted proportional odds models lend themselves to a general discussion of direction of response and do not have to focus on specific outcome categories" (p. 297). By extension, the same may be said for multilevel ordinal models, but it is incumbent on researchers to verify the reasonableness of any model they choose to apply to their data. Most software for the analysis of multilevel data fits the PO model, although other link options, such as the complementary log-log link for continuation ratio models, as well as methods for allowing partial-proportional odds, are available. Models in which some explanatory variables exhibit non-proportional odds are referred to as "partial" proportional odds models; fully non-proportional odds models are those in which all of the explanatory variables exhibit non-proportionality. Hedeker and Mermelstein (1998) provided the foundation for extensions of multilevel models for ordinal data that allow for non-proportional odds for all predictors as well as for subsets of the explanatory variables.

In a proportional odds analysis, the natural log of the odds of a response at or below each of the ordinal response categories form the quantities of interest. For example, with a six-category ordinal outcome such as *PROFMATH*, the $K - 1 = 5$ formulas shown in Table 7.2 can be used to compute the *cumulative probabilities* and, consequently, the *cumulative odds*. The cumulative probability is the probability that the response for the ith student in the jth school, which we can write as R_{ij}, is at or below a given proficiency level. As in logistic regression, the odds are a quotient comparing the probability of an event occurring to the probability of an event not occurring. Accordingly, the cumulative odds $[Y'_{kij}]$ represent the odds that the response for the ith child in the jth school would be, at most, in category k (rather than beyond category k). From Table 7.2 we see that the cumulative odds, in order, correspond to the probability of being in Proficiency Level 0 relative to all categories above it; the probability of being in Proficiency Level 0 or 1 relative to all above it; and so on until arriving at the probability of being in categories 0, 1, . . . , 4 relative to being in category 5. The Kth, or final, cumulative probability always would be 1.0 (probability of being at or below the highest possible level); therefore, its probability and associated odds are not included in the table. It is common to refer to the values marking each of these binary comparisons as "cutpoints" or cumulative splits. For example, the cutpoint for the first comparison is 0 (Proficiency Level 0

TABLE 7.2 Cumulative Odds Model for $K = 6$ ($k = 0, 1, \ldots, 5$), Where R_{ij} Represents the Proficiency Outcome (Response) for the ith Student in the jth School

Category	Cumulative Probability	Cumulative Odds $[Y'_{kij}]$	Probability Comparison
$k = 0$ (Proficiency 0)	$P(R_{ij} \le 0)$	$\frac{P(R_{ij} = 0)}{P(R_{ij} > 0)}$	Proficiency 0 versus all levels above
$k = 1$ (Proficiency 1)	$P(R_{ij} \le 1)$	$\frac{P(R_{ij} \le 1)}{P(R_{ij} > 1)}$	Proficiency 0 and 1 combined versus all levels above
$k = 2$ (Proficiency 2)	$P(R_{ij} \le 2)$	$\frac{P(R_{ij} \le 2)}{P(R_{ij} > 2)}$	Proficiency 0, 1, 2 combined versus 3, 4, 5 combined
$k = 3$ (Proficiency 3)	$P(R_{ij} \le 3)$	$\frac{P(R_{ij} \le 3)}{P(R_{ij} > 3)}$	Proficiency 0, 1, 2, 3 combined versus 4, 5 combined
$k = 4$ (Proficiency 4)	$P(R_{ij} \le 4)$	$\frac{P(R_{ij} \le 4)}{P(R_{ij} > 4)}$	Proficiency 0, 1, 2, 3, 4 versus proficiency 5

versus above 0); the cutpoint for the second comparison is 1 (Proficiency 0 and 1 versus above 1), and so on.

To better understand how the multilevel PO model works, imagine if the separate comparisons described in the last column of Table 7.2 were investigated using corresponding multilevel dichotomous-outcome logistic regressions for each of the associated cumulative splits. The *simultaneous* fitting of each of these separate $K-1$ logistic models represents the overall PO approach. The validity of this approach rests upon the critical assumption of proportional odds: The effects of the explanatory variables are assumed to be the same (i.e., not statistically different) across these separate comparisons. For single-level data, the assumption of equal slopes can be tested easily within statistical packages that provide either a score or likelihood ratio test for the proportional odds assumption (Agresti, 2019; Azen & Walker, 2011; for detailed discussion of the score test see Brant, 1990, as well as Agresti, 2013 and Long, 1997).

Within a multilevel context, the proportional odds assumption can be examined through a likelihood ratio test comparing the deviance for a model assuming proportional odds to the deviance for a model that relaxes this assumption and allows the effect of one or more explanatory variables to vary across the cumulative splits. Unfortunately, not all estimation strategies for ordinal outcomes can be used to provide a model deviance. Estimation methods that rely on numerical approximation such as adaptive quadrature (AQ) and Laplace do yield a reliable deviance statistic appropriate for likelihood ratio tests and model comparisons, but these approaches often prove difficult to converge as model complexity increases (Bauer & Sterba, 2011; Raudenbush et al., 2000; Snijders & Bosker, 2012). We discuss estimation options and related software considerations in later sections. In addition to a formal test of the proportional odds assumption, an ad hoc approach can be applied that investigates the consistency of slope estimates across the cumulative splits as described in Table 7.2. This approach, taken by Bell and Dexter (2000), is one we also demonstrate here.

For the ith student in the jth school, the multilevel proportional odds model is based on the cumulative logit link and fit according to the following equations (Raudenbush & Bryk, 2002):

$$\text{Level 1:}\quad \eta_{kij} = \ln(Y'_{kij}) = \ln\left(\frac{P(R_{ij} \le k)}{P(R_{ij} > k)}\right) = \beta_{0j} + \sum\nolimits_{q=1}^{Q} \beta_{qj} X_{qij} + \sum\nolimits_{k=2}^{K-1} D_{kij}\delta_k \tag{7.10}$$

$$\text{Level 2:}\quad \beta_{qj} = \gamma_{q0} + \sum_{s=1}^{S_q} \gamma_{qs} W_{sj} + u_{qj}$$

In these expressions, the level-1 equations represent the child-level models and the level-2 equations represent the school-level models. For the ith child in the *j*th school, Y'_{kij} represents the cumulative odds for each category *k*, with $k = 1, \ldots, K-1$ levels of the ordinal response and based on $q = 1, \ldots, Q$ child-level explanatory variables. The expression on the left side of the level-1 equation is the natural log of the cumulative odds at each category *k*, and is referred to as the logit for the cumulative distribution. The expression in the middle represents the two specific probability comparisons being made in the determination of the log of the odds, where R_{ij} represents the proficiency outcome (i.e., response) for the *i*th child in the *j*th school. The first set of terms on the far right side of the level-1 equation is similar to the regression coefficients in the multilevel logistic regression model, with β_{0j} representing the intercept for the *j*th school and the β_{qj} representing the collection of school-specific slopes for the *j*th school in terms of the expected change in the logit for each 1-unit change in the respective within-school explanatory variable, X_{qij}.

Across each of the $K-1$ underlying cumulative representations of the outcome data, these logit effects are assumed to be constant within each school; this is the proportional or cumulative odds assumption. For each school, the slope parameters, β_{qj}, are restricted to be constant for the *q*th explanatory variable across the separate possible cumulative splits derived according to the second column of Table 7.2. The level-2 equations model the differences between schools in these slope effects for each of the explanatory variables as well as for the intercepts at level 1. The random effects or residuals at level 2 are assumed to be multivariate normal and distributed as in Equation 7.9 for the multilevel logistic model, that is, $u_{qj} \sim N(\underline{0}, T)$.

The collection of estimates at the farthest right of the level-1 equation are referred to as "delta" coefficients, and they operate as deviations from the baseline intercept for each of the $K-1$ separate cumulative comparisons beyond the first, with β_{0j} as each school's intercept (i.e., the intercept logit marking the first comparison). D_{kij} is an indicator variable for the *k*th outcome category beyond the first (i.e., $k = 2$ to $K-1$). In other words, the effects of the explanatory variables are assumed to be constant across each comparison, but each cumulative comparison has its own intercept θ_k. For our model with $K = 6$, the $K-1$ intercepts could be written as $\theta_1 = \beta_{0j}$, $\theta_2 = \beta_{0j} + \delta_2$, $\theta_3 = \beta_{0j} + \delta_3$, $\theta_4 = \beta_{0j} + \delta_4$, and $\theta_5 = \beta_{0j} + \delta_5$. These intercepts are referred to as *thresholds* because they identify the category-specific relationship between an assumed underlying continuous response variable and the corresponding ordinal category, with the covariates set equal to 0 (Hedeker, 2015; Hox et al., 2013). Across schools, the intercepts for the first cumulative comparisons are allowed to vary at random (see Level 2 of Equation 7.10). Assuming a random effect of 0 we can write the overall intercepts or thresholds for each cumulative split in the model as $\theta_1 = \gamma_{00}$, $\theta_2 = \gamma_{00} + \delta_2$, $\theta_3 = \gamma_{00} + \delta_3$, $\theta_4 = \gamma_{00} + \delta_4$, and $\theta_5 = \gamma_{00} + \delta_5$.

The general level-2 equation presented in Equation 7.10 describes how the within-school effects (level-1 coefficients) vary based on selected school- or group-level characteristics. For example, γ_{q0} represents the intercept for each of the $Q + 1$ level-2 models (i.e., one for each of the q predictors from level 1 plus one for the intercept). These are the best estimates for each of the regression coefficients from level 1 when any school-level variables are set to 0, and "averaging" across all schools. For each specific level-2 equation, the γ_{qs}'s represent effects of each predictor for the qth equation at level 2. As in the multilevel logistic model, parameter estimates at level 2 are in the log-odds metric. The exponentiation of these logit-slopes provides the estimate of the effect of the associated group-level predictor on the cumulative odds.

In the PO model, the estimated odds of a child's response falling into category k or below is determined based on the collection of level-1 and level-2 explanatory variables. The multilevel PO analysis is based on a transformation of the odds, that is, the cumulative logit, which is the log of the cumulative odds. A predicted cumulative logit of zero corresponds to a cumulative odds of 1.0, which implies that there is no difference between the probability of a child being in a certain category (or below) and being above that category (e.g., .5/.5 = 1.0, and log(1.0) = 0). From Equation 7.10, a positive cumulative logit implies that the likelihood of being in lower categories is greater (e.g., .7/.3 = 2.33, log(2.33) = .85), and a negative cumulative logit implies that the likelihood of being in higher categories is greater (e.g., .3/.7 = .43, log(.43) = –.85). Estimated cumulative logits can be exponentiated and transformed to predicted cumulative probabilities using Equation 7.7. Forthcoming examples clarify this process.

Note on the Parameterization of the Multilevel Ordinal Model

The multilevel model can be presented in a variety of ways, similar to its single-level counterpart (for details see McCullagh, 1980 or O'Connell, 2006). The cumulative odds could be estimated in ascending order $\left[P(R_{ij} \leq k)\right]$ as shown here, or in descending order $\left[P(R_{ij} \geq k)\right]$. The cumulative odds for the complement could also be estimated; for the ascending option this is $\left[P(R_{ij} > k)\right]$. Further, the covariate effects represented by

$$\sum_{q=1}^{Q} \beta_{qj} X_{qij}$$

could be *subtracted* from rather than *added* to the intercept term. Model predictions will correspond to the ascending or descending order chosen by the analyst (or the software package); the intercept and threshold parameters will be adjusted to correspond to the intended direction of the cumulative odds. In either case, overall predictions for the cumulative odds requested are not affected by whether the covariate effects are added or subtracted. The `R mixor` package which we used for our demonstrations uses the ascending cumulative odds and offers a choice between adding or subtracting the covariate effects. Consistent with Equation 7.10, we used the additive approach—and our interpretation of variable effects corresponds to this parameterization. This is consistent with the approach taken in HLMv8.0 as well. The `R ordinal` package using the `clmm2` function (cumulative link mixed models) is based on the ascending approach but subtracts the covariate effects by default. Substantively, there is no difference in overall interpretation, but language regarding direction of effects must be carefully considered to reflect the cumulative probabilities ultimately being modeled.

Additionally, some software may report the separate intercepts or thresholds for each cumulative split, rather than report these as a departure from the first intercept as in Equation 7.10. Documentation should inform you about the format of results.

ESTIMATION AND MODEL COMPARISONS FOR GLMMs

GLMs require iterative maximum likelihood procedures for estimation of the regression parameters (Agresti, 2013). When these regression parameters are allowed to vary across contexts, as in GLMM analyses, we need to estimate not only the fixed effects portions of the model but also the random effects, that is, the variability between groups or contexts such as schools. Statistical software packages use different approaches that may vary in terms of computational intensity. The complexity of the model yields a reliance on approximation techniques to either numerically approximate the integral required to maximize the likelihood function (numerical integration via adaptive quadrature, or using Laplace transformations), or by approximating the non-linear model using Taylor series expansion, which creates a close linearization of the model for which parameter estimates are updated upon successive iterations (Bauer & Sterba, 2011; Diaz, 2007; Fielding, 2003; Hox et al., 2013; McCulloch & Searle, 2001; NcNeish et al., 2017, Raudenbush et al., 2000; Raudenbush & Bryk, 2002). The latter approach is the default taken in many statistical packages (i.e., penalized quasi likelihood or PQL), and this and its variants are referred to as quasi-likelihood strategies. However, deviance statistics are not available through quasi-likelihood methods and thus cannot be used for likelihood-based model comparisons. Hedeker (2015) and others have identified adaptive quadrature as the recommended approach for GLMMs, in part due to this limitation. However, optimal estimation given sample size and the nature and distribution of ordinal outcomes remains an active area of research (see e.g., Ali et al., 2016; Bauer & Sterba, 2011; Lo, 2020; and McNeish & Stapleton, 2016).

In this chapter, our emphasis is on intuitive understanding of analyses involving GLMM's, regardless of statistical package. For demonstration, our examples use adaptive quadrature estimation available through HLMv8.0 (Raudenbush et al., 2019) for the multilevel logistic models and the R package `mixor` (Archer et al., 2018) for multilevel ordinal models. Parameter estimates for the models we present are similar under PQL and Laplace methods. Using AQ, model fit can be assessed through comparison of deviances between nested models, as with deviance comparisons for standard HLMs (see Chapter 3, "Evaluation of Model Fit and Adequacy," of this volume, for additional details on model fit approaches).

Note on Population-Averaged Versus Unit-Specific Results

For GLMMs, a distinction can be made between unit- or cluster-specific (here, school-specific) results and population-average (or marginal) results. The unit-specific model is similar to the structure of a standard HLM in that "the goal is to provide inferences for covariates that can change within cluster[s]" (Hosmer et al., 2013, p. 317). Given that this is the aim of our presentation of GLMMs, we focus on the unit-specific perspective throughout this chapter. An alternative approach is to report population-average effects; these methods "account for clustering without explicitly splitting the model into multiple levels" (McNeish et al., 2017, p. 1). Population-average models tend to be preferred when the desired inferences focus on group-level variables rather than the investigations of varying effects of individual-level covariates; this is typically the case when population-level causal inferences are desired. Differences between the two types of results may be great. The population average model does not "condition on (or "hold constant") the random effect, u_{0j}" (Raudenbush et al., 2019, p. 122); rather, model estimates are averaged over all possible values of the level-2 random effects. On the other hand, the unit-specific model provides an expected value conditioning on the level-2 random effects. Depending on the software, output may provide both unit-specific and population-average results for GLMM analyses. Importantly, models estimated through generalized estimating equations (GEE) are providing marginal or population-averaged results (Hosmer et al., 2013). Researchers should carefully read the software manuals for their selected software to ensure the requested analyses are in the form anticipated.

DATA STRUCTURES FOR OUR EXAMPLES

Our examples focus on the prediction of proficiency in early numeracy. As described in the introduction, we used a collection of student-level and school-level predictors drawn from the public-use first-grade and follow-up third-grade sample of the ECLS-K. All of the data analyzed here are available through the NCES, and the subsample data created for our examples can be found on our book's website. Our examples were derived solely for the purpose of explicating the technical and methodological use of multilevel logistic and ordinal regression models. Although they are informative, these examples are not meant to provide a complete picture of early arithmetic skill for first-grade children.

We based our analyses on first-grade children who were not English-language learners and not repeating first grade nor retained in kindergarten. We selected only those schools that had at least five children included in the national ECLS-K study and that had no missing data on the school-level variables of interest. Our resulting data set had $N = 7{,}377$ children from $J = 569$ schools. Missing data for some of the student-level variables reduced our level-1 data set further to $N = 6{,}539$ for the multilevel analyses (note that when creating the MDM file for HLMv8, we also selected the option for deleting missing data when making the MDM file to facilitate model comparisons across a consistent sample). The average number of children per school in this sample was $M = 11.5$ ($SD = 4.2$).

The response variables included in the data file are PROFMATH based on Table 7.1, ORDPROF (a 5-level ordinal response variable which combined categories 0 and 1 shown in Table 7.1) and PROF45 (a dichotomization of PROFMATH, where students in the highest proficiency categories of 4 and 5 were considered "success"). Overall, 77.3% of the children in our sample had attained proficiency in levels 4 or 5 by the end of the first grade.

Explanatory variables at the student level included gender (MALE; 1 = male, 0 = female) and a count of the number of family risk factors a child had experienced prior to entering kindergarten (NUMRISK; 0 to 4 as described below). For our sample, the distribution of gender was approximately even (49% male). Earlier studies suggested that children entering kindergarten from families with particular characteristics (living in a single parent household, living in a family that receives welfare payments or food stamps, having a mother with less than a high school education, or having parents whose primary language is not English) tended to be at risk for poor performance (Zill & West, 2001). The NUMRISK variable is a sum of the number of these four risk characteristics present for each child in the analytic sample and ranges from 0 to 4. The mean number of risks for our sample was .39, with a standard deviation of .72. Table 7.3 contains

TABLE 7.3 Student-Level Descriptive Statistics for the Analytic Sample, N = 6,539

Proficiency level (*PROFMATH*)	**0**	**1**	**2**	**3**	**4**	**5**	**Total**
Student level	N = 9	N = 53	N = 192	N = 1,231	N = 3,188	N = 1,866	N = 6,539
Cumulative proportion[a]	.14%	.95%	3.88%	22.71%	71.46%	100%	—
Gender (MALE) % Male	33.33%	52.83%	48.44%	49.39%	46.55%	54.39%	49.4%
Number of risks (NUMRISKS) *M* (*SD*)	1.11 (1.17)	1.15 (1.06)	.90 (.97)	.55 (.82)	.40 (.72)	.19 (.50)	.39 (.72)
Proficiency level (*ORDPROF*[b])		**1**	**2**	**3**	**4**	**5**	**Total**
Student level		N = 62	N = 192	N = 1,231	N = 3,188	N = 1,866	N = 6,539
Cumulative proportion[c]		.95%	3.88%	22.71%	71.46%	100%	—
Gender (MALE) % Male		50.41%	48.44%	49.39%	46.55%	54.39%	49.4%
Number of risks (NUMRISKS) *M* (*SD*)		1.15 (1.07)	.90 (.97)	.55 (.82)	.40 (.72)	.19 (.50)	.39 (.72)

[a] P($R \leq$ cat. j), where R = proficiency level (response variable), and j = 0, 1, 2, . . . 5 possible values for proficiency.
[b] Categories 0 and 1 combined.
[c] P($R \leq$ cat. j), where R = proficiency level (response variable), and j = 1, 2, . . . 5 possible values for proficiency.

TABLE 7.4 School-Level Descriptive Statistics for the Analytic Sample, J = 569

Variable	Mean	*SD*	Minimum	Maximum
Neighborhood problems (NBHOODCLIM)	1.73	2.49	0.00	12.00
Private (PUBPRIV2)[a]	18%	—	—	—

[a] Public = 0; Private = 1.

descriptive information on our analytic sample regarding the dependent and predictor variables at the child level.

We used two school-level variables to demonstrate the development and interpretation of GLMMs. First, we used factor analysis to create a neighborhood climate variable, NBHOODCLIM, which represents a composite of principals' perceptions of the severity of six specific problems in the vicinity of their school, including extent of litter, drug activity and gang activity in the area, crime, violence, and existence of vacant lots or vacant homes. The resulting scale scores for NBHOODCLIM ranged from 0 to 12, with $M = 1.73$ ($SD = 2.49$). School sector was recorded as either public or private (PUBPRIV2; public = 0, private = 1) and used as a school-level predictor. In our sample, 18% of the schools were private schools, and no distinction was made among different types of private schools. Table 7.4 contains descriptive information on our analytic sample regarding the predictor variables at the school level.

We merged two data files corresponding to the levels of analysis of the data. The level-1 data file contained the child-level outcomes, MALE and NUMRISKS, with the original six-category ordinal numeracy proficiency scores (PROFMATH) recorded as a dichotomous (PROF45) variable (categories 0 through 3 versus categories 4 and 5 combined) and a reduced ordinal variable (ORDPROF; 5 categories with levels 0 and 1 combined due to small sample size for category 0), for each child at the end of first grade. The level-2 data file contained the school-level characteristics used for our demonstration, NBHOODCLIM and PUBPRIV2. The merged data file is provided on our website. We present data applications for 2-level models although 3-level models for dichotomous and ordinal outcomes are possible extensions to the 2-level configuration.

Model Building Process

In our dichotomous-outcome examples, success refers to a child having achieved proficiency by performing at levels 4 or 5 on the early numeracy measure. For the ordinal analyses, we used the ordinal variable ORDPROF with responses ranging from 1 through 5. We then ran a series of multilevel logistic and ordinal models on these data. The first five analyses are examples of multilevel logistic models for dichotomous data and used PROF45 as the outcome. The next two analyses used multilevel ordinal regression based on the proportional odds model and use ORDPROF as the outcome. We also demonstrate the test for the assumption of proportional odds. The collection of models examined for the analyses presented here include:

- logistic null (empty) model (no predictors at either level);
- logistic random coefficients model with MALE and NUMRISKS as level-1 predictors of proficiency;
- logistic reduced model, allowing only the intercepts to vary randomly across schools (the slopes for MALE and NUMRISKS were both fixed);
- logistic reduced model, with only NUMRISKS as a fixed level-1 predictor (MALE was removed from the model);
- logistic contextual model with NBHOODCLIM and PUBPRIV2 as school-level predictors of the intercepts;
- ordinal null (empty) model; and
- ordinal contextual model with NUMRISKS as a fixed level-1 predictor and NBHOODCLIM and PUBPRIV2 as school-level predictors of the intercepts.

DATA EXAMPLES FOR DICHOTOMOUS OUTCOMES

The Null Model

The dichotomous outcomes model uses PROF45 as the response variable, based on a child reaching proficiency on the arithmetic assessment in either category 4 or category 5: being able to solve simple addition and subtraction problems or simple multiplication and division problems (see Table 7.1). In multilevel logistic regression, the model constructed for the outcome is linear based on log-odds (the logits, or the natural log of the odds) and includes a random effect for the school. The null model, that is, the model with no predictors at either the child or school level, provides an overall estimate of the likelihood of proficiency for this sample, as well as information about the variability in the probability of proficiency between schools. The following equations define each level of the null model:

$$\begin{aligned} \text{Level 1:} \quad & \eta_{ij} = \beta_{0j} \\ \text{Level 2:} \quad & \beta_{0j} = \gamma_{00} + u_{0j} \end{aligned} \tag{7.11}$$

Combining equations across levels, the null model can be written: $\eta_{ij} = \gamma_{00} + u_{0j}$. With no explanatory variables in the model, the level 1 β_{0j} represents the log-odds of success for children in the *j*th school. These school-average, or unit-average (Raudenbush & Bryk, 2002), log-odds vary across schools, with $\text{var}(u_{0j}) = \tau_{00}$, where γ_{00} represents the average log-odds of attaining proficiency across all schools. The null model can be used to estimate the probability of students' attaining proficiency for a school with a value of zero for the random effect u_{0j}, that is, for a "typical" school. To find this estimated probability, Equation 7.7 can be applied as shown below.

Null Model Results

Adaptive Quadrature (AQ) estimation and HLMv8.0 was used for the multilevel logistic analyses. Results of the (unit-specific) null model in terms of estimated logits and corresponding odds ratios are provided in Table 7.5. Thus, the predicted logit for a typical school (referring to a school with a random effect of 0) in this sample is $\gamma_{00} = 1.36$ ($SE = .05$). The approach for significance testing for fixed and random effects corresponds to the standard HLM; thus, this average logit is statistically different from zero, $t(568) = 27.46$, $p < .001$. In addition, there was considerable variability around the intercepts for this collection of schools, $\tau_{00} = .67$ ($SE = .08$), *Wald's* $Z = 8.38$ (sig.). The column labeled "Odds Ratio" is the exponentiated value of the parameter estimate for each predictor in the model although for the intercept this exponentiated value is simply the odds (when all predictors and random effects are 0). Estimated odds of attaining proficiency for students within a typical school is $\exp(1.36) = 3.88$, and thus the estimated probability of attaining proficiency for these students is

$$\frac{\exp(1.36)}{1+\exp(1.36)} = \frac{3.88}{4.88} = .80.$$

This estimate is close to the overall proportion of children (ignoring school variability) who achieved proficiency in categories 4 or 5 for this sample (i.e.,

TABLE 7.5 Multilevel Logistic Null Model Results

Fixed Effects	Coefficient (*SE*)	Odds Ratio	*t*(*df*)	*p*
Model for numeracy proficiency (β_0)				
Intercept (γ_{00})	1.36 (.05)	3.88	27.46 (568)	<0.001
Random effects (var. components)	**Variance (*SE*)**		**Wald's *Z***	
Intercept variance (τ_{00}) (*SE*)	.67 (.08)		8.38 (sig.)	

Note: Deviance, under AQ(11) estimation: 18,804.677, parameters = 2; AQ(11) refers to 11 quadrature points

(3,188 + 1,866)/6,539 = .77; see descriptive statistics in Table 7.3). Under the model's assumption that the residuals (i.e., random effects) at level 2 follow a normal distribution, we can estimate that about 95% of schools in this sample have logits between 1.36 ± 1.96 * $(.67)^{1/2}$, or (–.24, 2.96). Again applying Equation 7.7, this corresponds to estimated probabilities between .43 and .95; this is the expected range of the proportions of students achieving proficiency across these schools. Thus, some schools have a considerable percentage of students who did not reach proficiency in early numeracy by the end of first grade.

Intraclass Correlation Coefficient

An appealing characteristic of nearly all multilevel models involves the determination and interpretation of the intraclass correlation coefficient (ICC) to assess degree of clustering in the data. Snijders and Bosker (2012) discuss two approaches to estimating an ICC for dichotomous outcomes. The first involves fitting a multilevel linear probability model by specifying an empty model and applying standard HLM to the dichotomous (0, 1) outcomes. Using this linear-based approach for our data, the

$$\text{ICC} = \frac{\tau_{00}}{\tau_{00} + \sigma^2}$$

and is approximately .11. The second method utilizes the variability estimated through the null logistic model and assumes that the outcome, Y, is a dichotomization of an unknown latent continuous variable, $\hat{Y}$, with a level-1 residual that follows the logistic distribution (Snijders & Bosker, 2012). The mean and variance of the logistic distribution are 0 and 3.29 (i.e., $\pi^2/3$), respectively (Evans et al., 2000). Accordingly, the ICC is scaled to the logit metric, and based on this approach is

$$\frac{\tau_{00}}{\tau_{00} + 3.29} = .17,$$

higher than the linear-based approach. In either case, the variability attributed to schools was larger than zero, indicating that a multilevel analysis is appropriate due to a strong clustering effect. That is, reaching proficiency is more similar within schools than an independence model (i.e., one ignoring the clustering effect) would assume. Readers are referred to Browne et al. (2005) for extensions of these methods when overdispersion may exist in the response variable.

Logistic Random Intercept and Random Coefficient Models

As in the standard HLM, level-1 coefficients in GLMM can be specified as fixed, randomly varying, or non-randomly varying. Thus the strategies used

TABLE 7.6 Logistic Random Coefficients Model with NUMRISKS, No Slope Variation

Fixed Effects	Coefficient (*SE*)	Odds Ratio	*t*(*df*)	*p*
Model for the intercepts (β_0)				
Intercept (γ_{00})	1.53 (.05)	4.62	30.52 (568)	<0.001
Model for NUMRISKS slope (β_1)				
Intercept (γ_{10})	–.43 (.04)	.65	–9.82 (5,969)	<0.001
Random effects (var. components)	**Variance (*SE*)**		**Wald's *Z***	
Intercept variance (τ_{00}) (*SE*)	.50 (.07)		7.14 (sig.)	

Note: Deviance, under AQ(11) estimation: 18,804.677, parameters = 2

to model the logit for proficiency are familiar and straightforward. We selected two child-level predictors, gender (MALE) and the number of family risk characteristics (NUMRISKS), to determine how these factors might relate to proficiency, PROF45. Table 7.6 contains the estimated coefficients and corresponding odds ratios for the resulting random intercepts model based on results from a series of random coefficient models fit to these data. For those models, we first specified MALE and NUMRISKS as child-level predictors of success with respect to proficiency. No between-schools variability in the slopes was found for either explanatory variable ($p > .50$ for both τ_{01} and τ_{02}), and the slope for MALE was not statistically different from zero. Thus, the results in Table 7.6 were derived from the following reduced model:

$$\begin{aligned} \text{Level 1:} \quad & \eta_{ij} = \beta_{0j} + \beta_{1j} \cdot NUMRISKS \\ \text{Level 2:} \quad & \beta_{0j} = \gamma_{00} + u_{0j} \\ & \beta_{1j} = \gamma_{10} \end{aligned} \tag{7.12}$$

The combined model is thus: $\eta_{ij} = \gamma_{00} + \gamma_{10} \cdot NUMRISKS_{ij} + u_{0j}$.

In our analysis, the estimated logit for a child from a typical school ($u_{0j} = 0$) and with no identified family risk factors (NUMRISKS = 0) is $\gamma_{00} = 1.53$ ($SE = .05$), which is statistically different from zero, $t(568) = 30.52$, $p < .001$. The value for the logit suggests that for a child with no family risks and from a typical school, the estimated odds of attaining proficiency is $\exp(1.53) = 4.62$ (note that the column labeled *Odds Ratio* provides an *OR* for each predictor but an odds for the intercept). The corresponding probability of attaining proficiency for a child with these characteristics is

$$\hat{p} = \frac{\exp(1.53)}{1 + \exp(1.53)} = .82.$$

The negative effect for NUMRISKS, $\gamma_{10} = -.43$ ($SE = .04$) indicates that as the number of risk factors increases by one, the estimated logit *decreases* by .43 units; this effect is statistically different from zero, $t(5{,}969) = -9.82$, $p < .001$. Exponentiating this slope, $\exp(\gamma_{10})$ yields $\exp(-.43)$, or 0.65. This *OR* value represents the effect of NUMRISKS on the odds of proficiency: As the number of risks increases by one, the odds of attaining proficiency is reduced by about one-third. For example, a child with NUMRISKS = 1 has a predicted logit $\hat{\eta}_{ij} = 1.53 + (-.43) * 1 = 1.10$. Exponentiating this value gives $\exp(1.10) = 3.00$, the odds of a child being proficient with a single NUMRISK. While the probability of being proficient with a single NUMRISK is still somewhat high $(3.00/(1 + 3.00) = .75)$, we saw while interpreting the intercept that the odds of attaining proficiency when NUMRISKS = 0 is much larger: $\exp(1.53) = 4.62$. The comparison of these two odds forms the *Odds Ratio* (see Equation 7.3):

$$\frac{Odds(NUMRISKS = 1)}{Odds(NUMRISKS = 0)} = \frac{3.00}{4.62} = .65,$$

which is $\exp(-.43)$. Thus, there is a 35% decrease in the odds as the number of risks increases by one (i.e., 100% * (OR – 1) = –35%). For a child with NUMRISKS = 4, the predicted logit is $\hat{\eta}_{ij} = 1.53 + (-.43) * 4 = -.19$, yielding a corresponding odds of $\exp(-.19) = .83$ and an estimated probability of attaining proficiency as $(.83/(1 + .83)) = .45$. Overall, children with the greatest number of family risks were at risk of not attaining proficiency in early numeracy (as defined here) by the end of the first grade.

Logistic Contextual Models

Although the previous models clarified the role of family risk on likelihood of proficiency in early numeracy and also indicated that gender was not a factor in numeracy proficiency at the end of first grade for this sample, the results in Table 7.6 also pointed to considerable residual variability in the intercepts across schools after adjusting for NUMRISKS, $\tau_{00} = .50$ ($SE = .07$), *Wald's* $Z = 7.14$ (significant at .05). The next set of analyses we conducted included several school-level variables to try to account for this residual variation. We included the measures of neighborhood climate (NBHOODCLIM) and school sector (PUBPRIV2) as predictors of the level-1 intercepts and the level-1 slopes for NUMRISKS. Since no between-school variability in the logit was found for the effect of NUMRISKS in the previous analysis, we anticipated no contribution from these level-2 variables as predictors of the NUMRISKS slopes; however, we included these effects here to illustrate the interpretation of cross-level interactions in a multilevel

TABLE 7.7 Results for the Contextual Logistic Models: Model I Uses Level 2a for the School-Level Model; Model II Uses Level 2b

Fixed effects	Model I[a] Coefficient (SE)	Model I[a] Odds ratio	Model II[b] Coefficient (SE)	Model II[b] Odds ratio
Model for the intercepts (β_0)				
Intercept (γ_{00})	1.60 (.06)	4.93**	1.57 (.06)	4.83**
NBHOODCLIM (γ_{01})	–.12 (.02)	.89**	–.11 (.02)	.90**
PUBPRIV2[c] (γ_{02})	.44 (.12)	1.55**	.50 (.12)	1.64**
Model for NUMRISKS slope (β_1)				
Intercept (γ_{10})	–.38 (.06)	.68**	–.34 (.04)	.71**
NBHOODCLIM (γ_{11})	.01 (.01)	1.01	—	—
PUBPRIV2[c] (γ_{12})	.30 (.22)	1.35	—	—
Random effects (var. components)	**Variance (se)**		**Variance (se)**	
Intercept variance (τ_{00}) (*SE*)	.40 (.06)**		.40 (.06)**	

** $p < .001$

[a] Deviance, under AQ(11): 18651.54, params. = 7

[b] Deviance, under AQ(11): 18653.98, params. = 5

[c] For public schools, PUBPRIV2 = 0; PUBPRIV2 = 1 for private schools.

logistic model. Table 7.7 contains the results of two analyses: The first included NBHOODCLIM and PUBPRIV2 as predictors of both the intercepts and the slopes (using Equation 7.13 with Level 1 and Level 2a models) and the second analysis was based on a reduced contextual model where only the intercepts are predicted from NBHOODCLIM and PUBPRIV2 (using Level 1 and Level 2b models from Equation 7.13). In both analyses, slope variability between-schools was constrained to zero, as supported through the previous analyses.

$$\begin{aligned}
\text{Level 1:}\quad & \eta_{ij} = \beta_{0j} + \beta_{1j} \cdot NUMRISKS \\
\text{Level 2a:}\quad & \beta_{0j} = \gamma_{00} + \gamma_{01} \cdot NBHOODCLIM + \gamma_{02} \cdot PUBPRIV2 + u_{0j} \\
& \beta_{1j} = \gamma_{10} + \gamma_{11} \cdot NBHOODCLIM + \gamma_{12} \cdot PUBPRIV2 \\
\text{Level 2b:}\quad & \beta_{0j} = \gamma_{00} + \gamma_{01} \cdot NBHOODCLIM + \gamma_{02} \cdot PUBPRIV2 + u_{0j} \\
& \beta_{1j} = \gamma_{10}
\end{aligned} \tag{7.13}$$

In combined form, Contextual Model I (Level 1 with Level 2a from Equation 7.13) can be written as:

$$\begin{aligned}
\eta_{ij} &= \gamma_{00} + \gamma_{01}(NBHOODCLIM)_j + \gamma_{02}(PUBPRIV2) + \gamma_{10}(NUMRISKS) \\
&\quad + \gamma_{11}(NBHOODCLIM * NUMRISKS) + \gamma_{12}(PUBPRIV2 * NUMRISKS) + u_{0j}
\end{aligned} \tag{7.14}$$

Model I is a random intercepts and slopes-as-outcomes logistic regression model. Similarly, Contextual Model II (Level 1 with Level 2b, above) can be written in combined form by eliminating the cross-level interaction terms from Equation 7.14 and is a random intercept logistic regression model. The parameter estimates for these contextual models are in log-odds (logits). For both models, the results in Table 7.7 show that the main effects of NBHOODCLIM (γ_{01}) and PUBPRIV2 (γ_{02}) were statistically different from zero, and as expected for Model I, the cross-level interactions between both of these variables and NUMRISKS were not statistically significant (γ_{11} and γ_{12}, respectively). Therefore, their effects on the NUMRISKS slope are close to zero, and Model I and Model II are substantively similar. We begin with an interpretation of Model II, and then, to illustrate interpretation of cross-level interactions, we refer to Model I despite a lack of statistical significance for these cross-level interaction effects.

From Table 7.7, the intercept for Model II is $\gamma_{00} = 1.57$ ($p < .001$). This value represents the expected log-odds for a child who has no family risks (NUMRISKS = 0) and is in a public school (PUBPRIV2 = 0) and in a neighborhood with no serious crimes or other problems (NBHOODCLIM = 0) and with corresponding random effect of 0. Thus, the estimated odds for a child with these individual and school characteristics, exp(1.57), is 4.83, which implies a predicted probability of proficiency of .83 (from Equation 7.7). To assist with clarifying the results of the multilevel logistic model, this odds value of 4.83 can be called the "referent odds," that is, the prediction for the odds in the case when all predictors (school- and child-level) and random effects are zero. According to these referent odds, it is more likely for a child with no family risks in a public school with no neighborhood problems to be proficient in early numeracy than not.

According to the results for Model II, the effect of NBHOODCLIM on the log-odds for the intercept is negative and statistically different from zero, controlling for both NUMRISKS and PUBPRIV2, $\gamma_{01} = -.11$ ($p < .001$). This effect suggests that when PUBPRIV2 and NUMRISKS are held constant, the odds of proficiency is expected to be lowered by a factor of .90 as the severity of neighborhood problems increases by one unit [$OR = \exp(-.11) = .90$]. The effect of attending private school (PUBPRIV2) on proficiency, holding NUMRISKS and NBHOODCLIM constant, is positive and statistically significant, $\gamma_{02} = .50$ ($p < .001$). Thus, controlling for NBHOODCLIM and NUMRISKS the odds of proficiency for children in private schools is 1.64 times greater than for children in public schools [$OR = \exp(.50) = 1.64$]. Overall, holding NBHOODCLIM and PUBPRIV constant, NUMRISKS has a negative and statistically significant effect on the log-odds for proficiency, $\gamma_{10} = -.34$ ($p < .001$), with an $OR = \exp(-.34) = .71$. With each one unit increase in the number of family risks, a child's odds of reaching proficiency in early numeracy decreases by nearly 29% (that is, 100% * ($OR - 1$) = 100% * (.71 – 1) = –29%), holding all other effects constant.

As with all multilevel models, variability in the level-1 slopes represents the between-schools variability in the relationships between level-1 predictors and the proficiency outcome. In both Model I and Model II, the child-level effect of NUMRISKS was fixed since it was not detected to vary between schools in the earlier analyses. However, the results of Model II show that there is considerable residual variability in the school intercepts (log-odds of proficiency when NUMRISKS is zero) even after accounting for NBHOODCLIM and PUBPRIV2, with $\tau_{00} = .40$ ($p < .001$). Although we do not pursue this analysis further here, additional child- or school-level predictors might help to explain some of this variability.

Now we turn to the interpretation of Model I, which allows for cross-level interactions between each of the school-level predictors and NUMRISKS, the child-level predictor, on proficiency. Although neither of the predictors of the NUMRISKS slope was statistically significant, this example illustrates the challenges in interpreting multilevel models where effects of level-1 explanatory variables may vary randomly or non-randomly. We will review these cross-level effects for demonstration purposes, although neither effect is statistically meaningful.

In Model I, NBHOODCLIM and PUBPRIV2 were included as predictors of the intercepts across schools as well as of the effect of a child's number of family risks characteristics (NUMRISKS) on child-level proficiency outcomes across schools. Thus, the intercepts and the slopes from the level-1 equations are treated as outcomes in a regression model at level 2. The fixed effects in the equation predicting the slopes are referred to as cross-level interactions. They represent the effect of a school-level variable on a child-level predictor of proficiency. In essence, these effects tell us how the relationship between NUMRISKS and the log-odds tends to vary across schools, based on the school variables NBHOODCLIM and PUBPRIV2. We focus on these cross-level interactions since interpretation of main effects should only be undertaken cautiously when interaction is present (although, as noted earlier, that is not the case here).

For children in a public school (PUBPRIV2 = 0) with no crime or other problems identified in their neighborhood (NBHOODCLIM = 0), the effect of NUMRISKS on proficiency is negative and statistically different from zero ($\gamma_{10} = -.38$ ($p < .001$)). Across schools, the effect of NUMRISKS is not moderated by NBHOODCLIM; note that this effect is very close to zero, $\gamma_{11} = .01$ (n.s.) and that the associated odds ratio is nearly one. The effect of attending a private school on the effect of NUMRISKS is also not statistically significant, but it is in the expected direction; the harmful effect of NUMRISKS on proficiency is lessened for private school students relative to those in public schools, $\gamma_{12} = .30$ (n.s.). That is, for public school students with NBHOODCLIM = 0, the effect of NUMRISKS on the log-odds is –.38; this becomes –.38 + .30 = –.08 for children in private schools. The odds ratio for PUBPRIV2, $\exp(\gamma_{12}) = 1.35$, indicates that for a specific value of NUMRISKS and holding NBHOODCLIM constant, the odds of achieving proficiency are 1.35 times greater for private-school children than for children in public schools.

Note that the residual variance in the level-1 intercepts is unchanged between Model I ($\tau_{00} = .40$) and Model II ($\tau_{00} = .40$), and both are significantly different from zero ($p < .01$). Additional level-2 predictors of the intercept may be useful in decreasing this variability.

Model Comparisons

Model estimates were based on AQ thus deviances for nested models can be compared directly. The deviances for the logistic models are provided in a note below each table and summarized below:

Logistic Model	Deviance	Parameters Estimated
Null	18,804.68	2
Random Coeffs	18,713.09	3
Contextual Model I	18,651.54	7
Contextual Model II	18,653.98	5

In general, a smaller deviance is preferred, and for nested models the difference in deviance can be compared statistically. When models are nested, the difference in the deviances follows a chi-square distribution with degrees of freedom determined by the difference in number of estimated parameters. According to the deviance estimates for our four models, the random coefficients model with NUMRISKS as a single level-1 predictor provides a better fit to the model than the null model, $\chi_1^2 = (18{,}804.68 - 18{,}713.09) = 91.59$, $p < .01$. (Note that the coefficient for NUMRISKS was treated as fixed). Similarly, both Contextual Models I and II provide a significantly smaller deviance than the null model. However, comparing Models I and II, the difference in deviances is negligible, $\chi_2^2 = 2.44$, $p > .05$. With fewer parameters estimated, Model II is reasonably a better choice for estimating PROF45 among these alternatives.

Probability Predictions

A convenient way to clarify the interpretation of GLMM's is to use the model to calculate estimated probabilities based on varying levels of the

TABLE 7.8 Model Predictions Based on Logistic Model II

Prediction			Logits	Odds	Estimated Probability
PUBPRIV2 = 0a	NUMRISKS = 0	NBHOODCLIM = 0	1.57	4.83	.83
	NUMRISKS = 2	NBHOODCLIM = 6	.24	1.28	.56
	NUMRISKS = 4	NBHOODCLIM = 12	−1.09	.34	.25
PUBPRIV2 = 1	NUMRISKS = 0	NBHOODCLIM = 0	2.07	7.93	.89
	NUMRISKS = 2	NBHOODCLIM = 6	.74	2.10	.68
	NUMRISKS = 4	NBHOODCLIM = 12	−.59	.55	.36

[a] For public schools, PUBPRIV2 = 0; PUBPRIV2 = 1 for private schools.

explanatory variables—while assuming random effects of 0, that is, for a typical cluster. Table 7.8 contains predictions for a selection of values of the explanatory variables based on estimates in Model II estimates. Variable effects are often more challenging to interpret for GLMMs than for standard HLMs due to the logit metric and the additional required steps of transforming log-odds predictions to odds and then to probability. Looking at specific predictions can assist in interpretation. Although the few predictions provided in Table 7.8 are helpful for understanding the direction of effects in the relatively simple model presented here, complex designs may warrant a more detailed look at the results of cross-level interactions. Tate (2004) presents some useful strategies for interpreting cross-level effects in multilevel logistic models based on approaches used in investigating simple effects for factorial analysis of variance designs.

Entries in Table 7.8 correspond with the results presented earlier. As children were exposed to greater numbers of family risks and increased severity of neighborhood problems, their likelihood of becoming proficient in early numeracy by the end of first grade diminished. Further, the likelihood of proficiency was lower for children in public schools.

DATA EXAMPLES FOR ORDINAL OUTCOMES

Ordinal Null Model and Threshold Interpretation

In the previous multilevel logistic model, the outcome, PROF45, was dichotomized from an underlying 6-level ordinal variable, PROFMATH (described in Table 7.1), and predictions based on the model could be used to estimate the probability of a child being "proficient" in early mathematics, as defined by having mastered at least through Proficiency Level 4. As with many variables in education and the behavioral sciences, dichotomizing an otherwise ordinal variable in this manner can result in a loss of information regarding individual- or context-level factors that overall might be related to a child's progression through these proficiency categories. The multilevel ordinal proportional odds model is based on the fuller ordinal progression of the response variable. The estimates from the resulting model summarize the effect of explanatory variables on the odds of being at or below any given proficiency category. Thus, the model provides for a simultaneous fitting of a series of multilevel logistic models corresponding to the different cumulative comparisons as described in Table 7.2 and under the assumption of proportional odds, which implies that the effect of explanatory variables across these separate simultaneous logistic models are constant and do not vary by level of response outcome.

The null model, with no level-1 or level-2 predictors, provides a convenient starting point for understanding and interpreting multilevel ordinal models. Results of the null model fit to the 5-level ordinal responses for early mathematics proficiency ORDPROF are presented in Table 7.9. We used adaptive quadrature estimation for the ordinal models via the R package mixor (Archer et al., 2018), and unit-specific estimates are presented. As with the logistic model, estimates are in log-odds and predictions based on the model can be transformed to probabilities using Equation 7.7. The

TABLE 7.9 Multilevel Ordinal Null Model (Model I) and Ordinal Contextual Model (Model II) for DV = ORDPROF, Using R mixor

	Null—Model I[a]		Contextual—Model II[b]	
Fixed Effects	**Coefficient (*SE*)**	**Odds Ratio**	**Coefficient (*SE*)**	**Odds Ratio**
Model for the intercepts (β_0)				
Intercept (γ_{00})	−4.898 (.120)	.0075**	−5.228 (.131)	.005**
NBHOODCLIM (γ_{01})			.123 (.015)	1.131**
PUBPRIV2[c] (γ_{02})			−.431 (.090)	.650**
Model for NUMRISKS slope (β_1)				
Intercept (γ_{10})			.399 (.039)	1.490**
For thresholds:				
δ_2	1.468 (.106)	4.341**	1.479 (.108)	4.389**
δ_3	3.580 (.113)	35.874**	3.610 (.116)	36.966**
δ_4	5.986 (.122)	397.820**	6.027 (.126)	414.470**
Random Effects (var. components)	Variance		Variance	
Intercept variance (τ_{00}) (*SE*)	.598** (.061)		.317** (.042)	

* $p < .05$; ** $p < .01$

[a] Deviance (AQ11) 14,903.11 (5 params.)

[b] Deviance (AQ11) 14,650.05 (8 params.)

[c] For public schools, PUBPRIV2 = 0; PUBPRIV2 = 1 for private schools.

probability being estimated in ordinal models under the assumption of proportional odds is $P(R_{ij} \leq \text{category } k|\underline{x})$, that is, the probability, conditional on the set of explanatory variables $\underline{x}$, that the proficiency level or response R for the ith student in the jth school is at or below category k, for $k = 1, \ldots, 5$. Since there are no predictors in the null model, these cumulative probabilities correspond to overall estimates across all students in the sample at each cut point k.

ORDPROF is a $K = 5$ level response variable; thus, there are $K - 1 = 4$ cumulative "splits" to the data, as described earlier in the section on the multilevel proportional odds model for ordinal outcomes. The results of the null model analysis presented in Table 7.9 include the series of δ estimates corresponding to the natural cumulative splits represented by the proportional odds model. These δ values can be used to estimate the thresholds or intercepts for each of the four underlying cumulative analyses in the overall ordinal model. Based on Equation 7.10 the null model can be written as follows:

$$\text{Level 1:} \quad \eta_{kij} = \ln(Y'_{kij}) = \ln\left(\frac{P(R_{ij} \leq k)}{P(R_{ij} > k)}\right) = \beta_{0j} + \sum\nolimits_{k=2}^{K-1} D_{kij}\delta_k \tag{7.14}$$

$$\text{Level 2:} \quad \beta_{0j} = \gamma_{00} + u_{0j}$$

Results indicate that the estimated log-odds of being at or below Proficiency Level 1 (the lowest category for ORDPROF) for a child in this sample is $\gamma_{00} = -4.898$, corresponding to an estimated odds of $\exp(-4.898) = .0075$ and estimated probability $P(R_{ij} \leq \text{category } 1)$ of $.0075/(1 + .0075) = .00744$ or approximately .74% (see Equation 7.7). This is holding the random effect at 0 or for a child from a "typical" school. Recall from the descriptive statistics in Table 7.3 that only 62 of the 6,539 students were in ORDPROF Proficiency Level 1 for this sample (.95%); thus the estimate from the model matches the actual data fairly well.

To estimate the log-odds, odds, and probability for the next cumulative comparison, $P(R_{ij} \leq \text{category } 2)$, we used the threshold estimate for $\delta_2 = 1.47$ in addition to γ_{00}. In general, the subscript refers to the specific cumulative comparison being made, not to the value of the response category. In this case, the first comparison is for $P(R_{ij} \leq \text{category } 1)$; the second comparison is for $P(R_{ij} \leq \text{category } 2)$. Similarly, the remaining estimates for the log-odds, odds, and probabilities can be determined for all four cumulative comparisons; these are provided in Table 7.10, where they also are compared with the actual proportions derived for our sample.

The cumulative probability estimates from the null model do appear to match the observed cumulative proportions. However, results of the analysis indicated a substantial amount of variation between schools in the level-1 intercepts, $\tau_{00} = .598$ ($SE = .061$), *Wald's* $Z = 9.083$ (sig.) revealing

TABLE 7.10 Model Predictions Based on the Ordinal Empty Model for ORDPROF

Model	Estimates for log odds	Estimates for odds	Estimates for $P(R_{ij} \leq \text{cat. } k)$	Actual Probability $P(R_{ij} \leq \text{cat. } k)$
$\log(Y'_{1ij}) = \beta_0$	−4.898	.0075	.0074	.0095
$\log(Y'_{2ij}) = \beta_0 + \delta_2$	−4.898 + 1.468 = −3.430	.0324	.0314	.0388
$\log(Y'_{3ij}) = \beta_0 + \delta_3$	−4.898 + 3.581= −1.317	.2678	.2113	.2271
$\log(Y'_{4ij}) = \beta_0 + \delta_4$	−4.898 + 5.986 = 1.088	2.9672	.7479	.7146

unaccounted for heterogeneity between schools in the probability of children being at or below a given proficiency level. Our next ordinal model was used to investigate how the individual- and school-level variables of NUMRISKS, NBHOODCLIM, and PUBPRIV2 may be associated with the probability of being at or below proficiency for any response category, k.

Intraclass Correlation Coefficient

The ICC for ordinal data can be defined in the same way as for dichotomous outcomes (Snijders & Bosker, 2012). We apply the second definition presented earlier:

$$\text{ICC} = \frac{\tau_{00}}{\tau_{00} + 3.29} = .154,$$

where 3.29 corresponds to $(\pi^2/3)$, the variance of the logistic distribution. This value for the ICC supports the utility of an ordinal multilevel analysis due to non-independence of the data.

Contextual Ordinal Model

The right-hand side of Table 7.9 provides the estimates for a contextual model of cumulative proficiency as follows:

$$\begin{aligned} \text{Level 1:} \quad & \eta_{kij} = \ln(Y'_{kij}) = \ln\left(\frac{P(R_{ij} \leq k)}{P(R_{ij} > k)}\right) = \beta_{0j} + \beta_{1j} NUMRISKS_{ij} \\ & + \sum_{k=2}^{K-1} D_{kij}\delta_k \\ \text{Level 2:} \quad & \beta_{0j} = \gamma_{00} + \gamma_{01} NBHOODCLIM_j + \gamma_{02} PUBPRIV2_j + u_{0j} \\ & \beta_{1j} = \gamma_{10} \end{aligned} \quad . \quad (7.15)$$

These equations represent a random intercepts model. Variability in the effect of NUMRISKS across schools was constrained to zero for purposes of this chapter and to correspond in structure to earlier models, but this constraint could easily be relaxed and tested.

With five parameters, the deviance of the null model was 14,903.11; the deviance for the more complex model with eight parameters was 14,650.05. The difference in deviances was statistically significant, $\chi^2(3) = 253.06$, $p < .001$. Thus the complex model has more explanatory value.

Tests of significance for the fixed effects and the variance components are similar methodologically to the logistic analyses presented earlier. That is, *t*-tests are used to test the fixed effects, and *Wald's Z*-tests are applied to the variance components. Although the between-schools variance was reduced by approximately 47% based on the predictors in the contextual model,

$$\left(\frac{\tau_{00}(\textit{empty model}) - \tau_{00}(\textit{contextual model})}{\tau_{00}(\textit{empty model})} = \frac{.598 - .317}{.598} = .47 \right),$$

there was still considerable residual variance remaining, $\tau_{00} = .317$ ($SE = .042$), *Wald's* $Z = 7.548$ (sig.).

Following Table 7.9, the effect of NUMRISKS on the cumulative log-odds of proficiency was positive and statistically different from zero, $\gamma_{10} = .399$, $p < .01$. Recall that in the cumulative model, the estimated probability is $P(R_{ij} \leq \text{cat. } k)$; thus, as the number of family risks increased, the probability of being at or below a given category, *rather than beyond that category*, tended to increase. Children with greater family risks were more likely to be at or below a given category. Although the parameter estimates have opposite signs between the logistic and cumulative logistic model, the interpretation is consistent and in the same direction.

School effects on the cumulative probability estimates were statistically significant and also in the same interpretive direction as the earlier logistic analysis. Children in schools with higher perceived severity of crime and related neighborhood issues were *more* likely to be at or below a given proficiency level, $\gamma_{01} = .123$, $p < .01$. Children in private schools were *less* likely to be at or below a given proficiency level, $\gamma_{02} = -.431$, $p < .01$; that is, by the rule of complements, children in private schools were more likely than their public school peers to be *beyond* a particular proficiency level.

To aid in interpretation of the model estimates, the predicted log-odds, odds, and probability can be determined for different values of the predictor variables. Each combination of predictor variables would provide four probability estimates for the cumulative set ranging from $P(R_{ij} \leq \text{cat. } 1)$ to $P(R_{ij} \leq \text{cat. } 4)$. Note that for the highest ordinal response category,

TABLE 7.11 Model Predictions for ORDPROF Based on Ordinal Contextual Model (Model II) for Selected Values of the Predictors

School Type	Number of Family Risks	Severity of Neighborhood Problems	$P(R_{ij} < \text{cat. } 1)$	$P(R_{ij} < \text{cat. } 2)$	$P(R_{ij} < \text{cat. } 3)$	$P(R_{ij} < \text{cat. } 4)$
Public	0	0	.0053	.0230	.1656	.6899
	0	6	.0111	.0471	.2938	.8235
	2	12	.0498	.1872	.6598	.9560
	2	0	.0118	.0498	.3061	.8318
	4	6	.0527	.1962	.6728	.9584
	4	12	.1044	.3386	.8117	.9797
Private	0	0	.0035	.0151	.1142	.5911
	0	6	.0073	.0311	.2128	.7519
	2	12	.0330	.1301	.5575	.9339
	2	0	.0077	.0329	.2228	.7627
	4	6	.0349	.1369	.5719	.9374
	4	12	.0704	.2496	.7369	.9691

$P(R_{ij} \leq \text{cat. } 5) = 1.0$, and thus is not estimated. When calculating the estimated log-odds, each of the four cumulative probabilities is determined using its respective threshold estimate. Table 7.11 presents a series of cumulative probability predictions using the same values for the predictors as chosen earlier for the logistic model. For space considerations, the log-odds and odds are not included in this table, but would be calculated as a first step in finding the cumulative probabilities. Most software packages provide estimated cumulative probabilities or class probabilities on request.

These probability estimates suggest that children in public schools who have the greatest number of family risks and are enrolled in schools in disadvantaged neighborhoods consistently have the highest predicted probabilities of being *at or below* any given proficiency value (see the 5th and 6th rows in the table). Thus, these children are least expected, relative to their peers, to achieve proficiency in mathematics *beyond* any given category.

Assumption of Proportional Odds

Estimates for the ordinal models presented above are based on the assumption of proportional odds, implying that the effects of the explanatory variables do not vary across the different cumulative comparisons being made. Thus, the parameter estimates summarize the effect of each variable across these splits. A likelihood ratio test on deviances of models that maintain and then relax this assumption can be constructed, although methods for estimating a non-proportional random-effects models are not available in all software packages; in this chapter our demonstrations are based on the R package `mixor` (Archer et al., 2018). An ad hoc approach to investigating the plausibility of the proportional odds assumption involves fitting the sequential series of binary logistic regressions underlying the cumulative model, such as the specific comparisons noted in the last column of Table 7.2. This strategy has been recommenced when analyzing both multilevel and single-level ordinal data, and can be useful if the likelihood ratio test for proportional odds fails (see, e.g., Bell & Dexter, 2000 and O'Connell, 2006).

With eight parameters, the deviance of the complex PO model was 14,650.05; the deviance for the model relaxing the PO assumptions (see Appendix) with 17 parameters was 14,600.09. The difference in deviances was statistically significant, $\chi^2(9) = 49.96$, $p < .0001$. A next step, then, would be trying to identify which of the predictors contributed to the non-proportionality.

Table 7.12 shows the parameter estimates and odds ratios for the four underlying multilevel logistic models corresponding to the cumulative odds

TABLE 7.12 Parameter Estimates and Odds Ratios (OR) for the Multilevel Logistic Models Corresponding to Cumulative Analyses for Model II in Table 7.9

Response Estimated	$R_{ij} < 1$	$R_{ij} < 2$	$R_{ij} < 3$	$R_{ij} < 4$
Fixed Effects	**Coeff (*SE*)** ***OR***	**Coeff (*SE*)** ***OR***	**Coeff (*SE*)** ***OR***	**Coeff (*SE*)** ***OR***
Model for the Intercepts (β_0)				
Intercept (γ_{00})	–5.78 (.40)** 0.003	–3.80 (.14)** 0.02	–1.57 (.06)** 0.21	.71 (.05)** 2.03
NBHOODCLIM (γ_{01})	.10 (.06) 1.11	.10 (.03)** 1.10	.11 (.02)** 1.11	.16 (.02)** 1.17
PUBPRIV2 (γ_{02})	–1.49 (.74)* 0.23	–1.36 (.34)** 0.26	–.50 (.11)** 0.61	–.38 (.09)** 0.68
Model for the NUMRISKS Slopes (β_1)				
Intercept (γ_{10})	.69 (.16)** 2.00	.58 (.08)** 1.79	.34 (.04)** 1.41	.53 (.05)** 1.70
Random Effects (variance components)				
Var. in intercepts[a] (τ_{00})	1.132	.416*	.400**	.267**
Deviance[b]	634.92	1,973.37	6,637.94	7,377.46

* $p < .05$; ** $p < .01$

[a] Whether the variance estimate is significant different from zero is tested through a Wald test

[b] Under Adaptive Gauss-Hermite quadrature with 11 quadrature points (the default in package R-`mixor`)

analysis. Of interest is the pattern of effects for each variable across these four logistic splits. The effects estimated in the proportional odds model summarize these effects, and thus, they should be quite similar, on average, across the five models if the assumption of proportional odds is feasible.

Reviewing the parameter estimates and their odds ratios for each of these separate logistic regressions, there seems to be similarity across splits for two of the variables: NUMRISKS and NBHOODCLIM. Averaging the logistic *OR*s for these two variables approximates the expected odds ratio under the proportional odds assumption for the ordinal model. For example, we find an "average" *OR* of 1.73 for NUMRISKS and an "average" *OR* of 1.12 for NBHOODCLIM. These are both fairly consistent with the values estimated in the ordinal GLMM: 1.49 and 1.13, respectively. The effects of PUBPRIV2 did not exhibit such consistency across the logistic regression results, with a broader range ("average" *OR* = .45 compared to GLMM estimate of .65), so the distribution of this variable across the proficiency categories, and the reasonableness of assuming a common odds ratio for the effect of attending a public or private school on proficiency, may be worth further investigation. We provide an example and interpretation of a partial proportional odds model for these data in the online materials for this chapter, relaxing the assumption of proportional odds for the PUBPRIV2 variable.

The assumption of proportional odds is an attractive one in terms of model parsimony, provided that the intent of the research is to estimate an overall effect of a predictor variable on an ordinal outcome. Although the proportional odds model is certainly the most widely used approach for analyzing ordinal data, there are popular alternatives, including continuation ratio GLMMs and mixed probit or multinomial models, in addition to mixed models that allow for non- or partial-proportional odds. Researchers should carefully consider the options available given their analysis goals and features of their data. Advances in statistical software, particularly through open-access R packages, serve to make these options widely accessible (e.g., Archer et al., 2018; Christensen, 2019).

CHAPTER SUMMARY

Multilevel logistic and ordinal models have intuitive appeal, given the kinds of variables often studied in education and in social or behavioral psychology. The examples we discussed in this chapter were selected to illustrate the potential of these kinds of models and to provide a starting point for researchers who may be new to these methods. Our objective was to lay the methodological foundation for researchers to investigate significant research questions involving dichotomous and ordinal outcomes across multiple and hierarchical contexts. We structured our examples around the four model-building elements for generalized models suggested by Dobson (2002): expressing the response as a collection of predictor variables, parameter estimation and interpretation, quality of fit, and statistical inference and hypothesis testing—particularly with regard to the assumptions made by the multilevel logistic and ordinal models. Understanding the impact of contexts such as schools, classrooms, organizations, or communities on individual outcomes continues to emerge as a critical feature of many research studies. The examples and discussions provided in this chapter should prepare researchers to take advantage of multilevel methods when their outcomes of interest are measured using dichotomous or ordinal scales.

In our examples, the ordinal outcome variable was proficiency level computed from student responses to a series of clustered test items. A related critical application for GLMMs occurs within educational testing, where interest often lies in the items that make up the reported scores. One issue of concern to educators and particularly to psychometricians is that of differential functioning of test items across different demographic groups, such as gender and ethnic, cultural, or racial groups. Logistic regression and ordinal logistic regression models have been used to model performance on dichotomously and ordinally scored test items as a function of student proficiency, group membership, and their interaction (Rogers & Swaminathan, 1993; Swaminathan & Rogers, 1994). In the case of items scored on an ordinal scale (such as where partial credit is assigned or responses to the item task are rated), a proportional odds or partial proportional odds model may be used. One of the strengths of logistic and ordinal regression procedures for investigating differential item functioning (DIF) is that they can be extended to take into account the hierarchical nature of student response data. Using multilevel models for assessing DIF not only allows for better estimation of DIF effects within schools or districts but also allows for a deeper analysis of factors that may explain or account for the presence of DIF.

Kamata (2001) provided an in-depth discussion of how models for item-response theory can be viewed as GLMMs and Rijmen et al. (2003) developed a unifying framework for IRT models and GLMMs. Segawa (2005) extended these concepts to growth models for item-level data (i.e., three-level GLMM). French, Finch, and Vo (Chapter 14, "Measurement and Multilevel Models," this volume) provide further demonstration of measurement models within a multilevel context. Rogers and Swaminathan (1998) developed Bayesian estimation procedures for fitting multilevel models for ordinal response data and applied them to investigate factors related to gender DIF in items from the National Assessment of Educational Progress (NAEP). Readers interested in methodology and applications for the analysis of item-level data within a multilevel framework would benefit from review of the few articles and chapters mentioned here.

SOFTWARE CONSIDERATIONS AND ONLINE APPENDIX

Software changes rapidly, particularly for the open-source R package as users are able to continuously improve, update, and develop new packages for complex analysis. All commercial or open-source software that we are aware of can fit mixed models for dichotomous outcomes, but differences across software packages seem to be more pronounced for analysis of ordinal data. Thus, as an aid to readers working with multilevel ordinal models, we provide a table in our online appendix that briefly describes estimation methods, approaches to assessing the proportional odds assumption, and capacity for fitting non- or partial-proportional odds models for six different software packages. Including this table in the online appendix allows us to keep the table updated over time. We hope you find this material helpful.

We note that data for this chapter, all code used for the chapter examples, and data for the "Try This!" section below can also be found in the online appendix (http://modeling.uconn.edu).

RECOMMENDED READINGS

Technical Readings

Bauer, D. J., & Sterba, S. K. (2011). Fitting multilevel models with ordinal outcomes: Performance of alternative specifications and methods of estimation. *Psychological Methods, 16*(4), 373.

Epasinghe, N., & Sooriyarachchi, R. (2017). A goodness of fit test for the multilevel proportional odds model. *Communications in Statistics-Simulation and Computation, 46*(7), 5610–5626.

Stroup, W. W. (2013). *Generalized linear mixed models: Modern concepts, methods and applications.* CRC Press.

Hedeker, D., & Mermelstein, R. J. (1998). A multilevel thresholds of change model for analysis of stages of change data. *Multivariate Behavioral Research, 33*(4), 427–455.

Applied Readings

Hedeker, D. (2015). Methods for multilevel ordinal data in prevention research. *Prevention Science, 16*(7), 997–1006.

Austin, P. C., & Merlo, J. (2017). Intermediate and advanced topics in multilevel logistic regression analysis. *Statistics in Medicine, 36*(20), 3257–3277.

Merlo, J., Wagner, P., Ghith, N., & Leckie, G. (2016). An original stepwise multilevel logistic regression analysis of discriminatory accuracy: The case of neighbourhoods and health. *PloS One, 11*(4), e0153778.

TRY THIS!

This exercise uses the 2015 OECD Programme for International Student Assessment (PISA) for which you can apply the skills, concepts, and approaches that we covered in this chapter. The dataset includes only the students sampled from the United States. The following variables are included in the data file:

- Student level variables—proficiency level in science; a dichotomous version of proficiency level in science; gender; science self-efficacy; and index of economic, social, and cultural status (ESCS).
- School level variables—school-level ESCS and schools' investment in science teaching.

Variable names and measurement information:

- CNTSCHID = school/cluster level identification number
- CNTSTUID = student level identification number
- PL1SCIE = ordinal science proficiency level (seven levels ranging from 0 to 6)
- D_PL1SCIE = dichotomous version of science proficiency level with levels 3 and beyond representing "intermediate to advanced" versus all lower levels as "basic"
- Female = gender with 1 = female and 0 = male
- SCIEEFF = science self-efficacy (continuous)
- ESCS = a scaled index of economic, social and cultural status (continuous, group-mean centered)
- MeanESCS = average school-level ESCS (continuous)
- SCFUND = schools' investment in science teaching (binary)

Three research questions to consider are:

Research Question 1: After controlling for school-level variables (school-level ESCS and schools' investment in science teaching), do student gender (Female), science self-efficacy, and ESCS relate to science proficiency?

Research Question 2: After controlling for student-level variables, do school-level ESCS and schools' investment in science teaching relate to science proficiency?

Research Question 3: After controlling for student gender (Female) and science self-efficacy and schools' investment in science teaching, does the relationship between student ESCS and science proficiency level vary by school-level ESCS?

You may consider running the logistic and ordinal models using the following steps:

1. Fit a logistic empty model (no predictors at either level) using D_PL1SCIE as the outcome.
2. Fit a logistic random coefficients model with Female, SCIEEFF, and ESCS as level-1 predictors of proficiency.
3. Find your best model(s) through model comparison: determine whether to keep the slopes of Female, SCIEEFF, and ESCS as random or eliminate the random effects for some predictors.
4. Fit a logistic contextual model with MeanESCS and SCFUND as school-level predictors of the intercepts.
5. Fit a logistic contextual model with MeanESCS as a school-level predictor of the slope for ESCS (MeanESCS*ESCS).
6. Fit ordinal models using PL1SCIE as the ordinal outcome by repeating the above steps and compare the results obtained from the logistic and ordinal models.
7. Don't forget to check the assumption of proportional odds for the best fitting model(s) used to answer the research questions.

Here are some guidelines that you will want to follow and/or consider as you complete this exercise.

1. Report the final estimates of your fixed effects, random effects, deviance and number of parameters for both logistic and ordinal models.
2. Include tables of results based on your model building process.
3. Transform the estimated log-odds for proficiency level into odds and estimated probabilities for selected values of the predictors.

Finally, summarize your findings in a few paragraphs with *user-friendly verbal descriptions* of the results of your final models. Be sure to interpret the fixed effects and explain the variance covariance components within your write-up of the results.

ACKNOWLEDGMENT

This chapter is an updated version of O'Connell, A. A., Goldstein, J., Rogers, J., & Peng, C. J. (2008). Multilevel logistic models for dichotomous and ordinal data. In A. A. O'Connell & D. B. McCoach (Eds.), *Multilevel modeling of educational data* (pp. 199–242). Information Age Publishing.

NOTES

1. Other options include probit models, although these are not widely used in education. The logit model tends to be simpler to interpret, "because it can be written as a linear model for the logodds" (Fox, 2017, p. 380).
2. Information on access to the public-use ECLS-K data is available from NCES online at https://nces.ed.gov/ecls/kindergarten.asp
3. The ECLS-K Users Guide explains that approximately 5% of children had response patterns that did not follow the Guttman model and attributes these patterns to guessing on multiple choice items.
4. Early releases of the ECLS-K data did not contain the highest proficiency score attained; these had to be derived from a series of dichotomous variables indicating pass/fail for each item cluster. Release of the third-grade data included highest proficiency scores for all earlier releases of the data, and those are the data used here.

REFERENCES

Ali, S, Ali, A., Khan, S. A., & Hussain, S. (2016). Sufficient sample size and power in multilevel ordinal logistic regression models. *Computational and Mathematical Models in Medicine, 2016*, Article ID = 7329158. http://dx.doi.org/10.1155/2016/7329158

Agresti, A. (2013). *Categorical data analysis* (3rd ed.). John Wiley & Sons.

Agresti, A. (2019). *An introduction to categorical data analysis* (3rd ed.). John Wiley & Sons.

Archer, K. J., Hedeker, D., Nordgren, R., & Gibbons, R. D. (2018). *Mixor: Mixed-effects Ordinal regression analysis.* R package version 1.0.4. https://CRAN.R-project.org/package=mixor

Armstrong, B. G., & Sloan, M. (1989). Ordinal regression models for epidemiologic data. *American Journal of Epidemiology, 129*, 191–204.

Azen, R., & Walker, C. M. (2011). *Categorical data analysis for the behavioral and social sciences.* Routledge.

Bauer, D. J., & Sterba, S. K. (2011). Fitting multilevel models with ordinal outcomes: Performance of alternative specifications and methods of estimation. *Psychological Methods, 16*(4), 373–390. https://doi.org/10.1037/a0025813

Bell, J. F., & Dexter, T. (2000). Using ordinal multilevel models to assess the comparability of examinations. *Multilevel Modeling Newsletter, 12*(2), 4–9.

Bender, R., & Benner, A. (2000). Calculating ordinal regression models in SAS and S-PLUS. *Biometrical Journal, 42*, 677–699.

Brant, R. (1990). Assessing proportionality in the proportional odds model for ordinal logistic regression. *Biometrics, 46*, 1171–1178.

Browne, W. J., Subramanian, S. V., Jones, K., & Goldstein, H. (2005). Variance partitioning in multilevel logistic models that exhibit overdispersion. *J.R. Statistical Society A, 168*(3), 599–613.

Christensen, R. H. B. (2019). *A tutorial on fitting cumulative link mixed models with* clmm2 *from the* ordinal *package.* https://cran.r-project.org/web/packages/ordinal/vignettes/clmm2_tutorial.pdf

Clogg, C. C., & Shihadeh, E. S. (1994). *Statistical models for ordinal variables.* SAGE.

Diaz, R. E. (2007). Comparison of PQL and Laplace 6 estimates of hierarchical linear models when comparing groups of small incident rates in cluster randomised trials. *Computations Statistics and Data Analysis, 51,* 2871–2888.

Dobson, A. J. (2002). *An introduction to generalized linear models* (2nd ed.). Chapman & Hall/CRC.

Evans, M., Hastings, N., & Peacock, B. (2000). *Statistical distributions* (3rd ed.). Wiley.

Fielding, A. (2003). Ordered category responses and random effects in multilevel and other complex structures. In Steven P. Reise and Naihua Duan (Eds.), *Multilevel modeling: Methodological advances, issues, and applications.* (pp. 181–208). Lawrence Erlbaum Associates.

Fox, J. (2017). *Applied regression analysis and generalized linear models* (3rd ed.). SAGE.

Fullerton, A. S. (2009). A conceptual framework for ordered logistic regression models. *Sociological methods & research, 38*(2), 306–347.

Heck, R. H., Thomas, S. L., & Tabata, L. N. (2012). *Multilevel modeling of categorical outcomes using SPSS.* Routledge.

Hedeker, D. (2015). Methods for multilevel ordinal data in prevention research. *Prevention Science, 16*(7), 997–1006.

Hedeker, D., du Toit, S. H. C., Demirtas, H., & Gibbons, R. D. (2018). A note on marginalization of regression parameters from mixed models of binary outcomes. *Biometrics, 74,* 354–361.

Hedeker, D., & Gibbons, R. D. (1994). MIXOR: A computer program for mixed-effects ordinal regression analysis. *Computer Methods and Programs in Biomedicine, 49,* 157–176.

Hedeker, D., & Gibbons, R. D. (2006). *Longitudinal data analysis.* Wiley.

Hedeker, D., & Mermelstein, R. J. (1998). A multilevel thresholds of change model for analysis of stages of change data. *Multivariate Behavioral Research, 33,* 427–455.

Hosmer, D. W., Lemeshow, S., & Sturdivant, R. X. (2013). *Applied logistic regression* (3rd ed.). Wiley.

Hox, J. J., Moerbeek, M., & van de Schoot, R. (2013). *Multilevel analysis: Techniques and applications* (3rd ed.). Routledge.

Kamata, A. (2001). Item analysis by the hierarchical generalized linear model. *Journal of Educational Measurement, 38,* 79–93.

Liao, T. F. (1994). *Interpreting probability models: Logit, probit, and other generalized linear models.* SAGE.

Lo, M.-T. (2020). *Alternative methods for modeling clustered ordinal data* [Doctoral dissertation, Ohio State University]. OhioLINK Electronic Theses and Dissertations Center. http://rave.ohiolink.edu/etdc/view?acc_num=osu1586792558628762

Long, J. S. (1997). *Regression models for categorical and limited dependent variables.* SAGE.

McCullagh, P. (1980). Regression models for ordinal data (with discussion). *Journal of the Royal Statistical Society, Series B, 42*(2), 109–142.

McCullagh, P., & Nelder, J. (1989). *Generalized linear models* (2nd ed.). Chapman & Hall.

McCulloch, C. E., & Searle, S. R. (2001). *Generalized, linear, and mixed models.* Wiley.

McCulloch, C. E., Searle, S. R., & Newhaus, J. M. (2008). *Generalized, linear, and mixed models* (2nd ed.). John Wiley Sons, Inc.

NcNeish, D. M., & Stapleton, L. M. (2016). The effect of small sample size on two-level model estimates: A review and illustration. *Educational Psychology Review, 28,* 295–214.

McNeish, D. M., Stapleton, L. M., & Silverman, R. D. (2017). On the unnecessary ubiquity of hierarchical linear modeling. *Psycholcgical Methods, 22*(1), 114–140.

Menard, S. (2000). Coefficients of determination for multiple logistic regression analysis. *The American Statistician, 54,* 17–24.

National Center for Education Statistics. (2002). *User's manual for the ECLS-K first grade public-use data files and electronic code book: NCES 2002-135.* U.S. Department of Education.

Nelder, J. A., & Wedderburn, R. W. (1972). Generalized linear models. *Journal of the Royal Statistical Society, Series A, 135,* 370–384.

O'Connell, A. A., (2000). Methods for modeling ordinal outcome variables. *Measurement and Evaluation in Counseling and Development, 33,* 170–193.

O'Connell, A. A. (2006). *Logistic regression models for ordinal response variables.* SAGE.

O'Connell, A. A., Yeomans-Maldonado, G., & McCoach, D. B. (2016). Residual diagnostics and model assessment in a multilevel framework: Recommendations toward best practice. In J. R. Harring, L. M. Stapleton, & S. Natasha Beretvas (Eds.), *Advances in multilevel modeling for educational research: Addressing practical issues found in real-world applications* (pp. 97–135). Information Age Publishing.

Peng, C. Y. J., & Nichols, R. N. (2003). Using multinomial logistic models to predict adolescent behavioral risk. *Journal of Modern Applied Statistical Methods, 2,* 1–13.

Pinto, I. V., & Sooriyarachchi, M. R. (2019). Comparison of methods of estimation for use in goodness of fit tests for binary multilevel models. *International Journal of Mathematical and Computational Sciences, 13*(4), 68–73.

R Core Team. (2021). *R: A language and environment for statistical computing.* R Foundation for Statistical Computing, Vienna, Austria. http://www.R-project.org/

Raudenbush, S. W., & Bryk, A. S. (2002). *Hierarchical linear models: Applications and data analysis methods* (2nd ed.). SAGE.

Raudenbush, S., Bryk, A., Cheong, Y. F., & Congdon, R., & du Toit (2019). *HLM 8: Hierarchical linear and nonlinear modeling.* Scientific Software International.

Raudenbush, S. W., Yang, M. L., & Yosef, M. (2000). Maximum likelihood for generalized linear models with nested random effects via high-order, multivariate Laplace approximation. *Journal of computational and Graphical Statistics, 9*(1), 141–157.

Rijmen, F., Tuerlinckx, F., De Boeck, P., & Kuppens, P. (2003). A nonlinear mixed model framework for item response theory. *Psychological methods, 8*(2), 185.

Rogers, H. J., & Swaminathan, H. (1993). A comparison of the logistic regression and Mantel-Haenszel procedures for detecting differential item functioning. *Applied Psychological Measurement, 17,* 105–116.

Rogers, H. J., & Swaminathan, H. (1998, August 14–18). *A multilevel approach for investigating DIF.* Paper presented at the annual APA meeting, San Francisco.

Segawa, E. (2005). A growth model for multilevel ordinal data. *Journal of Educational and Behavioral Statistics, 30,* 369–396.

Skrondal, A., & Rabe-Hesketh, S, (2004). *Generalized latent variable modeling: Multilevel, longitudinal, and structural equation models.* Chapman & Hall/CRC.

Snijders, T. A. B., & Bosker, R. J. (2012). *Multilevel analysis* (2nd ed.). SAGE.

Stroup, W. W. (2013). *Generalized linear mixed models: Modern concepts, methods, and applications.* CRC Press.

Swaminathan, H., & Rogers, H. J. (1994, April). Logistic regression procedures for detecting DIF in nondichotomous item responses. Paper presented at the annual NCME meeting, New Orleans.

Tate, R. (2004). Interpreting hierarchical linear and hierarchical generalized linear models with slopes as outcomes. *Journal of Experimental Education, 73,* 71–95.

West, J., Denton, K., & Germino-Hausken, E. (2000). *America's kindergartners: Findings from the Early Childhood Longitudinal Study, Kindergarten Class of 1998–1999: Fall 1998.* United States Department of Education, National Center for Education Statistics. (NCES 2000-070)

Wong, G. Y., & Mason, W. M. (1985). The hierarchical logistic regression model for multilevel analysis. *Journal of the American Statistical Association, 80,* 513–524.

Zill, N., & West, J. (2001). *Entering kindergarten: A portrait of American children when they begin school.* United States Department of Education, National Center for Education Statistics. (NCES 2001-035).

CHAPTER 8

SINGLE AND MULTILEVEL MODELS FOR COUNTS

Ann A. O'Connell
The Ohio State University

Nivedita Bhaktha
GESIS Leibniz-Institut für Sozialwissenschäften

Jing Zhang
Kent State University

Nearly all disciplines rely on count data to capture the frequency of specific events or occurrences of behaviors. In education settings, counts can be used to assess how often individual children in special education classrooms may be disruptive or may show prosocial behaviors. Other examples include counts of the number of words recalled during auditory exams, the number of volunteer experiences completed by high-school juniors and seniors, the number of times the "help" feature is accessed during an online algebra tutorial, or counts of violent crime events across different school zones or districts. Counts are common outcomes for intervention studies designed to reduce (or increase) the occurrence of particular events or behaviors, but analyzing such data brings unique challenges.

Multilevel Modeling Methods with Introductory and Advanced Applications, pages 255–304

Count data differ from more familiar continuous data. First, counts are discrete and are always non-negative integers. Thus, an "average" count may be helpful descriptively, but in conjunction with the observed standard deviation of the counts the interpretation of average counts can be problematic inferentially and may mask important patterns in the data (depending on degree of skew).

Second, while counts may follow a symmetric or close to normal distribution in some circumstances, most count data have a distribution described as "reverse J-shaped," with many occurrences of zeros, ones, or twos, tapering off in frequency as the number of observed events becomes larger. Figure 8.1 provides a simple illustration and shows the distribution of the number of inferential-type questions asked by pre-school teachers during interactive shared book-reading sessions. In this study, teachers submitted up to four videos of their shared book-reading sessions across the school year (Binici, 2014; Justice et al., 2009). Coded video data for 24 teachers indicated that the average number of inferential questions asked during shared book-reading was 9.8 (SD = 12.77). However, most teachers asked between 0 and 10 inferential questions, while a few teachers asked up to 40 or even 60 questions. To analyze this data without consideration of the

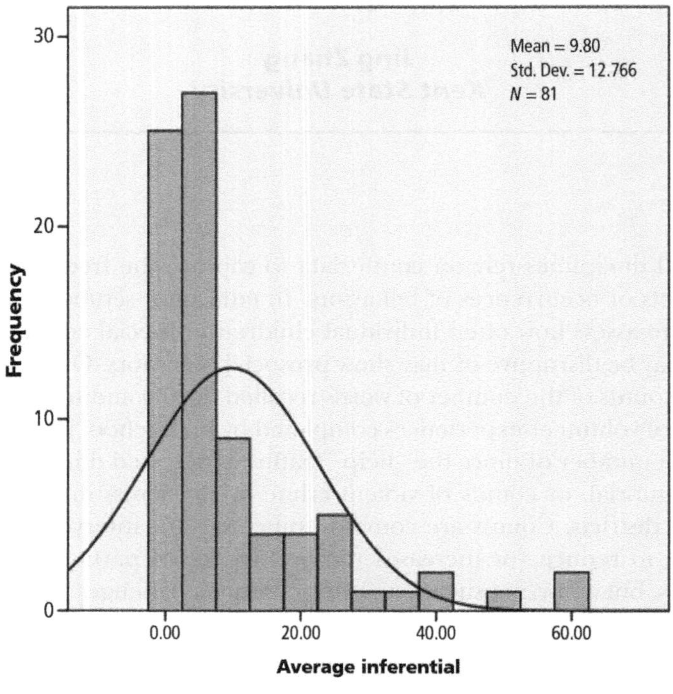

Figure 8.1 Familiar reverse J-shape for distribution of count data.

pattern and shape of the counts would provide a misleading assessment of variability within and between teachers in terms of how often inferential questions are used during shared book-reading, thus potentially masking factors that may be associated with teachers' use of inferential questions.

Third, counts of events of interest often occur within a bounded range such as a specific period of time or within a given geographic area. This feature makes rates of events more interesting than just raw numbers of occurrences. Rates can vary across individuals or cases within a sample, similar to how raw counts may vary. If the period of time for observation of the events or if the size of a geographic area varies within the sample, this variable of size or time—referred to as the "exposure" variable—must be incorporated into prediction models to obtain comparable assessments of the event rate. An exposure variable is included as an "offset" in count regression models (Cameron & Trivedi, 2013; Hilbe, 2014). An exposure variable can easily be specified as such in statistical software; alternatively, it can be incorporated into the model as a predictor with a constrained slope of 1.0, and with a log transformation to retain the same metric as the outcome (Hox et al., 2018).

Finally, researchers often seek to know if the variation in counts or rates can be explained by one or more predictors. When the distribution of counts is symmetric, ordinary least squares regression has sometimes been used to address these questions, but problems occur as a consequence of ignoring the nature of count data. Predictions of counts using an OLS model may be below zero or non-integer, which would clearly be incorrect; counts cannot be negative and must be a whole number. In addition, the residuals from an OLS analysis of counts are heteroskedastic, with the heteroskedasticity increasing to a greater degree as skew increases in the data. In this situation, heteroskedasticity affects precision of regression coefficient estimates by downwardly biasing their standard errors, and statistical inferences based on these models may be erroneous (Coxe et al., 2009; Gardner et al., 1995). A larger sample size does not improve the precision of regression estimates for counts based on the OLS model, as the standard errors themselves are inconsistent.

In addition to issues resulting from improper fitting of an OLS model to count data, many educational and community studies are hierarchical in nature with observations nested within clusters (e.g., classrooms, schools, neighborhoods) or nested within individuals over time (e.g., for repeated measures designs or assessing individual growth). Clustering induces a dependency in the data which leads to underestimated standard errors and inflated probability of Type I error if the dependency is ignored. Mixed models or multilevel models comprise a powerful analytical framework through which clustered and/or longitudinal data may be analyzed, and can be applied to clustered count data via extensions of single-level models for counts. Multilevel models for counts as well as for other types of discrete outcomes are typically referred to as generalized linear mixed models (GLMM; Hox et al., 2018; Raudenbush & Bryk, 2002; Stroup, 2013), multilevel counterparts to single-level generalized linear models.

There is a daunting array of distributions and models that can be used for count data and that adequately capture the shape and nature of diverse count distributions (Atkins et al., 2013; Cameron & Trivedi, 2013; Grimm & Stegmann, 2019; Hilbe, 2011, 2014). Our goal in this chapter is to review some common approaches to working with counts, starting with single-level data as a brief review and then moving to multilevel data. We demonstrate through example the nature of multilevel Poisson and negative binomial (NB) models, paying close attention to distributional differences, model fit, and parameter interpretation. Our demonstrations include examples in SAS (Statistical Analysis System Version 9.4), R (Version 4.3), and Mplus (Version 8). Syntax and results for all models are provided in the online appendix; we note for the reader that some software, such as HLM v8.0, does not currently have capacity for fitting NB models. We show how the intraclass correlation coefficient (ICC) can be estimated for multilevel count data, and briefly discuss extensions including the use of mixture models such as zero-inflated Poisson (ZIP) and zero-inflated negative binomial (ZINB), and their distinction from two-part Hurdle models.

Data for our Examples

For pedagogical purposes, we demonstrate count models using a widely analyzed epilepsy data set from Leppik et al. (1987) and reported and reanalyzed by Thall and Vail (1990) among others (e.g., Hox, 2010; Hox et al., 2018, and Skrondal & Rabe-Hesketh, 2004). The data are based on a clinical trial and include the number of seizures experienced by 59 epileptic patients (age range is 18 to 42) in the 2 week period prior to each of four clinic visits. Given the longitudinal nature of the data, the data were considered 2-level wherein observations obtained at the four visits are nested within individuals. The seizure counts show a high degree of overdispersion for these data, which allows us to consider several different approaches to modeling counts. In this research study, the treatment (placebo versus a drug-based intervention [progabide]) was randomly assigned after collection of baseline data, and study age and the patient's baseline number of seizures were log-transformed to address skewness and then centered at their mean. Table 8.1 provides a brief measurement description and descriptive statistics for the analysis variables. A fourth-visit indicator variable contained in the data set was used to mark the occasion for the final visit; however all 59 epilepsy patients in this dataset had completed the trial through their fourth visit. As in Hox et al. (2018) and Skrondal and Rabe-Hesketh (2004),

TABLE 8.1 Epilepsy Data: Analysis Variables

Variable Name	Description	Descriptive Statistics (n = 59)
C_logage	Log(age), mean-centered	M = 0.0, SD = .22
C_logbase	Log(0.25 * baseline seizures), mean-centered	M = 0.0, SD = .75
Seizures	Number of seizures	M = 8.26, SD = 12.36
Treat	Treatement, 1 = Probgabide, 0 = Placebo	53% Progabide 47% Placebo
Visit	Coded as –.30, –.10, .10, .30	M = 0.0, SD = .22

time of visit was coded as –.3, –.1, .1, and .3. This coding scheme is further clarified in the endnotes[1] and interpreted in later sections.

Our initial models will review the use of Poisson, overdispersed Poisson (i.e., quasi-Poisson), and NB models for single-level data. We then extend these models for multilevel data and illustrate statistical adjustments for clustered counts.

SINGLE-LEVEL MODELS FOR COUNTS

The traditional model for counts is based on the Poisson distribution, although more recently the NB distribution has become a more plausible alternative for many research situations, given the rigid assumptions required of the Poisson. A Poisson process counts the number of times that an event occurs, $Y = 0, 1, 2$, and so on. If the events occur during a given time span or other index of size, models can estimate the *rate* at which events occur. A Poisson distribution has just one parameter, λ, which represents the mean of the distribution (sometimes μ is used instead of λ). A distinctive feature of a Poisson distribution is that the variance is equal to the mean: $E(Y) = Var(Y) = \lambda$. However, in practice, the variance of samples of count data often exceeds the mean, leading to what is called "overdispersion." Overdispersion tends to underestimate standard errors of regression coefficients, which in turn leads to overstating results of statistical significance tests and an increase in Type I error. Poisson models can be adjusted for overdispersion in a somewhat ad hoc fashion, as will be demonstrated shortly, but the most commonly used model in the presence of overdispersion is the negative binomial (NB) model. The NB distribution has an additional parameter that specifically captures the extra variation present in the data (Hilbe, 2011; McCullagh & Nelder, 1989). The inclusion of this dispersion parameter makes the NB distribution capable of modeling a wide variety of count distributions. In fact, the Poisson distribution is a special case of the NB model when there is no overdispersion. Researchers should be careful regarding the parameterization of the NB model employed by their particular software choice. For example, Hilbe (2014) describes how two functions within the R package (`glm` (base R package)) and `glm.nb` (MASS package; R Core Team, 2020; Venables & Ripley, 2002) use an inverse of the dispersion parameter so that the model tends towards a Poisson as the dispersion parameter tends towards infinity (see p. 10, Hilbe, 2014). We draw attention to these differences in our examples and in our online Appendix.

For our demonstrations, we initially ignore the clustering of the panel data (four observations clustered within patients) and assume that there is no panel effect to violate the assumption of independence in the data, while still including *visit* as a predictor. The number of seizures in the initial 2-week period of the study (1st clinic visit) and the number of seizures in the 2-week period prior to the fourth clinic visit exhibit the characteristic reverse J-shape patterns for counts as seen in Figure 8.2 (data for both graphs ignore treatment assignment). At the initial clinic visit, the mean number of seizures was 8.95 for these 59 patients, with a standard deviation of 14.84 (variance = 220.23). At the 4th clinic visit (post intervention), the mean number of seizures was 7.31, and the standard deviation is 9.65 (variance = 93.12). For both variables, the variance of the data far exceeds the mean: At the initial clinic visit the ratio of variance to the mean is 220.23/8.95 = 24.61, and at the 4th clinic visit this ratio is 93.12/7.31 = 12.74. In practice, the variance often exceeds the mean for counts, as we see here—which calls into question the use of the Poisson distribution as an appropriate model for these data. Since this overdispersion leads to underestimation of the standard errors for regression coefficients, adjustments to the model are necessary. After reviewing and interpreting the Poisson results below, we then demonstrate two approaches to address the overdispersion: an over-dispersed Poisson model, and the NB model.

Poisson Regression Model

Poisson regression models the conditional distribution of Y, that is, the distribution that occurs when Y is conditional on the values of the predictors. The counts Y are non-negative integers which take on response values $y = 0, 1, 2$, and so on. The probability of observing a specific count, according to the Poisson probability function, is:

$$P(Y = y) = \frac{\lambda^y e^{-\lambda}}{y!} \quad \text{for } y = 0,1,2,\ldots \tag{8.1}$$

In this equation, the expected or mean count is $E(Y) = \lambda$, and $y! = y*(y-1)*(y-2)*\ldots*1$. Thus, once the mean is estimated, this formula can be used

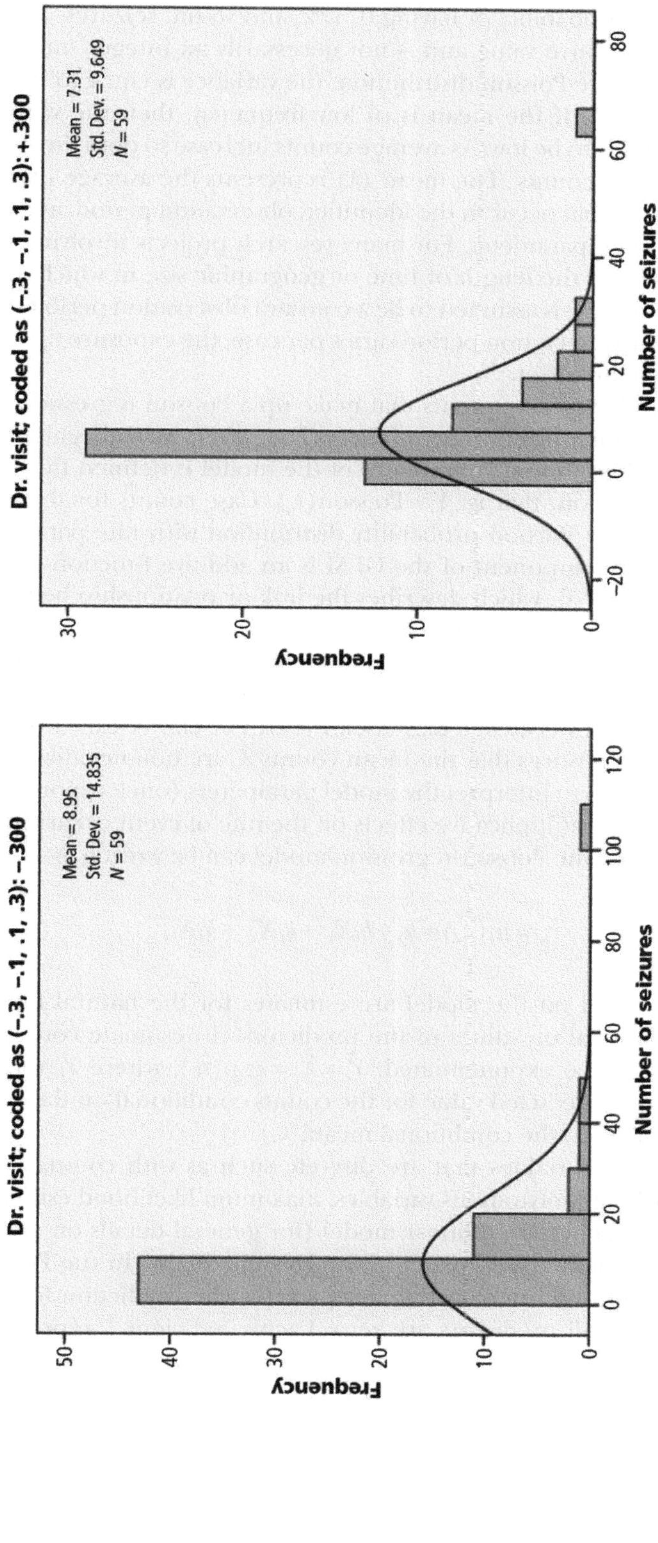

Figure 8.2 Distribution of number of seizures at baseline (left) and at fourth clinic visit (right).

to predict the probability of having 0, 1, 2, and so on, seizures. The mean λ can be any positive value and is not necessarily an integer since it is an average. For a true Poisson distribution, the variance is equal to the mean: $E(Y) = Var(Y) = \lambda$. If the mean is of low frequency, then the variance of counts also tends to be low. As average counts increase so does the expected variability of the counts. The mean (λ) represents the average number of events (counts) that occur in the identified observation period, and as such represents a rate parameter. For many research projects involving counts, the exposure—or the length of time or geographic size in which an event might be counted—is assumed to be a constant observation period for each case, but if the observation period varies per case, the exposure time can be adjusted for as an offset.

There are three components that make up a Poisson regression model, similar to all generalized linear models (Fox, 2017; McCullagh & Nelder, 1989). First, the random component of the model is defined through the Poisson distribution, that is: $Y_i \sim \text{Poisson}(\lambda_i)$. Case counts for the *i*th case are drawn from a Poisson probability distribution with rate parameter λ_i. The systematic component of the GLM is an additive function of the parameters: $\eta_i = \mathbf{X}_i^T \cdot \beta$, which describes the link or relationship between the predictors and the (transformed) response variable η_i. The choice for the transformed response variable is obtained through the canonical link function for Poisson regression models, which is the log link: $\log(\lambda_i) = \eta_i$, and thus $\lambda_i = \exp(\eta_i)$ (McCullagh & Nelder, 1989). The choice for a logarithmic transformation ensures that the mean counts λ_i are non-negative for each case and allows us to interpret the model parameters (once exponentiated) in terms of their multiplicative effects on the rate of event occurrence. For single-level data, the Poisson regression model can be written as:

$$\hat{\eta}_i = \ln(\hat{\lambda}_i) = b_0 + b_1 X_{1i} + b_2 X_{2i} + b_p X_{pi}. \qquad 8.2$$

Predictions based on the model are estimates for the natural log of the counts conditional on values of the predictors. To estimate counts, these predictions can be exponentiated: $\hat{Y}_i = \hat{\lambda}_i = \exp(\hat{\eta}_i)$, where $\hat{Y}_i$ represents the *i*th person's predicted value for the counts conditional on the values of the predictors (i.e., the conditional mean, $\hat{\lambda}_i$).

For response variables that are discrete such as with counts, dichotomous, or ordinal/polytomous variables, maximum likelihood estimation is used to fit the generalized linear model (for general details on maximum likelihood estimation see Fox, 2017, or Harrell, 2015). In the Poisson regression model, the intercept value represents the prediction for the log of counts when all predictors are zero. Each coefficient b_j represents the expected change in the natural log of the counts given a one-unit change in the predictor, holding other predictors constant. To represent change in actual counts (rather than change in the log-counts) each coefficient can be exponentiated as $\exp(b_j)$, which yields the rate ratio (*RR*) or incidence rate ratio (*IRR*; Hilbe, 2014). Similar to an odds ratio, each *RR* makes an explicit comparison of the rates of two values of the predictor as it increases by one unit. That is, for each increase of one unit in the predictor, the expected count will multiplicatively increase (or decrease, if the parameter is negative) by the value of the *RR*, holding other predictors constant. Our examples include clarification of these concepts.

For the epilepsy data we begin by using a person-period data set (59 subjects by 4 visits 2 weeks apart = 236 observations in total) and fit a single-level Poisson regression model with a linear coding scheme to model the passage of time over the four clinic visits (the time variable is "*visit*," coded as −.3, −.1, .1, and .3; see chapter notes for more detail). Additional predictors include the natural log of age, centered at its mean ("*c_logage*"); the natural log of the baseline number of seizures over the eight weeks prior to randomization to treatment, centered at its mean ("*c_logbase*"); and treatment ("*treat*": 0 = placebo, 1 = progabide). In our models, the interaction between treatment and (centered log of) baseline seizures ("*clbasXtrt*") was not statistically significant and thus was excluded from further consideration.

Trends for the two treatment groups are presented in Figure 8.3. We report results using SAS for our single level models based on Poisson, overdispersed Poisson and the NB model in Table 8.2.

Assessing Model Fit

In a model for counts, model fit is generally examined through a measure of discrepancy called the deviance. Deviance statistics capture badness of fit since they characterize the discrepancy between a model and the data. The larger the discrepancy, the worse the fit of a model to one that perfectly represents the data. Overdispersion in the data from a true Poisson distribution leads to greater discrepancy between the model and the data. It is important to consider factors contributing to the extra variance over that assumed in the Poisson model, which could possibly be corrected through better model specification, that is, by considering additional variables and/or interaction terms. Clustering of observations—an essential feature of the epilepsy data set, with clustering of repeated observations of seizure counts within persons—is a factor that contributes to overdispersion and is adjusted for in our multilevel models for counts in a later section of this chapter. Here, we continue with our single-level model to draw attention to important aspects of Poisson regression in general.

For Poisson models, the deviance residuals and Pearson residuals are two approaches to examining discrepancy for each case, and their individual components are squared and summed to form overall deviance G^2

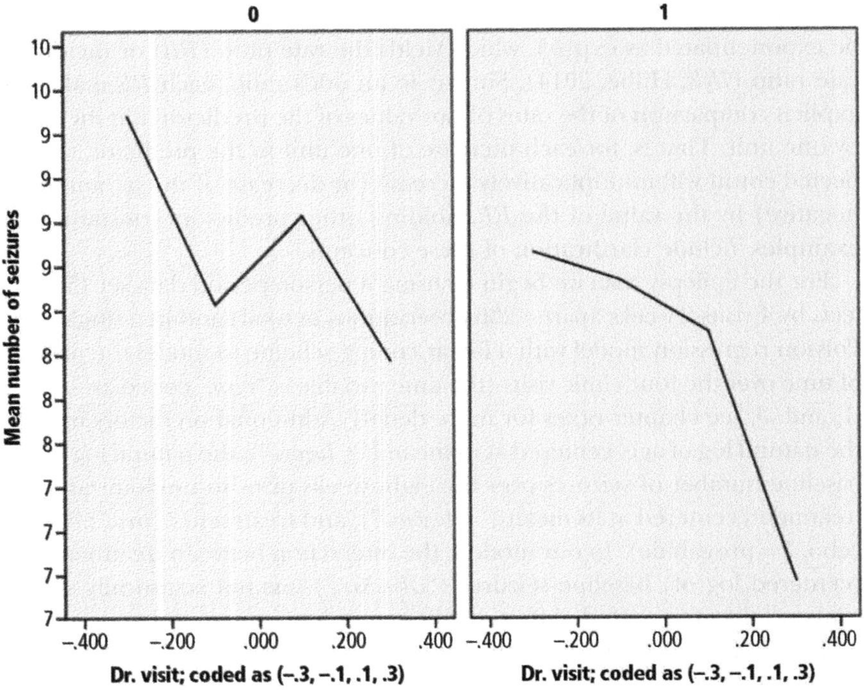

Figure 8.3 Trend over time in number of seizures for placebo (left) and intervention (right) patients.

and Pearson X^2 discrepancy measures. For each observation or case in the Poisson regression, Pearson residuals are defined as

$$r_i = \frac{y_i - \hat{\lambda}_i}{\sqrt{\hat{\lambda}_i}},$$

and deviance residuals are defined as

$$d_i = s_i \sqrt{2\left\{ y_i \cdot \ln\left(\frac{y_i}{\hat{\lambda}_i}\right) - (y_i - \hat{\lambda}_i)\right\}},$$

where s_i is the sign of $(y_i - \hat{\lambda}_i)$ and λ_i is the mean or rate parameter for the ith case. The use of residual or model plots to compare observed versus expected counts have been suggested for a visual examination of fit of the model to the data (Bilder & Loughin, 2016; Hilbe, 2014).

When count data are grouped (i.e., where counts form the cells or entries of a contingency table), and as long as the expected counts are reasonably

TABLE 8.2 Single-Level Poisson (P), Quasi/Overdispersed Poisson (QP), and Negative Binomial (NB) Models for the Epilepsy Data

	Poisson				Quasi (Overdispersed) Poisson				Negative Binomial			
	Model 0P		Model 1P		Model 0QP		Model 1QP		Model 0NB		Model 1NB	
	B	*SE*	*B*	*SE*	*B*	*SE*	*B*	*SE*	*B*	*SE*	*B*	*SE*
Intercept	2.11***	0.02	1.71***	0.04	2.11***	0.10	1.71***	0.09	2.11***	0.07	1.88***	0.07
visit			–0.29**	0.10			–0.29	0.22			–0.28	0.22
c_logage			0.59***	0.11			0.59*	0.24			0.37	0.23
c_logbase			1.23***	0.03			1.23***	0.07			1.06***	0.06
treat			–0.02	0.05			–0.02	0.11			–0.23**	0.10
Scale: ϕ	1.00		1.00		18.4781		4.7794					
$\sqrt{\phi}$					4.2986		2.1862					
NB Disp. α									1.11 (.110)		0.382 (.053)	
Criteria for Assessing Goodness of Fit												
LL(model)	–1643.87		–856.51		–1643.87		–856.51		–747.53		–650.96	
–2LL(model)	3287.74		1713.01		3287.74		1713.01		1495.06		1301.92	
Criterion	Value $df = 235$	Value/ df	Value $df = 231$	Value/ df	Value $df = 235$	Value/ df	Value $df = 231$	Value/ df	Value $df = 235$	Value/ df	Value $df = 231$	Value/ df
Deviance G^2	2521.75	10.73	947.02	4.0996	2521.75	10.73	947.02	4.0996	266.86	1.14	266.02	1.1516
Scaled G^2	2521.75	10.73	947.02	4.0996	136.47	0.58	198.15	0.8578	266.86	1.14	266.02	1.1516
Pearson X^2	4342.37	18.48	1104.04	4.7794	4342.37	18.48	1104.04	4.7794	426.06	1.81	269.89	1.1684
Scaled X^2	4342.37	18.48	1104.04	4.7794	235.00	1.00	231.00	1.0000	426.06	1.81	269.89	1.1684

Note: Models were fit using *SAS PROC GENMOD*. $^*p < .05$, $^{**}p < .01$, $^{***}p < .001$

large, G^2 and X^2 can be used to provide evidence of fit by comparing each to a χ^2 distribution with degrees of freedom equal to the number of cells less the number of model parameters. With individual level data (i.e., with continuous predictors) or data with many cell combinations, the asymptotic distribution of these statistics as χ^2 no longer strictly holds. In this case, and when the Poisson assumption of equidispersion (equal mean and variance) is valid, the deviance G^2 and the Pearson X^2 statistic should be close to their degrees of freedom (Cameron & Trivedi, 2013; Hilbe, 2014; Venables & Ripley, 2002). With continuous predictors, the degrees of freedom for each of these statistics are $n - p$, where p = the number of parameters in the model including the intercept. Accordingly, a diagnostic for goodness of fit is to divide the deviance G^2 and the Pearson X^2 statistics by their degrees of freedom (i.e., value/df) to see how much greater the result is from 1.0 (Cameron & Trivedi, 2013; Fox, 2017; Hilbe, 2011; McCullagh & Nelder, 1989; Powers & Xie, 2008). Given distributional properties, the Pearson X^2/df is typically used as the recommended assessment of overdispersion (lack of fit) in Poisson regression (Cameron & Trivedi, 2013; Hilbe 2011; Venables & Ripley, 2002).

In general, the deviance is also a form of likelihood ratio (LR) test that assesses how much the fitted model deviates from the fit of a perfect or saturated model (which replicates the data perfectly):

$$Dev_{fitted} = -2\ln\left(\frac{L_{fitted}}{L_{sat}}\right) = -2\ln(L_{fitted} - L_{sat}).$$

The deviance may also be written as $Dev_{fitted} = 2(LL_{sat} - LL_{fitted})$, where LL is the log-likelihood for the saturated and fitted model, respectively. As noted in the preceding paragraph, however, the deviance—even though by definition is a form of likelihood ratio test between the saturated and a fitted model—does not always follow the expected χ^2 distribution typically used to compare log-likelihoods for nested models (Cameron & Trivedi, 2013; Dobson & Barnett, 2018; McCullagh & Nelder, 1989; Stokes et al., 2000). To clarify, caution must be taken in using the deviance itself as a likelihood ratio test of goodness of fit relative to a "perfect" model.

A more informative statistical test using the deviance is via direct comparison of the fit of two competing nested models, such as a model with a predictor versus without, or as an overall assessment of a fitted model versus the null or intercept-only model. The likelihood ratio test comparing two nested models is twice the difference in model log-likelihoods, which is equivalent to a difference in deviances test, with the difference compared against a χ^2 with degrees of freedom equal to the difference in number of estimated parameters between the two nested models (reduced model being nested within the full(er) model): $\chi^2_{df} = -2[LL_{reduced} - LL_{full}] = -2LL_{reduced} - (-2LL_{full})$, or $\chi^2_{df} = [Dev_{reduced} - Dev_{full}]$ (Agresti, 2007, 2013; Azen & Walker, 2011; McCullagh & Nelder, 1989). We demonstrate this approach in our examples to follow, and in our comparison of Poisson to the NB model.

Results for the Poisson model. Model 0P and Model 1P in Table 8.2 are the null Poisson model (no predictors) and the four-variable Poisson model, respectively. The "Goodness of Fit" criteria in the lower section of Table 8.2 for Model 0P and Model 1P provide evidence for overdispersion, since the deviance statistics divided by their degrees of freedom $\left(df = n-(p+1)\right)$ are greater than 1.0 for the deviance G^2 and Pearson X^2 statistics. For Model 1P, these ratios are 4.0996 for G^2 and 4.7794 for X^2, each suggesting a high degree of lack of fit for the Poisson model and assumptions, even after adjusting for the contribution of the four predictors. The interpretation is that adjustments to the model are warranted to better align the model with the data; accordingly, our follow-up models will address the evident overdispersion.

Relative to the deviance of the null model, Model 1P with four predictors provides significant improvement:

$$\begin{aligned}\chi^2_4 &= -2[LL_{null} - LL_{fitted}] = -2\left[(-1643.87)-(-856.51)\right] \\ &= -2[-787.37] = 1574.72\end{aligned},$$

which is statistically different from zero, $p < .00001$. Using differences in deviances, we achieve the same result within rounding:

$$\chi^2_4 = [Dev_{null} - Dev_{fitted}] = [2521.75 - 947.02] = 1574.73.$$

Poisson parameter estimates. Although there is evidence for overdispersion in the Poisson model, it is useful to interpret the parameter estimates for later comparison with alternative count models. SAS uses Wald's χ^2 to assess statistical significance of parameter estimates in the Poisson; other software may use the same or a comparable approach. Parameter estimates are provided in the top section of Table 8.2. For Model 1P, the scale factor—identified as ϕ and also included in the top section of Table 8.2—is 1.00 which indicates no adjustment for overdispersion is being made in a Poisson regression model. Technically, this means that the standard errors are adjusted by or multiplied by $\sqrt{\phi} = 1.0$, reflecting no adjustment; in models adjusting for overdispersion this scale factor is changed. Three of the predictors in Model 1P are statistically significant (*c_logage*, *c_logbase*, and *visit*) with $p < .01$ or smaller; the treatment effect (*treat*) is not statistically significant.

Parameter estimates and model predictions are in terms of the natural log of expected counts; predictions are based on Equation 8.2. The intercept represents the estimated log(count) of seizures when all predictors are 0. Given the coding for our example, the intercept is the predicted log(count) of number of seizures at the midpoint of the study ($visit = 0$) for a placebo patient ($treat = 0$) of average age ($c\_logage = 0$) and with an average number of seizures experienced during the 8-week baseline period ($c\_logbase = 0$). To find estimated counts conditional on these same values, we exponentiate the model prediction. For Model 1P, this is an estimated $\exp(b_{intercept}) = \exp(1.71) = 5.53$ seizures. To help interpret the effect of treatment, we can use the regression model to estimate the log(count) of seizures for a similar patient in the treatment arm ($treat = 1$) holding the other variables constant (i.e., $visit = 0$, $c\_logage = 0$ and $c\_logbase = 0$). The estimated number of seizures is thus $\exp(1.71 + (-0.02)) = \exp(1.69) = 5.42$ and this difference ($b_{treat} = -0.02$) is not statistically different from zero, $p > .05$.

The slope for the treatment effect in the Poisson regression model can be used to determine the rate ratio (*RR*) for treatment: $\exp(-0.02) = .98$. As noted earlier, the *RR* for any predictor in the model is found by exponentiating the slope: $\exp(b_j)$. This value is also called the incidence rate ratio (*IRR*) and is the ratio of the predicted counts (or rates) for two ascending values of a predictor that captures a change of one unit, holding constant or controlling for the value of all other predictors. For the treatment variable, the

$$RR_{treat} = \frac{\hat{Y}_{treat=1}}{\hat{Y}_{treat=0}} = \frac{5.42}{5.53} = 0.98.$$

Given the one-unit change in *treat* (from 0 to 1), which represents the difference in the assignment variable between the placebo and the drug group, the estimated number of seizures in the drug group can be found by multiplying the count estimate for the placebo group by the RR_{treat}: this yields $5.53(0.98) = 5.42$, which is the same prediction as found in the previous paragraph (when $treat = 1$ and *visit*, *c_logage*, and *c_logbase* all $= 0$).

For the effect of *visit*, which for efficiency was coded in increments of two-tenths ($visit = -0.3, -0.1, +0.1, +0.3$), the log(counts) are expected to decrease by $b_{visit} = -0.29$ for each one-unit increment in clinic visits, which implies a decrease in the predicted counts of $\exp(-0.29) = 0.748$, or by about .75 for each one-unit increment in *visit*, holding all other predictors constant. However, we coded visits in increments of $\frac{2}{10}$, so we adjust the *RR* accordingly:

$$RR_{visit} = \exp\left(\tfrac{2}{10} \cdot (-0.29)\right) = \exp(-0.058) = 0.944. \tag{8.3}$$

To facilitate understanding of the *RR* for *visit*, we created a prediction chart in Table 8.3. Constructing the *RR* from the predictions for each of our four values of *visit* confirms that changes in the number of seizures decrease by a factor of .944 for each increasing value of *visit*. In the Poisson

TABLE 8.3 Predictions From the Single-Level Poisson Regression (Model 1P) for the Four Coded Values of Visit and Demonstration of Rate Ratio (*RR*) for Visit

Value of *visit*	log(counts)	$\hat{Y} = \exp(\ln(\text{counts}))$	*RR* for *visit* = $\exp(2/10 * (-0.29)) = .944$
–0.3	1.797	6.03	—
–0.1	1.739	5.69	5.69/6.03 = 0.944
0.1	1.681	5.37	5.37/5.69 = 0.944
0.3	1.623	5.07	5.07/5.37 = 0.944

Note 1: $\log(counts) = b_0 = b_{visit} * visit = 1.71 + (-0.29)visit$
Note 2: $\hat{Y}$ = expected number of seizures
Note 3: visit is coded in increments of two-tenths and the *RR* is adjusted accordingly

regression model, the *RR* for *visit* is statistically different from zero (corresponding to b_{visit}), $p < .01$.

We can also use the *RR* to estimate percent change in the predicted counts for a one-unit change in a predictor, with other predictors held constant, using $100(\exp(b) - 1)$. For the treatment effect, this is an estimated $100(0.98 - 1) = -2$ or 2% reduction in seizures (i.e., decrease or negative change). For *visit*, there is an estimated $100(0.94 - 1) = 6\%$ reduction in seizures for each additional increment in time.

Adjustment for Overdispersion via Quasi/Overdispersed Poisson

As noted previously, applications involving counts have a strong potential for overdispersion relative to a true Poisson distribution. The Poisson regression model itself has no error term (separate from the Poisson assumption of equal mean and variance), and overdispersion represents systematic variance that is not accounted for by the set of predictors, resulting in greater variability around predicted counts or rates than would be expected by a Poisson process; that is, $\sigma^2 > \lambda$. The presence of overdispersion in the data does not bias the Poisson regression coefficients, but it does under-estimate their standard errors and therefore affects statistical tests of the model estimates. The overdispersed Poisson uses a simple scaling or correction factor for the standard errors that is formed from the amount of overdispersion remaining after predictors have been accounted for. This correction factor, also called the dispersion parameter, is defined as ϕ = Pearson X^2/df, the same diagnostic measure discussed in the Poisson section to assess degree of overdispersion. Some software packages, including SAS, report the scaling factor as $\sqrt{\phi}$ in the output, so caution is necessary when identifying the correction used.

The model based on this adjustment to the standard errors is also known as a quasi-Poisson (QP) model, and parameter estimates are found through quasi-likelihood methods (Agresti, 2013; Hilbe, 2014). In the process of adding an additional parameter to capture the extra variance not explained by the Poisson model, the distributional family changes, which affects the availability of a likelihood-based approach to model comparison. Bilder and Loughin (2015) describe a quasi-likelihood function as a "likelihood function that has extra parameters added to it that are not a part of the distribution on which the likelihood is based" (p. 310). Here, the determination of ϕ is based, post-hoc, on the maximum likelihood estimation process used to obtain the Poisson model parameters, and thus doesn't affect the log-likelihood of the QP model. Consequently, estimates of the regression parameters are not affected, but inferences regarding these parameters and the fit of the overall model need to be adjusted. Note that in practice, Bilder and Loughin (2015) caution against the use of quasi-likelihood models without first giving due process to fully understanding potential sources of overdispersion, that is, through better model specification.

In QP models, the square root $\sqrt{\phi}$ is used to adjust the estimated standard errors for model coefficients from the corresponding Poisson models by multiplying each standard error by $\sqrt{\phi}$. Similarly, the Wald chi-square test statistics for each predictor are the corresponding Poisson model test statistics divided by ϕ. While it is possible to make these adjustments by hand, software packages do so automatically when the overdispersed or QP model is requested.

Model 1QP in Table 8.2 uses a scaling factor of

$$\phi = \frac{\text{Pearson } X^2}{df} = \frac{1104.04}{231} = 4.7794,$$

indicating a high degree of overdispersion even after adjusting for the explanatory variables in the model (if the model based on the Poisson distribution was a good fit for the data, this ratio—and thus the scaling factor ϕ—would be 1.0). A statistical test developed by Cameron and Trivedi (1990, 2013) available in R through the `AER` package assesses the degree to which the conditional variance exceeds the conditional mean (Kleiber & Zeileis, 2008). Unsurprisingly, the null hypothesis that $\phi = 1.0$ is rejected, Wald $Z = 3.35$, $p < .001$ (code provided in online appendix). Thus, alternate considerations to the Poisson such as the QP model with ϕ-corrected standard errors are warranted.

Results for the Overdispersed Poisson

Results for the null and four-variable overdispersed Poisson models are shown as Model 0QP and 1QP, respectively, in Table 8.2. In the "Goodness of Fit" section for Model 1QP, we first note that the deviance G^2 and its degrees of freedom are not changed from the Poisson model 1P: $G^2 = 947.02$ on 231 degrees of freedom. Adjusting the standard errors does not affect the parameter estimates and thus has no effect on the residuals or their summary deviance statistics. The value of the scaled deviance G^2 is 198.15, which can be found from the original Poisson G^2 by dividing by ϕ: $947.02/4.7794 = 198.15$. The ratio of the scaled G^2 to its df is now .8578 which is improved (i.e., closer to 1.0) in comparison to the Poisson model, illustrating the benefit of the correction made for overdispersion. The Pearson X^2 is likewise not affected by inclusion of ϕ, but the scaled Pearson X^2 is now $1104.04/4.7794 = 231$. Thus, the ratio of the scaled Pearson X^2 to its degrees of freedom in the model adjusted for overdispersion is $231/231 = 1.0$. This is expected, as ϕ was based on this ratio value from the original Poisson model. Since the overdispersed Poisson is based on quasi-likelihood methods, there is no direct comparison with the Poisson other than what can be provided through visual contrast of the scaled G^2 and X^2 statistics. Notice that the LL and –2LL values are the same for the P and QP models. Comparing the deviances for Models 1QP and 0QP yields the same result as earlier: $\chi_4^2 = -2[LL_{null} - LL_{fitted}] = -2[(-1643.87) - (-856.51)] = 1574.72$, a significant improvement favoring Model 1QP, $p < .00001$.

The parameter estimates and their interpretation for the overdispersed Poisson model are exactly the same as for the Poisson, although their standard errors have now been adjusted by $\sqrt{\phi} = \sqrt{4.7794} = 2.1862$. For example, the current standard error in Model 1QP for *visit* is 0.22. In the Poisson model 1P, the standard error for *visit* was 0.10. The adjusted value of the standard error correcting for overdispersion can be found from $0.10 * \sqrt{\phi} = 0.10 * 2.1862 = 0.22$ with rounding. Relative to the Poisson regression, results for Model 1QP indicate that the effect of *visit* is no longer statistically significant ($p > .05$), but the effect of age (*c_logage*, $p < .05$) and number of baseline seizures (*c_logbase*, $p < .001$) are still statistically different from 0 (although the obtained *p*-value for age has increased). There is no evidence for a treatment effect (*treat*, $p > .05$) in either the Poisson or overdispersed Poisson model.

Similar to the Poisson interpretation for *treat*, we would estimate that the number of seizures in the drug group is $\exp(-0.02) = 0.98$ times the estimated number of seizures in the placebo group holding all else constant, which is very close to 1.0 suggesting no difference between the two groups ($p > .05$). Similarly, for the effect of *visit* the $RR = \exp(-0.29) = 0.748$, and with time being coded in increments of two-tenths we have $RR = 0.944$ (see Equation 8.3). This is also not a statistically significant rate ratio for *visit* ($p > .05$).

Modeling the Overdispersion Directly via the Negative Binomial Model

The overdispersed Poisson corrects for bias resulting from inferences in statistical tests, but does not adjust parameter estimates for inconsistency that can arise from the presence of overdispersion. An alternative approach to predicting counts is to include a distinct source of dispersion in the model, relaxing the rigid Poisson assumption of equal mean and variance. The NB regression model is a flexible count estimation model for which the variance is allowed to differ from the mean, with a newly defined relationship between the variance and the mean. In the NB model, rates may vary across individuals even with the same values of the predictor variables. Instead of assuming equal mean and variance as in the Poisson, we now consider the mean as a random variable. Consequently, in order to represent the probability distribution for the counts, we need an additional distribution for the conditional means, capturing the overdispersion (Bilder & Loughin, 2015; Cameron & Trivedi, 2013; Dunn & Smyth, 2018; Hilbe, 2011; Long & Freese, 2014; McCullagh & Nelder, 1989).

The NB model that we discuss here is a type of mixed Poisson model for which the additional distribution for dispersion is the Gamma distribution (Agresti, 2013; Cameron & Trivedi, 2013; Fox, 2017; Hilbe, 2011, 2014). The Gamma distribution is a member of the exponential family of distributions and has the property that its variance increases with its mean. The Gamma distribution has two parameters—scale or location, and shape—and can be written, in general, as Gamma (μ_i, θ), where μ_i denotes the conditional mean of the distribution and θ captures the shape (Dunn & Smyth, 2018; Hilbe, 2011). Rather than relying on the Poisson distribution to describe the counts, with $E(y_i) = \lambda_i$ and $Var(y_i) = \lambda_i$, consider the λ_i as random variables from this Gamma distribution. For this mixture of the Poisson and Gamma distributions, the NB distribution has the same mean structure as the Poisson; that is, the conditional means or expected rates of the mixture distribution are still $E(y_i) = \lambda_i$, as in the Poisson (Dunn & Smyth, 2018; Hilbe, 2011, 2014; Long & Freese, 2014). The degree of overdispersion in the conditional means is governed by the θ parameter, and the variance of an NB distribution is a function of both the Poisson and the Gamma variances.

A distribution that is $\text{Gamma}(\lambda, \theta)$ has variance λ^2/θ (McCullagh & Nelder, 1989). In the Poisson-Gamma mixture, the variance is a combination of both the Poisson and the Gamma distributions: $Var(y) = \lambda + \lambda^2/\theta$ (Hilbe, 2011). The extra dispersion term tends to 0 as θ tends to infinity, representing an *indirect* relationship between θ and dispersion (Friendly & Myer, 2016; Hilbe, 2011, 2014). To capture a *direct* association between the parameter contributing to the increased dispersion and the Gamma variance, the NB model can be parameterized so that $\alpha = 1/\theta$ represents the factor by which the second term is increased; thus, this parameterization for the Poisson-Gamma mixture has variance represented by $\lambda + \alpha\lambda^2$. As α increases so does the additional variance, and as α tends towards 0, the Gamma distribution contributes no additional dispersion and the variance of the Poisson-Gamma mixture converges to that of the Poisson. Our interest is in estimating α, which is called the heterogeneity or NB dispersion parameter under this parameterization (Fox, 2017; Hilbe, 2011, 2014), and assessing whether its inclusion provides a better fit for the model to the data over that of the Poisson.

There are many ways in which an NB model can be expressed, and readers are referred to Hilbe (2011) for an extensive review. Starting from the Poisson distribution as a base, we can describe the conditional distribution of counts, y_i, as Poisson(λ_i), where only one parameter defines both the mean and the variance of the distribution: $E(y_i|\mathbf{x}_i) = \lambda_i$ and $V(y_i|\mathbf{x}_i) = \lambda_i$. Allowing for extra variability in the counts, a reasonable choice is the Poisson-Gamma mixture where the variance is a quadratic function of the mean: $V(y_i|\mathbf{x}_i) = \lambda_i + \alpha\lambda_i^2$. This is the standard and most common form of variance used in a direct model for overdispersion, and the model using this form is known as NB2. There are NB models with other functional forms between the mean and the variance that could be used to describe how the variance increases as the mean increases (the definition of overdispersion), and in general this relationship could be written as $V(y_i|\mathbf{x}_i) = \lambda_i + \alpha\lambda_i^p$. When $p = 2$, we have the quadratic form or NB2, i.e., $V(y_i|\mathbf{x}_i) = \lambda_i(1+\alpha\lambda_i)$; when $p = 1$, we are describing a linear relationship between the mean and the variance (known as NB1), that is, $V(y_i|\mathbf{x}_i) = \lambda_i(1+\alpha)$. Given the flexibility in the shape of the relationship between the mean and the variance, the quadratic form or NB2 is considered appropriate for most purposes and is the default NB model used in most statistical software (Cameron & Trivedi, 2013; Hilbe, 2011, 2014; Long & Freese, 2014). We note that the NB2 is the form of an NB model typically referred to simply as NB. In any case, however, the parameter α is known as the dispersion parameter. When $\alpha > 0$, the variance increases over that of the Poisson, indicating overdispersion. The Poisson regression model is a reduced form of the NB model when there is no overdispersion, and the NB model provides a general approach to adjusting count models in the presence of overdispersion. Given that the Poisson model is nested within the NB model and both models are estimated via maximum likelihood, a likelihood ratio test can be used to assess the comparative fit of these two models, as we demonstrate during our presentation of results below.

According to Hilbe (2011, 2014), the probability distribution for the NB can be written in many different ways. Based on our discussion above defining model parameters, the NB probability distribution can be written as follows (Hilbe, 2011, p. 189):

$$p(y|\lambda,\alpha)=\frac{\Gamma(y_i+\alpha^{-1})}{y!\Gamma(\alpha^{-1})}\cdot\left(\frac{1}{1+\alpha\lambda_i}\right)^{\frac{1}{\alpha}}\left(1-\frac{1}{1+\alpha\lambda_i}\right)^{y_i}. \quad (8.4)$$

In this expression $\Gamma(\cdot)$ is the gamma function (see Fox, 2017, p. 422 for details).

Results for the NB Regression Model

Models 0NB and 1NB in Table 8.2 refer to the null and four-predictor NB regression models, respectively (the names for our models should not be confused with the NB2 *form* of the model; all of our results follow the NB2 form as presented in the previous section). As with similar comparisons for the P and QP models, the four-predictor NB model fits better than the null model:

$$\chi_4^2=-2[LL_{Model0NB}-LL_{Model1NB}]=-2[(-747.53)-(-650.96)]=193.14.$$

This is a significant improvement favoring Model 1NB, $p < .00001$.

The NB regression model has the same structure as Equation 8.2 with an extra parameter estimated for additional dispersion over that of a Poisson, α. For Model 1NB, the dispersion parameter is included in the last row of the top section of Table 8.2 and is estimated as $\alpha = .382$. Estimated variance of the counts of number of seizures based on Model 1NB have conditional variance following the discussion in the previous section: $V(y_i|\mathbf{x}_i)=\lambda_i+\alpha\lambda_i^2$. The parameter estimates for Model 1NB account for the extra dispersion in the data. Notice that there are no scaling factors for the NB Models I in Table 8.2; rather, the overdispersion is captured through the inclusion of the NB dispersion parameter. In the "Goodness of Fit" section of the table, the reported diagnostic G^2 and X^2 values yield fit indices that are smaller than their counterparts in the Poisson models (i.e., value/df), indicating that the residuals from the NB models are also smaller overall with predicted counts that offer better fit to the data.

The degrees of freedom for the deviance and Pearson statistics are based only on the number of predictors in the model, $df = n - (4 + 1) = 231$; this is because these values represent the sums of the squared deviance or Pearson residuals, respectively, and the same predictors are included in both models. Modeling extra dispersion doesn't affect how residuals are formed for an NB model, as the residuals are based only on predictions from the systematic, and not the random, part of the model.

For Model 1NB, there is still some evidence of overdispersion given that the deviance $G^2 = 266.02$ is slightly greater than its degrees of freedom: $G^2/df = 1.1516$. However, Model 1NB is much improved over the Poisson regression analysis. A likelihood ratio test confirms the better fit of Model 1NB over Model 1P, with one parameter difference for the NB dispersion: $\chi_1^2 = -2(LL_{1P} - LL_{1NB}) = -2[(-856.5068) - (-650.9604)] = 411.0928, p < .00001$.

The addition of the NB dispersion parameter does not affect how the model's parameter estimates are interpreted. The intercept represents the predicted log(count) of number of seizures at the midpoint of the study for an average-aged patient with an average number of seizures and in the placebo arm. For Model 1NB, this is an estimated exp(1.88) = 6.55 seizures. For a similar patient in the treatment arm, the estimated number of seizures is exp (1.88 + (–0.23)) = exp(1.65) = 5.2 seizures. This difference is now statistically significant, $b_{treat} = -0.23$ ($p < .05$). Controlling for the other predictors in the model, we would estimate that the number of seizures in the drug group is exp(–0.23) = 0.79 times the estimated number of seizures in the placebo group. This *RR* represents a reduction of about 100(0.79 – 1) = 21% of seizures for patients in the drug group. The effect for *visit* is similar in value to both earlier analyses and is not statistically significant, $b_{visit} = -0.28$ ($p > .05$). As with the Poisson model, the *RR* for *visit* with time being coded in increments of two-tenths is *RR* = 0.944 (see Equation 8.3).

MULTILEVEL MODELS FOR COUNTS

Now that interpretation and differences in model results among single-level options for counts have been clarified, we can turn to models for analyzing clustered counts. We discuss and compare results for the epilepsy data using the multilevel Poisson and the multilevel negative binominal models. We omit presentation of the multilevel QP (overdispersed Poisson), because translating the crude adjustment of scaling standard errors is not straightforward in the multilevel case, and it becomes too simplistic to assume that a single scalar may correct the standard errors for effects of clustering across all random effects in a multilevel model. In addition, given the quasi-likelihood method used for estimation of the overdispersed Poisson, capacity for model comparisons is greatly diminished for QP models. We suggest Bolker et al.'s (2021) github webpage for more details on this issue. In the multilevel context when overdispersion is present, alternatives such as the NB are more flexible and reliable, and likelihood ratio tests, like those presented in the single-level section, can be used to compare competing models and test model parameters.

A multilevel approach accounts for variability that arises from clustered or correlated count data—such as counts for individual children (level 1) who are nested within a sample of classrooms or schools (level 2), or count data that is repeatedly collected on individuals over time (with the multiple responses [level 1] nested within persons [level 2]). Multilevel models for discrete outcomes are part of the generalized linear mixed model (GLMM) family. These models are mixed because they estimate both within- and between-cluster components of variability (i.e., within level-2 units and between level-2 units).

To facilitate presentation of multilevel models for counts, we begin with discussion of general estimation strategies for GLMMs, identifying estimation issues and decisions for working with count outcomes. We then focus specifically on modeling counts using a multilevel Poisson model and describe similarities and differences for multilevel NB models. Degree of clustering is an important issue for multilevel models in general, and we present recent advances in variance partitioning for multilevel count models including estimation of the intracluster or intraclass correlation coefficient, or ICC, following Leckie et al. (2020). Our analytic demonstrations are presented next, with results for the epilepsy data based on analysis in SAS, R, and Mplus. Syntax and data are provided in the online Appendix.

Estimation

Estimation processes for GLMMs are considerably more complicated than the normal mixed model case, since the likelihood function is now a nonlinear function of the fixed effects, and the residuals at level 1 are heteroscedastic and depend on the mean. Consequently, the integration required for solution of the parameter estimates in terms of maximizing the likelihood function becomes intractable (Brown & Prescott, 2006; McCulloch & Searle, 2001; Raudenbush et al., 2000). Thus, alternative strategies focused on approximating the likelihood function or approximating the model are generally relied on (Diaz, 2007; McNeish, 2016, 2019; SAS Institute, 2016; Rabe-Hesketh & Skrondal, 2009). We recommend McNeish (2019) for estimation details on these methods specifically related to multilevel Poisson.

Techniques that use numerical integration to approximate the likelihood function include Laplace approximation (Laplace) and adaptive quadrature (AQ). Since these methods yield an approximate likelihood function that can be optimized, likelihood ratio tests for model comparisons can be performed. Pinheiro and Bates (2006) note that the first-order Laplacian approximation is the same as a one-point adaptive quadrature solution. As the number of quadrature points used increases, parameter estimation will improve (Rabe-Hesketh & Skrondal, 2012). However, AQ with a larger number of quadrature points can experience convergence problems, and becomes progressively more difficult with model complexity, such as when the number of random effects per cluster increases (Raudenbush et al., 2000) or for models with crossed random factors or higher levels of clustering in the data (McCulloch & Searle, 2001).

Alternatively, techniques that approximate the model rather than the likelihood function rely on linearization methods such as Taylor series linearization, starting with an underlying linear mixed model for an approximated continuous "pseudo-outcome" and for which linearized estimates are iteratively updated. These are pseudo- or quasi-likelihood methods, and while they are often quicker to converge than AQ or Laplace, particularly for complex models, they cannot provide model deviance statistics for comparison across nested models. In addition, parameter estimates from multilevel quasi-likelihood methods may be biased (Rabe-Hesketh & Skrondal, 2009). Approximations that center on both the fixed and random components of the model are referred to as penalized quasi-likelihood (PQL) methods, and those focusing only on the fixed components are marginal quasi-likelihood methods (MQL; Snijders & Bosker, 2012; SAS Institute, 2016). We discuss these briefly in a later section on unit-specific versus population-averaged models, although in this chapter we are not focusing on marginal (or population-averaged) methods.

Finally, there are many options for characterizing within-person correlation for longitudinal data. One way to model this dependence is through the random-effects approach taken in this chapter, which assumes conditional independence among observations across occasions (i.e., the responses at the four occasions marked by *visit*). This structure is by default commonly applied in most statistical packages for multilevel models. Thus, we are accounting for the within-cluster correlation by including a person-specific (i.e., cluster-specific) random intercept. Other correlational structures are possible (i.e., autocorrelation), but given four time-points and our interest in illustrating mixed-effects models for counts in general, default options for the correlational structure were not changed in the software used.

In summary, PQL is simple and fast but cannot be used for likelihood-based comparisons across competing models. Estimation through AQ or Laplace overcomes the bias from PQL, but may have convergence problems for some data sets. Because accuracy of parameter estimates increases with a larger number of quadrature points, AQ is typically preferred but is often slower and harder to converge for some models than Laplace. Advantages and disadvantages of these approaches must be taken into account in estimation decisions for a particular data set and a particular GLMM analysis. For consistency across the multilevel count models we demonstrate here, numerical integration was used. In SAS and R we used Laplace estimation, as the quadrature methods suffered from convergence issues in the more complex models. Mplus models used numerical integration with adaptive quadrature. As is typically the case for GLMMs we assumed a normal distribution of random effects at level 2, but note that other options are possible (Hilbe, 2011).

Multilevel Poisson Models

Multilevel Poisson models extend the single-level Poisson models discussed earlier by allowing random effects in the model for the conditional mean. Looking back at Equation 8.2, the intercept and slopes were fixed values for all persons in the sample; they do not vary from cluster to cluster, or for the epilepsy data, from person to person. Multilevel models for clustered data allow for variability in effects—intercepts and slopes—between persons (or clusters). In essence, predictions are now considered cluster-specific. Depending on the context of the sample and study design, predictions and/or the multilevel models are sometimes referred to as "unit-specific" or "subject-specific."

By including an offset for varying exposure, multilevel Poisson models can also incorporate variable lengths of time in which counts are assessed or varying sizes of geographic regions in which counts occur. Similar to its single-level counterpart, however, the applicability of the multilevel Poisson is limited by the rigid equidispersion assumption (equal mean and variance in the data) which is exacerbated by the presence of clustering induced by the nesting of the data. While clustering can indeed lead to overdispersion if ignored, the multilevel Poisson does not directly accommodate other forms of overdispersion, and in these situations alternatives such as the NB model may be preferred.

In the multilevel Poisson model, the standard log link is used at level 1, such that the level-1 prediction equation is similar in form to the single-level Poisson regression equation reviewed earlier. But now we also incorporate information on the distribution of the random effects, which characterizes differences in level-1 intercepts and/or slopes. For GLMMs, these random effects are typically assumed to be normally distributed, as with multilevel models for continuous outcomes. For count outcomes in which the Poisson model is applied, variability at level 1 is not separately modelled, since an expected count always implies the equivalent value for the variance. This feature of the multilevel Poisson allows room for consideration of models that relax this equidispersion assumption while still maintaining an overall model that accommodates clustered data, such as the multilevel NB model.

We begin by assuming that level-1 data are nested within level-2 units or clusters. Let y_{ij} be the count response for level-1 case *i* within level-2 unit *j*, and $y_{ij} \sim \text{Poisson}(\lambda_{ij})$. Thus, λ_{ij} is the expected count of the $y_{ij}: \text{E}(y_{ij} | \lambda_{ij}) = \lambda_{ij}$. A *random intercepts* model is one in which intercepts from level 1 are allowed to vary *between* level-2 clusters or units, but effects of any predictors in the model are fixed to be the same regardless of cluster. The simplest form of random intercept model is the unconditional or null model, with no predictors. The unconditional model can be written as separate levels or in combined form as follows:

$$\begin{aligned} \text{Level 1:} \quad & \eta_{ij} = \log(\lambda_{ij}) = \beta_{0j} \\ \text{Level 2:} \quad & \beta_{0j} = \gamma_{00} + u_{0j} \\ \text{Combined:} \quad & \eta_{ij} = \gamma_{00} + u_{0j} \end{aligned} \tag{8.5}$$

In these expressions, η_{ij} is the log of expected counts for the ith person in the jth cluster (or the ith repeated measure for the j person). The β_{0j} are the level-1 randomly varying intercepts across the j clusters, and γ_{00} is the level-2 fixed overall or "average" intercept around which the j cluster-specific intercepts vary. The level-2 residuals or random effects, u_{0j}, capture the degree to which the cluster-specific intercepts vary around $\gamma_{00.}$ These residuals or random effects are assumed to be normally distributed with a mean of 0 and variance τ_{00}: $u_{0j} \sim N(0, \tau_{00})$.

There are two parameters estimated in the null or unconditional model, one for the intercept γ_{00} and one for the cluster-level variance around the intercept, τ_{00}. The null model is not as useful for count outcomes as with continuous responses, as variance partitioning and contributions to variation depend on the value of covariates forming the linear predictor (Austin, 2018). However, we can use the unconditional model to examine the impact of clustering as well as the presence of overdispersion in the data by partitioning model variance into its between-level and within-level components, as we demonstrate shortly.

The null model for the epilepsy data looks as in Equation 8.5; there are no predictors included in the model. Each person would have their own prediction for the log of expected number of seizures, and these predictions vary around the overall or "average" intercept, $\gamma_{00.}$

We can add level-1 predictors to this model while still allowing only the level-1 intercepts to vary at random. For p level-1 predictors, the corresponding level-1 and level-2 equations would be written as follows, for separate levels or in combined form:

$$
\begin{aligned}
\text{Level 1:} \quad & \eta_{ij} = \log(\lambda_{ij}) = \beta_{0j} + \sum\nolimits_{p=1}^{p} \beta_p X_{pij} \\
\text{Level 2:} \quad & \beta_{0j} = \gamma_{00} + u_{0j} \\
& \beta_1 = \gamma_{10} \\
& \beta_2 = \gamma_{20} \\
& \vdots \\
& \beta_p = \gamma_{p0} \\
\text{Combined:} \quad & \eta_{ij} = \log(\lambda_{ij}) = \gamma_{00} + \sum\nolimits_{p=1}^{p} \gamma_{p0} X_{pij} + u_{0j}
\end{aligned}
\tag{8.6}
$$

In Equation 8.6, η_{ij} is the log of expected counts for the ith person in the jth cluster (or the ith repeated measure for the jth person); β_{0j} are the randomly varying intercepts across the j clusters and if level-1 covariates are centered, the intercepts represent cluster-level predictions when covariates are at their average value. At level 2, γ_{00} is the fixed overall or "average" intercept around which the j cluster-specific intercepts vary. The level-2 random effects, u_{0j}, capture the degree to which the cluster-specific intercepts vary around γ_{00}. Formally, the expected value of the random intercepts is $E(\beta_{0j}) = \gamma_{00}$, and variation among the intercepts is estimated by $\tau_{00} = \text{Var}(u_{0j})$. The β_p or γ_{p0} are the fixed regression coefficients associated with each of the p level-1 covariates X_{pij}. Although we could simply refer to these as β_p since they are fixed, it is generally accepted to use gammas for fixed coefficients particularly when distinguishing effects at different levels, and we do so here. The values of the level-1 covariates may vary within clusters, but in the above model the effect (regression slope) for each covariate is constrained to be constant for all clusters or individuals in the sample; that is, unlike the intercept, these regression slopes do not vary by cluster. Overall there are $(p + 1) + 1$ parameters estimated in the Poisson random intercepts model, one for each fixed regression coefficient including the intercept, and one for the variance parameter τ_{00}. As in the single level Poisson model, there is no separately estimated variance at level 1.

For the epilepsy data, there are four observations for each person representing the number of seizures occurring prior to each clinic visit. Thus, observations are clustered within patients at level 1, with *visit* as the level-1 predictor or marker for time. Patients or individuals form the clusters at level 2. In the fixed-effects single-level model, the intercept represented the estimated log(count) of seizures for the entire sample when all predictors are zero. If we now allow the intercepts to be random rather than fixed for all patients, the following level-1 and level-2 models (Equation 8.7) are implied, with *visit* as the only level-1 predictor:

$$\begin{aligned}
\text{Level 1:}\quad & \eta_{ij} = \log(\lambda_{ij}) = \beta_{0j} + \beta_1(\mathit{visit}_{ij}) \\
\text{Level 2:}\quad & \beta_{0j} = \gamma_{00} + u_{0j} \\
& \beta_1 = \gamma_{10} \\
\text{Combined:}\quad & \eta_{ij} = \log(\lambda_{ij}) = \gamma_{00} + \gamma_{10}(\mathit{visit}_{ij}) + u_{0j}
\end{aligned} \tag{8.7}$$

The variable *visit* is coded as previously (–.3, –.1, .1, .3), so that $visit_{ij}$ represents the *i*th value of *visit* for the *j*th person. As noted above, the fixed-effects regression parameters are the gammas: γ_{00} and γ_{10} for the intercept and the level-1 predictor, *visit*, respectively. Only predictors that vary *within* level-2 clusters can have slopes that are allowed to randomly vary; but in this model we are constraining the linear change over time (slope for *visit*) for each patient to be the same. This constraint can be relaxed in a random-slopes model. Overall, there are three parameters estimated in the random-intercepts model in Equation 8.7; one for the regression parameter for *visit*, $\gamma_{10,}$ one for the intercept, γ_{00}, and one for the variance component, τ_{00}, which assesses variation between persons in their intercepts based on level 1, that is, controlling for *visit*.

With level-2 predictors to consider, a *conditional* random-intercepts Poisson model can be constructed with (or without) level-1 covariates but now also incorporating *Q* predictors at level 2, identified as W_{qj}. The values of these level-2 predictors vary *between* clusters but not *within* clusters. Thus the slopes for these covariates represent fixed regression coefficients. In Equation 8.8, the γ_{0q} are the *Q* regression coefficients associated with the level-2 covariates W_{qj}. Overall there are $(p + Q + 1) + 1$ parameters estimated in this model. The model is expressed as:

$$\begin{aligned}
\text{Level 1:}\quad & \eta_{ij} = \log(\lambda_{ij}) = \beta_{0j} + \sum\nolimits_{p=1}^{p} \beta_p X_{pij} \\
\text{Level 2:}\quad & \beta_{0j} = \gamma_{00} + \gamma_{01}W_{1j} + \gamma_{02}W_{2j} + \cdots \gamma_{0Q}W_{Qj} + u_{0j} \\
& \beta_1 = \gamma_{10} \\
& \beta_2 = \gamma_{20} \\
& \vdots \\
& \beta_p = \gamma_{p0} \\
\text{Combined:}\quad & \eta_{ij} = \gamma_{00} + \sum\nolimits_{p=1}^{p} \gamma_{p0}X_{pij} + \sum\nolimits_{q=1}^{Q} \gamma_{0q}W_{qj} + u_{0j}
\end{aligned} \tag{8.8}$$

For the epilepsy data, we have three predictors that are associated with patient j at level 2—*c_logage*, *c_logbase*, and *treat*. These predictors vary only between-patients rather than within-patients. For example, $treat_j$ represents the treatment group to which the jth person was assigned, and any particular patient was assigned to only one treatment group for the study. Similarly, $c\_logbase_j$ is the value of the jth person's baseline number of seizures; for a given patient in the study, this baseline variable doesn't change. Thus a conditional random-intercepts model for the epilepsy data, with $(p + Q + 1) + 1 = 6$ parameters ($p = 1$, $Q = 3$) can be written as Equation 8.9:

$$\begin{aligned}
\text{Level 1:}\quad & \eta_{ij} = \log(\lambda_{ij}) = \beta_{0j} + \beta_1(visit_{ij}) \\
\text{Level 2:}\quad & \beta_{0j} = \gamma_{00} + \gamma_{01}(c_logage_j) + \gamma_{02}(c_logbase_j) + \gamma_{03}(treat_j) + u_{0j} \\
& \beta_1 = \gamma_{10} \qquad (8.9) \\
\text{Combined:}\quad & \eta_{ij} = \log(\lambda_{ij}) = \gamma_{00} + \gamma_{10}(visit_{ij}) + \gamma_{01}(c_logage_j) \\
& \qquad + \gamma_{02}(c_logbase_j) + \gamma_{03}(treat_j) + u_{0j}
\end{aligned}$$

Finally, the most general form of multilevel model for counts is one with both random-intercepts and random-slopes. This model is more complex compared to the random-intercepts models above, as it relaxes the constraint of fixed regression coefficients for one or more covariates at level 1. Thus, there is an additional random effect for each randomly varying regression coefficient identified at level 1, each with its own variance, $\tau_{pp'}$, representing variability across the j clusters around the overall or "average" slope for that covariate, γ_{p0}, after controlling for other predictors in the model. In addition, covariances between all random effects in a model are also estimated, which quickly increases the overall number of parameters. There may be up to $[(p+1)*(p)]/2$ variance and covariance components to be estimated in addition to the number of fixed effects estimated in the multilevel model. The general model may also include cross-level interactions, allowing for level-2 predictors to moderate the effect (i.e., slope) of a level-1 covariate on the count outcome. These cross-level interactions arise by including one or more of the Q level-2 predictors in an equation for the slope of a predictor from level one. For example, a general model with cross-level interactions between the variable X_{1ij} and a set of level-2 predictors, W_{qj}, may be written as in Equation 8.10.

$$
\begin{aligned}
\text{Level 1:}\quad & \eta_{ij} = \log(\lambda_{ij}) = \beta_{0j} + \beta_{1j}X_{1ij} + \sum\nolimits_{p=2}^{P}\beta_p X_{pij} \\
\text{Level 2:}\quad & \beta_{0j} = \gamma_{00} + \sum\nolimits_{q=1}^{Q}\gamma_{0q}W_{qj} + u_{0j} \\
& \beta_{1j} = \gamma_{10} + \sum\nolimits_{q=1}^{Q}\gamma_{1q}W_{qj} + u_{1j} \\
& \beta_2 = \gamma_{20} \\
& \vdots \\
& \beta_p = \gamma_{p0} \\
\text{Combined:}\quad & \eta_{ij} = \gamma_{00} + \gamma_{10}X_{1ij} + \sum\nolimits_{q=1}^{Q}\gamma_{0q}W_{qj} + \left(\sum\nolimits_{q=1}^{Q}\gamma_{1q}W_{qj}\right)X_{1ij} \\
& \quad + \sum\nolimits_{p=2}^{P}\gamma_{p0}X_{pij} + u_{1j}X_{1ij} + u_{0j}
\end{aligned}
\tag{8.10}
$$

In this random-intercepts and random-slopes Poisson model, we have two random effects, u_{0j} and u_{1j}. The slopes for the level-1 covariate X_{1ij} have a deterministic part,

$$\gamma_{10} + \sum\nolimits_{q=1}^{Q}\gamma_{1q}W_{qj},$$

as well as a random part, u_{1j}. Similarly, the intercepts contain both a deterministic part,

$$\gamma_{00} + \sum\nolimits_{q=1}^{Q}\gamma_{0q}W_{qj},$$

and a random part, u_{0j}. The two residual or random components (u_{0j} and u_{1j}) follow a bivariate normal distribution with variances τ_{00} and τ_{11}, and covariance τ_{01}. The variance components are typically displayed as belonging to a tau, or T, matrix. In this illustration with two random effects, we would have a 2×2 T matrix, with variances on the diagonal, and covariance(s) off-diagonal.

A simpler form of random-intercepts and random-slopes model would include at least 2 random effects but have no level-2 predictors for any of the slope equations. Without cross-level interactions, the equations could be written as specified previously in Equation 8.8 with the inclusion of at least one random slope from level 1.

Specific research questions and theory should guide the inclusion of level-2 as well as level-1 predictors, along with anticipated cross-level interactions. Reliance on a solid theoretical framework is particularly compelling for discrete outcomes, given the complexity of models for counts or other types of limited dependent variables even prior to the inclusion of random effects.

The random intercepts and slopes model that we fit for the epilepsy data is given in Equation 8.11. No cross-level interactions were included because the sample size for our demonstration was small and the model complexity led to convergence issues. This multilevel Poisson model has eight parameters: five fixed effects (the level-2 γ's) and three covariance components: $\tau_{00} = \text{var}(u_{0j})$, $\tau_{11} = \text{var}(u_{1j})$, and τ_{01} or $\tau_{10} = \text{cov}(u_{0j}, u_{1j})$.

$$
\begin{aligned}
\text{Level 1:}\quad & \eta_{ij} = \log(\lambda_{ij}) = \beta_{0j} + \beta_1(\mathit{visit}_{ij}) \\
\text{Level 2:}\quad & \beta_{0j} = \gamma_{00} + \gamma_{01}(\mathit{c\_logage}_j) + \gamma_{02}(\mathit{c\_logbase}_j) + \gamma_{03}(\mathit{treat}_j) + u_{0j} \\
& \beta_1 = \gamma_{10} + u_{1j} \qquad (8.11) \\
\text{Combined:}\quad & \eta_{ij} = \log(\lambda_{ij}) = \gamma_{00} + \gamma_{10}(\mathit{visit}_{ij}) + \gamma_{01}(\mathit{c\_logage}_j) \\
& \quad + \gamma_{02}(\mathit{c\_logbase}_j) + \gamma_{03}(\mathit{treat}_j) + u_{1j}(\mathit{visit}_{ij}) + u_{0j}
\end{aligned}
$$

Multilevel Negative Binomial Models

Multilevel NB models are relatively straightforward extensions of the multilevel Poisson. The Poisson distribution is based on only one parameter, λ, which represents both the mean and variability. The NB distribution includes two parameters, λ and α, where α characterizes the overdispersion in the data, that is, beyond the variability imposed by the Poisson equidispersion assumption. Extra variability in the counts at level one could be due to unaccounted for clustering, omitted predictors, or other unaccounted for sources. Whatever the source, unaccounted for variability in the counts can be considered as an additional random effect in the model. In the Poisson-Gamma mixture, these level-1 random effects are assumed to come from a Gamma distribution.

In multilevel form, the NB model accounts for overdispersion as well as for clustering in the data. As noted previously, the level-2 random effects are typically considered to follow a normal distribution. Model building, estimation and parameter interpretation mirror that of the multilevel Poisson, with the addition of the NB dispersion parameter in accounting for extra variability in the counts. Thus, multilevel NB models are similar in form to the multilevel Poisson with the addition of the extra variability in counts characterized by the NB dispersion parameter. Accordingly, the null random intercepts model for the P and NB models compared side-by-side are shown in Equation 8.12 (Leckie et al., 2020).

$$
\begin{array}{ll}
\text{Multilevel Poisson} & \text{Multilevel NB} \\
y_{ij} \mid \lambda_{ij} \sim Poisson(\lambda_{ij}) & y_{ij} \mid \lambda_{ij} \sim Poisson(\lambda_{ij}) \\
\ln(\lambda_{ij}) = \gamma_{00} + u_{0j} & \ln(\lambda_{ij}) = \gamma_{00} + u_{0j} + e_{ij} \\
u_{0j} \sim N(0, \tau_{00}) & u_{0j} \sim N(0, \tau_{00}) \\
 & \exp(e_{ij}) \sim Gamma\left(\frac{1}{\alpha}, \alpha\right)
\end{array}
\tag{8.12}
$$

Here, the inclusion of e_{ij} at level 1 in the NB model refers to the variability in the counts that is not explained by predictors in the model, or by clustering or by the Poisson assumption that the mean and variance of the counts are equal. The e_{ij} are random effects for predictions of the natural log of the counts. Their distribution—or specifically, the distribution of $\exp(e_{ij})$—is consistent with the Poisson-Gamma mixture distribution of the NB model. Random draws from a Gamma distribution with location $1/\alpha$ and shape α have mean $E(Y) = (1/\alpha)(\alpha) = 1$; and $\mathrm{Var}(Y) = (1/\alpha)^2(\alpha) = 1/\alpha$ (see Fox, p. 422). The natural log of the mean of these residuals or random effects is 0. The goal is not to estimate these random effects, but to characterize their distribution and estimate the extra dispersion parameter α which guides the estimation of overall variability and quality of prediction for the NB model. With the estimation of α, the mean and variance of the Poisson-Gamma mixture follows the same form as our previous discussion for the NB2, with overall mean λ_i, and variance $\lambda_i + \alpha\lambda_i^2$.

Despite the seeming complexity of the multilevel NB, interpretation of parameter estimates are similar to those for the multilevel Poisson, although the regression coefficients and their standard errors are adjusted for the additional modeling of the extra-dispersion. Thus, Equations 8.5 through 8.11 are the same for the NB models as for the Poisson, with the addition of the scale parameter estimated for NB dispersion.

Note on Unit-Specific Versus Population-Averaged Models

With existing estimation issues induced by modeling a discrete rather than continuous outcome, the additional parameters required in a GLMM with random-intercepts and random slopes may limit the complexity of the model that a researcher is reasonably able to fit. This is the primary reason why random-intercepts models seem to be the predominant form for analyzing multilevel counts. Other options include model estimation through generalized estimating equations (GEE), which does not model the random effects but takes a marginal or population-average approach to account for clustering, treating the variability in the random effects as nuisance parameters rather than explicitly partitioning the variation across multiple levels (Gardiner et al., 2009; McNeish et al., 2017). Hedeker and Gibbons (2006) point out that "GEE models are not meant for situations in which scientific interest centers around the variance and/or covariance parameters" (p. 132). Further, when NB models are employed, GEE tends to overestimate large counts and leads to overdispersion estimates (i.e., the overall dispersion assessed for the GEE approach coupled with an overdispersion parameter from the NB distribution) that overlap in interpretation, affecting meaningful understanding of the NB model (Stroup, 2013). Accordingly, we highlight the application of unit-specific models (also called subject-specific) in this chapter, similar to the application of standard multilevel models used for continuous outcomes.

Partitioning Variance in Mixed Models for Counts

In a mixed model for continuous outcomes the intraclass correlation coefficient (ICC) is used to describe the proportion of variance in the outcome that can be attributed to between-cluster differences. As such, it is a variance partitioning coefficient as well as a correlation, as it also represents the expected correlation between two observations from the same cluster. The larger the ICC, the more similar observations are within clusters and the stronger the clustering effect. With continuous outcomes as in a traditional multilevel model, the ICC is easy to compute and interpret (e.g., Raudenbush & Bryk, 2000). The familiar formula is given by $ICC = \tau_{00} / (\tau_{00} + \sigma^2)$, where τ_{00} represents variability between clusters at level 2, and σ^2 represents variability within clusters at level 1. However, there are complications in the case of count outcomes, stemming from the equidispersion assumption required of the Poisson and because the level-1 and level-2 components of variance are modeled on different scales (count response scale at level 1 and continuous scale at level 2). There is also no reasonable argument to be made for the observed counts to arise as manifestations of some underlying latent variable, an argument that is sometimes utilized when estimating the ICC for multilevel dichotomous or ordinal outcomes (Austin et al., 2018; Rabe-Hesketh & Skrondal, 2012; Stryhn et al., 2006; & Chapter 7 this volume, "Multilevel Logistic and Ordinal Models"). Finally, a normal response approximation for the ICC is also not feasible for multilevel count data, since distributions of counts are typically highly skewed and reverse J-shaped.

In a traditional random-intercepts model for continuous outcomes, the inclusion of covariates doesn't change the form of estimation for the ICC; that is, a residual ICC can readily be calculated using the same expression noted above (Snijders & Bosker, 2012). However, for count outcomes the components of variance across level one and level two of a multilevel model depend on specific covariance patterns, and each covariate pattern would have a different ICC (Austin et al., 2018; Leckie et al., 2020), necessitating some adjustments in how we consider its estimation and interpretation. Austin et al. (2018) review several alternatives to establishing an ICC for count outcomes including an exact calculation (for each covariate pattern) and a simulation-based estimation, among other options. Of these, the two most feasible strategies for calculating ICC for count outcomes are the exact method and simulation, although the exact approach is recommended (Austin et al., 2018).

We follow recent derivations using the exact approach from Leckie et al. (2020) and based on the earlier work by Austin et al. (2018) to estimate the ICC for multilevel Poisson and NB models, focusing on variance partitioning for random-intercepts models with no covariates, that is, the null or unconditional model. The exact approach focuses on the variance partitioning coefficient, or VPC, which is equivalent to the ICC; the VPC represents the ratio of between-cluster variability to total variability (Leckie et al., 2020). Variance partitioning for counts requires a distinction between conditional or cluster-specific models and marginal or population-averaged models. Leckie and colleagues (2020) note that variance partitioning coefficients and intra-cluster correlation statistics are, in fact, marginal statistics (p. 790). In the case of multilevel models for normal outcomes these two approaches are the same, but with counts, marginal expectations are used to understand proportions of variance at each level of the model. One of the important results from the work of Leckie et al. and Austin et al. is that level-1 and level-2 variances for these marginal expectations can be found from the cluster-specific results.

ICC for Multilevel Poisson

Based on the unconditional random-intercepts model for the multilevel Poisson, the following formulas can be used to estimate the proportion of the marginal response variability that lies between clusters (Leckie et al., 2020). Recall that the essential assumption for a model based on the Poisson distribution is that of equidispersion; consequently, variability can be written in terms of expected values for the count outcome. The marginal expectation of the counts, λ_{ij}^{M}, is found by averaging over the random effects, with the result as shown in Equation 8.13. The marginal response variability is given by Equation 8.14; notice that if there is no clustering in the data, τ_{00} would equal zero and the marginal variance would reduce to λ_{ij}^{M}. Similarly, with no clustering effect, the marginal expectation for the counts would simply be γ_{00}. Based on these expressions, the ICC can then be calculated as shown in Equation 8.15. We use our chapter's notation in the equations below, and readers are urged to refer to the Leckie et al. (2020) paper for additional details and derivations.

$$\text{Marginal expectation of } y_{ij} \text{ (Poisson): } \lambda_{ij}^{M} \equiv E(y_{ij}) = \exp\left(\gamma_{00} + \frac{\tau_{00}}{2}\right) \quad (8.13)$$

$$\text{Marginal variance (Poisson): } Var(y_{ij}) = (\lambda_{ij}^{M})^2 \cdot \left[\exp(\tau_{00}) - 1\right] + (\lambda_{ij}^{M}) \quad (8.14)$$

$$\text{ICC(Poisson)} = \text{VPC}_{ij}(P) = \frac{(\lambda_{ij}^{M})^2 \cdot \left[\exp(\tau_{00}) - 1\right]}{(\lambda_{ij}^{M})^2 \cdot \left[\exp(\tau_{00}) - 1\right] + (\lambda_{ij}^{M})} \quad (8.15)$$

For the unconditional multilevel Poisson model, the numerator of $\text{VPC}_{ij}(\text{P})$ represents level-2 variance, and the denominator represents level-2 variance plus level-1 variance. We use (P) to distinguish this expression from the VPC for the NB model. Demonstration is provided for the epilepsy data in our next section.

ICC for Multilevel Negative Binomial

For NB models, similar equations can be derived for the VPC_{ij}(NB), although the marginal variance now also contains the additional source of variability given the NB dispersion parameter, α. The marginal expectation for y_{ij} is averaged over the random effects as well as over the distribution of the NB dispersion, with the equation the same as for the Poisson (Equation 8.13). As derived in Leckie et al. (2020), the variance partitioning coefficient for the NB model, VPC_{ij}(NB), can be determined from Equation 8.16 and is demonstrated for the epilepsy data in our next section. The numerator here represents level-2 variability, and the denominator represents level-2 plus level-1 variability.

$$\text{ICC(NB)} = \text{VPC}_{ij}(\text{NB}) = \frac{(\lambda_{ij}^{M})^2 \cdot \left[\exp(\tau_{00}) - 1\right]}{(\lambda_{ij}^{M})^2 \cdot \left[\exp(\tau_{00}) - 1\right] + \left[(\lambda_{ij}^{M}) + (\lambda_{ij}^{M})^2 \exp(\tau_{00})\alpha\right]} \quad (8.16)$$

DEMONSTRATION OF MULTILEVEL MODELS FOR CLUSTERED COUNTS

For our multilevel models, the cluster-level, or level 2, is the patient or person-level, and the number of seizures across occasions marked by *visit* are at level 1. Table 8.4 presents results of random intercepts models for the epilepsy data across three statistical packages, SAS, R, and Mplus. We

TABLE 8.4 Random Intercepts Models (n = 236 level-1 observations): Multilevel Poisson (P) and Negative Binomial (NB)

	Unconditional Models				Conditional Models (random intercepts)											
	SAS				SAS				R				Mplus			
	Model 2P		Model 2NB		Model 3P		Model 3NB		Model 4P		Model 4NB		Model 5P		Model 5NB	
	B	*SE*	*B*	*SE*	*B*	*SE*	*B*	*SE*	*B*	*SE*	*B*	*SE*	*B*	*SE*	*B*	*SE*
Within																
visit					–0.29**	0.10	–0.27	0.17	–0.29**	0.10	–0.27	0.17	–0.29**	0.10	–0.27	0.17
Between																
Intercept	1.62***	0.13	1.64***	0.13	1.79***	0.11	1.81***	0.11	1.79***	0.11	1.80***	0.11	1.79***	0.11	1.81***	0.11
c_logage[a]					0.34	0.34	0.33	0.34	0.34	0.34	0.33	0.34	0.34	0.34	0.33	0.34
c_logbase[a]					1.03***	0.10	1.03***	0.10	1.03***	0.10	1.03***	0.10	1.03***	0.10	1.03***	0.10
treat					–0.32*	0.15	–0.32*	0.15	–0.32*	0.15	–0.32*	0.15	–0.32*	0.05	–0.32*	0.15
Variability																
Intercepts: τ_{00}	0.89***	0.18	0.86***	0.18	0.27***	.06	0.23***	0.06	0.27		0.23		0.27***	0.06	0.23***	0.06
NB Disp., α $\theta = 1/\alpha$			0.14***	0.03			0.13***	0.03			7.464				0.13***	0.03
Model Fit																
Log(L)									–667.09		–625.94		–666.934		–625.868	
–2Log(L)	1402.50		1314.72		1334.02		1252.02		1334.00		1251.90					
Residual *df*	234		233		230		229		230		229		230		229	
AIC	1406.50		1320.72		1346.02		1266.02		1346.02		1265.88		1345.868		1265.736	
BIC	1410.65		1326.95		1358.48		1280.56		1366.80		1290.13		1366.651		1289.983	

[a] log(age), centered; log(baseline seizures), centered

* $p \le .05$; ** $p \le .01$; *** $p \le .001$

Note: `lme4` package in R provides the inverse of the dispersion parameter estimate and does not provide a standard error for the variance components.

present results for each of the three statistical packages to highlight important similarities and differences due to numerical estimation procedures, scaling of the NB dispersion parameter, and presentation on statistical output. Models 2P and 2NB are the unconditional (null) models for the multilevel Poisson and multilevel NB, respectively, with results presented only in SAS for space considerations. These two models were fit based on Equation 8.5 with inclusion of extra dispersion for the NB model. The remaining models in Table 8.4 were fit as random intercepts models with four predictors: *visit* at level 1, and *c_logage*, *c_logbase*, and *treat* at level 2. In these models, only the intercept was allowed to vary at random, and they correspond to Equation 8.9 with extra dispersion added for the NB models.

Results for random intercepts and slopes models for the epilepsy data are provided in Table 8.5 and were fit via SAS (6P and 6NB), R (7P and 7NB), and Mplus (8P and 8NB). In these models, the effect of *visit* was allowed to randomly vary across persons. The corresponding equation for these multilevel models was presented in Equation 8.11.

Guiding Questions

A series of questions can help guide the choice among options for modeling clustered counts. First, to what degree is clustering present in the data, and does adjusting for clustering improve model fit? Second, after adjusting for clustering, is the assumption of equidispersion reasonable, or should this assumption be relaxed? Third, does inclusion of covariates improve model fit, and should they be treated as fixed or random effects? Fourth, how are parameter estimates interpreted in multilevel models for counts? We work through each of these questions below in our demonstration, beginning with random intercepts models before examining models with random intercepts and slopes.

Random Intercepts Models

Degree of Clustering

A first step in multilevel modeling is to clarify the degree of clustering in the data, which is typically addressed prior to the inclusion of covariates in a model. We use the variance partitioning formulas presented in the previous section for Models 2P and 2NB, the unconditional random intercepts models for the epilepsy data. Following Equation 8.13, which can be applied to results for both the Poisson and NB unconditional models, the marginal expectation of number of seizures based on results of Model 2P is 7.89: $\lambda_{ij}^{M} = \exp(\gamma_{00} + \tau_{00} / 2) = \exp\big(1.62 + (0.89 / 2)\big) = \exp(2.065) = 7.89$. This

TABLE 8.5 Random Intercepts and Random Slopes (n = 236 level-1 observations): Multilevel Poisson (*P*) and Negative Binomial (*NB*)

	SAS				R				Mplus			
	Model 6P		Model 6NB		Model 7P		Model 7NB		Model 8P		Model 8NB	
	R	*SE*	*B*	*SE*	*B*	*SE*	*B*	*SE*	*B*	*SE*	*B*	*SE*
Within												
visit	–0.26	0.16	–0.26	0.17	–0.26	0.16	–0.25	0.18	–.26	0.16	–.27	0.18
Between												
Intercept	1.78***	0.11	1.81***	0.11	1.78***	0.11	1.80***	0.11	1.78***	0.11	1.81***	0.11
c_logage[a]	0.32	0.35	0.32	0.35	0.32	0.35	0.32	0.35	0.32	0.35	0.32	0.35
c_logbase[a]	1.03***	0.10	1.03***	0.10	1.03***	0.10	1.03***	0.10	1.03***	0.10	1.03***	0.10
treat	–0.31*	0.15	–0.32*	0.15	–0.31*	0.15	–0.32*	0.15	–0.32*	0.15	–0.31*	0.15
Variability												
Intercepts: τ_{00}	0.26***	0.06	0.23***	0.06	0.26		0.23		0.27***	0.06	0.23***	0.06
Slopes: τ_{11}	0.53*	0.23	0.001	.38	0.53		0.002		0.53*	0.23	0.02	0.33
NB Disp., α $\theta = 1/\alpha$			0.13***	0.04			7.464				0.13***	0.04
Model fit												
Log(L)					–657.09		–625.90		–657.02		–625.87	
–2LL	1314.17		1251.97		1314.20		1251.80					
Residual *df*	228		227		228		227		228		227	
AIC	1330.17		1269.97		1330.2		1269.8		1330.04		1269.75	
BIC	1346.79		1288.67		1357.9		1301		1357.75		1300.92	

[a] log(age), centered; log(baseline seizures), centered

* $p \le .05$; ** $p \le .01$; *** $p \le .001$

Note: `lme4` package in R provides the inverse estimate of the dispersion parameter for a Negative Binomial model. It also provides only estimates of standard deviation for the variance components and not the standard error.

estimated mean is slightly below the observed mean number of seizures for the data, $\bar{Y} = 8.26$, previously reported in Table 8.1. Applying Equation 8.14 for the results of Model 2P, the ICC(P) = $\text{VPC}_{ij}(\text{P}) = .9194$:

$$\left[7.89^2 * \left(\exp(0.89) - 1\right)\right] / \left\{\left[7.89^2 * \left(\exp(0.89) - 1\right)\right] + 7.89\right\}$$
$$= 89.34 / (89.34 + 7.89) = .9194$$

This is quite large but not unexpectedly so, given that our data represents repeated measures on the same subjects. To interpret, for the Poisson multilevel model 91.94% of the variability in counts of seizures is estimated to be between-subjects (level-2). This is after accounting for clustering by including a random-intercept.

For the unconditional NB model (Model 2NB), Wald's test can be used to assess the contribution of the extra dispersion parameter, which is found to be statistically different from zero, $\alpha = 0.14$ ($p < .001$). The dispersion parameter is incorporated into the assessment of variability for the ICC in the NB model. To determine the ICC based on estimates for Model 2NB, we first find $\lambda_{ij}^M = \exp(\gamma_{00} + \tau_{00}/2) = \exp\left(1.64 + (0.86/2)\right) = \exp(2.07) = 7.92$. For the numerator of the ICC(NB) formula provided in Equation 8.16, which represents level-2 variance, we have $(\lambda_{ij}^M)^2\left(\exp(\tau_{00}) - 1\right) = 7.92^2\left(\exp(0.86) - 1\right) = 62.73(1.36) = 85.51$. For the second half of the denominator of the ICC(NB) formula, representing level-1 variance, we have $\lambda_{ij}^M + (\lambda_{ij}^M)^2\left(\exp(\tau_{00})(\alpha)\right) = 7.92 + 7.92^2\left(\exp(0.86)(.14)\right) = 78.67$. Thus we have ICC(NB) = $\text{VPC}_{ij}(\text{NB})$ = 85.51/(85.51 + 78.67) = .5208. After accounting for clustering and extra dispersion via Model 2NB, 52.08% of the variance in seizures lies between-subjects. The ICC(NB) is smaller than the ICC(P). The Poisson model ignores overdispersion of the counts, and thus overestimates the importance of cluster-level or person-level contributions to variability in the numbers of seizures experienced per person.

A related question is: "For either the P or NB models, does adjusting for clustering improve model fit?" A likelihood ratio test comparing the unconditional multilevel Poisson Model 2P to the unconditional single-level Poisson Model 0P is statistically significant, $\chi^2(1) = [-2\text{LL}_{\text{Model }0P} - (-2\text{LL}_{\text{Model }2P})] = 3287.74 - 1402.50 = 1885.24$, $p < .0001$, based on a difference of 1 parameter for the random intercept. This indicates substantial variability between persons in number of seizures, affirming the need for accommodation of person-level clustering for the Poisson models. Similarly, we can compare the multilevel unconditional NB Model 2NB to the single-level unconditional NB Model 0NB, again with a difference of 1 parameter added for the random intercept, $\chi^2(1) = [-2\text{LL}_{\text{Model }0NB} - (-2\text{LL}_{\text{Model }2NB})] = 1495.06 - 1314.72 = 180.34$, $p < .0001$, which indicates that the addition of the random intercept in the NB model provides significantly better fit. Overall, both the Poisson and the NB models are improved when clustering is taken into account.

Contribution of Overdispersion

Our second guiding question was: "After adjusting for clustering in the data, does allowing for overdispersion improve model fit?" Given that clustering is substantial in this data, as evidenced by the size of the VPC's for the unconditional multilevel Poisson and NB models, and the results of the likelihood ratio tests above, a next step is to clarify the presence of overdispersion in the multilevel models. We can compare the multilevel unconditional NB Model 2NB to the multilevel unconditional Poisson Model 2P, which differ by the 1 parameter added for the NB dispersion. The likelihood ratio test yields $\chi^2(1) = [-2LL_{\text{Model } 2P} - (-2LL_{\text{Model } 2NB})] = [1402.50 - 1314.72] = 87.78$, $p < .0001$, which confirms the need to allow for overdispersion even when adjusting for clustering. Thus, modeling overdispersion in these multilevel models via the NB distribution improves the fit relative to the Poisson assumption of equidispersion.

Overall, results of these comparisons support the need to accommodate for overdispersion in addition to clustering for these data. The unconditional multilevel NB model would be preferred over the unconditional multilevel Poisson. This comparison can be revisited for more complex models once covariates are considered.

Inclusion of Covariates in the Random-Intercepts Models

Our next step is to compare the random-intercepts NB model with full covariates (Model 3NB) to the unconditional model (Model 2NB). In the full model, *visit* is treated as a fixed effect, as are *c_logage, c_logbase,* and *treat.* Using a likelihood ratio test we find $\chi^2(4) = [-2LL_{\text{Model } 2NB} - (-2LL_{\text{Model } 3NB})] = [1314.72 - 1252.02] = 62.7$, $p < .0001$, confirming that adding the four predictors to the model as fixed effects improves the fit. With covariates included, the overdispersion parameter in Model 3NB is significantly different from zero, $\alpha = 0.13$ ($p < .001$) and changed only slightly from Model 2NB ($\alpha = .14$). Although not necessary given our preference for the NB model, we note a similar conclusion regarding the inclusion of covariates for the Poisson models, comparing Model 3P to Model 2P: $\chi^2(4) = [-2LL_{\text{Model 2P}} - (-2LL_{\text{Model } 3P})] = [1402.50 - 1334.02] = 68.48$, $p < .0001$.

Since adding predictors can reduce variability, we can also assess the need for the NB model once covariates are added by comparing the fit of the four-predictor NB model to that of the corresponding Poisson. We find that Model 3NB provides significant reduction in deviance relative to Model 3P: $\chi^2(1) = [-2LL_{\text{Model } 3P} - (-2LL_{\text{Model } 3NB})] = [1334.02 - 1252.02] = 82.00$, $p < .0001$, on a difference of 1 parameter added for the extra dispersion parameter.

This step isn't completely necessary, but it does offer additional support for Model 3NB.

Interpretation of Model Parameters

Before shifting to consideration of whether random slopes are warranted, we focus our initial interpretations on the full random-intercept models, Model 3NB and 3P. While the NB model would be preferred based on the comparisons above, we discuss both models to clarify similarities and differences. The predicted outcome for the multilevel models, $\ln(Y_{ij})$, is the log of the number of seizures for the jth person at the ith visit. The values of the parameter estimates themselves do not visually vary too much between Models 3NB and 3P, but the standard error for the effect of *visit* does; in fact, the slope for *visit* is not significantly different from zero in Model 3NB, while it is significantly different from 0 for Model 3P. With the inclusion of extra variability captured at level 1 in the NB model, the effect of linear change (in the log of the counts) over time by *visit* is no longer significant.

In Model 3P, exponentiating the intercept can be interpreted in the same way as the single-level models, with the additional consideration of the random effects. For example, $\exp(\gamma_{00}) = \exp(1.79) = 5.99$, which is the estimated number of seizures for a participant at the midpoint of the study (*visit* = 0), in the placebo group (*treat* = 0), and with average values for log(age) and log(baseline number of seizures) (i.e., *c_logage* = 0 and *c_logbase* = 0), given a random effect of 0. For Model 3NB, this estimate is $\exp(1.81) = 4.05$. The difference is due to the inclusion of overdispersion allowed by the NB model. Specifically, there is more unaccounted for variability in the number of seizures based on the Poisson model. Note that the intercept variance is slightly larger in Model 3P than in Model 3NB: For Model 3P, $\tau_{00} = .27$ ($p < .001$) and for Model 3NB, $\tau_{00} = .23$ ($p < .001$).

Tests of the variance components for the random effects are often applied using one-sided tests, since they represent variance which is a positive quantity. Consider Model 3NB as an example. The fixed effect estimate for an individual (i.e., cluster) with a value of zero for all covariates was found to be $\gamma_{00} = 1.81$, in the metric of ln(counts). The variance component for the intercept, τ_{00}, represents cluster-level or person-level variation around this estimate. Taking the square root of the variance component, $\sqrt{\tau_{00}} = .48$, gives an idea of the range of plausible values for the intercept for these data. Recall that GLMM's assume that the random effects are normally distributed (see discussion following Equation 8.5); thus a 68% plausible range for the individuals in the sample for the ln(counts) when all covariates are zero (i.e., for a placebo patient) is 1.33 to 2.29 ($\gamma_{00} \pm 1(.48)$, i.e., ± 1 standard deviation). When exponentiated to provide predicted counts, the predictions range from 3.78 seizures to 9.87. This is quite a salient range for persons experiencing seizures and gives a substantive interpretation to the size of the variance component. Further, the contribution of the extra dispersion parameter implies there is additional variability in the counts.

In either the multilevel Poisson or NB models, fixed effect estimates for the predictors can be exponentiated and interpreted as rate ratios, as was previously demonstrated for the single-level models. Here, the effect of *treat* is coincidentally the same for Models 3NB and 3P, $\gamma_{00} = -0.32$, $p < .05$. To interpret the effect of *treat*, we exponentiate the coefficient to obtain the *RR* between the drug and placebo groups: $\exp(-.32) = .73$. Controlling for the other predictors in the model and for a "typical" participant (i.e., with a random effect of 0), the number of seizures in the drug group is estimated to be about .73 times the number of seizures in the placebo group, a reduction of $100(0.73 - 1)$ or 27%. This is a statistically significant difference ($p < .05$).

It is instructive to examine differences in slope parameters between the single-level (Models 1P and 1NB) and the multilevel models (Models 3P and 3NB). Specifically, in contrast to the single-level Poisson model (1P), the effect of treatment is statistically significant in Model 3P ($b_{treat} = -0.32$, $p < .05$), while it was not statistically significant in Model 1P ($b_{treat} = -0.02$, $p > .05$). For the NB models, the effect of treatment was statistically significant in Model 3NB ($b_{treat} = -0.32$, $p < .05$), as well as for Model 1NB ($b_{treat} = -0.23$, $p < .05$). Further, the values for the coefficients for *treat* are larger (more negative) in both multilevel models relative to their single-level counterparts. Thus, accounting for clustering as well as overdispersion allows the intervention effect to be more clearly recognized. Interpretation of other regression parameters in the multilevel models can proceed similar to earlier discussions, with the additional consideration that model predictions are referenced to a "typical" cluster (here, person) with a random effect of zero.

Random Intercepts and Slopes Models

Including Effects as Fixed or Random

We now consider the effect of *visit* in our models and examine its possible inclusion as a random effect (Models 6P and 6NB). This allows the *visit* slopes to vary by person. Results for these models follow Equation 8.11 with the overdispersion parameter estimated for the NB model.

Comparing the multilevel Poisson models with and without random slopes (Models 6P and 3P, respectively), we see that including the random effect for *visit* provides a significant reduction in the deviance and improved model fit, $\chi^2(2) = [-2LL_{\text{Model 3P}} - (-2LL_{\text{Model 6P}})] = [1334.02 - 1314.17] = 19.85$, $p < .0001$. Degrees of freedom are based on the two additional parameters, τ_{11} and τ_{01} (covariance estimates are not shown in Table 8.5 but they are nonetheless estimated in the analysis). Note that the slope for *visit* is no longer statistically significant in Model 6P, $\gamma_{00} = -0.26$, $p > .05$; however, the variability in slopes is statistically greater than 0 based on Wald's test, $\tau_{11} = .53$, $p = .0106$ (one-sided). Although the overall slope for *visit* in the random slopes model may be 0, there is evident variability between patients in their individual predictions of seizures based on *visit.*

In contrast, there is essentially no improvement from the addition of a random effect for *visit* in the NB models. For the NB models with and without random slopes for *visit* (Model 6NB and 3NB, respectively), the −2LL for Model 6NB is not changed ($-2LL_{\text{Model 6NB}} = 1251.97$). Note also that the slope variance for Model 6NB, $\tau_{11} = .001$, is quite small and not statistically significant, suggesting no variability across persons in the effect of *visit.* For the multilevel NB, a model with a fixed-effect regression slope for *visit* is preferred. Overall, the Poisson model is extremely vulnerable to overdispersion, and as we've seen in our model comparisons above there is still overdispersion present in the data even after adjusting for clustering via the multilevel Poisson. Thus a flawed interpretation of the contribution of *visit* as a random effect may arise from ignoring this overdispersion when the Poisson model is applied.

Parameter Interpretation for the Random Slopes Model

Interpretations of slopes for all predictor variables in the random-slopes models are consistent with our previous discussions. Parameter estimates represent the expected change in the log of the number of seizures as that variable increases by 1 unit, controlling for other predictors in the model; and exponentiating the coefficient yields a rate ratio for that respective variable. The parameter estimates are interpreted the same way in Poisson or NB models. Given that Model 6NB was not preferred over Model 3NB, we will interpret selected parameter estimates for the Poisson random slopes model, Model 6P. Variance components have been previously discussed so we will focus on the fixed effects.

For the intercept in Model 6P, exp(1.78) = 5.93. This is the expected number of seizures for a participant at the midpoint of the study (*visit* = 0), in the placebo group (*treat* = 0), and with average values for log(age) and log(baseline number of seizures) (i.e., *c_logage* = 0 and *c_logbase* = 0), given random effects of 0. *Visit* is not a statistically significant predictor of log(seizures). However, recall that *visit* was coded in in increments of two-tenths (*visit* = −0.3, −0.1, +0.1, +0.3), thus for interpretation of the slope for *visit,* we note that the log(counts) are expected to decrease by $\gamma_{10} = -0.26$ for each one-unit increment in clinic visits, which implies a decrease of exp(−0.29) = 0.748, or by .75 for each one-unit increment in *visit,* holding all other predictors constant. Given the coding scheme in increments of $\frac{2}{10}$, we adjust the RR according to Equation 8.3: $RR_{visit} = \exp\left(\frac{2}{10} \cdot (-0.26)\right) = \exp(-0.052) = 0.949$. This corresponds to change of about 100(0.95 − 1) = −.05 or 5% reduction in seizures for each additional increment in time. The RR for *c_logage* is exp(.32) = 1.377 (n.s.) and for *c_logbase* is exp(1.03) = 2.80 ($p < .001$), representing positive multiplicative effects on the expected number of seizures as these variables increase by 1 unit. Finally, for the treatment effect, exp(–0.31) = .73 ($p < .05$), representing approximately an estimated 27% reduction in seizures for those in the progabide group, which is a statistically significant difference.

SUMMARY OF THE MULTILEVEL MODELS

To complete our comparison of the multilevel Poisson and NB models, we note that the model with the smallest information criteria for both AIC and BIC is Model 3NB (AIC = 1266.02). This is the NB model with only the intercept allowed to vary at random, and this finding is consistent with the many results and comparisons described above that support the preference for Model 3NB. For the epilepsy data, the choice for a multilevel model was indicated by design (counts of seizures nested within patients) but also through our examination of how clustering affects the interpretation of our model. The multilevel Poisson and NB models were both superior to their single-level counterparts at estimating number of seizures. Yet in comparing results between the multilevel Poisson versus multilevel NB, we observed how variability appeared to be redistributed at level two of the Poisson model, particularly for its indication of significant variability in the random effects for *visit*. With the inclusion of the dispersion parameter in the NB model—which relaxes the strict assumption of equal mean and variance in seizure counts that the Poisson induces—variability in the effect of *visit* is reflected in a very different way, with no significant slope for *visit* (Model 3NB and Model 6NB), and no significant variability in the slopes for *visit* (Model 6NB). Further we saw how the NB models enhanced the detection of a treatment effect. Thus, overall, Model 3NB would be the preferred model. While these results and their implications are clearly specific to the data set we used for our demonstration, we have highlighted model interpretation and a model comparison process that should be useful to others confronted with the complex choice of how best to model clustered count data.

Extensions: Two-Part Models

Overdispersion may arise from the presence of "extra" zeros in the data. Both zero-inflated (Lambert, 1992) and hurdle models (Mullahy, 1986) are used to analyze count data with excess zeros (Hu et al., 2011; Rose et al., 2006; Zeileis et al., 2008). These two types of models are distinct from each other in their way of interpreting and treating zero observations. A zero-inflated model assumes that zero observations are from two distinct sources: one structural source, for which no positive counts are observed, and one sampling source, for which both zero and positive counts are observed. The zero-only part, that is, structural zeros, are modeled through a binomial distribution, while the remaining zero and non-zero observations, that is, sampling zeros, are simultaneously modeled through a Poisson or NB distribution (Hu et al., 2011; Hester & Hartman, 2017). In contrast, hurdle models assume that all zero observations come from the structural source. The assumption is that there is a threshold, or "hurdle," and a positive observation occurs once the "hurdle" is crossed (Mullahy et al., 1986). A hurdle model analyzes the data through two separate processes, a binomial distribution governing the modeling of non-zero observations, and a truncated Poisson or NB distribution governing the non-zero or positive observations. Research indicates that results from zero-inflated and hurdle models are similar. The theoretical considerations about the source of zero observations are key to decide which approach is more appropriate (Hester & Hartman, 2017).

Software Considerations

Data and syntax in SAS, R, and Mplus can be found in the online Appendix for the models demonstrated in this chapter. We also highlight important differences in estimation and modeling across these packages, but remind the reader to carefully consult the software manuals or online support materials for their software of choice, as updates and options can change rapidly.

RECOMMENDED READINGS

Technical Readings

Hilbe, J. M. (2011). *Negative binomial regression* (2nd ed.). Cambridge University Press.

McNeish, D. (2019). Poisson multilevel models with small samples. *Multivariate Behavioral Research, 54*(3), 444–455.

Leckie, G., Browne, W. J., Goldstein, H., Merlo, J., & Austin, P. C. (2020). Partitioning variance in multilevel models for count data. *Psychological Methods, 26*(6), 787–801.

Applied Readings

Heck, R. H., Thomas, S. L., & Tabata, L. N. (2012). *Multilevel modeling of categorical outcomes using IBM SPSS.* Routledge.

Grimm, K. J., & Stegmann, G. (2019). Modeling change trajectories with count and zero-inflated outcomes: Challenges and recommendations. *Addictive Behaviors, 94*, 4–15.

Aiken, L. S., Mistler, S. A., Coxe, S., & West, S. G. (2015). Analyzing count variables in individuals and groups: Single-level and multilevel models. *Group Processes & Intergroup Relations, 18*(3), 290–214.

TRY THIS!

Data for the "Try This!" example are modeled after proficiency data from the ECLS-K (https://nces.ed.gov/ecls/kindergarten.asp) and have been simulated/modified to fit a count outcome and correspond to models demonstrated in Chapter 8. Data match a study in which proficiency counts were collected from a sample of $n = 11,301$ kindergarten children sampled from within $J = 720$ schools. Proficiency was scored based on six early reading skill items and six early math skill items, and summed for a total count of number of items correct (range is 0 to 12). Variables at the child level (level one) and school level (level 2) are described below. Consider the guiding questions posed in this chapter to decide on either multilevel Poisson or multilevel NB models, and address the following questions. In this example, focus on the treatment of the three level-1 predictors as random or fixed before adding the level-2 predictors. Syntax and annotation are provided in the online appendix.

Child Level

Y = profcount = proficiency count
X1 = male (0 = female; 1 = male)
X2 = c_momed (centered version of mother's highest level of education)
X3 = foodinsec (coded 0 for child is food secure; 1 for child is food insecure)

School Level (level-2 identifier is "schoolid")

W1 = public (0 = private school; 1 = public school)
W2 = nbhocdprobs (range is from 0 to 22, with higher value indicating greater severity; summed score on severity of 11 items (racial tension, litter, drug activity, etc., each rated from 0 to 2) summed to represent severity of community problems in the neighborhood of the school).

1. To what degree is clustering present in the data, and does clustering improve model fit?
   - Begin by looking at descriptive statistics and a histogram for the proficiency outcome.
   - For comparison with the multilevel models, fit the single-level null and the three-variable Poisson and NB models. Examine the dispersion parameter for the NB model. What can you conclude regarding the fit of the Poisson versus the NB model?
   - Next, fit the null multilevel Poisson and null multilevel NB models. Determine and interpret the ICC for each model.
   - Finally, how do each of these models compare to their single-level counterpart models? What decision would you make regarding the presence and degree of clustering in the data?
2. After adjusting for clustering, is the assumption of equidispersion reasonable, or should this assumption be relaxed?
   - How do the null multilevel Poisson and NB models compare to each other? Which model would you choose for further development?
3. Does inclusion of covariates improve model fit, and should any level-1 predictors be treated as random or fixed?
   - Based on your decision for Step 2, add the level-1 predictors to a random-intercepts model. Compared to the null model, what can you conclude about the addition of the covariates?
   - Next, build a random coefficients model and allow the level-1 predictors to vary at random.
   - Make a decision on how the level-1 predictors should be incorporated into the model. Since this is a very large sample, this decision may be based on substantive/practical grounds as well as statistical.
   - Based on results, make a decision on your final model.
4. For your final model from Step 3, interpret the parameter estimates, and write a brief results section summarizing your findings.

NOTE

1. *Description of coding scheme*: The number of seizures occurring over a 2-week span was collected at clinic visits every 2 weeks for a total of four clinic visits. There are many options available for coding the passage of time (Biesanz et al., 2004). One option for coding the linear passage of time would be 0, 1, 2, 3, with a change of one unit corresponding to a 2-week time difference. Another option for coding linear time is –3, –1, 1, 3. Here, a change of 2 units corresponds to the number of weeks between visits, and the midpoint or center of the coding scheme is 0, which makes the intercept interpretable

as the prediction at the midpoint of the study. Finally, coding time as –.3, –.1, .1, .3 divides the passage of time into tenths; in the current analysis this increases the size of the regression coefficients by 10, and since the coefficients are somewhat small in either of the previous coding schemes, this adjustment helps make the coefficients more efficient for table form. All three of these examples are linear; thus predictions for each consecutive visit are the same regardless of coding scheme use. We interpret the coefficient for *visit* as the expected change for each consecutive clinic visit in the study, with the difference between coded values as two-tenths.

REFERENCES

Agresti, A. (2007). *An introduction to categorical data analysis* (2nd ed.). John Wiley & Sons.

Agresti, A. (2013). *Categorical data analysis* (3rd ed.). John Wiley & Sons.

Atkins, D. C., Baldwin, S. A., Zheng, C., Gallop, R. J., & Neighbors, C. (2013). A tutorial on count regression and zero-altered count models for longitudinal substance use data. *Psychology of Addictive Behaviors, 27*(1), 166.

Austin, P. C., Stryhn, H., Leckie, G., & Merlo, J. (2018). Measures of clustering and heterogeneity in multilevel Poisson regression analyses of rates/count data. *Statistics in Medicine, 37*(4), 572–589.

Azen, R., & Walker, C. M. (2011). *Categorical data analysis for the behavioral and social sciences.* Routledge.

Biesanz, J. C., Deeb-Sossa, N., Papadakis, A. A., Bollen, K. A., & Curran, P. J. (2004). The role of coding time in estimating and interpreting growth curve models. *Psychological Methods, 9*(1), 30.

Bilder, C. R., & Loughin, T. M. (2015). *Analysis of categorical data with R.* CRC Press.

Binici, S. (2014). *Preschool teachers' inferential questions during shared reading and their relation to low-income children's reading comprehension at kindergarten and first grade* [Unpublished doctoral dissertation]. The Ohio State University.

Bolker, B. (and others). (2021, January 20). *GLMM FAQ*. https://bbolker.github.io/mixedmodels-misc/glmmFAQ.html

Brown, H., & Prescott, R. (2006). *Applied mixed models in medicine* (2nd ed.). John Wiley & Sons.

Cameron, A. C., & Trivedi, P. K. (1990). Regression-based tests for overdispersion in the Poisson model. *Journal of econometrics, 46*(3), 347–364.

Cameron, A. C., & Trivedi, P. K. (2013). *Regression analysis of count data* (2nd ed.). Cambridge University Press.

Coxe, S., West, S. G., & Aiken, L. S. (2009). The analysis of count data: A gentle introduction to Poisson regression and its alternatives. *Journal of personality assessment, 91*(2), 121–136.

Diaz, R. E. (2007). Comparison of PQL and Laplace 6 estimates of hierarchical linear models when comparing groups of small incident rates in cluster randomized trials. *Computational Statistics and Data Analysis, 51*, 2871–2888.

Dobson, A. J., & Barnett, A. G. (2018). *An Introduction to Generalized Linear Models* (4th ed.). CRC Press.

Dunn, P. K., & Smyth, G. K. (2018). *Generalized linear models with examples in R.* Springer.

Fox, J. (2017). *Applied regression analysis and generalized linear models* (3rd ed.). SAGE.

Friendly, M., & Meyer, D. (2016). *Discrete data analysis with R: Visualization and modeling techniques for categorical and count data.* CRC Press.

Gardiner, J. C., Luo, Z., & Roman, L. A. (2009). Fixed effects, random effects and GEE: What are the differences? *Statistics in Medicine, 28*(2), 221–239.

Gardner, W., Mulvey, E. P., & Shaw, E. C. (1995). Regression analyses of counts and rates: Poisson, overdispersed Poisson, and negative binomial models. *Psychological Bulletin, 118*(3), 392.

Grimm, K. J., & Stegmann, G. (2019). Modeling change trajectories with count and zero-inflated outcomes: Challenges and recommendations. *Addictive Behaviors, 94,* 4–15.

Harrell, F. (2015). *Regression modeling strategies with applications to linear models, logistic and ordinal regression, and survival analysis* (2nd ed.). Springer.

Heck, R. H., Thomas, S. L., & Tabata, L. N. (2012). *Multilevel modeling of categorical outcomes using IBM SPSS.* Routledge.

Hedeker, D., & Gibbons, R.D. (2006). *Longitudinal data analysis.* Wiley.

Hester, R., & Hartman, T. K. (2017). Conditional race disparities in criminal sentencing: A test of the liberation hypothesis from a non-guidelines state. *Journal of Quantitative Criminology, 33*(1), 77–100.

Hilbe, J. M. (2011). *Negative binomial regression.* Cambridge University Press.

Hilbe, J. M. (2014). *Modeling count data.* Cambridge University Press.

Hox J. (2010). *Multilevel modeling: Techniques and applications* (2nd ed). Routledge.

Hox, J. J., Moerbeek, M., & Van de Schoot, R. (2018). *Multilevel analysis: Techniques and Applications* (3rd ed.). Routledge.

Hu, M. C., Pavlicova, M., & Nunes, E. V. (2011). Zero-inflated and hurdle models of count data with extra zeros: Examples from an HIV-risk reduction intervention trial. *The American Journal of Drug and Alcohol Abuse, 37*(5), 367–375.

Justice, L. M., Kaderavek, J. N., Fan, X., Sofka, A., & Hunt, A. (2009). Accelerating preschoolers' early literacy development through classroom-based teacher–child storybook reading and explicit print referencing. *Language, Speech, and Hearing Services in Schools, 40*(1), 67–85.

Kleiber, C., & Zeileis, A. (2008). *Applied econometrics with R.* Springer Science & Business Media.

Lambert, D. (1992). Zero-inflated Poisson regression, with an application to defects in manufacturing. *Technometrics, 34*(1), 1–14.

Leckie, G., Browne, W. J., Goldstein, H., Merlo, J., & Austin, P. C. (2020). Partitioning variation in multilevel models for count data. *Psychological methods.*

Leppik, I. E., Dreifuss, F. E., Porter, R., Bowman, T., Santilli, N., Jacobs, M., . . . & Gutierrez, A. (1987). A controlled study of progabide in partial seizures: methodology and results. *Neurology, 37*(6), 963–963.

Long, J. S., & Freese, J. (2014). *Regression models for categorical dependent variables using Stata* (3rd ed). Stata Press.

McCullagh, P., & Nelder, J. A. (1989). Generalized Linear Models (Vol. 37). CRC Press.

McCulloch, C. E., & Searle, S. R. (2001). *Generalized, linear, and mixed models.* Wiley.

McNeish, D. (2016). Estimation methods for mixed logistic models with few clusters. *Multivariate Behavioral Research, 51*(6), 790–804.

McNeish, D. (2019). Poisson multilevel models with small samples. *Multivariate Behavioral Research, 54*(3), 444–455.

McNeish, D., Stapleton, L. M., & Silverman, R. D. (2017). On the unnecessary ubiquity of hierarchical linear modeling. *Psychological Methods, 22*(1), 114–140.

Mullahy, J. (1986). Specification and testing of some modified count data models. *Journal of econometrics, 33*(3), 341–365.

Pinheiro, J. C., & Bates, D. M. (2006). Approximations to the log-likelihood function in the nonlinear mixed effects model. *Journal of Computational and Graphical Statistics, 4,* 12–35.

Powers, D., & Xie, Y. (2008). *Statistical methods for categorical data analysis.* Emerald Group Publishing.

R Core Team (2021). *R: A language and environment for statistical computing.* R Foundation for Statistical Computing. Vienna. https://www.R-project.org/

Rabe-Hesketh, S., & Skrondal, A. (2009). Generalized linear mixed-effects models. In G. Fitzmaurice, M. Davidian, G. Verbeke, & G. Molenberghs (Eds.), *Longitudinal Data Analysis* (pp. 79–106). CRC Press.

Rabe-Hesketh, S., & Skrondal, A. (2012). *Multilevel and longitudinal modeling using stata: Volume II: Categorical responses, counts, and survival.* Stata Press.

Raudenbush, S. W., & Bryk, A. S. (2002). *Hierarchical linear models: Applications and data analysis methods* (2nd ed.). SAGE.

Raudenbush, S. W., Yang, M. L., & Yosef, M. (2000). Maximum likelihood for generalized linear models with nested random effects via high-order, multivariate Laplace approximation. *Journal of computational and Graphical Statistics, 9*(1), 141–157.

Rose, C. E., Martin, S. W., Wannemuehler, K. A., & Plikaytis, B. D. (2006). On the use of zero-inflated and hurdle models for modeling vaccine adverse event count data. *Journal of Biopharmaceutical Statistics, 16*(4), 463–481.

SAS Institute. (2016). *SAS/STAT 13.1 User's Guide: The GLIMMIX Procedure.* SAS Institute.

Skrondal, A., & Rabe-Hesketh, S. (2004). *Generalized latent variable modeling: Multilevel, longitudinal, and structural equation models.* Chapman & Hall/CRC.

Snijders, T. A. B., & Bosker, R. J. (2012). *Multilevel analysis: An introduction to basic and advanced multilevel modeling* (2nd ed.). SAGE.

Stokes, M. E., Davis, C. S., & Koch, G. G. (2000). *Categorical data analysis using the SAS system* SAS Institute.

Stroup, W. W. (2013). *Generalized linear mixed models: Modern concepts, methods and applications.* CRC Press.

Stryhn, H., Sanchez, J., Morley, P., Booker, C., & Dohoo, I. R. (2006, August). Interpretation of variance parameters in multilevel Poisson regression models. In *Proceedings of the 11th International Symposium on Veterinary Epidemiology and Economics* (Vol. 702), Cairns, Australia.

Thall, P. F., & Vail, S. C. (1990). Some covariance models for longitudinal count data with overdispersion. *Biometrics,* 657–671.

Venables, W. N., & Ripley, B. D. (2002). *Modern Applied Statistics with S.* Springer.
Zeileis, A., Kleiber, C., & Jackman, S. (2008). Regression models for count data in R. *Journal of Statistical Software, 27*(8), 1–25.

SECTION II

LONGITUDINAL DATA

CHAPTER 9

INDIVIDUAL GROWTH CURVE MODELS FOR LONGITUDINAL DATA

D. Betsy McCoach
University of Connecticut

Bethany A. Bell
University of Virginia

Aarti P. Bellara
The Ohio State University

This chapter introduces individual growth curve modeling within a multilevel framework and provides an overview of three of the simplest and most commonly used individual growth models: the linear growth model, the piecewise linear growth model, and the polynomial growth model. We include an applied example using NLSY97 data (U.S. Bureau of Labor Statistics, 2019) on frequency of alcohol consumption from ages 18 to 25 to illustrate the development and interpretation of individual growth models within a multilevel modeling framework.

Multilevel Modeling Methods with Introductory and Advanced Applications, pages 307–354

CONCEPTUAL INTRODUCTION TO GROWTH MODELING

Research questions about growth (or decline)[1] implicitly involve the measurement of systematic change over time. How does the outcome variable of interest change across time? Does it systematically increase or systematically decrease across time? Is the rate of growth steady or does the growth rate change over time? Do people grow in the same way or is there variability between people in their rates of change? What predicts inter-individual differences in rates of change? Individual growth curve models allow us to answer such questions. Conceptually, individual growth models allow for a separate growth trajectory for each person in the sample, thereby acknowledging that participants start at different levels and change at different rates.

The Importance of Theory

The development of growth curve models should be theoretically driven. Growth modeling requires a priori specification of the general shape of the growth trajectory. Do we expect growth to be constant across time throughout the data collection period? If so, then a linear growth model is appropriate. However, if the rate of change is not constant across the data collection period, then some sort of nonlinear growth model is more appropriate.

Imagine that we were interested in studying how children's gross motor skills develop during the first 3 years of life, using an assessment tool that would yield ratio level data. What shape would the expected trajectory be? When babies are born, they do not have gross motor skills. Over the first year of life, these skills grow slowly. Most typically developing children can crawl and/or pull up into the standing position by age 1. Between 12 and 18 months, many children are standing alone or walking holding onto surfaces and some are drinking from cups. Between 18 and 23 months many children are dancing and eating with utensils, at 24 months children are able to run, jump, and throw balls. By age 3, or 36 months, children are often able to catch balls thrown to them, slide down slides, and run and jump with ease. Over the first 3 years, gross motor skills do not grow at a constant rate; motor skills accelerate. Further, there is a great deal of variability between children in terms of how quickly these skills are acquired. Some children engage in gymnastic tumbling at age 3 whereas others are just learning to walk up the stairs unassisted. A model of gross motor skills development needs to reflect all this information. A reasonable model might hypothesize that at birth, children have no gross motor skills (i.e., a value of 0 on the assessment of skills), and the between-person variability in gross motor skills is also 0 at birth. The rate of change in babies' gross motor skills is not steady: Babies grow slowly at first, but then the rate of skill development accelerates across time. Finally, not all children grow at the same rate. Some grow more slowly, and some grow more quickly. In other words, there is considerable variability between a child in the development of gross motor skills. This basic theory provides valuable information with which to design a study and conduct analyses.

Know Your Data

Before fitting growth curve models, it is essential to examine the data. Plotting means on the outcome variable across time provides a sense of the aggregate growth across time. However, individual growth trajectories may look very different from the average growth trajectory. Therefore, it is also important to examine individual growth trajectories for a sample of cases. Figure 9.1 depicts a set of individual growth trajectories for our example data. Such graphs plot time on the *x*-axis and the outcome variable on the *y*-axis. In this figure, the outcome variable is the number of days in the previous 30 days that a participant consumed alcohol and the values of 0, 2, 4, and 6 correspond to participant ages of 18, 20, 22, 24.

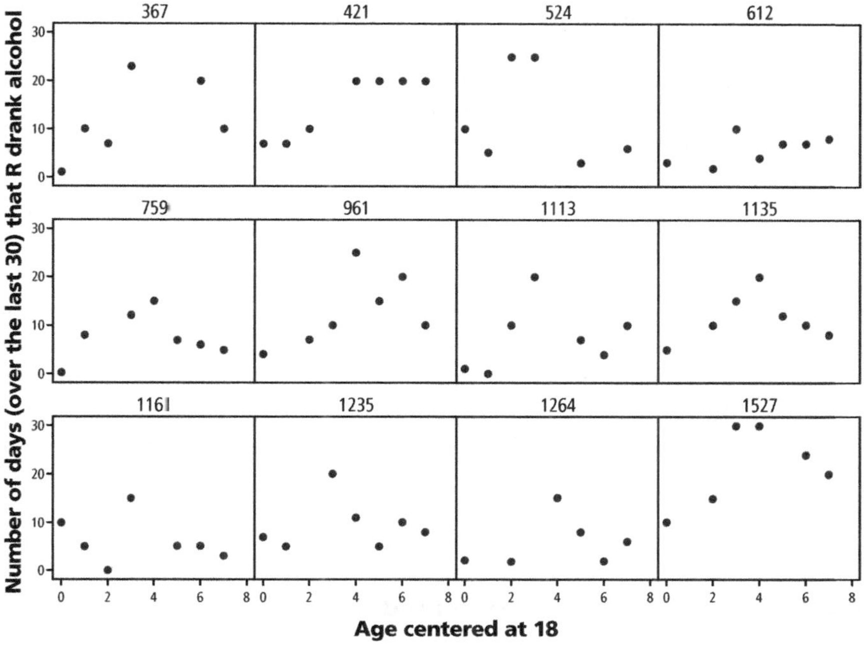

Figure 9.1 Individual growth plots for 12 individuals within the NLSY sample.

A review of the plots displayed in Figure 9.1 reveals multiple patterns of participants' self-reported alcohol consumption. For example, in earlier time periods, which correspond to ages 18, 19, and 20, participant 421 reported drinking on 10 or fewer days per month however, starting with age 21, this participant's drinking behaviors increased dramatically and then plateaued; participant 421 reported consuming alcohol on approximately 20 of the last 30 days from age 21 to age 25. Conversely, participant 759 reported steadily increasing rates of consuming alcohol between ages 18 to 21 and then a steady decrease in alcohol consumption after age 21. There does not appear to be a consistent, or linear, relationship between age and alcohol consumption for many of the participants (e.g., participants 367, 524, and 1161).

Comparing the growth plots across individuals provides important insights into similarities and differences in individuals' growth trajectories. Accurately modeling the within-person growth trajectories and understanding between-person differences in the within-person growth trajectories are key issues in growth modeling. Therefore, these individual growth plots offer invaluable guidance on building the growth model, and they also preview the likely results.

Advantages of Individual Growth Models Over Traditional Repeated Measures Designs

Traditional repeated measures designs such as Repeated Measures ANOVA do not accommodate missing data. If a person's data is not collected on the same schedule as the other participants, then the researcher must either (a) ignore that fact or (b) ignore all the observations for that individual. In addition, individuals who are missing observations at any of the data collection points are excluded from the analysis.[2] Eliminating incomplete cases from the data file has two adverse consequences. First, it decreases statistical power and lowers the precision of the estimates of growth. Second, it introduces a potential selection bias: People with missing data are likely to be systematically different from people with complete data. Therefore, eliminating cases from the analysis is likely to introduce bias into the estimates of growth. Luckily, individual growth models have the flexibility to (a) retain cases even when some time points are missing and (b) allow for the collection of time unstructured data (McCoach & Rambo-Hernandez, 2018).

Change Scores Versus Growth Models

Multi-wave data yield better estimates of individual change than two wave data (Rogosa et al., 1982). The simplest type of change measure, a difference score, is literally the difference in two scores between pretest and posttest. Although difference scores are simple to calculate, they are inadequate measures of growth (Cronbach & Furby, 1970). The difference between values at two time points confounds "true" change and measurement error. When measurement error is intertwined with the pre-test or post-test scores (or both), then "true" change and measurement error become confounded, thus the observed change between two scores may either overestimate or underestimate the degree of "true" change. Therefore, to separate the true change trajectory from time specific measurement error requires at least three data points (Singer & Willett, 2003). In addition, with only two time points, all change must be linear (and perfectly so). Thus, examining the shape of the change trajectory across time requires multiple observations, and more complex trajectories require more observations across time than simple linear trajectories.

REQUIREMENTS FOR LONGITUDINAL MODELS

First, estimating even the simplest linear longitudinal model requires observations from at least three time points. The goal of the growth curve is to model the true growth trajectory. Values on any observed variable represent a combination of true scores and error. With at least three data points, it is possible to estimate a linear trajectory, which represents the rate of change as well as time specific errors, which capture the occasion specific measurement error (McCoach & Rambo-Hernandez, 2018). In general, more complex growth trajectories require more repeated observations. Therefore, increasing the number of observations per person affords more flexibility and provides opportunities to fit more complex growth functions. When designing longitudinal studies, researchers should carefully consider both the number and the spacing of data collection points necessary to accurately capture the change process of interest. Planning the data collection schedule for a longitudinal study requires careful consideration of the hypothesized model for change. Determining the optimal collection schedule is not a trivial matter, and it is essential to consider these issues carefully at the outset of the study (McCoach & Cintron, 2021).

Second, correctly modeling the functional form of growth across time requires a reliable measure of how much time has elapsed between repeated measurements. A growth plot is a two-dimensional graph that displays scores on the outcome variable (*y*-axis) as a function of time (*x*- axis). To accurately measure the distance between measurement occasions to plot scores at the correct locations on the time continuum (*x*-axis), multilevel growth models include at least one "time" variable. A time variable is a level-1 predictor (covariate) that "clocks" the passage of time, quantifying how much time has elapsed between observations. Common metrics for time variables include the number of weeks/months/years between measurement occasions or the individual's age in months (or years) for each administration of the assessment (McCoach & Rambo-Hernandez, 2018).

Third, the outcome variable must measure the same underlying construct, and scores on the outcome variable must be equatable or comparable across time (Singer & Willett, 2003). Two ways to meet these requirements are (a) to use the same assessment at each administration or (b) to use a vertically scaled assessment (Singer & Willett, 2003). Conceptually, vertical scaling places all scores on the same equal-interval "ruler" so that growth can be measured on the same metric, even if examinees of different ages, grades, or ability levels complete different assessments. Scores on two different, unlinked scales cannot measure growth or change. Finally, scores on the outcome variable of interest should be psychometrically sound. The scores should demonstrate evidence of reliability, both at a given time point and across time. Likewise, the assessment should be sensitive to changes in the outcome variable and appropriate to measure the outcome across the entire timespan of data collection (Singer & Willett, 2003).

Time-Structured Versus Time-Unstructured Data

Longitudinal data can be either time-structured or time-unstructured. In time-structured data, all individuals are measured on the same data collection schedule, so the intervals between data collection points are identical across the entire sample (Kline, 2015). Data are time-unstructured if time intervals vary both within and across people (Singer & Willett, 2003). Time-unstructured designs allow the data collection schedule to be completely different for every person in the sample (Skrondal & Rabe-Hesketh, 2008). Forcing time-unstructured data to follow a time-structured pattern can decrease precision and increase bias in the parameter estimates.

Multilevel growth models contain (at least) one time variable to clock the passage of time. In MLM, time is an independent variable. Therefore, each individual can have their own unique values on the time variable, enabling multilevel growth models to easily accommodate time unstructured data. The ability to accommodate time unstructured data means that multilevel growth models can also easily accommodate missing observations. Rather than listwise deleting all cases with incomplete data on the outcome variable, we use full information maximum likelihood (FIML, see Chapter 3, "Evaluation of Model Fit and Adequacy," for additional details) to estimate growth models using all available data, regardless of the number of observations across time.

THE TWO-LEVEL MULTILEVEL MODEL FOR LINEAR GROWTH

In a multilevel growth model, observations across time (the level-1 units) are nested within people (the level-2 units). In a 2-level growth curve model, level 1 represents the within-individual level, and level 2 represents the between-individual level. The level-1 model captures the shape of an individual's growth trajectory over time and includes variables that vary across occasions, within individuals. The level-2 model captures inter-individual variability in the level-1 growth parameters.

Unconditional Linear Growth Model

The unconditional linear growth model (Equation 9.1) contains an independent variable to clock the passage of time, but no other covariates (at level 1 or level 2):

Level 1:

$$y_{ti} = \pi_{0i} + \pi_{1i}(time_{ti}) + e_{ti}$$

Level 2:

$$\pi_{0i} = \beta_{00} + r_{0i} \tag{9.1}$$

$$\pi_{1i} = \beta_{10} + r_{1i}$$

Combined:

$$y_{ti} = \beta_{00} + \beta_{10}(time_{ti}) + r_{1i}(time_{ti}) + r_{0i} + e_{ti}$$

The dependent variable (y_{ti}), the outcome for person i at time t, is a function of π_{0i} (i.e., the intercept for person i), $\pi_{1i} * time_{ti}$ (i.e., the time slope for person i), and time-specific individual error, e_{ti}. In the level-1 equation, the intercept (π_{0i}) represents the predicted value on the outcome variable (y_{ti}) for person i when time = 0. The time slope, π_{1i} represents the unit change in y per unit change in t, which is the linear rate of change in the outcome (y_{ti}) over time.

The level-2 equations in the unconditional linear growth model predict the expected initial level and average growth trajectory (across people). Specifically, the randomly varying intercept (π_{0i}) for each individual (i) is a function of the fixed effect for the overall intercept (β_{00}), and r_{0i}, the level-2 residual for the intercept, which represents the difference between the model-predicted intercept (β_{00}) and person i's intercept (π_{0i}). Likewise, the randomly varying linear growth slope (π_{1i}) for each individual (i) is a function of the intercept for the slope (β_{10}) and r_{1i}, the level-2 residual for the linear growth slope, which represents the difference between the overall model-predicted linear growth slope and person i's growth slope.

Imagine a simple scenario in which the time variable, age, centered at age 18, so that time = 0 at age 18, time = 1 at age 19, time = 2 at age 20, and so on. The outcome variable is drinking behavior, measured by the number of days per month that the individual drinks alcohol. The intercept (π_{0i}) represents the predicted drinking behavior of person i when time equals 0 (at age 18), and β_{00} is the expected value of the intercept. Therefore, β_{00} represents the expected (average or model predicted) drinking behavior at age 18, and each person's r_{0i} denotes the difference between average (or expected) drinking behavior and their individual drinking behavior. If r_{0i} is positive, then individual i's intercept is higher (more positive) than the overall intercept, indicating that individual i drinks more frequently than the average person in the sample at time = 0. If r_{0i} is negative, then individual i drinks less frequently than the average person at time = 0.

In a simple linear growth model, the time slope, β_{10}, is the expected change in Y per unit change in time. It represents the constant (linear) rate of change in the outcome over time, measured in the units of the time variable. Time can be measured in any interpretable, equal interval metric: milliseconds, hours, days, months, weeks, years, and so on. In our example, as time increases by one unit, expected drinking behavior increases by β_{10} units. The linear growth rate for person i is π_{1i}. The r_{1i} residual represents the discrepancy between the expected growth rate (β_{10}) and individual i's growth rate (π_{1i}). If r_{1i} is positive, then person i's growth slope is more positive than the overall expected growth slope, indicating that person i's drinking frequency grows faster than average. If r_{1i} is negative, then person i's rate of growth is more negative than the overall (expected) growth rate, indicating that person i's drinking frequency grows more slowly than average.

Centering Time in Growth Curve Models

Centering time variables requires special consideration. Remember, the intercept reflects the expected value of the outcome when time equals zero, which may or may not be a reasonable (or realistic) value. Because the intercept represents the predicted value on the outcome variable when time = 0, it is important to code or center time so that the intercept is an interpretable parameter of substantive interest. There are many possible ways to center time around some meaningful value that occurs during the study. One common approach is to center time at the beginning of the study period (or at the first data collection point), as in the previous example. In that case, the intercept represents the expected value on the outcome at the first time point (at the beginning of the study). When centering time at the final time point in the study, the intercept represents the expected value at the end of the study. It is also possible to center the level-1 predictor for time around a particular time point (e.g., Year 3 of a 5-year study) that represents a specific meaning within a particular research context.

Alternatively, each participant's age at each timepoint can serve as the time variable. This approach has the added advantage of controlling for age in addition to centering time. For example, imagine a longitudinal study examining school aged children's math achievement in which data are collected in kindergarten, first, and second grades. Centering time in the fall of kindergarten produces an intercept representing expected math achievement in fall of kindergarten. However, some students may be 4 years old, some may be 5 years old, and some may be 6 years old in the fall of kindergarten. One alternative is to use age as the time variable, centering age at 6 years. In this case, the intercept represents expected mathematics achievement at age 6, regardless of the student's grade level. To center age at age 6, subtract 6 from each age, so that the centered age variable is –1 at age 5, –.5 at 5.5, 0 at age 6, .5 at age 7.5, 1 at age 7, and so on. Using uncentered age is clearly problematic, as it represents expected math achievement when age = 0 (i.e., when the child is first born). Generally, it is best to center the time variable so that there is at least some between-person variance in the intercept $\left[var(r_{0i}) = \tau_{00}\right]$. For example, in our prior example, centering age at birth would result in $\left[var(r_{0i}) = \tau_{00} = 0\right]$ because there is no between-person variability in mathematics skills when children are first born.

Variance Components

Both the slope and intercept terms contain the subscript, *i*, indicating that each individual (*i*) has a separate slope and intercept. Conceptually, the error term, (e_{ti}) captures the deviation of a particular observation at time *t* for person *i* from the model-predicted growth trajectory. Thus, e_{ti} represents the within-person measurement error associated with individual *i*'s data at time point *t*. The variance of e_{ti} is the pooled within-person error variability around individuals' growth trajectories ([$Var\ e_{ti} = \sigma^2$]; Raudenbush & Bryk, 2002). This pooled within-person error variance ([$Var\ e_{ti} = \sigma^2$]) is generally assumed to be constant across repeated observations. In other words, the amount of within-person prediction error should not vary as a function of time.

The variance of r_{0i} ([$Var\ r_{0i} = \tau_{00}$]) represents between-person variability around the fixed effect for the intercept. When centering the intercept at the initial time point in the level-1 equation, τ_{00} represents the between-person variability in initial status. In that case, all other things being equal, the greater the between-person variance in the intercept (i.e., the larger τ_{00}), the more people differ from each other on the dependent variable in terms of where they start. Centering time affects both the expected value of the intercept (β_{00}) as well as the between-person variance in the intercept (τ_{00}). In our math achievement example, centering time at age 6, τ_{00} represents the between-person variability in math achievement at age 6. Centering math scores at age 5 would decrease τ_{00}, whereas centering math scores at age 7 would increase τ_{00}. Why? There is much less between-person variance in math scores at age 5 (when most students have very few math skills) than there is at age 7 (when some students have mastered basic arithmetic and some students are still learning basic arithmetic concepts).

The variance of r_{1i} ([$Var\ r_{1i} = \tau_{11}$]) represents between-person variability in the time slope (Raudenbush & Bryk, 2002). All other things being equal, greater between-person variance in the slope (i.e., larger τ_{11}) indicates greater variability in people's growth rates. In linear growth models, because the growth rate is constant across time, the centering of time has no effect on the expected growth slope (β_{10}), or the between-person variance in the growth slope (τ_{11}).[3] If the growth rate varies across time (as is the case in higher order polynomial models, a topic covered later in this chapter), then the choice of centering does impact both the growth slope and the between-person variance in the growth slope (τ_{11}).

Inclusion of r_{0i} and r_{1i} in the level-2 equations allows for between-person variability in peoples' intercepts (τ_{00}) and growth slopes (τ_{11}). As a general rule, we allow the randomly varying intercepts and slopes to co-vary, (i.e., τ_{01} is freely estimated).[4] Standardized τ_{01} estimates the correlation between the intercept (i.e., the expected value of the outcome when $t = 0$) and the growth slope. To compute the correlation between the slope and the intercept from τ_{01}, simply divide τ_{01} by the square root of the product of τ_{00} and τ_{11} (Rabe-Hesketh & Skrondal, 2008):

$$r_{\tau_{01}} = \frac{\tau_{01}}{\sqrt{\tau_{00} \cdot \tau_{11}}} \tag{9.2}$$

If time is centered at initial status, then, τ_{01} quantifies the relationship between where people start and how fast they grow. Again, the choice of centering affects the estimate of τ_{01}. For instance, the correlation between math scores at age 6 and math growth (τ_{01} when the time variable is centered at age 6) is not equal to the correlation between math scores at age 10 and math growth (τ_{01} when the time variable is centered at age 10.)

Regardless of the number of time points, a standard unconditional linear growth model estimates 6 parameters: 2 fixed effects (the expected intercept (β_{00}) and the expected growth slope (β_{11})) and 4 variance components (the between-person variance in the intercept (τ_{00}), the between-person variance in the slope (τ_{11}), the covariance between the slope and the intercept (τ_{01}), and the pooled within-person residual (σ^2; McCoach & Cintron, 2021). The T matrix includes the variances and covariances of the level-2 random effects:

$$\tau = \begin{bmatrix} \tau_{00} & \\ \tau_{10} & \tau_{11} \end{bmatrix} \tag{9.3}$$

Incorporating Time-Invariant Covariates Into a Linear Growth Model

As mentioned earlier, fitting a linear growth curve model with a randomly varying intercept and a randomly varying slope requires a minimum of three observations across time. For simplicity, imagine a dataset that contains only three observations per person, with time centered at the initial timepoint. After fitting a linear model to the data (by including a time variable at level-1), we might wish to determine whether a person-level variable (such as sex or some other stable person-level characteristic) can help explain between-person variability in people's initial status (intercept) and/or their growth rate (slope). To do so, we include a time-invariant covariate (TIC) into our simple 2-level linear growth model. The TIC can predict the randomly varying intercept (π_{0i}) and/or the randomly varying growth slope (π_{1i}). Equation 9.4 illustrates a model in which sex predicts both the varying intercept (π_{0i}) and the randomly varying growth slope (π_{1i}).

Level 1:

$$y_{ti} = \pi_{0i} + \pi_{1i}(time_{ti}) + e_{ti}$$

Level 2:

$$\pi_{0i} = \beta_{00} + \beta_{01}(male) + r_{0i} \tag{9.4}$$

$$\pi_{1i} = \beta_{10} + \beta_{11}(male) + r_{1i}$$

Combined:

$$y_{ti} = \beta_{00} + \beta_{01}(male) + \beta_{10}(time_{ti}) + \beta_{11}(male)(time_{ti}) + r_{1i}(time_{ti}) + r_{0i} + e_{ti}$$

For this example, sex is dummy coded such that male = 1, female = 0 and time is coded 0, 1, and 2. The fixed effect for the intercept (β_{00}) now represents the expected value on the outcome variable when both time and male equal zero, so β_{00} is now the expected initial score for females. The fixed effect of male on the intercept, β_{01}, represents the expected difference between males and females on the outcome variable at initial status or when time = 0. For example, if the student is male, the intercept for person i (π_{0i}) is predicted from the expected value of female students on the initial measure (β_{00}) and the expected differential between males and females in initial scores (β_{01}). Conceptually, the random effect, r_{0i}, captures the difference between the model predicted intercept ($\beta_{00}+\beta_{01}(male)$), and the person i's intercept (π_{0i}). Likewise, linear growth for person i (π_{1i}) is predicted from the expected growth of all female students (β_{10}) and the expected differential in growth between males and females (β_{11}). Again, conceptually, the random effect (r_{1i}) is the difference between the model predicted slope, based on the level-2 model ($\beta_{10}+\beta_{11}(male)$) and person i's slope (π_{0i}). The variance of $r_{0i}\left[var(r_{0i})=\tau_{00}\right]$ now represents residual between-person variability in the intercept, holding sex constant at 0. The variance of $r_{1i}\left[var(r_{1i})=\tau_{11}\right]$ represents residual between-person variability in the time slope, holding sex constant at 0 (Bryk & Raudenbush, 1988; Raudenbush & Bryk, 2002). The standardized covariance between the slope and the intercept (τ_{01}) indicates the relationship between the slope and the intercept, holding sex constant at 0.

Centering Level-2 Predictors

Researchers must also consider how to center time-invariant (person-level) predictors. We recommend centering level-2 continuous predictors at the grand-mean (assuming no higher-level groups exist). Again, the interpretation of the intercept changes depending on whether and how predictors are centered. Finally, it is best to calculate sample means using person-level data to avoid inadvertently weighting people with more waves of longitudinal (level-1) data more heavily (see Singer & Willett, 2003, p. 114 for details).

APPLIED EXAMPLE: SIMPLE LINEAR GROWTH MODEL

Data for our applied example are from the National Longitudinal Youth Survey 1997 (NLSY97; U.S. Bureau of Labor Statistics, 2019) which is an ongoing longitudinal project that follows the lives of a sample of Americans who were born between 1980 and 1984. NLSY97 contains data on a variety of topics including employment, educational attainment, physical and mental health, relationships, and crime and substance use (U.S. Bureau of Labor Statistics, 2019). Data are currently available for Round 1 (1997–1998) through Round 18 (2017–2018) and 75% of those interviewed in

Round 1 also participated in Round 18. Our example uses a subsample of participants aged 18 to 25 who had reported alcohol consumption data for the majority of time points between ages 18 and 25. The outcome variable is the number of days during the previous 30 days that participants reported drinking at least one alcoholic beverage. Our example also includes sex as a TIC and average weekly hours worked within a given year as a time-varying covariate (which we discuss later in this chapter).

Our first two models are a simple unconditional linear growth model of alcohol consumption (Model 1), and a conditional linear growth model, in which sex predicts both initial alcohol consumption at age 18 (the intercept) and yearly growth in alcohol consumption from ages 18–25 (the slope). Our model allows both the intercept and slope to randomly vary across people. Although Figure 9.1 indicates that growth in alcohol consumption appears to be nonlinear, for now, we ignore this complication and fit an unconditional simple linear growth model to the data to explain the interpretation of the simple linear model, both without and with a time invariant covariate, sex. Later in the chapter, we return to the same dataset and fit a more appropriate, nonlinear growth model. For substantive studies, it would not make sense to add predictors to a growth model without first establishing the pattern for trajectories of individual change. Thus, if we were conducting these analyses as part of a research study, we would never fit the conditional linear growth model with sex (Model 2), given the inability of the linear model to adequately capture growth in alcohol consumption from ages 18–25 as shown in Figure 9.1. However, pedagogically, presenting and interpreting a simple linear growth model provides the most accessible entry into individual growth models.

Model 1: Unconditional Linear Growth Model

Model 1, an unconditional linear growth model, assumes that frequency of drinking increases at a constant rate from ages 18 to 25. The time variable, age, is centered at age 18. Therefore, the time variable is coded 0 at age 18, 1 at age 19, 2 at age 20, 3 at age 21,... and 7 at age 25. In the unconditional linear model, the fixed effect for the intercept (β_{00}) represents the predicted score on the outcome variable when time = 0. Because the time variable is centered at age 18, the intercept (β_{00}) represents the expected number of days per month that an 18-year-old drinks alcohol. The fixed effect for the linear growth slope (β_{10}) is the expected yearly change in the number of days of alcohol consumption per month. Table 9.1 contains the parameter estimates for the unconditional linear growth model (Model 1). The estimated intercept is 5.19 ($\beta_{00} = 5.19$), and the estimated growth slope is .33 ($\beta_{10} = .33$). So, on average, 18-year-olds drink alcohol

TABLE 9.1 Comparison Table for Five Longitudinal Models of Number of Days Participants Reported Drinking Alcohol Each Month

	Linear Model			Linear Model w/ Male			Piecewise Model			Piecewise Model w/ TVC			Piecewise Model w/ TVC and Male		
Fixed	**Est.**	***SE***	***p***	**Est.**	***SE***	***p***	**Est.**	***SE***	***p***	**Est.**	***SE***	***p***	**Est.**	***SE***	***p***
Intercept	5.194	0.120	<0.001	4.266	0.166	<0.001	4.737	0.129	<0.001	4.920	0.136	<0.001	4.214	0.189	<0.001
Age_18	0.327	0.025	<0.001	0.248	0.035	<0.001	—	—	—	—	—	—	—	—	—
Male	—	—	—	1.901	0.237	<0.001	—	—	—	—	—	—	1.409	0.272	<0.001
Age_18*Male	—	—	—	0.160	0.050	0.001	—	—	—	—	—	—	—	—	—
Pre_21	—	—	—	—	—	—	0.558	0.091	<0.001	0.499	0.093	<0.001	0.283	0.129	0.029
Age_21	—	—	—	—	—	—	1.209	0.172	<0.001	1.155	0.174	<0.001	1.551	0.250	<0.001
Post_21	—	—	—	—	—	—	−0.034	0.047	0.464	−0.073	0.048	0.128	0.011	0.067	0.874
Hrswk	—	—	—	—	—	—	—	—	—	0.018	0.004	<0.001	0.017	0.006	0.002
Mean_Hrswrk	—	—	—	—	—	—	—	—	—	0.044	0.009	<0.001	0.018	0.014	0.185
Hrswk*Mean_Hrswk	—	—	—	—	—	—	—	—	—	−0.000	0.000	0.121	—	—	—
Male*Pre_21	—	—	—	—	—	—	—	—	—	—	—	—	0.437	0.186	0.019
Male*Age_21	—	—	—	—	—	—	—	—	—	—	—	—	−0.791	0.349	0.023
Male*Post_21	—	—	—	—	—	—	—	—	—	—	—	—	−0.159	0.096	0.097
Male*Hrswk	—	—	—	—	—	—	—	—	—	—	—	—	−0.002	0.008	0.796
Male*Mean_Hrswk	—	—	—	—	—	—	—	—	—	—	—	—	0.024	0.017	0.168

(continued)

TABLE 9.1 Comparison Table for Five Longitudinal Models of Number of Days Participants Reported Drinking Alcohol Each Month (continued)

Random	Variance	Variance	Variance	Variance	Variance
Intercept	15.396	14.489	12.960	12.859	12.313
Age_18	0.514	0.508	—	—	—
Pre_21	—	—	1.586	1.641	1.593
Post_21	—	—	1.269	1.253	1.238
Age_21	—	—	8.119	7.876	7.768
Residual	24.420	24.420	21.743	21.511	21.526
Est. Pars	6	8	15	18	23
Deviance	79,362.486	79,221.590	79,057.982	76,345.038	76,212.516
AIC	79,374.186	79,237.590	79,087.982	76,381.038	76,258.516
BIC-L1	79,419.057	79,297.018	79,199.410	76,514.140	76,428.591
BIC-L2	79,407.899	79,282.141	79,171.515	76,481.269	76,386.589

Note: The total number of observations for the models that include hours worked (models 4 and 5) as a time varying covariate is lower than the number of observations for models 1–3. Therefore, it is inappropriate to use Likelihood Ratio Tests or Information Criteria to compare models 4 and 5 to models 1–3.

5.19 days per month, and alcohol consumption increases by approximately one-third of a day per month each year from ages 18 to 25. Therefore, this model predicts that an average 21-year old consumes alcohol 6.18 days per month ($6.18 = 5.19 + 3 * .33$) and an average 25-year old consumes alcohol 7.5 days per month ($5.19 + 7 * .33 = 7.5$).

The between-person variances in the intercept (τ_{00}) and slope (τ_{11}) indicate how much people differ from each other in terms of alcohol consumption at age 18 (τ_{00} = variance in the randomly varying intercepts, π_{0i}) and yearly growth in alcohol consumption (τ_{11} = variance in the randomly varying growth slope, π_{1i}). For example, the between-person variance in frequency of alcohol consumption at age 18 is 15.40 ($\tau_{00} = 15.40$). The square root of 15.40 is 3.92, which is the standard deviation of the intercepts, the predicted frequency of drinking at age 18. Assuming the randomly varying intercepts are normally distributed, 68% of intercepts should fall within one standard deviation, placing them within the interval $5.19 \pm 3.92 = (1.27, 9.11)$. In other words, the model predicts that 68% of 18-year-olds drink between 1.27 days per month and 9.11 days per month. Likewise, 95% of intercepts should fall within 1.96 standard deviations: $5.19 \pm (1.96 \cdot 3.92) = 5.19 \pm 7.68 = (-2.49, 12.87)$. The 95% interval of predicted intercepts is quite wide. Although, the model predicts that on average, 18 year-olds drink alcohol approximately 5 days per month, the 95% plausible interval for intercepts extends below 0 (which isn't actually possible in the data) to 12.87, suggesting that some 18 year-olds do not drink alcohol at all, and other 18 year-olds drink alcohol almost 13 days per month.

We can construct a similar plausible interval for the yearly growth slope in alcohol consumption. The average growth slope is .33 ($\beta_{10} = .33$), and the variance in the growth slope is .51 ($\tau_{11} = .51$). The standard deviation of the yearly growth slope is the square root of .51, which is .71. Although on average, alcohol consumption increases by approximately one-third day per month each year from ages 18 to 25, there is a great deal of between-person variability in the yearly growth slope as well. We can again create 68% and 95% intervals of plausible growth slopes.

The 68% plausible interval of growth slopes is $.33 \pm .71$, which is –.38 to 1.04. In other words, we expect 68% of individuals' yearly growth slopes to be between –.38 and +1.04. The 95% plausible interval of growth slopes is –.88, 1.73, so we expect 95% of the growth slopes to be between –.88 and 1.73. Again, these results indicate that there is a great deal of between-person variability in the predicted growth slopes for alcohol consumption. Although the model predicts that on average, alcohol consumption increases by approximately .33 points per year from 18 to 25, some people's drinking frequency increases across time whereas other people's drinking frequency decreases during this time period.

The negative covariance between the slope and the intercept ($\hat{\tau}_{01} = -0.568$) indicates that those who drink more at age 18 tend to increase their alcohol consumption more slowly and those who drink less at age 18 tend to increase their drinking more quickly from ages 18–25. To compute the correlation between the slope and the intercept, we can standardize the covariance by dividing it by the square root of the product of the variance of the intercept and the variance of the slope, as previously shown in Equation 9.2. Therefore, the correlation between the slope and the intercept equals –.20, as shown below.

$$r_{\tau_{01}} = \frac{\tau_{01}}{\sqrt{\tau_{00} \cdot \tau_{11}}} = \frac{-.568}{\sqrt{.51 * 15.40}} = \frac{-.568}{2.802} = -.20 \tag{9.5}$$

Model 2: Adding Sex as a Time-Invariant Covariate

Does a person's sex help to predict between-person variability in drinking behavior at age 18 or between-person variability in the change in drinking behavior from age 18 to age 25? To answer these questions, Model 2 includes sex as a time invariant predictor of both the intercept (predicted alcohol consumption at age 18) and the linear growth slope (yearly change in drinking behavior from ages 18 to 25). In these analyses, sex is dummy coded: Female is the reference group (coded 0) and male is coded 1. Recall, the intercept (β_{00}) is the predicted/expected value when all predictors are held constant at 0. Therefore, in Model 2, the intercept, ($\beta_{00} = 4.27$), represents the expected number of days per month that females (sex = 0) drink at age 18. The fixed effect of male on the intercept ($\beta_{10} = 1.90$) is the difference between males' and females' drinking behavior at age 18. In other words, on average, males drink 1.90 days more per month than females at age 18, so the expected number of days per month that males drink alcohol at age 18 is 6.17 ($\beta_{00} + \beta_{10} = 4.27 + 1.90$). The estimated yearly growth slope for females is .25 ($\beta_{10} = .25$). The difference between males and females in terms of yearly growth is .16 ($\beta_{11} = .16$), so the yearly growth slope for males is .41 ($\beta_{10} + \beta_{11} = .25 + .16$). This means drinking frequency increases more quickly for males, who are growing at a rate of .41 days per month than it does for females, who are growing at a rate of .25 days per month. Using parameter estimates from the linear growth model, we can generate model predicted estimates of drinking behavior at any age from 18 to 25. For example, at age 21, on average, males are expected to drink 7.40 days per month ($\beta_{00} + \beta_{01} + 3 \cdot (\beta_{10} + \beta_{11}) = 4.27 + 1.90 + 3 * (.25 + .16) = 6.17 + 3 * .41 = 6.17 + 1.23$), whereas females are expected to drink 5.02 days per month at age 21 ($4.27 + 1.90(0) + 3 * (.25 + .16 * 0) = 4.27 + .75 =$

5.02). By age 25, males are expected to drink an average of 9.04 days per month ((4.27 + 1.90) + 7 * (.25 + .16 * 1) = 9.04) whereas females are expected to drink an average of 6.02 days per month (4.27 + .25 * 7).

Importantly, there is still a great deal of variability in both the intercepts (predicted drinking behavior at age 18) and the growth slopes (yearly change in drinking behavior from ages 18 to 25), even after controlling for sex. For example, the standard deviation for the predicted intercepts in Model 2 is 3.81 ($\sqrt{14.489} = 3.81$). Therefore, 68% of females are predicted to have intercepts between .46 and 8.08 (4.27 +/− 3.81) and 95% of females are predicted to have intercepts between −3.2 and 11.74 (4.27 ± 7.47). In comparison, 68% of males have predicted intercepts between 2.36 and 9.35 (6.17 ± 3.81) and 95% of males have predicted intercepts between −1.3 and 13.64 (6.17 ± 7.47). Even though males do tend to drink more than females at age 18, there is a considerable degree of overlap in the distributions of intercepts for males and females. The same is true for their growth slopes. The standard deviation for the predicted growth slopes is .71 ($\sqrt{.508} = .71$). Therefore, although the average yearly growth slope for females is .25, 68% of females have predicted growth slopes between −.46 and .96, and 95% of females have predicted growth slopes between −1.14 and 1.64. In contrast, although the average male yearly growth slope is .41, 68% of males have predicted growth slopes between −.30 and 1.12 and 95% of males have growth slopes between −.98 and 1.80. Again, although males tend to have more positive growth slopes than females, there is considerable between-person variability in the growth slopes, even after controlling for sex, and there is a considerable degree of overlap in the distributions of growth slopes for males and females.

Although sex is a statistically significant predictor of both drinking frequency at age 18 and growth in alcohol consumption from ages 18 to 25, it explains a relatively small amount of the variability in terms of alcohol consumption at age 18 and growth in alcohol consumption from 18–25. The intercept variance (τ_{00}) in the unconditional (or baseline, B) growth model is 15.396 ($\tau_{00B} = 15.396$). The residual intercept variance (τ_{00}) in the full (F) model that includes sex is 14.489 ($\tau_{00F} = 14.498$). We can compute the proportion reduction in intercept variance using the following formula:

$$\frac{\tau_{00B} - \tau_{00F}}{\tau_{00B}} \tag{9.6}$$

where B indicates the baseline (simpler) model and F indicates the full model (Raudenbush & Bryk, 2002). In this example, adding sex reduces the intercept variance by

$$\frac{15.396 - 14.489}{15.396},$$

or .0589. This can be interpreted similarly to an R-square measure in multiple regression, although this is not truly an R-square. This means that sex explains approximately 5.9% of the variance in drinking frequency at age 18. Similarly, we can compute the proportion of variance in the growth slope that is explained by sex using the formula

$$\frac{\tau_{11B} - \tau_{11F}}{\tau_{11B}}.$$

Adding sex reduces the slope variance by

$$\frac{.514 - .508}{.514} = .01167.$$

In other words, sex explains only 1.17% of the between-person variance in alcohol consumption growth slopes.

As we mentioned earlier, Figure 9.1 indicates that a simple linear growth model would not adequately capture the growth process of interest. Our simple example of a linear growth model with a time invariant covariate served an important pedagogical purpose: It allowed us to introduce the basics of individual growth modeling. Now we turn our attention to modeling more complex level-1 models, which are more appropriate given the growth patterns in Figure 9.1.

MORE COMPLEX LEVEL-1 MODELS

The linear growth model is the simplest growth model. However, the level-1 model can be extended through the incorporation of time varying covariates, piecewise linear terms, or polynomial terms to model nonlinearity that occurs in the growth trajectory. We briefly consider the use of time varying covariates, piecewise regression models, and polynomial growth models.

Incorporating Time-Varying Covariates

Time-varying covariates (TVCs) are level-1 variables: They can vary across time as well as across people. Thus, TVC can have different values for the same person at different time points throughout the study period. Although the time variable itself is technically a TVC, the term TVC usually refers to a variable other than time that varies both within and across people. The slope parameter (regression coefficient) for the time varying covariate indicates the expected change in the outcome variable for each one unit increase in the TVC, and this slope can randomly vary across people.

Equation 9.7 displays the multilevel equations and the combined equation for a linear growth model with a TVC.

Level 1:

$$y_{ti} = \pi_{0i} + \pi_{1i}(time_{ti}) + \pi_{2i}(TVC_{ti}) + e_{ti}$$

Level 2:

$$\begin{aligned} \pi_{0i} &= \beta_{00} + r_{0i} \\ \pi_{1i} &= \beta_{10} + r_{1i} \\ \pi_{2i} &= \beta_{20} + r_{2i} \end{aligned} \tag{9.7}$$

Combined:

$$y_{ti} = \beta_{00} + \beta_{10}(time_{ti}) + \beta_{20}(TVC_{ti}) + r_{1i}(time_{ti}) + r_{2i}(TVC) + r_{0i} + e_{ti}$$

In Equation 9.7, TVC_{ti} is a time-varying covariate for person i at time t. The coefficient for the TVC slope for person i is π_{2i}, which represents the direct effect of the TVC on y_{ti} for person i, controlling for any other variables in the model. Including r_{2i} as a random effect for π_{2i} indicates that the TVC slope may randomly vary across people. In other words, for some people the effect of the TVC on the dependent variable could be quite strong and positive, whereas for others it could be weak, zero, or even negative. The variance in r_{2i} (τ_{22}) is the between-person variance in the slope of the TVC on the outcome variable, after controlling for other variables in the model.

Adding a TVC changes the interpretation of both the intercept (β_{00}) and the growth slope (β_{10}). In Equation 9.7, the intercept (β_{00}) is now the expected value of the outcome variable when both time *and* the TVC are 0. Similarly, the growth slope is now conditional on the TVC, so β_{10} is the expected growth per unit change in time, holding the TVC constant at 0. In other words, β_{10} is the expected growth rate, after controlling for the effect of the TVC.

Centering Time-Varying Covariates

The interpretation of the growth parameters also depends on the centering or coding of the TVC. It is common to dummy code dichotomous TVCs. With continuous variables, it is essential to carefully consider the best approach to centering the TVC.

Centering a TVC that does not systematically change over time. If TVC does not systematically increase or decrease across time,[5] centering decisions are relatively straightforward. There are several possibilities for centering the TVC: (a) grand mean centering; (b) group (i.e., person) mean centering; (c) time-point specific centering; or (d) centering around an inherently meaningful constant.

Time-varying predictors change across time points, so calculating the grand mean uses the entire data file (i.e., it averages both across time and people). Therefore, a grand mean centered TVC represents an individual's deviation from the overall average, both across time and across people. As such, when centering the TVC around the grand mean, the intercept (β_{00}) represents the expected value of the outcome variable at time 0 when the TVC is at the overall mean, both across time and across people.

Group mean centering involves subtracting the cluster (level-2) mean of the TVC from each cluster's score. In growth models, observations across time are clustered within people, so the cluster mean is the individual's person-specific average on the TVC across all time points. Therefore, a group mean centered TVC represents an individual's deviation from their own personal average. With a group mean centered TVC, the intercept (β_{00}) represents the expected value of the outcome variable at time 0, assuming each person is at their own personal mean (across time). With a group mean centered TVC, the slope (β_{10}) represents the expected growth rate after controlling for the TVC (within person, but not between people).

As an alternative, a predictor could also be centered around its initial value (i.e., its value at the first time point), as opposed to the mean across all time points (see Singer & Willett, 2003, p. 176 for more details). If time is centered at initial status, then the intercept (β_{00}) is the expected value at the initial time point. The slope of the TVC (β_{10}) is now the average growth rate, holding the TVC constant at its initial value. Finally, in some cases, centering around an inherently meaningful constant, may enhance interpretability. Singer and Willett's (2003) example using high school dropout wage earnings centers time-varying unemployment rates around a value of 7 (a common rate found during the timeframe under study), so that average wages are for someone in an area with a 7% unemployment rate.

How do we interpret a group mean centered TVC? Imagine a growth curve model of drinking behavior across time that includes depression as a TVC. Time is measured in years and centered at age 18. To group mean center depression, compute each person's mean depression: This is average depression across time for person *i*. Then, subtract person *i*'s mean depression from each of their time specific depression scores. Group mean centered depression for person *i* at time *t* represents the deviation in depression from the person's own average depression level. Therefore, the coefficient for the depression slope indicates the expected change in drinking if depression increases by one unit. Because depression is group mean centered, π_{0i} represents predicted drinking at age 18 when person *i*'s depression is at their own personal average across time, and β_{00} is expected drinking behavior at age 18. The growth slope for individual *i*, π_{1i}, is the expected change in drinking each year, assuming individual *i*'s depression is at their own personal average (i.e., it is the increase in drinking, holding depression constant at individual *i*'s own mean), and β_{10} is the expected yearly change in drinking, holding depression constant within-person, but not between people. In contrast, when grand mean centering depression at level 1, the intercept (β_{00}) is predicted drinking at age 18 when depression is at the overall average, both across time and across people. The growth slope (β_{10}) is the expected yearly growth rate when depression is at the overall mean.

The TVC slope (β_{20}) is the predicted change in drinking as depression increases by one unit. In both the group and grand mean centered equations, the TVC slope represents the expected change in drinking for a unit change in depression. However, using group mean centering, the change in depression is the deviation from a person's average level whereas in grand mean centering, the change in depression is the deviation from the overall mean.

Because group mean centering eliminates the between cluster variance in the predictor variable, if the average level of the TVC differs across people, it is generally advisable to include the mean value of the TVC as a person level predictor at level-2. For instance, in Equation 9.8, depression is group-mean centered, so we include mean depression as a grand-mean centered time-invariant predictor at level-2 (Dep_c). A person's average level of depression is a predictor of people's initial drinking (β_{01} predicts between-person differences in the randomly varying intercept, π_{0i}). In addition, a person's average level of depression is a predictor of change across time in drinking behavior (β_{11} predicts between-person differences in the randomly varying growth slope, π_{1i}), and occasion specific changes in drinking as a function of intrapersonal changes in depression (β_{21} is the effect of mean depression on the individually-varying slope of the within-person changes in depression on drinking, π_{2i}).

Level 1:

$$y_{ti} = \pi_{0i} + \pi_{1i}(time_{ti}) + \pi_{2i}(Depression_{ti} - Depression_{.i}) + e_{ti}$$

Level 2:

$$\begin{aligned} \pi_{0i} &= \beta_{00} + \beta_{10}(MeanDepression) + r_{0i} \\ \pi_{1i} &= \beta_{10} + \beta_{11}(MeanDepression) + r_{1i} \\ \pi_{2i} &= \beta_{20} + \beta_{21}(MeanDepression) + r_{2i} \end{aligned} \tag{9.8}$$

Combined model:

$$\begin{aligned} Y_{ti} = {} & \beta_{00} + \beta_{01}(MeanDep) + \beta_{10}(time_{ti}) + \beta_{11}(MeanDep)\,(time_{ti}) \\ & + \beta_{20}(Dep\_c_{ti}) + \beta_{21}(MeanDep)\,(Dep\_c_{ti}) + r_{0i} + r_{1i}(time_{ti}) \\ & + r_{2i}(Dep\_c) + e_{ti} \end{aligned}$$

Centering a TVC that systematically changes across time. It is possible that the value of the TVC may itself systematically increase or decrease across time. Imagine a scenario in which time is centered at initial status and both the outcome variable and the TVC increase across time. For example, we might want to predict growth in a student's writing skills across time, and we plan to use reading skills as a time varying covariate. Given what we know about how reading skills develop as children get older, we expect reading skills (our TVC) to increase across time and we expect reading skills to be related to writing skills at each time point. If we expect the TVC itself to systematically increase or decrease across time, how should we center the TVC? If we grand mean center the TVC, the intercept is no longer the expected value of the outcome at the initial time point. Instead, it is the predicted value of the outcome at the initial time point for someone whose score on the TVC is at the overall mean, a value that is likely to occur around the middle of the data collection period. If we group mean center the TVC, the intercept is the predicted value of the outcome at the initial time point for someone whose score on the TVC is at their own personal mean, a value that once again is likely to occur around the middle of the data collection period. Neither strategy seems optimal when the TVC itself systematically changes across time.

The optimal analysis in such a scenario is a multivariate longitudinal model, in which the time varying predictor becomes a second outcome variable (Hoffman, 2015). To specify a multivariate longitudinal model using standard MLM software, we can fit a 3-level model in which outcomes (level 1) are nested within occasions (level 2), which are nested within-persons (level 3). Such an approach requires reorganizing the data into a double stacked format so that the data for both outcome variables (the original outcome variable and the TVC) are in a single variable (column). Hoffman (2015) provides an excellent discussion of issues related to centering predictors (TVC) that systematically change across time, and she provides additional information about modeling and interpreting multivariate longitudinal models within a standard MLM framework.

TVC Variance Components

In Equation 9.6, if we fit an unstructured **T** matrix, the slope of the TVC randomly varies and it covaries with both the time slope and the intercept. Therefore, the **T** matrix has six parameters: three variances (the intercept variance, growth slope variance, and TVC slope variance) and three covariances (the covariance between the growth slope and the intercept (τ_{10}), the covariance between the intercept and the TVC slope outcome (τ_{20}), and the covariance between the growth slope and the TVC slope (τ_{21}); see Equation 9.9). Standardizing these covariances produces correlations between initial status and growth, initial status and the TVC slope, and growth and the TVC slope.

$$\tau = \begin{bmatrix} \tau_{00} & & \\ \tau_{10} & \tau_{11} & \\ \tau_{20} & \tau_{21} & \tau_{22} \end{bmatrix} \tag{9.9}$$

Can the impact of a TVC vary across time? The TVC_{ti} is a level-1 variable, and the value of this covariate changes across time; however, for a given individual *i* the effect of the TVC on the dependent variable (π_{2i}) is *constant* across time. In other words, although the effect of the TVC varies across people (i.e., the random slope for the TVC), for each person, the effect of the TVC is constant across time. In our earlier example, both the dependent variable (drinking), and the time varying covariate (the depression) are measured each year. An individual's depression level can change each year, and the effect of depression on drinking can differ across individuals, but the estimated within-person relationship between depression and drinking remains constant across the entire study.

However, we may wish to allow the passage of time to moderate the effect of the TVC on the outcome variable (i.e., to examine if the impact of the TVC does vary across time). One simple way to relax the assumption that the effect of the TVC remains constant across time is to create an interaction between time and the TVC. To do so, we create a new observed variable that equals the product of time and the centered TVC (Singer & Willett, 2003). Centering continuous TVCs makes parameter estimates for the interaction and the TVC more easily interpretable.

Level 1:

$$y_{ti} = \pi_{0i} + \pi_{1i}(time_{ti}) + \pi_{2i}(TVC_{ti}) + \pi_{3i}(time_{ti} * TVC_{ti}) + e_{ti}$$

Level 2:

$$\begin{aligned} \pi_{0i} &= \beta_{00} + r_{0i} \\ \pi_{1i} &= \beta_{10} + r_{1i} \\ \pi_{2i} &= \beta_{20} + r_{2i} \\ \pi_{3i} &= \beta_{30} + r_{3i} \end{aligned} \tag{9.10}$$

Combined:

$$\begin{aligned} y_{ti} = {} & \beta_{00} + \beta_{10}(time_{ti}) + \beta_{20}(TVC_{ti}) + \beta_{30}(time_{ti} * TVC_{ti}) + r_{1i}(time_{ti}) \\ & + r_{2i}(TVC_{ti}) + r_{3i}(time_{ti} * TVC_{ti}) + e_{ti} \end{aligned}$$

In Equation 9.10 the level-2 parameter for the interaction term (β_{30}) captures the expected differential effect of the TVC across time. If β_{30} is positive, then the effect of the TVC becomes increasingly positive across time; in contrast, if β_{30} is negative, the effect of the TVC becomes more negative as time passes (assuming time is centered at initial status). If the interaction term randomly varies across people (includes r_{3i}), the slope of the time-varying effect of the TVC may differ across people. In other words, including r_{3i} allows between-person variance in the degree to which the effect of the TVC on the outcome varies as a function of time ($Var(r_{3i}) = \tau_{33}$).

Although introducing a randomly varying interaction between time and a given TVC provides great flexibility, it also increases the number of estimated parameters in the model. For example, excluding the randomly varying interaction yields a model with one random intercept, two random slopes, and three covariance parameters (see Equation 9.10). The addition of the randomly varying slope (r_{3i}) for the interaction effect adds four additional covariance parameters to the model: one variance parameter (τ_{33}) and three covariance parameters (τ_{03}, τ_{13}, and τ_{23}). Therefore, the model in Equation 9.7 has a total of 15 parameters: four fixed effects (β's), σ^2, and ten Tau's (four variances and six covariances, as shown in the $\underline{\mathbf{T}}$ matrix below [Equation 9.11]) and σ^2.

$$\underline{\mathbf{T}} = \begin{bmatrix} \tau_{00} & & & \\ \tau_{10} & \tau_{11} & & \\ \tau_{20} & \tau_{21} & \tau_{22} & \\ \tau_{30} & \tau_{31} & \tau_{32} & \tau_{33} \end{bmatrix} \quad (9.11)$$

Although the inclusion of TVCs creates complexities in the growth model building process, incorporating TVCs can be critical to understanding the process of change. Thoughtfully coded time-varying variables can more accurately reflect the process of change, as well as correlates of that change. Incorporating TVCs can also be an effective strategy for modeling discontinuities or nonlinearity in growth trajectories (McCoach & Kaniskan, 2010), as we discuss next.

Piecewise Linear Growth Models

Often, growth trajectories are nonlinear. In one type of nonlinear model, the overall growth trajectory consists of multiple growth segments corresponding to fundamentally different patterns of change (Collins, 2006). In other words, growth occurs in multiple phases, each of which has a different growth rate. Piecewise linear growth models (also called spline growth models) allow researchers to determine whether growth rates are the same or different across time periods and to investigate differences in substantive predictors of growth across different growth phases. Piecewise models divide the growth trajectory into separate linear components (Raudenbush & Bryk, 2002). Each piece (linear component) has a different slope parameter, and these growth slopes may randomly vary across people. For example, imagine researchers collect multiple observations before and after implementing a new intervention. Conceptually, they hypothesize that the intervention changes the growth rate. Therefore, their model must accommodate one growth rate (slope) prior to the intervention and another growth rate (slope) during the intervention.

How Many Individually Varying Slopes Can My Data Support?

To accommodate individual variation in the growth parameters, such as the slope and the intercept (as well as individual variation in the slopes of TVC), we include random effects (r_{0i}, r_{1i}, etc.) in our model. The variances of these random effects (τ_{00}, τ_{11}, etc.) capture the between-person variability in the growth parameters. In general, MLM can accommodate one fewer random effect than there are time points ($r_{max} = t - 1$). For example, a linear growth model with a randomly varying slope and a randomly varying intercept requires two random effects (one for the intercept and one for the growth slope). Thus, fitting a linear model with a randomly varying slope and a randomly varying intercept requires at least three time points. Why? To answer this question, we consider the number of parameters being estimated.

Under the assumption of homogeneity, a model with a randomly varying slope and a randomly varying intercept contains four variance components: σ^2, τ_{00}, τ_{01}, τ_{11}. (The model also contains two additional parameters, the fixed effects for the slope (β_{11}) and the intercept (β_{00}), but this exposition focuses on the identification of the variance components.) We use those four variance components (σ^2, τ_{00}, τ_{01}, τ_{11}) to reproduce or recreate the elements of the covariance matrix of composite residuals (Σ_r), which has the same dimension as the variance covariance matrix of the outcome variable across time.[6] With three time points, the observed variance covariance matrix has six unique elements (three variances and three covariances). A model estimating the four variance component parameters is overidentified: We use four parameters to reproduce a variance covariance matrix with six unique elements. Thus, the standard linear growth model estimates two fewer variance covariance components than there are observed variances and covariances.

Assuming an unstructured Tau matrix, a model with three random effects contains seven variance components: σ^2, τ_{00}, τ_{01}, τ_{02}, τ_{11}, τ_{12}, τ_{22} (three variances for the three level-2 residuals, three covariances among the three level-2 residuals and the level-1 residual variance). However, with three time points, the composite variance covariance matrix still contains only six unique pieces of information. The number of parameters to estimate (unknowns = 7) exceeds the number of unique elements in the variance/covariance matrix (knowns = 6); therefore, there is –1 degrees of freedom (*df*). Thus, a model with three time points and three random effects is underidentified. However, when there are four time points, the composite variance covariance matrix contains ten unique elements. So, with four time points, the model with three random effects is overidentified; it has three *df* (10 – 7). Again, it is not possible to estimate a model that contains four random effects with four time points because a model with four random effects contains ten elements in the variance covariance (**T**) matrix (4 variances and 6 covariances) plus the level-1 residual variance parameter, for a total of 11 parameters. So, the model is underidentified: it has –1 *df*.

As a general rule, growth models can accommodate one fewer random effect (factor) than there are time points (observations). There are ways to reduce the number of variance covariance components, either by constraining one or more of the covariances or by constraining the level-1 residual variance in some way. However, such approaches require additional assumptions. McCoach et al. (2006) provides an example of constraining the level-1 residual variance.

A piecewise model that contains two randomly varying slopes (and a randomly varying intercept) requires at least four observations (time points) per person, and a piecewise model that contains three randomly varying slopes (and a randomly varying intercept) requires at least five observations per person.[7] These guidelines indicate the *minimum* number of observations required for identification, not the *optimal* number of observations for such a model. Additional time points allow for more thorough examination of the adequacy of the specified functional form. For instance, although three time points are adequate for a linear model, it is not possible to compare the linear trajectory to a more complex growth trajectory to determine whether the linear model adequately captures the shape of the growth. Allowing the slopes for TVC to randomly vary also requires additional random effects, necessitating additional time points. For these reasons, we recommend collecting more than the minimal number of time points needed to estimate a hypothesized model.

The Mechanics of Coding Time for the Piecewise Model

Figure 9.2 illustrates a growth trajectory for mathematics achievement collected across six time points during an intervention study. The intervention is hypothesized to change the growth rate in the outcome variable. The first three observations represent baseline data; they occur during the baseline (pre-intervention phase). Treatment begins at the third time point

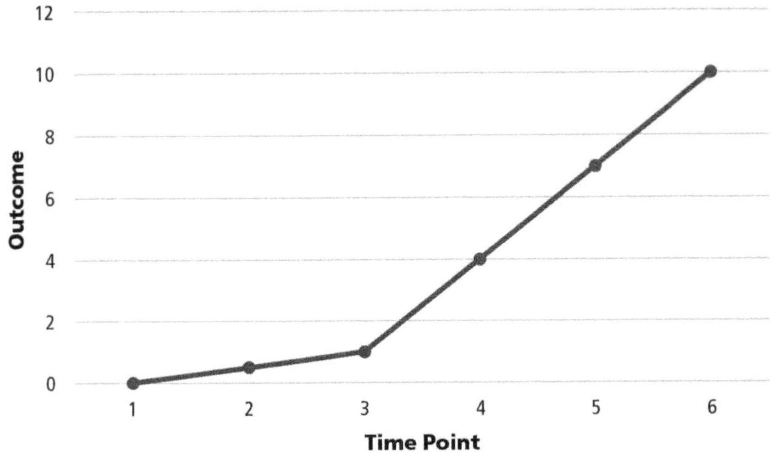

Figure 9.2 Piecewise growth example trajectory with a change in slope that occurs at/after time point 3.

(immediately after collecting the third observation), and the intervention continues through the subsequent three time points. In Figure 9.2, the linear slope changes in the middle of data collection period. How can we code time to allow for two separate growth slopes?

Centering time at the beginning of the study, in a linear model with six equally spaced observations, time is coded 0, 1, 2, 3, 4, 5. To capture multiple linear growth slopes, the piecewise linear growth model includes multiple time variables. A two-piece linear growth model contains two separate time variables as shown below:

Level 1:

$$y_{ti} = \pi_{0i} + \pi_{1i}(time\_piece1_{ti}) + \pi_{2i}(time\_piece2_{ti}) + e_{ti}$$

Level 2:

$$\begin{aligned} &\pi_{0i} + \beta_{00} + r_{0i} \\ &\pi_{1i} + \beta_{10} + r_{1i} \\ &\pi_{2i} + \beta_{20} + r_{2i} \end{aligned} \tag{9.12}$$

Combined:

$$\begin{aligned} y_{ti} = {} & \beta_{00} + \beta_{10}(time\_piece1_{ti}) + \beta_{20}(time\_piece2_{ti}) + r_{0i} \\ & + r_{1i}(time) + r_{2i}(time\_piece2_{ti}) + e_{ti} \end{aligned}$$

The "magic" of the piecewise model occurs in the coding of time, and the interpretations of the growth slopes depend upon the coding of time. Moving forward, think of these systems of time variables as "stopwatches": each "stopwatch" captures the passage of time and can be turned on or off, as needed. Using multiple stopwatches provides an easy method to produce separate growth rates.

Creating two *time* variables allows the growth rate to differ across the two phases of the study. To estimate two distinct growth rates, each stopwatch turns on, captures the growth rate during its respective period, and turns off at the end of its period. Table 9.2 shows the time coding to estimate the growth trajectory in Figure 9.2 as two separate slopes. The first time variable is coded 0, 1, 2, 2, 2, 2. The second time variable is coded 0, 0, 0, 1, 2, 3. Therefore, the first slope captures the baseline growth rate (from waves 1–3) and the second slope captures the growth rate during the intervention (from waves 3–6). Wave three is the *joint*: It is both the last time point included in the first growth slope and the first (baseline) time point for the second growth slope. The piecewise coding in Table 9.2 has two important features:

1. Only one stopwatch is turned on at each time point.
2. Summing the time codes for piece 1 and piece 2 produces the time coding for a linear growth model.

Using this coding scheme, the intercept (β_{00}) is the expected value when both clocks 1 and 2 are 0, at wave 1. The first growth slope (β_{10}) captures the growth rate during the first time period (from waves 1–3), and the second growth slope (β_{20}) captures the growth rate during the second time period (from waves 3–6).

Alternatively, we can model a baseline growth trajectory (slope) and a change in the slope. Think of the change in slope as a *deflection* parameter: It captures the increase or decrease in the slope during the second time period. To model a baseline slope and a change in slope using this coding

TABLE 9.2 Coding Scheme for Figure 9.2 to Produce Two Separate Slopes

Wave	Piece 1	Piece 2	Sum (P1 + P2)
1	0	0	0
2	1	0	1
3	2	0	2
4	2	1	3
5	2	2	4
6	2	3	5

TABLE 9.3 Coding Scheme for Figure 9.2 to Produce Slope and Change in Slope

Wave	Piece 1	Piece 2	Sum (P1 + P2)
1	0	0	0
2	1	0	1
3	2	0	2
4	3	1	4
5	4	2	6
6	5	3	8

scheme, the coding of the first-time variable is identical to the time coding for a standard linear growth model. The coding of the first time variable is 0, 1, 2, 3, 4, 5. The coding for the second time variable is identical to the coding of the second time variable for the 2 piece model (shown in Table 9.3). The second time variable would still be coded 0, 1, 2, 2, 2, 2. Table 9.3 contains the coding scheme for the baseline slope plus change in slope coding scheme.

Although coding for the second time variable is identical across the two coding schemes, the meaning of the second growth slope changes. The second growth slope (β_{20}) now captures the change in the growth rate between the first and second time periods (from waves 3–6). The interpretation of the intercept (β_{00}) and the first growth slope (β_{10}) remain constant across the two coding schemes. In either case, the intercept (β_{00}) is still the expected value at wave 1 (when time 1 and time 2 are 0) and the first growth slope (β_{10}) is the baseline growth rate across the first phase, from waves 1–3. However, when coding the first time variable as 0, 1, 2, 2, 2, 2, β_{20} captures the entire growth rate during the second phase of the study. When coding the first time variable as 0, 1, 2, 3, 4, 5, β_{20} captures the *change* in the growth rate between the first and second phases of the study. These two models are statistically equivalent: They produce the same deviance. To test the hypothesis that the slope during the second time period is 0, use the coding scheme in Table 9.2. To test the hypothesis that the change in slope between the first and second phases of the study equals 0, use the coding scheme in Table 9.3.

Polynomial Growth Models

Polynomial growth models represent the outcome variable (y_{ti}) as an nth degree polynomial of time. Simple linear growth models (and intercept-only models) are part of the family of polynomial growth models. Higher order polynomial growth models (i.e., polynomial models of second order or above) enable the modeling of nonlinear growth trajectories. Below, we briefly describe zero-order (intercept-only), first-order (linear), second-order (quadratic) and third-order (cubic) growth models using MLM notation.

Intercept-Only Model

An intercept only model is a zero-order growth model in which the outcome variable is a function only of $\pi_{0i} * (time_{ti})^0$. Because $time^0 = 1$, we generally drop the $(time_{ti})^0$ portion of the level-1 intercept term from the equation, yielding:

$$\text{Level 1:} \quad y_{ti} = \pi_{0i} + e_{ti}$$

$$\text{Level 2:} \quad \pi_{0i} = \beta_{00} + r_{0i} \tag{9.13}$$

$$\text{Combined:} \quad Y_{ti} = \beta_{00} + r_{0i} + e_{ti}$$

The zero-order growth model specifies that the outcome variable at time t, for individual i, is a function of the intercept (which randomly varies across people) and a within-person residual, e_{ti}, which captures the time-specific deviation of the outcome variable from the individual's intercept. The intercept represents the person's average level on the outcome variable across time. An intercept-only model specifies a mean (an average level) on the outcome variable, and variability around that average level, but does not allow for systematic change across time.

The zero-order growth model may seem overly simplistic, but it can actually be quite useful. Imagine that a researcher conducts a daily diary study of mood and alcohol consumption in a sample of middle-aged adults who are daily drinkers. As part of the study, participants report on their daily stressors and alcohol consumption every day for a 1-month period. The researcher wishes to examine how much daily alcohol consumption varies across people (i.e., the between-person variance in daily drinking), how much people's alcohol consumption varies across time (i.e., the within-person variance in daily drinking), and whether a person's average daily stressors predict their overall (average) drinking behavior, and whether daily fluctuations in daily stressors predict within-person changes in drinking. The research hypothesis is that people's daily drinking does *not* systematically increase or decrease across the one-month period. Instead, people have a homeostatic[8] level of alcohol consumption. This homeostasis represents a dynamic state of equilibrium rather than a constant, unchanging state. An additional hypothesis is that daily stressors can disrupt an individual's equilibrium; higher daily stressors increase daily drinking behavior.

To test these hypotheses, we could model a zero-order growth trajectory that includes daily stressors as a group mean centered time varying (level-1) covariate and person's average daily stressors as a time-invariant (level-2) covariate, as shown in Equation 9.13. β_{00} represents expected daily drinking (for a person with average stressors), β_{01} represents the average effect of stressors on daily drinking, β_{10} represents the effect of within-person changes in stressors on daily drinking for a person with average overall stress levels. In other words, how much do daily changes in stress affect daily drinking (for a person whose daily stressors are average)? The β_{11} parameter represents the effect of overall stressors on the within-person effect of daily fluctuations in stress on daily drinking. How do we interpret these parameters? If β_{10} is positive, an average person drinks more on more stressful days and drinks less on less stressful days. If β_{11} is positive, then π_{1i} is more positive for people with higher average daily stressors, so the effect of daily changes in stressors on drinking is more pronounced for people who report higher overall daily stressors. In other words, the higher a person's overall level of daily stressors is, the more changes in their daily stressors affect their drinking behavior: People with higher average daily stressors are more susceptible to fluctuations in (within-person) drinking caused by (within-person) fluctuations in daily stressors. In contrast, if β_{11} is negative, then π_{1i} is more negative for people with higher average daily stressors. In this case, the drinking behavior for people with higher overall daily stressors is less affected by daily fluctuations in stressors. In other words, the higher a person's overall daily stressors, the less within-person fluctuations in daily stressors lead to within-person fluctuations in alcohol consumption.

Level 1:

$$Drinks_{ti} = \pi_{0i} + \pi_{1i}(Stress_{.j}) + e_{ti}$$

Level 2:

$$\begin{aligned} \pi_{0i} &= \beta_{00} + \beta_{01}(Stress_{..}) + r_{0i} \\ \pi_{1i} &= \beta_{10} + \beta_{11}(Stress_{..}) + r_{1i} \end{aligned} \tag{9.14}$$

Combined:

$$\begin{aligned} Y_{ti} = {} & \beta_{00} + \beta_{10} + \beta_{01}(Stress_{..}) + \beta_{10}(Stress) + \beta_{11}(Stress_{..}) + r_{0i} \\ & + r_{1i}(Stress) + e_{ti} \end{aligned}$$

Linear Model

Linear growth models are first-order polynomial growth models: The outcome variable (y_{ti}) is a function of $(time_{ti})^0$ and $(time_{ti})^1$.

Level 1:

$$y_{ti} = \pi_{0i} + \pi_{1i}(time_{ti}^1) + e_{ti}$$

Level 2:

$$\pi_{0i} = \beta_{00} + r_{0i} \quad (9.15)$$

$$\pi_{1i} = \beta_{10} + r_{1i}$$

Combined:

$$y_{ti} = \beta_{00} + \beta_{10}(time_{ti}) + r_{1i}(time_{ti}) + r_{0i} + e_{ti}$$

In the level-1 equation, y_{ti} is a function of $\pi_{0i} * time_{ti}^0$ and $\pi_{1i} * time_{ti}^1$. The intercept (π_{0i}) represents individual *i*'s level on the outcome variable when *time* = 0. The time slope (π_{1i}) represents the predicted change in the outcome variable (y_{ti}) for individual *i*, at time *t*, for every one-unit change in time. In a linear growth model, although the level of the outcome variable changes across time, the *rate of change* remains constant over time.

Quadratic Model

Building on linear growth models, in the second-order polynomial model (aka quadratic growth model), y_{ti} is a function of $\pi_{0i} * time_{ti}^0$, $\pi_{1i} * time_{ti}^1$, and $\pi_{2i} * time_{ti}^2$. The quadratic model is:

Level 1:

$$Y_{ti} = \pi_{0i} + \pi_{1i}(time_{ti}) + \pi_{2i}(time_{ti})^2 + e_{ti}$$

Level 2:

$$\pi_{0i} = \beta_{00} - r_{0i}$$

$$\pi_{1i} = \beta_{10} + r_{1i} \quad (9.16)$$

$$\pi_{2i} = \beta_{20} - r_{2i}$$

Combined:

$$Y_{ti} = \beta_{00} + \beta_{10}(time_{ti})^1 + \beta_{20}(time_{ti})^2 + r_{0i} + r_{1i}(time_{ti})^1 + r_{2i}(time_{ti})^2 + e_{ti}$$

In a linear growth model, as $(time_{ti})^1$ increases by one unit, the outcome variable (y_{ti}) increases by π_{1i} units, and this rate of change is constant across the time continuum. However, in a quadratic model, the rate of change is no longer constant across time. Because $(time_{ti})^2$ increases more quickly than $(time_{ti})^1$ does, the coefficient for the quadratic parameter (π) has an increasingly large impact on the outcome variable across time.

Quadratic growth models include parameters for the level when $time = 0$ (the intercept, π_{0i}), the instantaneous rate of change when $time = 0$ (π_{1i}), and the change in the instantaneous rate of change across time (π_{2i}). When time is coded so that $t = 0$ at initial status, then π_{0i} represents individual i's initial level on the outcome variable, and the first-order parameter (π_{1i}) represents individual i's instantaneous rate of change at initial status. The instantaneous rate of change can be positive (indicating a positive growth slope when $t = 0$), negative (indicating a negative growth slope when $t = 0$), or zero (indicating a flat trajectory when $t = 0$).

In a quadratic model, the instantaneous rate of change (π_{1i}) differs across time; the quadratic parameter (π_{2i}) captures the change in this rate of change. Other names for this second-order parameter (π_{2i}) include the acceleration parameter or the curvature parameter. The quadratic parameter essentially describes the degree and direction of the curve in the growth slope over time. The quadratic term can be positive, indicating that the growth slope becomes more positive across time or the quadratic term can be negative, indicating that the growth slope becomes more negative across time. Notably, negative quadratic terms produce convex curves, whereas positive quadratic terms produce concave curves. It is imperative to consider the combination of first-order (linear) and second-order (quadratic) slopes; they determine the ultimate shape of the growth trajectory in combination—no single parameter can adequately capture the quadratic model for change.

Figure 9.3 illustrates the interplay of linear and quadratic slope directionality. Both the first order and second order parameters determine the shape of the quadratic growth trajectory. Figure 9.3 contains four hypothetical quadratic trajectories, each of which is defined by a positive or negative π_{1i}, and a positive or negative π_{2i}. The upper-left quadrant depicts a trajectory with positive π_{1i} and a positive π_{2i} (+/+). When both π_{1i} and π_{2i} are positive, growth is positive when $time = 0$ and becomes increasingly positive across time. The upper-right quadrant depicts a trajectory in which π_{1i} is positive π_{2i} is negative (+/−). This trajectory exhibits positive growth at the start. However, the negative π_{2i} means that the growth slope becomes increasingly negative over time. Therefore, the growth slope becomes less and less positive and eventually becomes more and more negative. The bottom-left quadrant illustrates a trajectory with a negative π_{1i} and a positive π_{2i} (−/+) This depicts a trajectory that begins as negative but becomes

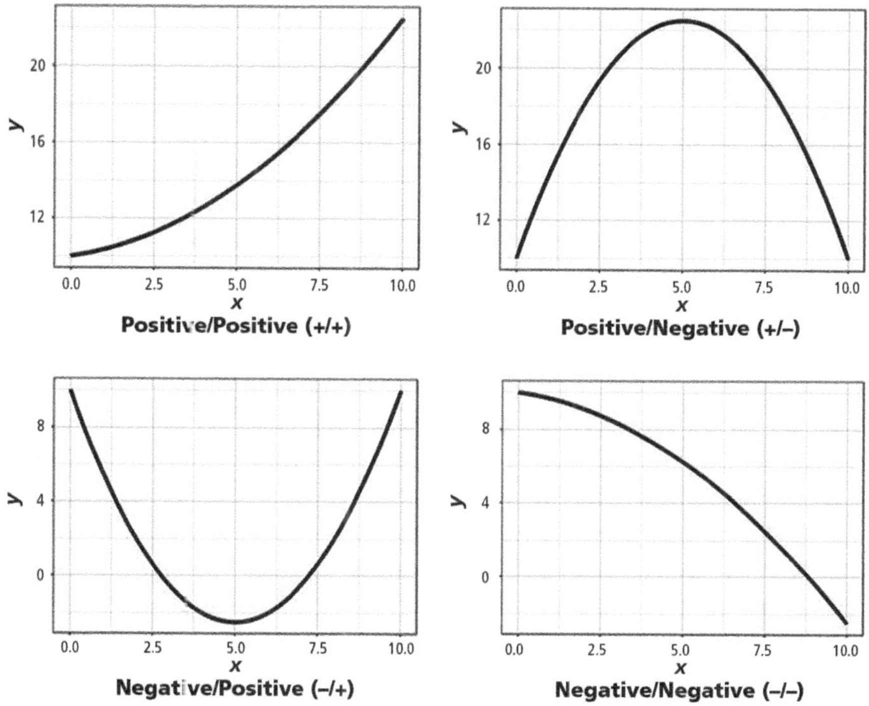

Figure 9.3 Four different quadratic growth trajectories.

increasingly positive across time. At first, when the slope is negative, the positive π_{2i} slows the rate of negative growth. Eventually, the positive π_{2i} overtakes the negative π_{1i}, and the growth rate becomes positive, resulting in a u shape. The bottom-right quadrant (–/–) with a negative π_{1i} and a negative π_{2i}, illustrates initially negative growth (or decline) that becomes increasingly more negative across time over time.

Cubic Model

In a quadratic model, the rate of change is not constant across time. The quadratic parameter captures the change in the rate of change, or the acceleration (deceleration) in the rate of change. However, a quadratic model does not allow for change in the acceleration parameter: The trajectory becomes increasingly positive or increasingly negative (and the rate of change changes at the same rate). Thus, a quadratic model can only have one curve. To model a polynomial growth trajectory with more than one curve, it is necessary to fit a higher-order polynomial function. In a cubic model, the acceleration (curvature) parameter is not constant; it changes across time. The cubic parameter in a polynomial model captures the change in the acceleration parameter: It represents the rate of change in the *changing rate of change.*

A cubic model does allow for change in the quadratic parameter across time; therefore, cubic models may contain two curves. To fit a cubic model, we add an additional term to the growth model: time^3.

Level 1:

$$Y_{ti} = \pi_{0i} + \pi_{1i}(time_{ti}) + \pi_{2i}(time_{ti})^2 + \pi_{3i}(time_{ti})^3 + e_{ti}$$

Level 2:

$$\begin{aligned}\pi_{0i} &= \beta_{00} + r_{0i}\\ \pi_{1i} &= \beta_{10} + r_{1i}\\ \pi_{2i} &= \beta_{20} + r_{2i}\\ \pi_{3i} &= \beta_{30} + r_{3i}\end{aligned} \tag{9.17}$$

Combined:

$$\begin{aligned}Y_{ti} = {} & \beta_{00} + \beta_{10}(time_{ti})^1 + \beta_{20}(time_{ti})^2 + \beta_{30}(time_{ti})^3 + r_{0i} + r_{1i}(time_{ti})^1\\ & + r_{2i}(time_{ti})^2 + r_{3i}(time_{ti})^3 + e_{ti}\end{aligned}$$

This cubic parameter (π_{3i}) represents the rate of change in the quadratic parameter across time. Therefore, the shape of a cubic model is determined by the parameters for first-, second-, and third-order time variables (remember that only the highest-order coefficient is invariant to the centering decision). Parameter estimates for the cubic model must be interpreted in the context of the intercept and three growth slopes: They work together as a team to determine the shape of the growth trajectory. Therefore, as was the case for quadratic models, computing predicted values and graphing the growth trajectories is necessary to interpret the results of a cubic model.

To allow all coefficients (the intercept and the three growth slopes) in the cubic model to randomly vary across people requires at least five time points (although more time points are desirable for a model of this complexity). One advantage of modeling a multilevel cubic model with randomly varying intercept and slope coefficients is that conceptually, every person in the sample has their own intercept, linear, quadratic, and cubic parameters. Thus, in a cubic model with random effects for all four growth parameters, each individual trajectory can have a different shape, as long as the trajectory has no more than two curves. For example, one person could have a negative linear, negative quadratic, and positive cubic parameter whereas another person could have a positive linear, positive quadratic, and negative cubic parameter. Although the trajectories of these two people would look very different, a cubic model that allows all growth parameters to randomly vary across people could accommodate both trajectories. Polynomial models are very flexible, much more so than piecewise models. However, the gain in flexibility does come at some cost. Generally, polynomial models become increasingly difficult to interpret as we add higher order polynomial terms. In addition, in some cases, it may be difficult to make a strong theoretical justification for higher order polynomial models (McCoach & Cintron, 2021).

Centering Higher Order Polynomial Models

The choice of centering has implications for the interpretation of more complex growth functions that extend beyond a simple linear trend. Interpretation of model parameters including polynomial functions such as quadratic and cubic change are dependent upon the centering of time: The highest-order coefficient's interpretation does not change with choice of centering, but the lower-order effect coefficients are dependent upon centering choice. Thus, in a cubic model (which includes time-cubed), the choice of centering impacts the interpretation of the intercept, linear, and quadratic terms. The intercept is still the level (value) of the outcome when time equals zero ($t = 0$), but the linear term represents the instantaneous rate of change at the point where time equals zero and the quadratic term represents the changing rate of change (i.e., curvature) when time equals zero. The cubic term indicates the change in the curvature parameter. In other words, the cubic parameter captures the change in the changing rate of change. In physics, this is sometimes referred to as *jerk* or *jolt.*

Polynomial models using different centering schemes are statistically equivalent, thus parameter estimates can be transformed from one centering scheme to another. However, the coefficients for the intercept and the first order growth slope lower-order parameters may be quite different in magnitude and/or sign. Therefore, centering decisions are especially important in higher-order polynomial models. When fitting polynomial models, it is helpful to generate predicted values across time for proto-typical members of the sample and graph these results. Plotting prototypical growth trajectories aids in the interpretation of the results and helps convey the findings to a variety of audiences (McCoach & Cintron, 2021)

BUILDING GROWTH MODELS

Having discussed the mechanics of growth models, we offer some general recommendations for individual growth models. Generally, building growth curve models requires a sequential approach. First, focus on the within-person model of growth to ensure that the shape of the growth trajectory has been adequately specified. After selecting a growth model that adequately reproduces the observed data, introduce TVC into this model, including any conceptually important interactions among time variables and TVC. After determining the shape of the growth trajectory and modeling within-person growth, then add person-level, TICs to explain the between-person differences (variability) in within-person growth. TICs, such as race/ethnicity or other stable characteristics, differ across individuals but remain constant across occasions (within a given individual).

Including TICs should help explain individual differences (between-person variability) in the intercept and/or the growth slopes. Conditional growth models estimate residual intercept and slope variances, the variances in the intercept and slope(s) that are not explained by the covariates. Ideally, if person-level covariates help to explain inter-individual variability in the intercept and slope, including those predictors in the model should decrease the (residual) variances for the slope and intercept. Therefore, one way to evaluate the necessity/utility of a given predictor is to determine what proportion of the variance in initial status (intercepts) or growth (slopes) a given predictor can explain, as demonstrated in our description of Model 2 (See Equation 9.6).

In summary, to fit multilevel growth curve models, we recommend the following four-step procedure (McCoach & Cintron, 2021):

1. (Optional.) Fit unconditional linear growth model as a baseline model. Often, when the model is nonlinear, this model simply serves as a straw man. Even so, fitting the unconditional linear growth model provides a baseline model to compare to other, more realistic models.
2. Model the shape of the growth trajectory without additional covariates. The unconditional growth model should capture the shape of the unconditional growth trajectory and should provide reasonable predictions of individual growth data. Often, correctly modeling the trajectory requires the use of piecewise, polynomial, or other types of nonlinear growth models.
3. Add other within-person (time varying) covariates to the model (if there are any).
4. Add between-person (time invariant) covariates to the model.

EXAMPLE: MORE ADEQUATELY MODELING ALCOHOL CONSUMPTION FROM AGES 18 TO 25

Piecewise Linear Growth Model

For now, pay attention to the average number of days of alcohol consumption from ages 18 to 25 displayed in Figure 9.4. The solid line displays the sample means across time. The growth in alcohol consumption is clearly nonlinear: Frequency of drinking appears to increase at a constant rate from ages 18 to 20. However, at age 21, there is a large spike in the frequency of drinking. At age 21, reported drinking increases by a full day per month, as compared to age 20. Conceptually, this makes sense because the legal drinking age is 21 years old. However, from age 21 on, drinking behavior appears to stabilize; there appears to be little or no growth in drinking frequency from ages 21 to 25. Therefore, a linear model, which assumes a constant growth rate in drinking frequency, would fail to capture the shape of the growth trajectory.

To more accurately capture the shape of the growth trajectory, we fit a piecewise model with a change in slope and a change in level. Conceptually, we need to fit a growth trajectory in which the outcome grows linearly from ages 18 to 20 and spikes at age 21, and then flattens from ages 21 to 25. In fact, it appears that drinking frequency may not actually grow at all after age 21. We can statistically test whether alcohol consumption systematically

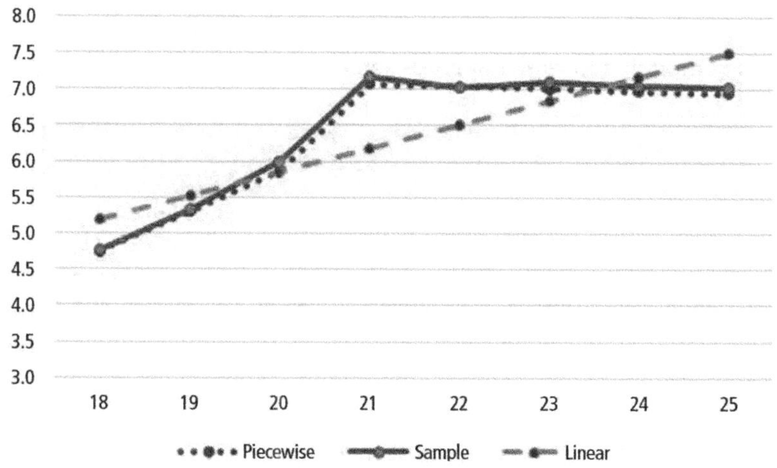

Figure 9.4 Model predicted values for the linear model (dashed line) and the piecewise model (dotted line), compared to the sample means (solid line)

changes between ages 21–25. We do this by fitting a linear trajectory from ages 21–25. We can then test to see if the age 21–25 growth slope is statistically different from 0.

To fit our theoretical model, we need to create three separate pieces (time clocks) to capture our hypothesized trajectory.

$$y_{ti} = \beta_{00} + \beta_{10}(pre21_{ti}) + \beta_{20}(age21_{ti}) + \beta_{20}(post21_{ti}) + r_{0i} \\ + r_{1i}(pre21_{ti}) + r_{2i}(age21_{ti}) + r_{2i}(post21_{ti}) + e_{ti} \tag{9.18}$$

Table 9.4 contains the coding for the time variables (clocks), which estimate three separate pieces. Together these three pieces/clocks allow us to model the nonlinear trajectory, with differing growth rates from ages 18–20 and 21–25, and a change in level (jump) at age 21. The first time-clock (*pre21*) starts at 0 at age 18 and increases by one unit per year until age 20 (so age 19 = 1 and age 20 = 2). At age 20, this first clock stops: From age 20 onward, the first time variable is coded as 2. To capture the discontinuity at age 21, we add a second clock (*age21*) that is coded 0 until age 21 and coded 1 from age 21 onward. Using this coding, the second time variable is only active to capture the jump (change) that occurs between ages 20 and 21. The third piecewise slope (*post21*) captures linear growth from age 21 onward. To do so, at age 21, the third clock starts at 0 and increases by one unit per year until age 25. This coding scheme captures the actual shape of the growth trajectory. It also allows us to statistically test whether the linear growth slope from ages 21–25 is statistically different from 0.

The third model in Table 9.1 contains the parameter estimates for the piecewise model. The *pre21* slope is .56 ($\beta_{10} = 0.56$), indicating that from

TABLE 9.4 Coding Scheme for Linear Growth Model and Piecewise Growth Model for the NLSY97 Example

Coding	Linear Model	Model 3: 2-Piece Model With Change in Level (Discontinuity) at Age 21		
Age	(Age 18)	Slope 1	Slope 2	Change in Level at Age 21
18	0	0	0	0
19	1	1	0	0
20	2	2	0	0
21	3	2	1	0
22	4	2	1	1
23	5	2	1	2
24	6	2	1	3
25	7	2	1	4

ages 18–20, drinking frequency increases by .56 days per month. The *age21* parameter indicates that drinking frequency increases by 1.21 days per month from age 20 to age 21 ($\beta_{20} = 1.21$). The *post21* slope captures growth from age 21 onward. The parameter estimate for growth from age 21 onward is slightly negative ($\beta_{30} = -0.03$), but it is not statistically significantly different from 0. This means that after age 21, on average, drinking frequency remains constant. However, the variance in the *post21* growth slope (τ_{33}) is 1.27. Taking the square root of this variance ($\sqrt{1.27} = 1.33$), we can compute the likely range of *post21* growth slopes. Approximately 68% of the sample have growth slopes between −1.16 and 1.10 and approximately 95% of the sample have growth slopes between −2.29 and +2.23: Some people have negative *post21* growth slopes and some people have positive *post21* growth slopes. Therefore, even though on average, growth in drinking frequency is approximately 0, some people's drinking frequency increases after age 21 and some people's drinking frequency decreases after age 21. In other words, on average, drinking behavior plateaus at age 21, but there is between-person variability in how drinking behavior changes from ages 21 to 25. There is also considerable variability in drinking frequency at age 18 ($\tau_{00} = 12.96$). On average, 18-year-olds drink 4.74 days per month. However, the standard deviation around the intercept is 3.6 ($\sqrt{12.96} = 3.6$), indicating substantial between-person variance in drinking behavior at age 18. Likewise, the variance of the *age21* bump is also quite large ($\tau_{22} = 8.12$). So, although the average change from age 20 to age 21 is 1.21 days per month, the standard deviation is 2.85. Therefore, for 68% of the sample, the age 21 bump falls between −1.64 and 4.06 days per month.

Using the same parameter estimates, we can generate model predicted drinking frequency for ages 18 to 25. In the linear model, drinking increases at a constant rate of 0.33 per month per year. In the piecewise model, the three piecewise parameters capture growth from ages 18–20, the change from 20–21, and growth from ages 21 onward. To compute the predicted score at each age, we multiply the coding scheme for each of the three slopes at a given age by the parameter estimate for each of the slopes at that age, then sum these values and the intercept. For example, predicted drinking frequency at age 25 is $(4.73) * 1 + (0.56) * 2 + (1.21) * 1 + (-0.03) * 4 = 6.95$. In contrast, using the linear model, the predicted drinking fluency at age 25 is $5.19(1) + 0.33(7) = 7.5$. Using the coding scheme in Table 9.4, we can easily reproduce the predicted values for both the linear and the piecewise model in Table 9.5. Table 9.5 also includes the sample means at ages 18–25, and Table 9.6 compares the predicted values for the linear model and the piecewise model to the sample means at each age. Figure 9.4 graphs the sample means, the predicted means from the linear model, and the predicted means from the piecewise model.

TABLE 9.5 Estimated Means for the Linear Growth Model and the Piecewise Linear Growth Model for the NLSY97 Example

Model	Fixed Effects	Parameter Estimate	Age							
			18	19	20	21	22	23	24	25
Linear Growth	β_{00}	5.19								
	β_{10}	0.33								
	Estimated Means		5.19	5.52	5.85	6.18	6.51	6.84	7.17	7.50
Piecewise Linear Growth	β_{00}	4.74								
	β_{10}	0.56								
	β_{20}	–0.03								
	β_{30}	1.21								
	Estimated Means		4.74	5.30	5.86	7.07	7.04	7.01	6.98	6.95

Note: Here we calculated the predicted monthly alcohol consumption at each age using the coding scheme for the linear model and the piecewise model listed in Table 9.4. The predicted values at each age are a function of the intercept estimate, the slope(s) estimates, and the time coding scheme.

TABLE 9.6 Predicted Values for the Piecewise and Linear Models Compared to the Sample Means at Each Time Point

Age	Piecewise	Sample	Linear
18	4.74	4.77	5.19
19	5.30	5.33	5.52
20	5.86	6.00	5.85
21	7.07	7.18	6.18
22	7.04	7.03	6.51
23	7.01	7.11	6.84
24	6.98	7.06	7.17
25	6.95	7.03	7.50

Why do we prefer the piecewise model over the linear model? As shown in Figure 9.4, the piecewise model is better able to reproduce the sample means than the linear model is. Additionally, comparing the piecewise model to the linear model, the likelihood ratio test (LRT) and information criteria such as the AIC, the BIC_2, and the BIC_1 all favor the piecewise model. The deviance of the linear model is 79362.5 with 6 parameters. The deviance of the piecewise model is 79058 with 15 parameters. The difference in deviance is 304.5; the difference in the number of parameters is 9. The critical value of chi-square with 9 *df* is 21.666 at alpha = .01. Therefore, using the LRT, we favor the piecewise model.

Having established that the piecewise model best captures the growth trajectory, we move to a model that includes TVCs (Model 4 in Table 9.1). The TVC, hours of work per week, is the average number of work hours per week during a given year and is measured at all eight timepoints, from age 18–25. We person-mean centered weekly work hours for each individual in the study and included the person's mean work hours across the 7-year period as a level-2 (time-invariant) covariate. We also included a cross-level interaction between mean work hours and person mean centered work hours.

The coefficient/slope for the TVC of within-person centered work hours is 0.02. For every additional hour worked per week, drinking frequency increases by 0.02 days per month. So, on average, working 10 additional hours per week increases a person's drinking frequency by 0.2 days per month. The effect of mean hours worked on the intercept was 0.04 points, indicating that individuals who work more on average also tend to drink more frequently at age 18. Finally, the interaction between average work hours and within-person centered work hours was –0.0005. This interaction was not statistically significantly different from 0. However, for pedagogical reasons, we review the interpretation of the cross-level interaction between the TVC and its aggregate at level 2. The coefficient is negative, meaning that the effect of within-person change in number of work hours on drinking frequency is more negative for people who work more hours overall. Because the within-person work hours slope is positive, this means the effect of within-person increases in work hours has less of a positive effect on drinking behavior for people who work longer hours overall. In this example, we did not allow the TVC (work hours) to randomly vary across people. This means that we assume that the effect of within-person changes in work hours on drinking behavior is constant across the entire sample.

In the final model (Model 5 in Table 9.1), we add sex as time-invariant predictor of drinking behavior at age 18 (the intercept), the *pre21* growth slope, the discontinuity at age 21, the *post21* growth slope, and the group mean centered workhours slope. We also include a same level interaction between sex and the average number of work hours to examine whether the effect of average work hours on drinking frequency varies by sex.

In this model, the females who work an average number of hours per week drink an average of 4.21 days per month at age 18; comparable males drink an average of 5.62 days per month at age 18 (4.21 + 1.41). Holding work hours constant at the mean, from ages 18–20 drinking frequency increases by 0.28 days per month for females and 0.72 (.28 + .44) days per month for males. In other words, holding work hours constant, males drink more at age 18 and their drinking frequency grows more from ages 18 to 20. By age 20, holding work time constant, females drink 4.77 days per month at age 20, whereas males drink an average of 7.06 days per month at age 20. However, from age 21 onward, males have a negative effect on increases in drinking (holding work constant). From age 20 to 21, holding work constant, drinking frequency increases by 1.55 days per month for females, but 0.76 days per month for males. The growth slope from ages 21–25 is 0.01 for females (and is not statistically significantly different from 0). The *post21* growth slope for males is –0.15 (–0.16 + 0.01). The differential in the *post21* drinking slope (–0.16) is not statistically significantly different from 0. However, this does not mean that the *post21* growth slope for males is not statistically significantly different from 0. The easiest test of whether the *post21* growth slope for males is statistically significant is to create a new dummy coded variable where male is coded 0 and female is coded 1. These results are not included in Table 9.1. However, reversing the dummy coding of *sex* so that male = 0 and female = 1 produces the following results. The coefficient for the *post21* slope is the growth slope from *age21* onward for males (holding work hours constant at the mean.) The coefficient is –0.15 ($z = -2.17$, $p = .03$), so the male *post21* growth slope is statistically different from 0 (at $\alpha = .05$). Sex does not moderate the within-person effect of work hours on drinking frequency (–0.002), nor does it moderate the effect of average work hours on drinking frequency (0.024).

CHAPTER SUMMARY

This chapter introduced a series of models that can be used to examine change in longitudinal models with continuous outcomes: simple linear growth models, piecewise models, and polynomial growth models. Theory plays a guiding role in the model building process. As demonstrated through the monthly alcohol consumption example, it is also crucial to examine the data prior to fitting growth curve models to ensure that the growth model adequately captures the shape of the individual growth trajectories. Plotting the sample data, it was apparent that changes in monthly alcohol consumption were not linear. Our piecewise model with a discontinuity at age 21 more accurately reflects the change process that we observe than the linear model. Although we introduced a wide variety of models that provide a great deal of flexibility, some forms of growth cannot be adequately captured by any of these models. In the next chapter, Harring and Blozis provide an introduction to truly nonlinear growth models within an MLM framework.

RECOMMENDED READINGS

Technical Readings

Curran, P. J., & Bauer, D. J. (2011). The disaggregation of within-person and between-person effects in longitudinal models of change. *Annual Review of Psychology, 62*, 583–619. https://doi.org/10.1146/annurev.psych.093008.100356

Hoffman, L. (2015). *Longitudinal analysis: Modeling within-person fluctuation and change.* Routledge.

Singer, J. D., & Willett, J. B. (2003). *Applied longitudinal data analysis: Modeling change and event occurrence.* Oxford University Press.

Applied Readings

McCoach, D. B., O'Connell, A. A., Reis, S. M., & Levitt, H. A. (2006). Growing readers: A hierarchical linear model of children's reading growth during the first 2 years of school. *Journal of Educational Psychology, 98*(1), 14–28. https://doi.org/10.1037/0022-0663.98.1.14

Nærde A, Ogden T, Janson H, & Zachrisson H. D. (2014). Normative development of physical aggression from 8 to 26 months. *Developmental Psychology, 50*(6), 1710-1720. https://doi.org/10.1037/a0036324.supp

Rambo, K., & McCoach, D. B. (2014). Using summer growth patterns to assess the impact of schools on high achieving and gifted students' reading skills. *Journal of Educational Research.* https://doi.org/10.1080/00220671.2013.850398

TRY THIS!

The "Try This!" data are from the National Youth Survey (NYS), a longitudinal study sponsored by the National Institute of Mental Health that explored conventional and deviant behavior in adolescents (https://www.icpsr.umich.edu/icpsrweb/ICPSR/series/88). Using these data, we can examine adolescents' antisocial attitudes from ages 11–15 to answer three main research questions: (a) "How antisocial are adolescents' attitudes at age 11?" (b) "How do adolescents' antisocial attitudes change across time from ages 11 to 15?" ("What is the shape of the growth trajectory—How do adolescents change across this developmental period?"); and (c) "Does sex moderate either initial antisocial attitudes (at age 11) or growth in antisocial attitudes?" Questions 1 and 2 focus on the modeling of the trajectory; neither question requires any person-level covariates. In contrast, research Question 3 examines inter-individual differences in adolescents' initial status and rate of change as a function of sex, a person-level covariate.

For this example, our primary variable of interest is attitudes toward (i.e., tolerance of) antisocial/deviant behavior (*ATTIT*) across five time points (from age 11 to age 15), for 239 adolescents. Given that *ATTIT* were collected 5 times for each person, these data include a total of 1,079 observations of the *ATTIT* variable. Remember, multilevel growth models require a long dataset, containing a row for each observation and a single outcome variable (*ATTIT*). The data are already restructured to be a person-period (long) data file.

1. First, create individual growth plots and a mean growth plot (comparing the average antisocial attitudes of boys and girls from ages 11–15) to examine the shape of the growth trajectory. Does it look linear? If it looks nonlinear, how could you model the shape of the growth trajectory using a piecewise linear model? How could you model the shape of the growth trajectory using a polynomial model?
2. To model the growth in antisocial attitudes from ages 11 to 15, we recommend that you run the following unconditional growth models and compare the fit using information criteria such as AIC and/or BIC (not all of the models below are nested):
   a. an unconditional linear growth model;
   b. a piecewise linear growth model; or
   c. a polynomial growth model (Should you fit a quadratic model or a cubic model, given the shape of the growth trajectory you graphed in Question 1? If you are unsure, you could fit both models and then compare the fit of the two models.)

   Plot the predicted values from the unconditional linear growth model, the piecewise linear growth model, and the polynomial growth model and compare those predicted values to the sample means at ages 11–15. Which of the models appears to best reproduce the sample means? Also, which of the models fits the best, according to the AIC/BIC?
3. Select your preferred growth model from Question 2, then add sex as a time invariant covariate, predicting both antisocial attitudes at age 11 and growth in antisocial attitudes from ages 11 to 15. Be sure that you can interpret the results. Using the parameter estimates, create a growth plot that compares boys and girls antisocial attitudes at age 11 and their growth in antisocial attitudes from ages 11 to 15.

NOTES

1. The term *growth curve modeling* is used to refer to models in which observations systematically increase (positive growth) or systematically decrease (negative growth) across time.

2. Researchers could employ modern missing data techniques such as multiple imputation to salvage such cases.
3. This is true as long as the centering decision is the only thing that changes. If a researcher moves from using a time structured time variable to a time unstructured time variable, the linear growth slope may differ. However, the change is not due to the change in centering. Instead, it is due to the change in the nature of the time variable itself.
4. To do so, we fit a model with an unstructured *T* matrix.
5. In other words, if the TVC does not grow or decline across time.
6. See Chapter 11 of this volume, "Within-Subject Residual Variance/Covariance Structures in Longitudinal Data Analysis" by Kim et al. for more information about the composite residual matrix.
7. There can be missing data, so not every person must have that number of data points. But your design should include at least that many data collection points, and at least some people in the sample should have that many observations.
8. Homeostasis refers to stability or equilibrium. Homeostasis represents a dynamic state of equilibrium rather than a constant, unchanging state.

REFERENCES

Bryk, A., S., & Raudenbush, S. W. (1988). Toward a more appropriate conceptualization of research on school effects: A three-level hierarchical linear model. *American Journal of Education, 97*(1), 65–108. http://www.jstor.org/stable/1084940

Collins, L. M. (2006). Analysis of longitudinal data: The integration of theoretical model, temporal design, and statistical model. *Annual Review of Psychology, 57,* 505–528. https://doi.org/10.1146/annurev.psych.57.102904.190146

Cronbach, L. J., & Furby, L. (1970). How should we measure "change": Or should we? *Psychological Bulletin, 74*(1), 68–80. https://doi.org/10.1037/h0029382

Hoffman, L. (2015). *Longitudinal analysis: Modeling within-person fluctuation and change.* Routledge.

Kline, R. B. (2015). *Principles and practice of structural equation modeling* (4th ed.). Guilford Press.

McCoach, D. B., & Cintron, D. (2021). *Introduction to modern modeling methods.* SAGE.

McCoach, D. B., & Kaniskan, B. (2010). Using time-varying covariates in multilevel growth models. *Frontiers in Psychology, 1*(17), 1–12. https://doi.org/10.3389/fpsyg.2010.00017

McCoach, D. B., O'Connell, A. A., Reis, S. M., & Levitt, H. (2006). Growing readers: A hierarchical linear model of children's reading growth over the first 2 years of school. *Journal of Educational Psychology, 98*(1), 14–28. https://doi.org/10.1037/0022-0663.98.1.14

McCoach, D. B., & Rambo-Hernandez, K. E. (2018). Issues in the analysis of change. In C. Secolsky & D. B. Denison (Eds.), *Handbook of measurement, assessment, and evaluation in higher education* (2nd ed.; pp. 527–544). Routledge.

Rogosa, D., Brandt, D., & Zimowski, M. (1982). A growth curve approach to the measurement of change. *Psychological Bulletin, 92*(3), 726–748. https://doi.org/10.1037/0033-2909.92.3.726

Rabe-Hesketh, S., & Skrondal, A. (2008). *Multilevel and longitudinal modeling using Stata* (2nd ed.). Stata Press.

Raudenbush, S. W., & Bryk, A. S. (2002). *Hierarchical linear models: Applications and data-analysis methods.* SAGE Publications.

Singer, J. D., & Willett, J. B. (2003). *Applied longitudinal data analysis: Modeling change and event occurrence.* Oxford University Press. https://doi.org/10.1093/acprof:oso/9780195152968.001.0001

Skrondal, A., & Rabe-Hesketh, S. (2008). Multilevel and related models for longitudinal data. In J. de Leeuw & E. Meijer (Eds.), *Handbook of multilevel analysis* (pp. 275–299). Springer Science + Business Media. https://doi.org/10.1007/978-0-387-73186-5_7

U.S. Bureau of Labor Statistics. (2019). *National Longitudinal Survey of Youth 1997: Index to the NLSY97 cohort* [Data set]. https://www.nlsinfo.org/content/cohorts/nlsy97

CHAPTER 10

MODELING NONLINEAR LONGITUDINAL CHANGE WITH MIXED EFFECTS MODELS

Jeffrey R. Harring
University of Maryland, College Park

Shelley A. Blozis
University of California, Davis

Investigations of cognitive and noncognitive developmental processes in education and in the social and behavioral sciences necessitate that individuals be measured repeatedly over time or other analytic conditions. Such longitudinal data are typically comprised of continuous or categorical response scales collected from direct observation, experimental manipulation, self-reports, survey instruments and assessment batteries. The scores from these data sources reflect individual differences in performance (e.g., reading comprehension) or gradations of individual attributes or attitudes (e.g., motivation, self-efficacy). These differences can be marked by dissimilar patterns of change, such as differences in performance levels at

Multilevel Modeling Methods with Introductory and Advanced Applications, pages 355–387

different ages, as well as different rates of change as individuals refine their skills, abilities and attitudes. Notably, change is not uniform but rather faster during some periods and slower in others.

In a study of early childhood reading skill development, for example, a child's performance may show gradual slow gains giving way to a period of rapid improvement followed by deceleration as a higher level of proficiency is achieved. Although development may occur in the same general manner across children, the timing of change as well as the amount gained is specific to each child. Moreover, there is no requirement for a child's development to conform to the trajectory for the average, or typical, child. To help cement these ideas, consider the repeated measures data of 9 children displayed in the trellis plot in Figure 10.1. These data are due to Conner et al. (2006) who investigated the effects of preschool instruction on academic gains. The larger sample contains assessment information on 383 children (195 girls and 188 boys) from an economically and ethnically diverse community located on the urban fringe of a major Midwestern city in the United States.[1] The response variable used here is letter–word identification (LWID) subtest scores (maximum of 75 on the raw score scale) from the Woodcock-Johnson III Test of Achievement (designed to measure letter and word recognition; McGrew et al., 1991). The children were assessed on 10 occasions in the fall and spring of each school year from preschool through second grade (e.g., FK denotes the fall assessment in kindergarten).

Interestingly, individual profiles implied by the black dots follow the generic pattern characteristic of the behavior but vary on particular features. Performance differs at the initial assessment, gains are shallower or steeper, points of inflection are reached earlier or later, and potential (or asymptotic) performance has just been realized or attainment occurs earlier. Variability within children over time is evident and the scatter of responses differs from child to child. Conspicuously, the number of data points measured on the children is unequal. Some children began the study and then dropped out whereas others arrived for the study at the moment when they entered school. To adequately account for these data and profile characteristics, the statistical modeling framework needs to accommodate them all: missing data, nonlinear change patterns, intraindividual variability, and individual differences in response.

Nonlinear mixed-effects (NLME) models (Davidian & Giltinan, 1995; Pinheiro & Bates, 1995, 2000; Vonesh & Chinchilli, 1997) provide a flexible analytic framework for describing individuals' continuous repeated measures data that exhibit complex, curvilinear patterns of change, such as those presented in Figure 10.1. As Davidian and Giltinan (2003) argued, an NLME model is an appropriate statistical framework for scenarios in which (a) the response-time functional relation is similar across individuals yet idiosyncratic differences are evident; (b) there exists a scientifically

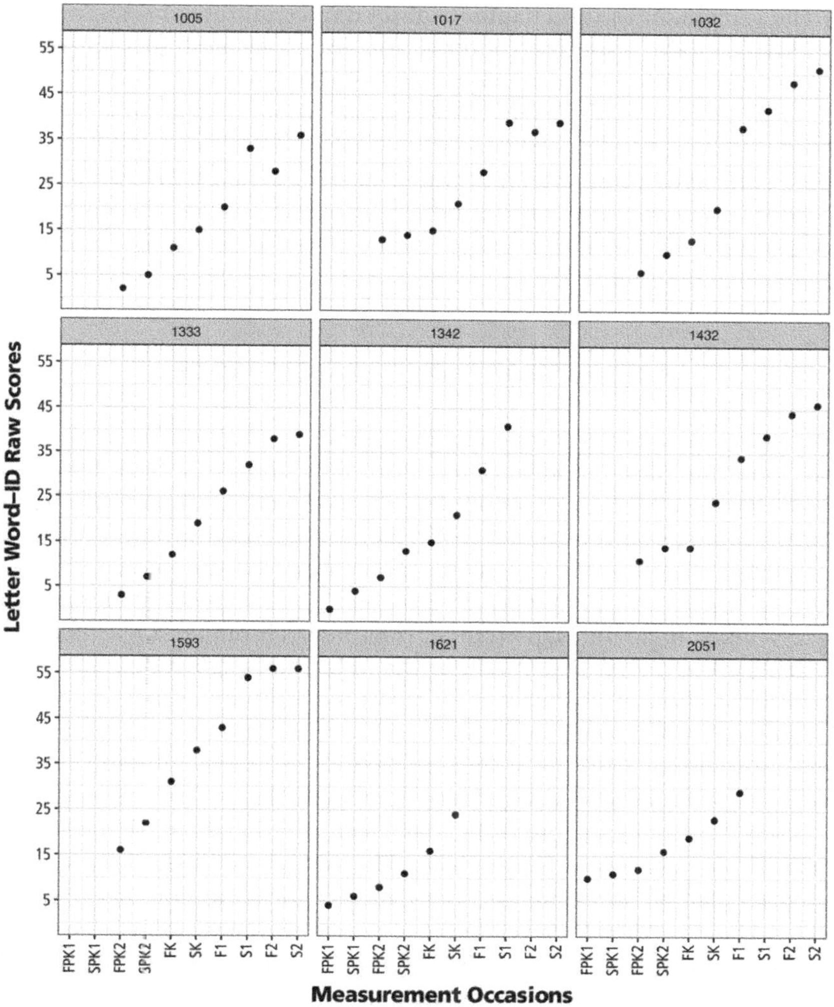

Figure 10.1 Letter Word–ID raw scores across ten occasions for a random sample of nine children.

defensible nonlinear function that effectively characterizes individual behavior in terms of meaningful parameters that are linked to interesting, salient features of the underlying change process; and (c) model parameters vary across individuals giving rise to within- and between-individual variation in time-response. Of primary interest is in accurately describing the typical behavior of the phenomenon, but once this task is completed, the analytic objectives are to understand the extent to which the model parameters vary across individuals and investigate whether some of this variation can be explained by individual attributes or characteristics.

The remainder of this chapter is divided into four major sections. First, the NLME model is specified and the notation explicated. In a subsequent section, data analytic and modeling issues associated with NLME models are highlighted and discussed. Next, an analysis of the data will ensue with an emphasis on those moments when analytic choices must be decided upon. A detailed presentation of the results, focusing on interpretation of model parameters and corresponding graphical representations of typical and individual behavior, follow. Lastly, the chapter is summarized with some concluding remarks.

A NONLINEAR MIXED-EFFECTS MODEL

Like linear mixed-effects models (Fitzmaurice et al., 2011; Laird & Ware, 1982) and multilevel models (Hox, 2010; Raudenbush & Bryk, 2002; Snijders & Bosker, 2012), a common way to present an NLME model is hierarchically noting that repeated measurements are nested within individuals. Let $y_{i1}, y_{i2}, \ldots, y_{in_i}$ be n_i measurements on the response variable y for individual $i\,(i = 1, \ldots, m)$ taken at the jth design point, t_{ij}. In many longitudinal studies, t_{ij} is taken to be the elapsed time from the beginning of the experimental protocol to the jth assessment or the individual's age at occasion j. The n_i subscript allows individuals to be measured at unique times and for the real possibility that some participants will have dropped out of the study; or have intermittent patterns of missing data under the assumption that the missingness is random (MAR). More on missing data is forthcoming.

Given the basic structure of the data, a conventional NLME model consists of three interconnected modules. The first is a nonlinear function that describes change over time or other data analytic condition (see, e.g., Cudeck & Harring, 2010 for an example of non-time design points). This function is parameterized at the individual level in which coefficients are individual specific and so may vary between individuals. The second element is the covariance structure of the residuals—time-specific deviations from an individuals' data to their fitted function. The third component is the set of models for the regression coefficients that summarize differences among individuals. Each of these is further explicated below.

Individual-Specific Function for Nonlinear Change

As its name implies, an NLME model is well-suited to summarize nonlinear patterns of change—typically by using a simple algebraic function that adequately describes individual-specific behavior yet differs from its linear mixed-effects modeling counterparts in that at least one parameter with a random effect must enter the expectation function nonlinearly (Cudeck, 1996; Seber & Wild, 1989). The functional form characterizing the within-individual behavior is the primary focus of the individual level (or level 1) model,

$$y_{ij} = f(t_{ij}, \boldsymbol{\beta}_i, \mathbf{z}_i) + e_{ij}, \quad j = 1, \ldots, n_i \tag{10.1}$$

Time-specific residuals are e_{ij}. Nonlinear function $f(\cdot)$ is dependent on t_{ij}, a set of p individual regression coefficients, $\boldsymbol{\beta}_i = (\beta_{1i}, \beta_{2i}, \ldots, \beta_{pi})'$, and a set of individual-specific attributes, ($\mathbf{z}_i$; age, gender, SES, treatment condition), that do not vary across time.

A logistic function (see Browne, 1993) of the form

$$f(t_{ij}, \beta_{1i}, \beta_{2i}, \beta_{3i}) = \frac{\beta_{1i}\beta_{2i}}{\beta_{1i} - (\beta_{1i} - \beta_{2i})\exp\{-\beta_{3i}t_{ij}\}}, \tag{10.2}$$

may provide good model-data fit for the letter-word identification data presented earlier given that learning in general tends to follow nonlinear monotonic trajectories that level off at later measurement occasions. Parameters of the logistic function in Equation 10.2 describe key features of performance. Specifically, β_{1i} is a child's expected LWID raw score at $t_{ij} = 0$ (or at the beginning of the assessment protocol), β_{2i} is a child's expected potential performance (i.e., personal maximum) achieved at later times (i.e., $f(t_{ij}) \to \beta_2$ as $t_{ij} \to \infty$), and β_{3i} is the child's nonlinear growth parameter that reflects how quickly change occurs from initial to potential performance. To make this more concrete, suppose an individual child's logistic growth coefficients were $\beta_{1i} = 5$, $\beta_{2i} = 30$, and $\beta_{3i} = 0.25$, respectively. Then in the fall of the second year of kindergarten (i.e., when $t_{ij} = 3$ given that the fall of kindergarten was coded as $t_{ij} = 0$), the child's expected LWID raw score based on the logistic function in Equation 10.2 would be

$$f(t_{ij} = 3, \beta_{1i} = 5, \beta_{2i} = 30, \beta_{3i} = 0.25) = \frac{5 \cdot 30}{5 - (5 - 30)\exp\{-0.25 \cdot 3\}} = 8.92,$$

while this same child in the fall of second grade (i.e., $t_{ij} = 9$) would have an expected model-based LWID raw score of 19.65.

Other parameterizations of the logistic function, even algebraically equivalent functions, could be employed as could other non-logistic functions. Reparameterization is often possible with nonlinear functions (see, e.g., Preacher & Hancock, 2015), yet more importantly is the idea that nonlinear functions can frequently be tailored so that parameters have a natural, physical interpretation which correspond to substantive characteristics of the phenomenon under study (Blozis & Harring, 2015; Cudeck & Harring, 2010). As is true in many problems involving repeated measures, several nonlinear functions may perform similarly. The decisive factor in selecting a final functional form may well be guidance from practitioners and applied researchers who understand the science underlying the change process.

Residuals and the Level 1 Covariance Structures

The residuals, e_{ij}, on the right-hand side of the model in Equation 10.1 are random variables that describe lack-of-fit of the model, $e_{ij} = y_{ij} - f(t_{ij}, \boldsymbol{\beta}_i, \mathbf{z}_i)$, in accounting for the responses within an individual. The statistical assumptions about these residuals relate to what is called the level-1 covariance structure. First, the residuals are assumed to be normally distributed and satisfying $E(e_{ij}|\boldsymbol{\beta}_i) = 0$ for all i and j. The variances of and covariances among the residuals are modeled through an $n_i \times n_i$ covariance matrix, $\boldsymbol{\Lambda}_i(\boldsymbol{\beta}_i, \boldsymbol{\lambda})$, whose elements are functions of individual-specific regression coefficients, $\boldsymbol{\beta}_i$, and fixed-effects, $\boldsymbol{\lambda}$.[2]

Considerable flexibility can be accommodated in $\boldsymbol{\Lambda}_i(\boldsymbol{\lambda})$, yet it is frequently relegated to taking on a conditionally independent structure with homogeneous time-specific variances (i.e., $\boldsymbol{\Lambda}_i(\boldsymbol{\lambda}) = \sigma^2 \mathbf{I}_{n_i}$ where $\mathbf{I}_{n_i}$ is an identity matrix of dimension n_i). However, greater complexity than a conditional-independence structure may be necessary to describe intricate patterns of time-specific variances and serial correlation likely to arise with nonlinear repeated measures data. Davidian and Giltinan (2003) discussed a general structure that uncouples the time-specific variances (i.e., diagonal elements of $\boldsymbol{\Lambda}_i(\boldsymbol{\lambda})$) and the serial correlation among the repeated measures (i.e., off-diagonal elements of $\boldsymbol{\Lambda}_i(\boldsymbol{\lambda})$) so that each could be modeled separately. This can be written as

$$\boldsymbol{\Lambda}_i(\boldsymbol{\lambda}) = \boldsymbol{\Omega}_i^{1/2}(\boldsymbol{\omega}) \cdot \mathbf{P}_i(\boldsymbol{\rho}) \cdot \boldsymbol{\Omega}_i^{1/2}(\boldsymbol{\omega}), \tag{10.3}$$

where $\boldsymbol{\Omega}_i^{1/2}(\boldsymbol{\omega})$ is either a scalar or an $n_i \times n_i$ diagonal matrix depending on fixed-effects, $\boldsymbol{\omega}$. For example, consider a balanced design with equally-spaced measurements. If heterogeneity is suspected in the variances then $\boldsymbol{\Omega}_i^{1/2}(\boldsymbol{\omega})$ could be parameterized such that its $[j, j]$th element would be σ_j. That is,

$$\boldsymbol{\Omega}_i^{1/2}(\boldsymbol{\omega}) = \begin{pmatrix} \sigma_1 & & & \\ & \sigma_2 & & \\ & & \ddots & \\ & & & \sigma_{n_i} \end{pmatrix},$$

where $\boldsymbol{\omega}' = (\sigma_1^2, \sigma_2^2, \ldots, \sigma_{n_i}^2)$.

Serial correlation in pairs of residuals may be modeled through the $n_i \times n_i$ correlation matrix, $\mathbf{P}_i(\rho)$, which depends on fixed effects, ρ. If residuals were hypothesized to follow a first-order autoregressive process, then off-diagonal elements of $\mathbf{P}_i(\rho)$ would be parameterized as $\rho^{|j-k|}$ for all $j \neq k$. In many instances, the data collection times t_{ij} and t_{ik} are far enough apart relative to the *locality* of the process that correlation, as Davidian and Giltinan (1995) put it, has "dampened out" sufficiently as to be practically insignificant. In this case for the particular times involved, $\mathbf{P}_i(\rho) = \mathbf{I}_{n_i}$ may be a reasonable approximation.

As Harring and Blozis (2014) pointed out, specifying a model that adequately summarizes level-1 variation is frequently an afterthought for many researchers, not only because the primary focus may be on parameters defining the change process itself, but also because the literature has not been very informative of how to examine this type of variation. Yet, gross misspecification can impact the analysis in nontrivial ways. Cudeck and Harring (2007) made a compelling argument justifying the statistical investment in the residual covariance structure suggesting that (a) the choice of covariance structure impacts the fit of the model and can influence decisions made about the adequacy of the response function, and as they stated so concisely; (b) "the precision of estimation of all parameters in the composite model is affected by the covariance structure. This implies, for example, that the significance of a particular parameter is affected by choice of the within-subject, level 1 specification" (p. 623).

Individual's Coefficients and the Level 2 Covariance Structure

The other set of random variables on the right-hand side of the model in Equation 10.1 are the individual regression coefficients, $\boldsymbol{\beta}_i = (\beta_{1i}, \beta_{2i}, \ldots, \beta_{pi})'$ in $f(\cdot)$. In most scenarios, each coefficient is not measured directly but represents a type of latent variable that symbolizes an interesting, scientifically important aspect of the phenomenon evident in the underlying change process. The coefficients are distinguished by their role in determining the shape of the trajectory. A secondary goal of many NLME analyses is to attempt to understand individual differences in these parameters, and toward that end, a submodel may be specified for each coefficient. The degree of complexity of these submodels directly corresponds to the features inherent in the data, the accessibility of individual-level covariates, and the nature of the questions articulated in the research hypotheses.

A basic form of the model for an individual's coefficients is a simple additive structure, $\beta_{ki} = \beta_k + b_{ki}$. Thus as a starting point, an individual's coefficients are comprised of population parameters, β_k, known as *fixed effects*, and individual deviations, b_{ki}, known as *random effects*. Cleverly, an individual's coefficient is related to the population value but offset by an amount, b_{ki}, to allow the curve to conform to the individual's data. More generally, an individual's coefficients is modeled through function $d(\cdot)$ which depends on a $p \times 1$ vector of fixed parameters, $\boldsymbol{\beta}' = (\beta_1, \beta_2, \ldots, \beta_p)$, covariate vector $\mathbf{z}_i$, and a $q \times 1$ vector of random effects, $\mathbf{b}_i' = (b_{1i}, \ldots, b_{qi})$. For individual coefficient k, this is

$$\beta_{ki} = d_k(\mathbf{z}_i, \beta_k, b_{ki}).$$

Here, $d_k(\cdot)$ is a flexible function of the arguments and could very well be distinct for each coefficient. Uncoupling the dimensions of the fixed and random effects (i.e., p and q) allows for optimal control in the form the coefficients take. For example, consider the previously mentioned preschool instruction study and logistic model expressed in Equation 10.2. Consider the situation in which the functions for the three individual regression coefficients are

$$\begin{aligned}
\beta_{1i} &= \beta_1 + \gamma_1 z_{1i} + b_{1i} \\
\beta_{2i} &= \beta_2 + \gamma_2 z_{1i} + \gamma_3 z_{2i} + b_{2i} \\
\beta_{3i} &= \beta_3.
\end{aligned}$$

Excluding b_{3i} implies that only two of the three coefficients vary across individuals. Note that the regressions for β_{1i} and β_{2i} use different time-invariant covariates, z_{1i} and z_{2i}, to explain individual differences in initial and potential performance, respectively. These variables might be a categorical variable indicating a child's gender or a continuous variable representing another reading-related skill thought to impact the development of letter and word recognition.

In the population of individuals, it is generally assumed that the random effects follow a q-dimensional normal distribution with zero mean vector and covariance matrix, $\boldsymbol{\Phi}$. That is,

$$\mathbf{b}_i \sim N(\mathbf{0}, \boldsymbol{\Phi}).$$

This is often referred to as the level-2 covariance structure. To make this idea concrete, the covariance matrix among random effects for individual coefficients above $(q = 2)$ would be

$$\text{cov}(\mathbf{b}_i) = \boldsymbol{\Phi} = \begin{pmatrix} \text{var}(b_{1i}) & \\ \text{cov}(b_{2i}, b_{1i}) & \text{var}(b_{2i}) \end{pmatrix} = \begin{pmatrix} \varphi_{11} & \\ \varphi_{21} & \varphi_{22} \end{pmatrix},$$

where the diagonal elements of $\boldsymbol{\Phi}$ are variance components that summarize the extent that the random effects vary around zero, whereas the off-diagonal elements quantify the degree to which the random effects are linearly associated. A positive value of φ_{21} would suggest that children with higher initial LWID scores are associated with higher potential performance. Lastly, it is assumed that the random effects and level-1 residuals are uncorrelated (i.e., $\text{cov}(\mathbf{b}_i, \mathbf{e}_i') = \mathbf{0}$).

ANALYTIC ISSUES AND DECISION POINTS

Longitudinal Designs and Data Gathering Protocols

From a longitudinal design perspective, in the NLME modeling framework, the number of measurements taken over time, as well as the intervals between assessments, can be unique to each individual. Thus, unlike traditional analytic approaches like repeated measures analysis of variance, this statistical approach permits missing data and it attends to the actual timing of assessments that may be unique to the individual. This makes the method highly amenable to studies in which individuals may not have complete data as prescribed in a study plan or have identical times of assessments. These last points have important implications for practitioners as it offers a very flexible data collection protocol.

First, in many environments, especially schools, data collection is challenging due, in part, to adhering to a schedule that permits "visits" sparingly as to not hinder the regular workings of the classroom yet adequately captures change in the phenomenon of interest (e.g., math self-efficacy, reading comprehension). In this same context, data collection protocols such as planned missingness designs can be very convenient from the experimenter's viewpoint because it lessens the burden of over-assessing students and keeping individuals from becoming exhausted and unmotivated. For example, in a data collection protocol designed to assess students at weeks 1, 2, 3, 4, 5, 6, and 7, some could be measured at weeks 1, 3, 5, and 7 while others could be measured at completely different times such as weeks 2, 4, and 6.

Secondly, many developmental studies examine phenomena tied to age—meaning that the age distinction is an important substantive feature. Allowing the observations to be anchored to the chronology metric rather than the order in which the observations were obtained is critical to understanding when development in the phenomena changes most rapidly or dramatically slows. From a data collection design standpoint, this means that participants do not need to be assessed at the same time. If the protocol calls for each individual to be measured four times then subject 1 could be measured at months 43, 55.5, 60, 64.3, while data could be gathered for subject 2 at months 45, 53, 62.5, and 66, and so on. The preschool instruction data used throughout the chapter has the age of evaluation for each child. This will need to be examined empirically against treating the assessment occasions as discrete points in time.

Handling Missing Data

Missing data in longitudinal studies can occur because a participant misses an assessment or discontinues their participation at some point during the study. Estimation of NLME model parameters is typically carried out using maximum likelihood estimation or Bayesian estimation. With such estimation methods, complete data for each individual is not required, and so an individual may have, for example, data for only one occasion and still be included in the analysis. This is useful in longitudinal data analysis where missing data are common and it is important to retain all available information regarding the variables under investigation.

Inferences drawn from an NLME modeling analysis is considered to be valid if the missingness (i.e., values on variables in which data to be gathered on the individuals were not realized) is independent of the missing data (see, e.g., Enders, 2010, and Enders & Hayes, Chapter 15, this volume). This means that missingness can be related to observed data, such as the response measures observed up to or possibly after the missing assessment or to any of the observed covariates that are also included in the analysis. More generally, inference from a NLME model is considered valid if the missingness is not informative which essentially means that no special steps are needed to address the missing data. If, on the other hand, the missingness is informative (i.e., in the missing data literature this is referred to as *nonignorable* missingness), then this indicates that special attention must be taken to address the missing data to obtain valid statistical inference. This can include extending a model to include correlates of the variables with missing data or possibly by extending a longitudinal model to include a submodel that directly characterizes the missing data process. There are both technical as well as many readable treatments on handling missing data more generally (see, e.g., Enders, 2010; Chapter 16 this volume; Graham, 2009; Little & Rubin, 2002; van Buuren, 2011), and especially for longitudinal designs (the interested reader is directed to Blozis, 2012; Hedeker & Gibbons, 1997; Laird, 1988; Xu & Blozis, 2011). In the subsequent analysis we assume a MAR mechanism for any missingness that arises.

Estimation

Estimation for NLME models is more complicated than with their linear mixed-effects model counterparts. Inference within NLME models is carried out on the marginal distribution of the data. Two prominent estimation strategies for estimating parameters are within a Bayesian framework and via maximum likelihood. Here we briefly review how maximum likelihood estimation occurs in practice and its implications for the types of research questions to be addressed.

The marginal distribution of the data, $\mathbf{y}_i$, is computed as a multidimensional integral of the joint distribution of $\mathbf{y}_i$ and $\mathbf{b}_i$,

$$h(\mathbf{y}_i) = \int p(\mathbf{y}_i|\mathbf{b}_i)p(\mathbf{b}_i)d\mathbf{b}_i,$$

where $p(\mathbf{y}_i|\mathbf{b}_i)$ is the distribution of the data conditioned on the random effects and $p(\mathbf{b}_i)$ is the marginal distribution of the random effects. Unlike linear models in which the integration can be done algebraically leading to a closed-form solution for the marginal distribution of $\mathbf{y}_i$, when at least one random effect enters the function $f(\cdot)$ in Equation 10.1 nonlinearly, this nice analytic arrangement no longer holds, and in the vast majority of circumstances the integral will be analytically intractable (cannot be computed algebraically). Let $\xi = vech(\mathbf{\Phi})' = (\varphi_{11}, \varphi_{21}, \varphi_{22}, \ldots, \varphi_{rr})$, where the $vech(\cdot)$ operator creates a column vector of a symmetric matrix (i.e., $\mathbf{\Phi}$) by stacking successive row-wise elements of the lower triangle below one another. Combining the unknown, non-stochastic parameters into one vector, $\boldsymbol{\theta}$ where $\boldsymbol{\theta} = (\boldsymbol{\beta}', \boldsymbol{\gamma}', \xi, \boldsymbol{\omega}')'$, maximum likelihood estimates for $\boldsymbol{\theta}$ can be found by maximizing the loglikelihood in $\boldsymbol{\theta}$

$$\begin{aligned} l(\boldsymbol{\theta}) = \ln L(\boldsymbol{\theta}) &= \ln \prod_{i=1}^{m} \int p(\mathbf{y}_i|\mathbf{b}_i)p(\mathbf{b}_i)d\mathbf{b}_i \\ &= \sum_{i=1}^{m} \ln \int p(\mathbf{y}_i|\mathbf{b}_i)p(\mathbf{b}_i)d\mathbf{b}_i. \end{aligned} \tag{10.4}$$

From a frequentist perspective as Wall (2009) suggested, there exist two general classes of computational methods for carrying out the integration embedded in the loglikelihood. The first is accomplished by approximating the integrand prior to integration through linearizing nonlinear function $f(\cdot)$ via a Taylor series polynomial (Beal & Sheiner, 1982). The second general approach attempts a direct approximation of the integral with a finite sum (e.g., numerical integration; see Skrondal & Rabe-Hesketh, 2004).[3] These two approaches lead to analyses that produce estimates that have population-average and subject-specific interpretations, respectfully (Harring & Blozis, 2016).

Population-Average Interpretations

Linearization leads to parameter estimates in $f(\cdot)$ that have a population average interpretation. Research questions are focused on the average response profile (i.e., $E(\mathbf{y}_i|\mathbf{z}_i)$) among individuals summarized by regression coefficients $\boldsymbol{\beta}$ and covariates, $\mathbf{z}_i$. The function for the average individual using typical values of the regression coefficients is the same as the curve of the means, $f(t_{ij}, \beta_1, \ldots, \beta_p) = \mu_j$ across all j. The emphasis is clearly on the population average response profile as an accurate representation of mean change whereas understanding individual change is of less importance.

Subject-Specific Interpretations

In contrast, numerical integration (i.e., Gaussian quadrature or importance sampling) of the loglikelihood leads to parameter estimates that have a subject-specific interpretation. The fixed effects of the regression coefficients $\beta_1, \ldots, \beta_p$ are mean values of the individuals' coefficients—$E(\beta_{pi}) = \beta_p$. Zeger et al. (1988) referred to these fixed effects as *typical values* and when these coefficients are used to evaluate the nonlinear function, the result is a hypothetical fitted curve for a *typical* or *average* individual, a subject whose coefficients are equal to the population mean values. This theoretical trajectory is not the same as the curve of the means obtained when using fixed effects from estimating the NLME model with a linearization method. Several authors have discussed these issues in more detail (see, e.g., Davidian & Giltinan, 2003; Fitzmaurice et al.; 2011) and provided a graphical representation of fitted functions stemming from estimating a NLME model using both linearization and numerical integration (Harring & Blozis, 2016). From this point forward, all NLME models will be estimated via numerical integration producing parameter estimates with a subject-specific interpretative orientation.

PRELIMINARY ANALYSIS

An NLME model analysis is carried out in much the same manner as many other analyses of repeated measures data using different statistical frameworks. However, because estimation of model parameters is more complicated, it is prudent to initially explore the functional form of the trajectory and variability in the data across time in a pre-analysis phase. These exploratory activities are outlined below.

Exploring the Data

Since the majority of longitudinal analyses address the relation of time-response, plotting individual data can be very informative regarding individual differences in the change process. These plots come in two main flavors: (a) spaghetti plots (time plots) where, for a sample of individuals, data are plotted, connected by lines, or just the lines; while (b) each individual could be plotted in a separate graphing window.

Spaghetti Plots

Every good statistical analysis begins with an "ocular test," that is, a good look at the data. A spaghetti plot can be used to get a general sense of the change patterns of individuals and identify their salient characteristics. Do individual trajectories show general improvement or cessation of the behavior? Does the behavior begin or end gradually? Where do periods of rapid change occur? Answers to these types of questions can suggest a particular nonlinear function or a class of nonlinear functions. A 10% random sample of children's LIWD scores is displayed in Figure 10.2.

The trends over time exhibited in Figure 10.2 suggest a logistic functional form that allows asymptotic behavior at both the beginning and at the end of the study period. Individual difference in profiles is evident as is variability within each child across time.

Scatterplots With Superimposed Summaries

There are several summaries that can be superimposed on a plot of individuals' data. One could draw a scatterplot with averages at the time points superimposed. This can also be done separately for covariate groups (e.g., gender).

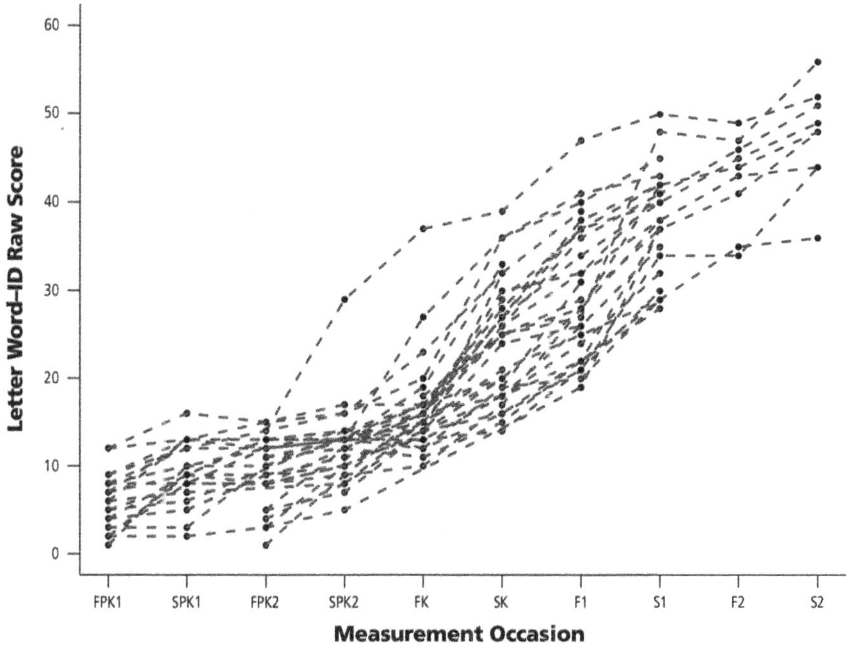

Figure 10.2 Spaghetti plot of LWID raw scores over ten occasions for a 10% random sample of children.

Figure 10.3 shows a spaghetti plot of LWID scores for a random sample of children by gender. The fact that these gender-specific profiles show differences—females perform better than males initially and reach greater potential performance levels—suggest that gender may be useful in explaining growth characteristics in subsequent analyses. Showing the same type of information for continuous covariates, while more challenging, simply

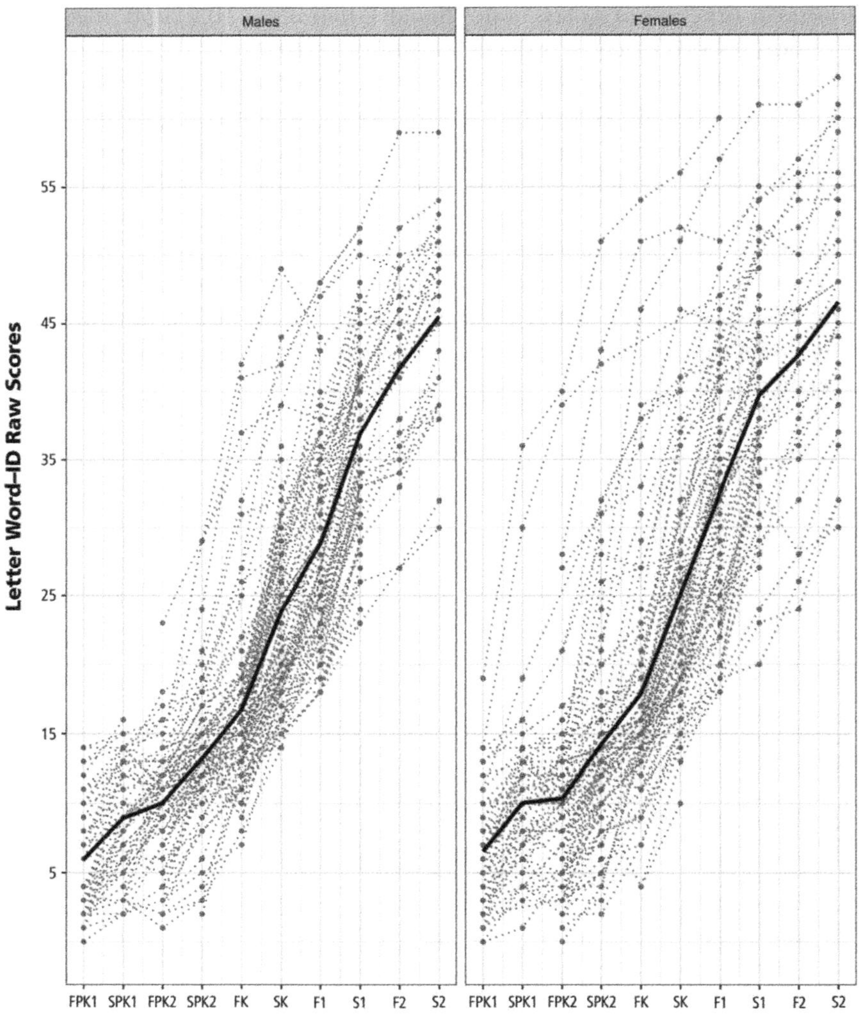

Figure 10.3 Spaghetti plot of LWID raw scores across ten occasions for males and females with a superimposed black solid line connecting the empirical means for each gender.

requires that the continuous scale of the covariate be discretized and then graphed in a similar fashion as variables representing true categories.

Choosing a Growth Function

Central to an NLME model is a theoretically justifiable or empirically derived user-specified nonlinear function that effectively summarizes the repeated measures and one that can be parameterized to highlight important features of the latent developmental process as captured in the data. Initial scientific research questions often focus on determining a particular functional form. The function that is ultimately decided upon must also provide good fit to the data in the most parsimonious manner possible. Empirically deriving a functional form relies on information garnered from graphs, summary statistics, and preliminary analyses.

Exploring Different Functional Forms Using NLS

Once an initial list of potential functions (or class of functions) has been enumerated, fitting the functions to the empirical means from the dataset via nonlinear least squares (NLS) seems like a reasonable course of action. Fit can be assessed using the residual sum of squares (RSS) or a variance accounted for measure such as R^2. The exploratory plots of the data suggest that the general class of logistic functions would be a sensible place to start since the basic function accommodates lower and upper asymptotes in addition to a period of rapid growth near the middle of the study.

The logistic growth curve (Verhulst, 1845) is one of the simplest of the S-shaped growth curves. We start with an adaptation to the original model developed by Browne (1993) shown in Equation 10.2—also Model 1 in Table 10.1. A second function to be tried is a parameterization of the Gompertz function (Browne, 1993) that allows functional response to follow a sigmoidal or S-shaped curve but with an asymmetric inflection point (change in curvature),

$$f(t_j,\beta_1,\beta_2,\beta_3)=\beta_2\exp\left[\log(\beta_1/\beta_2)\cdot\exp(-\beta_3-t_j)\right].$$

This is Model 2 in Table 10.1. The parameterization leads to the same interpretations of coefficients as the logistic function in Equation 10.2. The last two logistic growth functions come from a reparameterization of the model originally introduced by Verhulst. The first of these (Model 3 in Table 10.1), is a three-parameter model,

$$f(t_j,\beta_1,\beta_2,\beta_3)=\frac{\beta_2}{\left[1+\exp(-\beta_3)\cdot(t_j-\beta_1)\right]},$$

TABLE 10.1 Four Alternative Functions Fit by Nonlinear Least Squares

Function	Model	Number of Parameters	RSS
Logistic (Browne, 1993)	$f(t_j,\beta_1,\beta_2,\beta_3)=\dfrac{\beta_1\beta_2}{\beta_1-(\beta_1-\beta_2)\exp\{-\beta_3 t_j\}}$	3	10.037
Gompertz	$f(t_j,\beta_1,\beta_2,\beta_3)=\beta_2\exp\left[\log(\beta_1/\beta_2)\cdot\exp(-\beta_3-t_j)\right]$	3	16.744
Reparameterized Logistic	$f(t_j,\beta_1,\beta_2,\beta_3)=\dfrac{\beta_2}{\left[1+\exp(-\beta_3)\cdot(t_j-\beta_1)\right]}$	3	10.037
Generalized Logistic	$f(t_j,\beta_1,\beta_2,\beta_3,\beta_4)=\beta_1+\dfrac{\beta_2-\beta_1}{\left\{1+\exp\left[(-\beta_3-t_j)/\beta_4\right]\right\}}$	4	3.615

Note: The RSS for models 1 and 3 are identical. This is not surprising as one is algebraically equivalent to the other (i.e., one is a reparameterization of the other), but with one parameter that has a different interpretation. The algebraic equivalence is presented in the Appendix.

where β_1 is the time where the inflection point (change in curvature) occurs, β_2 is the upper asymptote, and β_3 is the nonlinear growth rate. The second is a four parameter logistic curve (Model 4 in Table 10.1) reminiscent of a generalized logistic or Richard's curve (1959),

$$f(t_j,\beta_1,\beta_2,\beta_3,\beta_4)=\beta_1+\frac{\beta_2-\beta_1}{\left\{1+\exp\left[(-\beta_3-t_j)/\beta_4\right]\right\}}, \tag{10.5}$$

where β_1 is the lower asymptote, β_2 is the upper asymptote, β_3 is the time needed to realize half the gain [i.e., when $t_j=\beta_3$, $f(t_j)=\beta_1+\frac{1}{2}(\beta_2-\beta_1)$], and β_4 is a shape (smoothing) parameter that determines the steepness of the rising curve with larger positive values corresponding to less steep curves. Figure 10.4 displays a generic version of the four-parameter logistic function. Of course, other logistic functions are possible with alternate parameterizations highlighting different aspects of the change process (see, e.g., Choi et al., 2009). The results from the NLS analyses are presented in Table 10.1.

The RSS for Model 1 and Model 3 are the same, which indicates that these two models fit equally well. It also indicates that the two are in fact equivalent, and so this equivalence in model fit is not surprising. Since the data-model fit is identical, choosing one parameterization over another is typically decided upon based on an increase in interpretability of model parameters (see, e.g., Cudeck & du Toit, 2002) that reflect meaningful aspects of change corresponding to substantive research hypotheses (Preacher & Hancock, 2012, 2015).

Checking the Final Form With Individuals

Once the list of candidate models is whittled down to a single final function, that function should be tried with a random sample of individuals to assure

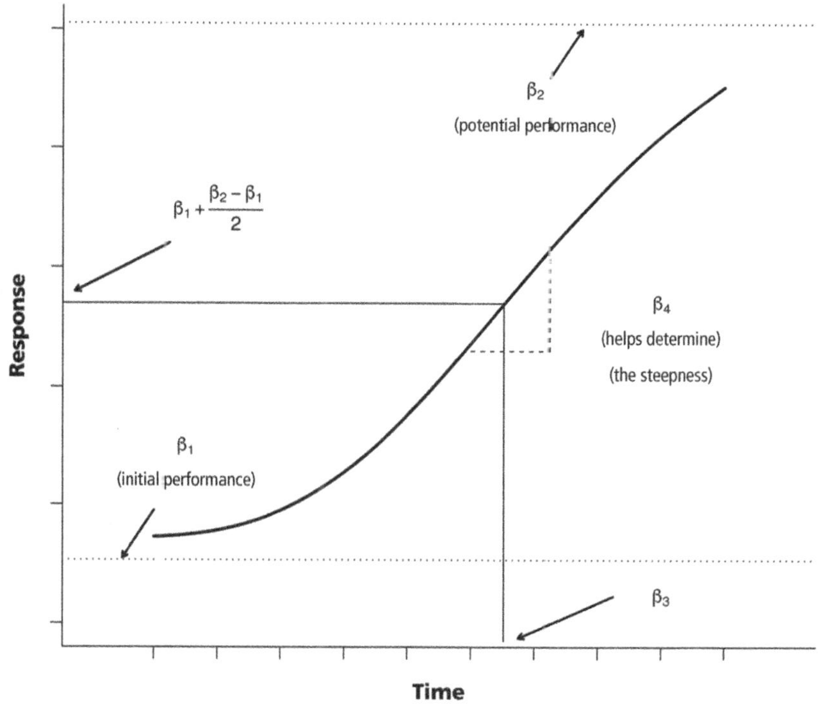

Figure 10.4 Generic schematic of a variant of the four-parameter logistic function.

that it also fits adequately at this level. This can be accomplished by again relying on NLS used on individuals' data, one at a time, and taking the fitted functions and graphing them against individuals' data. Based on the fit statistics in Table 10.1 and the interpretability of the parameters, the analysis now moves forward using the four-parameter logistic function in Equation 10.5.

FITTING THE NLME MODEL

In contrast with linear mixed-effects models where estimation is usually straightforward, even when every fixed effect has a corresponding random effect, estimating NLME models is more complex and relies heavily on reasonable starting values for the fixed effects, and more importantly, the variances and covariances of the random effects in $\mathbf{\Phi}$. In fact, programs that fit NLME models[4] insist that the user provide starting values that are in the neighborhood of the maximum of the loglikelihood surface to improve the prospect that the algorithm will converge to a proper and sensible solution. So, how are starting values to be used in an NLME modeling analysis obtained?

Starting Values

Summary statistics computed from regression coefficient estimates obtained from fitting individuals' data by NLS to a particular nonlinear function under consideration can be used to generate starting values for an analysis of an NLME model. In Table 10.2, for example, are generic coefficient estimates from fitting the four-parameter logistic function in Equation 10.5 from each individual. Starting values for $\boldsymbol{\beta}$ and $\boldsymbol{\Phi}$ could be obtained from calculating the means, variances, and covariances from the coefficient estimates (see Table 10.2). For a conditionally independent, homogeneous covariance structure at level 1, a value for σ^2 could be computed as a pooled estimate of the residual variances obtained across the m, nonlinear least squares analyses.

If a subject-specific analysis is sought, another strategy to finding good starting values might first begin by running the analysis using a conditional linearization method (Lindstrom & Bates, 1990; operationalized in the `nlme` package in R or the SAS macro, `NLINMIX`, with the `eblup=expand` option) and take those estimates and use them for starting values when using a direct approach to the multidimensional integration like Gaussian quadrature. This is an important data proposition as the ultimate success of an analysis may very well rest on finding reliable starting values for model parameters. Thus, thoughtful care and diligence at this stage is strongly recommended.

TABLE 10.2 Individual Estimates From Running Nonlinear Least Squares

	Coefficients			
Individual	β_{1i}	β_{2i}	β_{3i}	β_{4i}
1	$\hat{\beta}_{11}$	$\hat{\beta}_{21}$	$\hat{\beta}_{31}$	$\hat{\beta}_{41}$
2	$\hat{\beta}_{12}$	$\hat{\beta}_{22}$	$\hat{\beta}_{32}$	$\hat{\beta}_{42}$
⋮	⋮	⋮	⋮	⋮
m	$\hat{\beta}_{1m}$	$\hat{\beta}_{2m}$	$\hat{\beta}_{3m}$	$\hat{\beta}_{4m}$
Means	$E(\hat{\beta}_{1i}) \approx \beta_1$	$E(\hat{\beta}_{2i}) \approx \beta_2$	$E(\hat{\beta}_{3i}) \approx \beta_3$	$E(\hat{\beta}_{4i}) \approx \beta_4$
Covariance matrix	$\text{cov}\begin{pmatrix} \hat{\beta}_{1i} \\ \hat{\beta}_{2i} \\ \hat{\beta}_{3i} \\ \hat{\beta}_{4i} \end{pmatrix} = \begin{pmatrix} \text{var}(\hat{\beta}_{1i}) & & & \\ \text{cov}(\hat{\beta}_{2i},\hat{\beta}_{1i}) & \text{var}(\hat{\beta}_{2i}) & & \\ \vdots & \vdots & \ddots & \\ \text{cov}(\hat{\beta}_{4i},\hat{\beta}_{1i}) & \text{cov}(\hat{\beta}_{4i},\hat{\beta}_{2i}) & \cdots & \text{var}(\hat{\beta}_{4i}) \end{pmatrix} \approx \boldsymbol{\Phi}$			

Are Random Effects Necessary?

The next analytic step would be to determine whether each of the four regression coefficients in the logistic function in Equation 10.5 should be allowed to vary across children or should some of these coefficients be fixed. A nonlinear random coefficient model (Cudeck & Harring, 2007) would be a good starting point in which all coefficients are allowed to vary and covary. Initially, a Wald test statistic (i.e., an estimate divided by its standard error) can be computed for variance and covariance components of the random effects in order to adjudicate whether any of them are statistically different than zero.[5] P-values from the hypothesis tests involving these Wald test statistics greater than the nominal 0.05 significance level would indicate some zero elements of $\mathbf{\Phi}$, from which a more formal model comparison test could be utilized. Testing whether elements of $\mathbf{\Phi}$ are zero can be accomplished vis-à-vis the likelihood ratio test (or LRT). The test statistic, D, is computed as

$$D = -2\left[\ln L_R(\hat{\theta}_v) - \ln L_F(\hat{\theta}_w)\right]$$

where $L_R(\hat{\theta}_v)$ is the loglikelihood for the model with fewer variance and covariance parameters (i.e., the *reduced* model with some elements of $\mathbf{\Phi}$ set to zero) with v parameters, and $L_F(\hat{\theta}_w)$ is the loglikelihood for the model with all elements in $\mathbf{\Phi}$ estimated—the *full* model with w parameters. If only covariance parameters in $\mathbf{\Phi}$ are set to zero in the reduced model (i.e., random intercepts and slopes do not covary), and for sufficiently large sample sizes, D is distributed as a χ^2 with degrees of freedom equal to the difference in dimensionality of θ between competing models, $(w - v)df$. However, when testing whether variances along with corresponding covariances in $\mathbf{\Phi}$ are zero in the reduced model, the asymptotic null distribution of D is no longer χ^2 because the null hypothesis of zero variances places the population value of the variance on the boundary of the parameter space of the alternative hypothesis (i.e., variances must be non-negative). In this scenario, several researchers have suggested that the appropriate null distribution with which to compare the LRT test statistic is a mixture of chi-square distributions (Self & Liang, 1987; Stoel et al., 2006; Verbeke & Molenberghs, 2000). The chi-square distributions used in the mixture are determined by the number of variance components that are in the *full* and *reduced* models. Again, using the four parameter logistic function in Equation 10.5 as an example, suppose through the Wald test that of the four coefficients, only $\text{var}(b_{1i}) = \varphi_{11}$, $\text{var}(b_{2i}) = \varphi_{22}$, and $\text{var}(b_{3i}) = \varphi_{33}$ were found to be statistically different than zero. The random effects covariance structures for the *full* and *reduced* models would be, respectively

$$\boldsymbol{\Phi}=\begin{pmatrix}\varphi_{11} & & & \\ \varphi_{21} & \varphi_{22} & & \\ \varphi_{31} & \varphi_{32} & \varphi_{33} & \\ \varphi_{41} & \varphi_{42} & \varphi_{43} & \varphi_{44}\end{pmatrix} \text{ and } \boldsymbol{\Phi}=\begin{pmatrix}\varphi_{11} & & & \\ \varphi_{21} & \varphi_{22} & & \\ \varphi_{31} & \varphi_{32} & \varphi_{33} & \\ 0 & 0 & 0 & 0\end{pmatrix}.$$

The correct null distribution would be a mixture of χ^2_{4df} and χ^2_{3df} drawing values from each distribution with equal weights of 0.5 (Shapiro, 1985; Verbeke & Molenberghs, 2000). Other scenarios in which more than one random effects variance is zero is covered in sufficient detail by Stoel et al. (2006). Since tables of critical values of various mixture chi-square distributions are not readily available (for an exception, see Appendix C in Fitzmaurice et al., 2011), researchers are relegated to computing the empirical mixture distributions themselves, which may be seen as unrealistic in practice (although Stoel et al. [2006] provide software script files to facilitate these computations). For more complex comparisons among nested random effects models for the variances and covariances in $\boldsymbol{\Phi}$ where the null distribution of the likelihood ratio test is not well understood (e.g., comparison of nested models with q correlated random effects and $q + k$ correlated random effects, where $k > 1$), Fitzmaurice et al. (2011) suggest simply using a more stringent significance level (i.e., $\alpha = 0.01$) with the conventional chi-square distribution accompanying the LRT to prevent selection of a model that is too parsimonious.

Level-1 Covariance Structure

Given that all four coefficients in the logistic function in Equation 10.5 were found to vary randomly (see Table 10.2 in the unconditional model), attention turns to choosing an appropriate level 1 covariance structure. Coupled with the complexity of $\boldsymbol{\Phi}$, only two simple and parsimonious level 1 covariance structures were fit to the data: (a) a conditionally independent homogenous variance structure (i.e., $\text{var}(e_{ij}) = \sigma^2$ for all j, and $\text{cov}(e_{ij}, e_{ik}) = 0$); and (b) a first-order autoregressive structure. A criterion based on information theory, such as AIC (Akaike, 1974), is frequently used as an arbiter for deciding on the relative merits of different covariance structures where the index accounts for model-data fit but penalizes models that are over-parameterized. The AIC indices for the conditionally independent structure and the first-order autoregressive structure were (AIC = 12465.8) and (AIC = 12468.2), respectively.[6] Based on the values of AIC where relatively better fitting models produce smaller values, the best fitting error structure was the one where the level 1 residuals were conditionally independent with equal variances across time. All subsequent analyses will integrate this level 1 covariance structure into the analyses.

Conditional NLME Model

Typically, NLME model analyses proceed in two stages. In the first stage, substantial time and effort is spent exploring the data and finalizing all of the components making up what is referred to in the literature as the *unconditional* growth model (i.e., in this case, an NLME model without added time-varying or time-invariant covariates). These components include deciding on a functional form for individual and average or typical change, the structures capturing both within- and between-level variation. Upon completion of these important analytic activities, focus then shifts to the second stage of the analysis—adding covariates to the model.

Additional insights into the developmental behavior captured by the repeated measures can be garnered when relations between individual attributes and regression coefficients are posited to account for the response. For the preschool instructional study, an indicator-coded variable denoting a child's gender (0 = males and 1 = females) was collected and will be added as a time-invariant covariate (a variable that does not change during the course of the study) to the analysis to help explain why children differ in logistic growth characteristics. An individual's coefficients will follow the same general regression function, d_k, which takes on the form across $k = 1,\ldots,4$,

$$\beta_{ki} = d_k(\mathbf{z}_i, \beta_k, b_{ki}, \gamma) = \beta_k + \gamma_k z_{1i} + b_{ki}.$$

That is, each regression coefficient will have the same single predictor, gender. Maximum likelihood estimates and corresponding standard errors from both the unconditional and conditional NLME models are presented in Table 10.3.

In terms of the regression coefficients for the conditional model, because of the indicator coding of gender, the fixed effects, $\hat{\beta}_1 - \hat{\beta}_4$, represent logistic growth parameter estimates for children who were male. The regression coefficients, $\hat{\gamma}_1 - \hat{\gamma}_4$, represent the difference in corresponding growth parameter estimates between females and males. As it turned out, none of the coefficients representing gender differences were statistically significant. Thus, interpretation of the model will resume based on the unconditional model as variables associated with non-significant estimates in many analytic scenarios would be dropped from the analysis and the model re-estimated.

The results indicate that the initial performance level for the typical individual is a little greater than 3 raw score points on the LWID scale ($\hat{\beta}_1 = 5.06$, $se(\hat{\beta}_1) = 0.30$). Potential performance (performance at later grades) for the typical individual is approximately 53 raw score scaled points ($\hat{\beta}_2 = 53.15,\ se(\hat{\beta}_2) = 0.84$). The time that the typical child took to reach half their gains relative to initial performance is about 6 ($\hat{\beta}_3 = -5.77,\ se(\hat{\beta}_3) = 0.07$). Due to the coding of time, this translates to roughly the start of first grade. Note that potential performance for the typical individual is not realized by the end of the study or period of data collection (spring of second grade). This would seem to indicate that growth in letter and word identification continues on through the middle primary grades. For many children, their growth in letter and word identification is not steep, but quite gradual. It is not surprising then that the magnitude of the parameter estimate for the typical child that controls the rate of increase is rather large and statistically significant ($\hat{\beta}_4 = 1.65,\ se(\hat{\beta}_4) = 0.05$).

Individual differences in the four aspects of letter and word identification development can be ascertained by examining the variances and covariances of the random effects. From $\hat{\mathbf{\Phi}}$, the variance estimates of the random coefficients were all statistically significant. The correlation between potential performance and the steepness of an individuals' curve is

$$\hat{\rho}_{b_{4i},b_{2i}} = \frac{1.19}{\sqrt{79.15 \cdot 0.16}} = 0.334,$$

and points to a weak to moderate positive trend that those children who have greater potential performance levels tend to also have smoothing parameters that are larger, indicative of more gradual gains than do those children with lower potential performance in identifying letters and words.

Individual Coefficients

For NLME models, graphs appear to be indispensable at every stage of an analysis. The parameter estimates from the unconditional model in Table 10.3 have a subject-specific interpretation because of the decision made earlier to keep the focus of the analysis on individual children. This was guaranteed by choosing to use a numerical integration routine instead of a linearization method to facilitate the estimation of model parameters. As such, the fixed effects define the logistic trajectory of the typical child. Given this orientation, it is also instructive to illustrate how well the model works for individual children. This is accomplished by predicting individual children's regression coefficients and using these to plot their fitted functions against their data. So, how are individual regression coefficients obtained?

To compute children's regression coefficients, predictions of random effects vector $\mathbf{b}_i$ must be obtained for each child. The expected value of $\mathbf{b}_i$ given their data $\mathbf{y}_i$ is computed as

TABLE 10.3 Maximum Likelihood Estimates and Standard Errors for Analyses of the Unconditional and Conditional Four-Parameter Logistic NLME Models

Parameter	Unconditional		Conditional	
β_1	5.06	(0.30)	4.86	(0.42)
β_2	53.15	(0.84)	53.99	(1.10)
β_3	–5.77	(0.07)	–5.82	(0.09)
β_4	1.65	(0.05)	1.71	(0.07)
γ_1	n/a		0.31*	(0.56)
γ_2	n/a		–1.29*	(1.10)
γ_3	n/a		0.06*	(0.10)
γ_4	n/a		–0.09*	(0.08)
φ_{11}	9.39	(1.34)	9.29	(1.38)
φ_{21}	8.81	(0.95)	8.80	(0.94)
φ_{22}	79.15	(8.63)	76.77	(7.57)
φ_{31}	0.69	(0.10)	0.68	(0.10)
φ_{32}	–2.23	(0.61)	–2.10	(0.56)
φ_{33}	1.37	(0.08)	1.37	(0.07)
φ_{41}	–0.63	(0.13)	–0.62	(0.14)
φ_{42}	1.19	(0.36)	1.08	(0.34)
φ_{43}	–0.26	(0.03)	–0.25	(0.03)
φ_{44}	0.16	(0.02)	0.16	(0.02)
σ^2	6.86	(0.26)	6.87	(0.26)

Note: Standard errors are in parentheses. Both models were estimated using a non-adaptive Gaussian quadrature method with $Q = 5$ quadrature points. All estimates except those with (*) were statistically significant at $\alpha = 0.05$.

$$\hat{\mathbf{b}}_i = E[\mathbf{b}_i|\mathbf{y}_i] = h^{-1}(\mathbf{y}_i)\int_{\mathbf{b}_i} \mathbf{b}_i p(\mathbf{y}_i|\mathbf{b}_i)p(\mathbf{b}_i)d\mathbf{b}_i,$$

using Bayes' theorem. Fortunately, carrying out these calculations does not require special software modules and can be achieved with a minimum of additional programming. Once an individual child's random effects have been obtained, their regression coefficients can be calculated for the unconditional model as $\hat{\beta}_i = \beta + \hat{\mathbf{b}}_i$ from which a fitted trajectory for each child could be graphed and inspected. However, unless the sample is small, it is seldom possible to present every individual. In many ways this graph is the focus of the analysis so every effort should be made to examine as many as

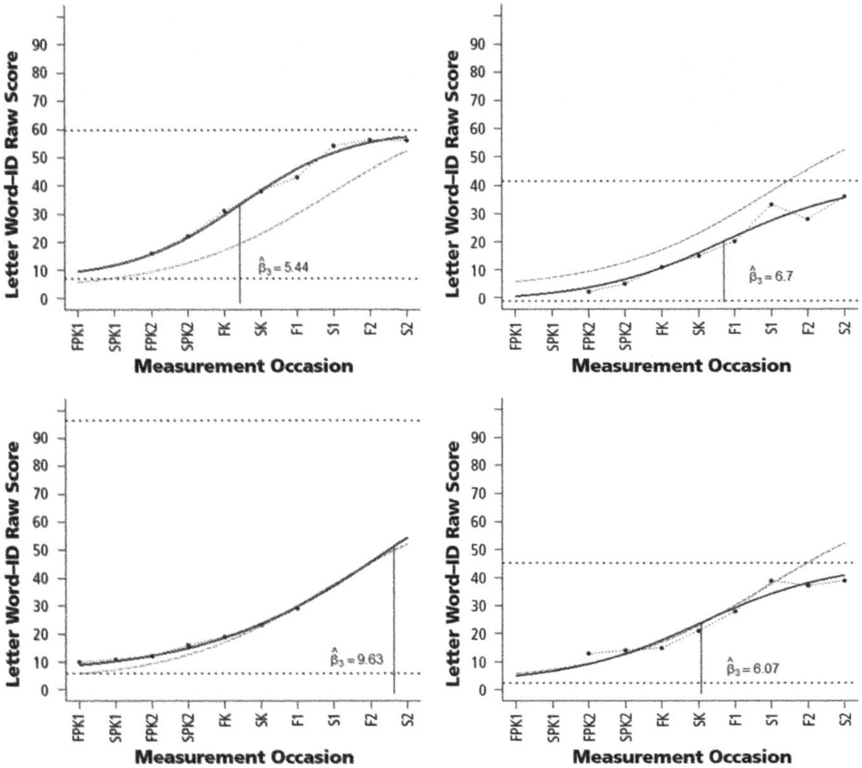

Figure 10.5 Fitted logistic function for 4 selected individuals (—) superimposed on individuals' data. The trajectory for the typical individual (- - -) was also superimposed. Three salient features of each individual's trajectory are also plotted: initial performance (· · ·), potential performance (· · ·), and the time when individuals realized half their gains emphasized with their predicted coefficient and a vertical line (—).

possible and display a selected few that illustrate the range of the response patterns and interesting anomalies. A trellis display in Figure 10.5 presents individual data with individual fitted functions and the typical child's trajectories superimposed.

CONCLUDING REMARKS

Nonlinear patterns of change arise frequently in longitudinal investigations of real-world phenomena found in education, and in social and behavioral science research. The primary feature of nonlinear development is that change is not constant, but rather change is more rapid in some time periods than in others. Although it is assumed that individual profiles follow the same general curvilinear pattern, characteristically, there are substantial differences in individual behavior and also variation within individuals.

NLME models have become popular choices for adequately describing such data mainly because they can be customized so that parameters have scientifically relevant interpretations which correspond to substantive characteristics of the underlying developmental process. Covariation arising from measuring the same individuals on multiple occasions can be partitioned into within- and between-individual components. Time-invariant covariates can be incorporated that account for variability in individual coefficients. Provisions for missing data stemming from an ignorable missingness process can be accommodated, as well as highly flexible longitudinal measurement schedules—even those in which individuals are measured idiosyncratically.

An empirical example using language development data from a preschool instruction study was presented to demonstrate the utility of the modeling approach. A general analytic process was presented highlighting a logical progression of investigative activities to work through in order to successfully execute an analysis. Modeling issues inherent when working with NLME models were discussed stressing the conceptual understanding of their potential impact while also providing practical ways to address some challenging technical aspects of an analysis.

Throughout the chapter, various graphics have been presented and their importance emphasized. Arguably, these graphical displays are the most important parts of any longitudinal analysis, especially for NLME models. Raw data plots of individuals with and without superimposed statistical summaries are crucial. Spaghetti plots and trellis plots may be two of the more important and revealing types. Graphs are also invaluable when choosing a function, when assessing the accuracy of individual fits, and when evaluating whether assumptions hold.

Lastly, choosing to approach a longitudinal analysis utilizing an NLME model is not without costs. As has been previously discussed, an NLME model can be more difficult to specify and estimate than a simple linear model. Interpretation of parameters depends on which estimation scheme is chosen, and should be reflective of the researcher questions needing to be addressed. Despite these possible apprehensions and concerns, the increased overhead when using NLME models is more than compensated by the increased utility of the approach.

RECOMMENDED READINGS

Technical Readings

Blozis, S. A., & Harring, J. R. (2016). On the estimation of nonlinear mixed-effects models and latent curve models for longitudinal data. *Structural Equation Modeling: A Multidisciplinary Journal, 23,* 904–920. http://dx.doi.org/10.1080/10705511.2016.1190932

Cudeck, R. (1996). Mixed-effects models in the study of individual differences with repeated measures data. *Multivariate Behavioral Research, 31,* 371–403.

Davidian, M., & Giltinan, D. M. (2003). Nonlinear models for repeated measurement data: An overview and update. *Journal of Agricultural, Biological, and Environmental Statistics, 8,* 387–419.

Applied Readings

Burke, C. T., Shrout, P. E., & Bolger, N. (2007). Individual differences in adjustment to spousal loss: A nonlinear mixed model analysis. *International Journal of Behavioral Development, 31,* 405–415.

Cudeck, R., & Harring, J. R. (2007). The analysis of nonlinear patterns of change with random coefficient models. *Annual Review of Psychology, 58,* 615–637.

Harring, J. R., & Blozis, S. A. (2016). A note on recurring misconceptions when fitting nonlinear mixed models. *Multivariate Behavioral Research, 51,* 805–817. http://dx.doi.org/10.1080/00273171.2016.1239522

TRY THIS!

The data file in this chapter's online appendix (ECLSK-Math-Example.dat) contains mathematics scores from a random sample of $N = 400$ students from the full sample of the ECLS-K 1998–1999 cohort (Tourangeau et al., 2009). Vertically equated math achievement scores were obtained at seven unequally spaced occasions from kindergarten to eighth grade in the fall and spring: fall-k, spr-k, fall-1st, spr-1st, spr-3rd, spr-5th, and spr-8th. The current sample has only 6 of the 7 repeated measures as the fall-1st scores were dropped because the sample was refreshed at this juncture and the new sample did not have the same data characteristics as the cohort that began in the fall of their kindergarten year. Names of the variables are on the first line of the dataset. There are four time-invariant covariates to be used as predictors at a secondary stage of the analytic process. A full description of the variables can be found in the accompanying codebook (ECLSK-Math-Codebook.pdf).

Some of the data-analytic challenges awaiting you include the appropriate handling of missing data, accommodating unequally spaced measurement occasions, deciding on a time metric, and software implementation among others. The data analysis should proceed in four phases: (a) an exploratory data analytic phase, (b) modeling the repeated measures, (c) adding time-invariant covariates to explain differences in subject-specific growth parameters, and (d) the checking of assumptions and examining diagnostic information.

Exploratory Data Analysis

1. First, consider what kind of nonlinear trajectories might make sense given the data described above? What data characteristics are important features to capture and what might the shape of the growth trajectory look like and why? Graph the sample data in three ways so that you can empirically examine the likely shape of the growth trajectory. First, create spaghetti plots of several random samples of size $n = 30$. Secondly, create a graph using different fitted trajectories (e.g., logistic, exponential) superimposed on the empirical means (a nonlinear least squares program will be helpful in estimating parameters for this step). Remember that the time axis should reflect the unequally-spaced measurements. Also, create individual plots for at least 50 of the students.
2. Compute descriptive statistics (i.e., means, variances, correlations) of the math scores across time as well as the time-invariant covariates.

Model Fitting

3. Prepare the data file for longitudinal modeling using `PROC NLMIXED` in SAS (or the `nlmer()` function in the `lme4` package or the `nlme()` in the `nlme` package in R). This can be done manually as a pre-processing data step in SAS or using `PROC TRANSPOSE`.
4. In SAS (or R), run and compare four different unconditional models:

   (1) an exponential growth model,

$$y_{ij} = \beta_{2i} - (\beta_{2i} - \beta_{1i})\exp(-\beta_{3i}t_{ij}) + e_{ij},$$

   (2) a variant of Michaelis-Menten model,

$$y_{ij} = \beta_{1i} + \frac{(\beta_{2i} - \beta_{1i})t_{ij}}{\beta_{3i} + t_{ij}} + e_{ij},$$

(3) a linear-linear piecewise growth model where the two segments join at the changepoint,

$$y_{ij} = \left[\begin{cases} \beta_{1i} + \beta_{2i} t_{ij} & t_{ij} \le \tau_i \\ \beta_{1i} + \tau_i(\beta_{2i} - \beta_{3i}) + \beta_{3i} t_{ij} & t_{ij} > \tau_i \end{cases} \right] + e_{ij}, \text{ and}$$

(4) a reparameterized quadratic growth model (Cudeck & du Toit, 2002),

$$y_{ij} = \beta_{2i} - (\beta_{2i} - \beta_{1i}) \left(\frac{t_{ij}}{\beta_{3i}} - 1 \right)^2 + e_{ij}.$$

All function growth parameters should have random effects and a homogeneous mutually uncorrelated covariance structure should be assumed for the residuals. Organize your results in a table which might include columns denoting the model, number of parameters, degrees of freedom, deviance, and BIC values. Use the first-order linearization method (`method = FIRO` in SAS PROC NLMIXED) to fit each model.

Based on the Bayesian Information Criterion (BIC), which model best fits the data? For the best fitting model, were all random effects needed? Explain. If your final decision is to let some parameters be fixed (not random) show that these random effects are not needed by executing the appropriate likelihood ratio test comparing a model with all the random effects included versus a model with the random effects for some parameters eliminated.

5. Next, briefly interpret each parameter in the best-fitting model remembering that your interpretations must reflect the timing of measurements and the description of the response.
   a. Change the timing of measurements for the best fitting model so that the last measurement occasion (spr-8th) is coded at zero. Refit the model. Have the parameters changed their interpretations? Explain.
   b. SAS PROC NLMIXED will compute predicted values of the random effects for each subject and write estimates to a new file. It is always informative to see the individuals. Unless the sample is small, however, it is generally not possible to draw results for all individuals on a single display. But even a few records are helpful to get a sense of how variable results are and to be sure the model is appropriate for individuals. In one graph (called a lattice plot), plot the data and individual functions of three boys and three girls. There should be six individuals on this figure.

Label your graph clearly so it is transparent which graphs belong to which individuals.

Adding Covariates

6. Using the best fitting NLME model from Question 4, now add **sex**, **ses**, **race** (dummy-coded), and **gk1** (general knowledge scaled score) as level-2 predictors of the growth parameters. You do not have to eliminate non-significant fixed effects.

   Interpreting only the significant parameter estimates for the covariates, write-up the results from the analysis in at most two paragraphs. Your write-up of your final level 2 model should be for a non-technical, but knowledgeable audience. Be sure to explain the variance and covariance components within your write-up of the results.

APPENDIX

Model 1 and Model 3 from Table 10.1 are reproduced here for ease of explication.

$$f(t_j,\beta_1,\beta_2,\beta_3)=\frac{\beta_1\beta_2}{\beta_1-(\beta_1-\beta_2)\exp\{-\beta_3 t_j\}} \qquad \text{(Model 1)}$$

$$f(t_j,\beta_1,\beta_2,\beta_3)=\frac{\beta_2}{\left[1+\exp(-\beta_3)\cdot(t_j-\beta_1)\right]} \qquad \text{(Model 2)}$$

The inflection point in Model 1 occurs at that time when the curve changes concavity (i.e., the first derivative of the function with respect to time is a maximum or the second derivative with respect to time is zero). In other words, when

$$\frac{df}{dt}=\frac{\beta_1\beta_2(\beta_2-\beta_1)\beta_3\exp(\beta_3 t_j)}{\left[\beta_2-\beta_1+\beta_1\exp(\beta_3 t_j)\right]^2}$$

$$\frac{d^2 f}{dt^2}=-\frac{\beta_1\beta_2(\beta_2-\beta_1)\beta_3^2\exp(\beta_3 t_j)\left[\beta_1\exp(\beta_3 t_j)-\beta_2+\beta_1\right]}{\left[\beta_2-\beta_1+\beta_1\exp(\beta_3 t_j)\right]^3}$$

$$-\frac{\beta_1\beta_2(\beta_2-\beta_1)\beta_3^2\exp(\beta_3 t_j)\left[\beta_1\exp(\beta_3 t_j)-\beta_2+\beta_1\right]}{\left[\beta_2-\beta_1+\beta_1\exp(\beta_3 t_j)\right]^3}=0$$

$$\beta_1\beta_2(\beta_2-\beta_1)\beta_3^2\exp(\beta_3 t_j)\left[\beta_1\exp(\beta_3 t_j)-\beta_2+\beta_1\right]=0, \text{ solve for } t_j$$

$$t_j=\frac{\ln\left(\dfrac{\beta_2}{\beta_1}-1\right)}{\beta_3}.$$

One way to show equivalence is to set the two equations equal to one another and plug in a common value (i.e., time where the inflection of the logistic curve occurs). If they are equivalent

$$f_{\text{model 1}}\left(t_j=\left(\ln\left\{\left[\beta_2/\beta_1\right]-1\right\}\right)/\beta_3;\beta_1,\beta_2,\beta_3\right)=f_{\text{model 3}}(t_j=\beta_1^*;\beta_1^*,\beta_2,\beta_3),$$

the function values evaluated at the time when the inflection occurs should be equal.

$$\frac{\beta_1\beta_2}{\beta_1-(\beta_1-\beta_2)\exp\left\{-\beta_3\left[\dfrac{\ln\left(\dfrac{\beta_2}{\beta_1}-1\right)}{\beta_3}\right]\right\}}=\frac{\beta_2}{\left[1+\exp(-\beta_3)\cdot(\beta_1^*-\beta_1^*)\right]}$$

$$\frac{\beta_1\beta_2}{\beta_1-(\beta_1-\beta_2)\dfrac{\beta_1}{(\beta_2-\beta_1)}}=\frac{1}{2}\beta_2$$

$$\frac{1}{2}\beta_2=\frac{1}{2}\beta_2 \quad \therefore$$

AUTHOR NOTE

The authors wish to thank Dr. Kevin Grimm for facilitating access to the preschool instruction study data used throughout the chapter.

Correspondence concerning this book chapter should be addressed to Jeffrey R. Harring, Human Development and Quantitative Methodology, University of Maryland, College Park, MD, 20742-1115. Email: harring@umd.edu

NOTES

1. This dataset will be used throughout the chapter.

2. Davidian and Giltinan (1995, 2003) refer to the level-1 covariance structure that permits dependence on i through an individual's response, given β_i. While this does indeed provide the most flexibility for modeling time-specific variances, the empirical example used throughout this chapter will not exploit this dependency, and therefore, we will suppress the β_i notation and assume that the level-1 covariance structure may only vary across children in terms of its dimension (i.e., allowing for missing data), but not otherwise.
3. Other estimation procedures are possible including MCMC algorithms, such as Gibbs sampling or Metropolis-Hastings within a Bayesian framework (Davidian & Giltinan, 1995, 2003).
4. The most popular software programs or modules to fit NLME models are SAS `PROC NLMIXED` (SAS, 2017) and the `nlmer()` function from the R package `lme4` or the `nlme()` function from the R package `nlme`. Of course there are other alternatives. In this chapter we use PROC NLMIXED to fit the NLME models while R is used for graphing. Annotated code for fitting the models in this chapter can be found on the multilevel modeling book website.
5. Some have argued that since the sampling distribution of variances are non-normal that the Wald test is inappropriate. However, we believe in a good initial test so that a model comparison can eventually be executed.
6. Greater detail of fitting level 1 covariance structures in SAS PROC NLMIXED can be found in Harring and Blozis (2014).

REFERENCES

Akaike, H. (1974). A new look at the statistical model identification. *IEEE Transactions on Automatic Control, AC-19*, 716–723.

Beal, S. L., & Sheiner, L. B. (1982). Estimating population kinetics. *CRC Critical Reviews in Biomedical Engineering, 8*, 195–222.

Blozis, S. A. (2012). Sensitivity analysis of mixed-effects models when longitudinal data are incomplete. In J. R. Harring & G. R. Hancock (Eds.), *Advances in longitudinal methods in the social and behavioral sciences* (pp. 137–157). Information Age Publishing.

Blozis, S. A., & Harring, J. R. (2015). Understanding individual-level change through the basis functions of a latent curve model. *Sociological Methods and Research*, 1–28. https://doi.org/10.1177/0049124115605341

Browne, M. W. (1993). Structured latent curve models. In C. M. Cuadras & C. R. Rao (Eds.), *Multivariate analysis: Future directions 2* (pp. 171–198). North-Holland.

Choi, J., Harring, J. R., & Hancock, G. R. (2009). Latent growth modeling for logistic response functions. *Multivariate Behavioral Research, 44*, 620–645.

Conner, C. M., Morrison, F. J., & Slominski, L. (2006). Preschool instruction and children's emergent literacy growth. *Journal of Educational Psychology, 98*, 665–689.

Cudeck, R. (1996). Mixed-effects models in the study of individual differences with repeated measures data. *Multivariate Behavioral Research, 31*, 371–403.

Cudeck, R., & Harring, J. R. (2007). The analysis of nonlinear patterns of change with random coefficient models. *Annual Review of Psychology, 58*, 615–637.

Cudeck, R., & Harring, J. R. (2010). Developing a random coefficient model for nonlinear repeated measures data. In S.-M. Chow, E. Ferrer, & F. Hsieh (Eds.), *Statistical methods for modeling human dynamics: An interdisciplinary dialogue.* Routledge.

Cudeck, R., & du Toit, S. H. C. (2002). A version of quadratic regression with interpretable parameters. *Multivariate Behavioral Research,* 37, 501–519.

Davidian, M., & Giltinan, D. M. (1995). *Nonlinear models for repeated measurement data.* Chapman & Hall/CRC.

Davidian, M., & Giltinan, D. M. (2003). Nonlinear models for repeated measurement data: An overview and update. *Journal of Agricultural, Biological, and Environmental Statistics, 8,* 387–419.

Enders, C. K. (2010). *Applied missing data analysis.* The Guilford Press.

Fitzmaurice, G. M., Laird, N. M., & Ware, J. H. (2011). *Applied longitudinal analysis* (2nd ed.). Wiley.

Graham, J. W. (2009). Missing data analysis: Making it work in the real world. *Annual Review of Psychology, 60,* 549–576.

Harring, J. R., & Blozis, S. A. (2014). Fitting correlated residual error structures in nonlinear mixed-effects models using SAS PROC NLMIXED. *Behavioral Research Methods, 46,* 372–384.

Harring, J. R., & Blozis, S. A. (2016). A note on recurring misconceptions when fitting nonlinear mixed models. *Multivariate Behavioral Research.* https://doi.org/10.1080/00273171.2016.1239522

Hedeker, D., & Gibbons, R. D. (1997). Application of random-effects pattern-mixture models for missing data in longitudinal studies. *Psychological Methods, 2,* 64–78.

Hox, J. (2010). *Multilevel analyses: Techniques and applications* (2nd ed.). Erlbaum.

Laird, N. M. (1988). Missing data in longitudinal studies. *Statistics in Medicine, 7,* 305–315.

Laird, N. M., & Ware, J. H. (1982). Random-effects models for longitudinal data. *Biometrics, 38,* 963–974

Lindstrom, M. J., & Bates, D. M. (1990). Nonlinear mixed effects models for repeated measures data. *Biometrics, 46,* 673–687.

Little, R. J. A., & Rubin, D. B. (2002). Bayes and multiple imputation. In R. J. A. Little & D. B. Rubin (Eds.), *Statistical analysis with missing data* (pp. 200–220). Wiley.

McGrew, K. S., Werder, J. K., & Woodcock, R. W. (1991). Woodcock-Johnson technical manual. DLM.

Pinheiro, J. C., & Bates, D. M. (1995). Approximations to the log-likelihood function in the nonlinear mixed-effects model. *Journal of Computational and Graphical Statistics, 4,* 12–35.

Pinheiro, J. C, & Bates, D. M. (2000). *Mixed-effects models in S and S-PLUS.* Springer Science & Business Media.

Preacher, K. J., & Hancock, G. R. (2012). On interpretable reparameterizations of linear and nonlinear latent growth curve models. In J. R. Harring & G. R. Hancock (Eds.), *Advances in longitudinal methods in the social and behavioral sciences* (pp. 25–58). Information Age Publishing.

Preacher, K. J., & Hancock, G. R. (2015). Meaningful aspects of change as novel random coefficients: A general method for reparameterizing longitudinal models. *Psychological Methods, 20,* 84–101.

Raudenbush, S. W., & Bryk, A. S. (2002). *Hierarchical linear models: Applications and data analysis methods* (2nd ed.). SAGE.

Richards, F. J. (1959). A flexible growth function for empirical use. *Journal of Experimental Botany, 10,* 290–300.

Seber, G. A. F., & Wild, C. J. (1989). *Nonlinear regression.* Wiley.

Self, S. G., & Liang, K. Y. (1987). Asymptotic properties of maximum likelihood estimators and likelihood ratio tests under nonstandard conditions. *Journal of the American Statistical Association, 82,* 605–610.

Shapiro, A. (1985). Asymptotic distribution of test statistics in the analysis of moment structures under inequality constraints. *Biometrika, 72,* 133–144.

Skrondal, A., & Rabe-Hesketh, S. (2004). *Generalized latent variable modeling: Multilevel, longitudinal, and structural equation models.* Chapman & Hall.

Snijders, T. A. B., & Bosker, R. J. (2012). *Multilevel analysis: an introduction to basic and advanced multilevel modeling* (2nd ed.). SAGE.

Stoel, R. D., Garre, F. G., Dolan, C., & van den Wittenboer, G. (2006). On the likelihood ratio test in structural equation modeling when parameters are subject to boundary constraints. *Psychological Methods, 11,* 439–455.

Tourangeau, K., Nord, C., Lê, T., Sorongon, A. G., & Najarian, M. (2009). *Early childhood longitudinal study, kindergarten class of 1998–99 (ECLS-K): Combined user's manual for the ECLS-K eighth-grade and K–8 full sample data files and electronic codebooks.* NCES 2009-004. National Center for Education Statistics.

van Buuren, S. (2011). Multiple imputation of multilevel data. In J. J. Hox & J. K. Roberts (Eds.), *Handbook for advanced multilevel analysis* (pp. 173–196). Routledge/Taylor & Francis Group.

Verbeke G., & Molenberghs G. (2000). *Linear mixed models for longitudinal data.* Springer-Verlag.

Verhulst, P. (1845). Mathematical researches into the law of population growth increase. *Nouveaux mémoires de l'Académie Royale des Sciences et Belles-Lettres de Bruxelles, 18,* Article 1.

Vonesh, E., & Chinchilli, V. M. (1997). *Linear and nonlinear models for the analysis of repeated measurements.* CRC.

Wall, M. (2009). Maximum likelihood and Bayesian estimation for nonlinear structural equation models. In R. Millsap (Ed.), *The SAGE handbook of quantitative methods in psychology* (pp. 540–567). SAGE.

Xu, S., & Blozis, S. A. (2011). Sensitivity analysis of a mixed model for incomplete longitudinal data. *Journal of Educational & Behavioral Statistics, 36,* 237–256

Zeger, S. L., Liang, K.-Y., & Albert, P. S. (1988). Models for longitudinal data: A generalized estimating equations approach. *Biometrics, 44,* 1049–1060.

CHAPTER 11

WITHIN-SUBJECT RESIDUAL VARIANCE–COVARIANCE STRUCTURES IN LONGITUDINAL DATA ANALYSIS

Minjung Kim
The Ohio State University

Hsien-Yuan Hsu
University of Massachusetts, Lowell

Oi-man Kwok
Texas A&M University

This chapter demonstrates how to specify the alternative within-subject residual variance–covariance (V–CV) structures in a 2-level growth model. We begin with a brief literature review on how existing studies have specified the within-subject V–CV structures in their analyses using multilevel modeling (MLM). We provide an overview of the consequences of misspecifying

Multilevel Modeling Methods with Introductory and Advanced Applications, pages 389–419

the within-subject V–CV structures that have been identified in previous simulation studies. Then we describe the model formulation in terms of the decomposition of the between- and within-subject V–CV components in a simple linear growth model in the MLM framework. Following that, we extend the model formulation to be in matrix algebra form under the generalized linear modeling framework, which elaborates the variance components of the growth models. Then we introduce a set of alternative V–CV structures for modeling the within-subject residual components that are offered in popular statistical software packages. Finally, we demonstrate how to select the within-subject residual V–CV that fits the data better in MLM using an empirical dataset from the National Longitudinal Study of Youth 97 (NLSY97) project.

As discussed in Chapter 9 of this volume (McCoach et al.), MLM is popular for analyzing longitudinal data to investigate individual change over time in educational and social science studies (Laird & Ware, 1982; Raudenbush & Bryk, 2002). In the MLM framework, longitudinal data sets are in long (or univariate) format, so that each subject (e.g., each student) has multiple rows of observations measured over multiple time points. Long format data sets require a variable that indicates the time when each observation is made (e.g., age, year, or time of measurement). Because the data contains time information for each occasion of measurement, MLM does not require repeated measures to be equally spaced in time or to contain an equal number of repeated measures for all individuals (Raudenbush & Bryk, 2002).

Although the time structure can be more flexible in MLM, substantive researchers tend to focus mainly on specifying the mean and variability of the growth trajectory (i.e., the mean structure and the between-subject V–CV structure, respectively) but ignore the within-subject residual V–CV structure. Generally, the within-subject residual V–CV structures are assumed to be independently and identically distributed in the model building process, even though this assumption is often violated (Kwok et al., 2007). The *standard* MLM usually specifies the saturated level-2 V–CV structure (i.e., freely estimating the random intercept- and slope-variances and covariance between them), while constraining the level-1 variances to be equal across repeated measures with zero covariance. However, misspecifying the within-subject residual V–CV structure leads to biased standard errors of the fixed effects, which in turn interfere with statistical significance testing and construction of confidence intervals for the fixed effects (Singer & Willett, 2003; Weiss, 2005).

More importantly, the within-subject V–CV structure contains unique information specific to the measures at each time point, that is the change in the variation of the average trend over time. For example, in a simple linear growth model, the diagonal elements of the within-subject residual V–CV matrix are the residual variances of repeated measures that cannot be explained by the time-specific means of the repeated measures. The magnitudes of the residual variances indicate whether time-varying predictors (if any) are needed to further explain the within-subject variability at each of the time points. After adding a time-varying predictor, the change in the size of the response variability across time can be computed to indicate the predictive power of the time-varying predictor (akin to the R-square change in regression analysis). On the other hand, the relationships among the within-subject residual variances depict the unique feature of repeated measures studies in which the subjects' responses at different time points are usually dependent on one another to some degree (e.g., measures from adjacent time points are more similar to each other). Substantive researchers can use this information to understand and confirm the data quality before the main analysis.

In summary, oversimplification of the within-subject residual V–CV structure results in biased standard errors of the fixed effects. In addition, the within-subject residual V–CV matrix contains useful information, including the unexplained variances of repeated measures and relationships between these unexplained variances, which are relevant for including time-varying predictor(s) and understanding the characteristics of repeated measures data, respectively. Therefore, it is important to put greater effort into optimally specifying the within-subject residual V–CV structure. Although the importance of optimally specifying the within-subject residual V–CV structure has been identified during the past decade (Ferron et al., 2002; Kim et al., 2016; Kwok et al., 2007), guidance on how to specify the alternative V–CV structures in multilevel growth modeling has been limited.

LITERATURE REVIEW

We conducted a literature review to understand how researchers analyze longitudinal data using growth modeling in MLM. We specifically focused on whether the within-subject V–CV structure was considered for the growth model specification. We used two different strategies to search for the relevant literature for this review. First, we selected two APA psychology journals (*Child Development* and *Developmental Psychology*) published between 2015 and 2017. Using the keywords ["multilevel" or "hierarchical"] and ["growth," "trajectory," "change," or "longitudinal"], we identified 23 applied studies that used MLM to investigate individual trajectories. Only three studies tested the alternative level-1 error variance structures; the majority of studies did not address modification of the within-subject residual V–CV structures.[1] Conversely, most studies included discussion of the level-2 random effects (83%), supporting our claim that researchers generally focus on between-subject differences when analyzing longitudinal data using MLM.

We used a second search strategy to review empirical studies that use the MLM for growth analysis. We found studies that cited Singer (1998) using Google Scholar. We chose Singer's paper as the starting point for the search because it is a widely cited foundational article for understanding MLM in general, and it also specifically addresses level-1 error V–CV structures in longitudinal MLM. At the time of this writing, Google Scholar listed 3,243 citations for Singer (1998). We further restricted our search to contemporary articles (2010 to present) and applied the search keywords ["longitudinal data" and "linear growth"] or ["longitudinal data" and "quadratic growth"], which left us with 59 results. After eliminating methodological papers, dissertations, and papers that did not employ MLM, we identified 30 empirical studies to review. Not surprisingly, most of the reviewed studies did not consider modifying the structure of the within-subject variance components in the growth analysis. Among the 30 reviewed studies, only 2 (7%) compared models by changing the level-1 error variance structures. Again, although most studies did not test alternative level-1 V–CV structures, the majority of studies specified the random-slopes (at level-2) component in the model (71%).

Perhaps many researchers ignore the within-subject V–CV structure because they simply adopt the default or embedded structure provided by their statistical software of choice, rather than altering the syntax to fit the alternative V-CV structures. Because the primary interest in model interpretation is the overall growth shape, the within-subject variance component is often disregarded in the model specification. We note that the default structures are usually too simple in the basic 2-level model setting (e.g., the identity structure in SAS, or SPSS) to capture the reality of within-subject residual variance structures in longitudinal data, which represents a model misspecification.

CONSEQUENCES OF MISSPECIFYING THE WITHIN-SUBJECT RESIDUAL V–CV STRUCTURES IN GROWTH ANALYSIS

Previous simulation studies have shown that misspecifying the within-subject residual variance structure introduces bias into the estimated standard errors of the fixed effects, which our statistical inferences. Specifically, Ferron et al. (2002) investigated the effects of oversimplifying the within-subject V–CV structure (using the identity structure) in 2-level growth models and found that the misspecification led to substantially biased estimates of variance parameters. Similarly, Kwok et al. (2007) found that oversimplifying (underspecifying) the within-subject covariance structure resulted in overestimated standard errors of the growth-related parameters, which in turn reduced the statistical power to detect the significance of growth parameters. On the other hand, overspecifying the within-subject V–CV, which might produce a structure that is unnecessarily too complex, resulted in smaller estimates of the random effect variances, and this in turn increased the likelihood of type I error inflation when testing the significance of growth parameters (Ferron et al., 2002; Kwok et al., 2007).

Therefore, applied researchers should choose the most optimal structure for the within-subject V–CV after comparing models with alternative V–CV structures. It is hard to provide a clear answer regarding how many and which error structures should be compared. However, previous research shows that starting with the unstructured (saturated) pattern can be a reasonable option, given that it makes no particular assumptions about the residual structure (Liu et al., 2012). Similarly, a limited number of studies within the latent growth modeling framework suggest using the unstructured V–CV as the starting point for searching for growth trajectories (Kim et al., 2016; Wu & West, 2010). The estimated results with the unstructured residuals can be a useful diagnostic to indicate which other alternatives should be considered to better estimate the model. Previous simulation studies (Liu et al., 2012; Kim et al., 2016) have recommended using the Bayesian information criterion (BIC; Sclove, 1978) and the Akaike information criterion (AIC) to compare models with different V–CV structures. This chapter demonstrates the use of the BIC and the AIC to compare models with different V–CV structures.

MODEL FORMULATION

In this chapter, we use the MLM 2-level linear growth model in which repeated measures (level 1) are nested within subjects (level 2). The 2-level linear growth model can be written as follows.

$$
\begin{aligned}
&\text{Level 1:}\\
&\quad y_{ti} = \pi_{0i} + \pi_{1i} Time_{ti} + e_{ti}\\
&\text{Level 2:}\\
&\quad \pi_{0i} = \beta_{00} + u_{0i},\\
&\quad \pi_{1i} = \beta_{10} + u_{1i}\\
&\text{Combined (mixed):}\\
&\quad y_{ti} = \beta_{00} + \beta_{10} Time_{ti} + u_{0i} + u_{1i} Time_{ti} + e_{ti}
\end{aligned}
\tag{11.1}
$$

In these equations, y_{ti} is the repeated measure y at time t $(1, \ldots, t)$ for individual i $(1, \ldots, i)$, β_{00} is the mean intercept of y_{ti}, β_{10} is the average slope of the linear growth across all individuals, u_{0i} is the deviation of the individual intercepts from the mean intercept β_{00}, and u_{1i} is the deviation of the individual slopes from the mean slope β_{10}.

The mixed model in Equation 11.1 can be divided into two components: the fixed effects (i.e., $\beta_{00} + \beta_{10} Time_{ti}$) and the random effects (i.e., $u_{0i} + u_{1i} Time_{ti} + e_{ti}$). The random effects can be further decomposed into two components: the between-subject V–CV structure for the level-2 model [i.e., $Var(u_{0i} + u_{1i} Time_{ti})$] and the within-subject V–CV structure for the level-1 model [i.e., $Var(e_{ti})$]. Next, we rewrite the mixed model in matrix algebra form to further elaborate the V–CV components (Kwok et al., 2007):

$$\begin{bmatrix} y_{11} \\ \vdots \\ y_{t1} \\ \vdots \\ \vdots \\ \vdots \\ y_{1i} \\ \vdots \\ y_{ti} \end{bmatrix} = \begin{bmatrix} 1 & TIME_1 \\ \vdots & \vdots \\ 1 & TIME_t \\ \vdots & \vdots \\ \vdots & \vdots \\ \vdots & \vdots \\ 1 & TIME_1 \\ \vdots & \vdots \\ 1 & TIME_t \end{bmatrix} \begin{bmatrix} \beta_{00} \\ \beta_{10} \end{bmatrix} + \begin{bmatrix} 1 & TIME_1 & 0 & \ldots & 0 & 0 & 0 \\ \vdots & \vdots & \vdots & \ldots & \vdots & \vdots & \vdots \\ 1 & TIME_t & \vdots & \ldots & \vdots & 0 & 0 \\ \vdots & \vdots & \vdots & \ldots & \vdots & 0 & 0 \\ \vdots & \vdots & \vdots & \ldots & \vdots & \vdots & \vdots \\ \vdots & \vdots & \vdots & \ldots & \vdots & 0 & 0 \\ 1 & 0 & \vdots & \ldots & \vdots & 1 & TIME_1 \\ \vdots & \vdots & \vdots & \ldots & \vdots & \vdots & \vdots \\ 0 & 0 & 0 & \ldots & 0 & 1 & TIME_t \end{bmatrix} \begin{bmatrix} u_{01} \\ u_{11} \\ u_{02} \\ u_{12} \\ \vdots \\ \vdots \\ \vdots \\ u_{0i} \\ u_{1i} \end{bmatrix} + \begin{bmatrix} e_{11} \\ \vdots \\ e_{t1} \\ \vdots \\ \vdots \\ \vdots \\ e_{1i} \\ \vdots \\ e_{ti} \end{bmatrix} \tag{11.2}$$

which again can be generalized to:

$$\underline{y} = \mathbf{X}\underline{\beta} + \mathbf{Z}\underline{u} + \underline{e}, \tag{11.3}$$

where $\underline{y}$ is a $[t * i]$ column vector containing the t repeated measures for all i individuals, X is a $[(t * i)$ by $2]$ matrix containing the intercept (i.e., 1) and the predictor variable *TIME*, $\underline{\beta}$ is a [2] column vector containing the unknown linear growth parameters β_{00} and β_{10}, Z is a $\left[(t * i) \text{ by } (2i)\right]$

design matrix, $\underline{u}$ is a [2 * i] column vector containing the random effects representing between-subject variation (individual differences), and $\underline{e}$ is a [t * i] column vector containing the within-subject random errors. More specifically, the variance of the random effects in Equation 11.3 is equal to: $\text{VAR}(\underline{y}) = \text{VAR}(\boldsymbol{Z}\underline{u} + \underline{e}) = \text{VAR}(\boldsymbol{Z}\underline{u}) + \text{VAR}(\underline{e}) = \boldsymbol{ZTZ}^T + \Sigma$, which equals the following.

$$\begin{bmatrix} Z_1 & 0 & . & . & 0 \\ 0 & Z_2 & . & . & 0 \\ \vdots & 0 & . & \vdots & \vdots \\ \vdots & \vdots & \vdots & . & \vdots \\ 0 & . & . & . & Z_i \end{bmatrix} \begin{bmatrix} T & 0 & . & . & 0 \\ 0 & T & . & . & 0 \\ \vdots & \vdots & \vdots & . & \vdots \\ 0 & . & . & . & T \end{bmatrix} \begin{bmatrix} Z_1 & 0 & . & . & 0 \\ 0 & Z_2 & . & . & 0 \\ \vdots & 0 & . & \vdots & \vdots \\ \vdots & \vdots & \vdots & . & \vdots \\ 0 & . & . & . & Z_i \end{bmatrix}^2 + \begin{bmatrix} \Sigma_1 & 0 & . & . & 0 \\ 0 & \Sigma_2 & . & . & 0 \\ \vdots & 0 & . & \vdots & \vdots \\ \vdots & \vdots & \vdots & . & \vdots \\ 0 & . & . & . & \Sigma_i \end{bmatrix}$$

$$= \begin{bmatrix} Z_1TZ_1^T + \Sigma_1 & 0 & . & . & 0 \\ 0 & Z_2TZ_2^T + \Sigma_2 & . & . & 0 \\ \vdots & \vdots & \vdots & \vdots & \vdots \\ 0 & . & & . & Z_iTZ_i^T + \Sigma_i \end{bmatrix} \tag{11.4}$$

where the elements on the diagonals of the block diagonal matrices are defined as follows:

$$Z_i = \boldsymbol{Z} = \begin{bmatrix} 1 & TIME_1 \\ 1 & TIME_2 \\ \vdots & \vdots \\ 1 & TIME_t \end{bmatrix}$$

$$\boldsymbol{T} = \begin{bmatrix} \tau_{00} & \tau_{01} \\ \tau_{10} & \tau_{11} \end{bmatrix} = \begin{bmatrix} Var(u_{0i}) & Cov(u_{0i}, u_{1i}) \\ Cov(u_{1i}, u_{0i}) & Var(u_{1i}) \end{bmatrix} \tag{11.5}$$

$$\Sigma_i = \Sigma = \begin{bmatrix} \sigma_{11}^2 & \sigma_{12} & \cdots & \cdots & \sigma_{1t} \\ \sigma_{21} & \sigma_{22}^2 & \cdots & \cdots & \sigma_{2t} \\ \vdots & \vdots & \cdots & \cdots & \vdots \\ \vdots & \vdots & \cdots & \cdots & \vdots \\ \sigma_{t1} & \sigma_{t2} & \cdots & \cdots & \sigma_{tt}^2 \end{bmatrix}$$

In this linear growth model, T is a [2 by 2] between-subject V–CV matrix associated with the two growth parameters (i.e., intercept and linear slope). Specifically, τ_{00} and τ_{11} correspond to the variance of the intercepts (u_{0i}) and linear slopes (u_{1i}), respectively, and τ_{01} (or τ_{10}) is the covariance between them. Σ is a [t by t] within-subject V–CV matrix containing the variances and covariances of random errors related to within-subject random errors. Both T and Σ are square and symmetric matrices, meaning that the off-diagonal elements are mirrored by the variances on the diagonal (e.g., $\sigma_{12} = \sigma_{21}$). Equation 11.5 assumes that the between- and within-subject variance components are mutually exclusive (i.e., $COV(u_{0j}, e_{ij}) = 0$), so that they can be modeled separately.

The combined covariance structure of the four repeated measures can be written as below under the *standard* MLM framework, which allows the level-2 V–CV to be saturated with the equal level-1 variance and zero covariance:

$$ZTZ^T + \Sigma = \begin{bmatrix} 1 & \\ 1 & \\ \vdots & \\ \vdots & \\ 1 & \end{bmatrix} * \begin{bmatrix} \tau_{00} & \tau_{01} \\ \tau_{10} & \tau_{11} \end{bmatrix} * \begin{bmatrix} 1 & 1 & \cdots & \cdots & 1 \\ TIME_1 & TIME_2 & \cdots & \cdots & TIME_t \end{bmatrix} + \begin{bmatrix} \sigma^2 & 0 & 0 & 0 \\ 0 & \sigma^2 & 0 & 0 \\ 0 & 0 & \sigma^2 & 0 \\ 0 & 0 & 0 & \sigma^2 \end{bmatrix} \tag{11.6}$$

$$= \begin{bmatrix} \tau_{00} + \sigma^2 & & & \\ \tau_{00} + \tau_{01} & \tau_{00} + 2\tau_{01} + \tau_{11} + \sigma^2 & & \\ \tau_{00} + 2\tau_{01} & \tau_{00} + 3\tau_{01} + 2\tau_{11} & \tau_{00} + 4\tau_{01} + 4\tau_{11} + \sigma^2 & \\ \tau_{00} + 3\tau_{01} & \tau_{00} + 4\tau_{01} + 3\tau_{11} & \tau_{00} + 5\tau_{01} + 6\tau_{11} & \tau_{00} + 6\tau_{01} + 9\tau_{11} + \sigma^2 \end{bmatrix}$$

Types of Longitudinal Data

As shown in Equation 11.6, Σ in a 2-level growth model is a [t by t] within-subject residual V–CV matrix, where $\underline{t}$ is the number of time points. The time metric in the longitudinal data can be either balanced or unbalanced (Raudenbush & Bryk, 2002). Being *balanced on time* means that each subject is observed at the same fixed set of time points. For example, suppose that researchers want to study a change in engagement in family routines across 1997, 1998, 1999, and 2000 for adolescents who are 12 years old in 1997. Table 11.1 presents the different time formats for the corresponding data. Analytically, the time variable in the growth model is commonly coded as 0, 1, 2, and 3 (Column 4), representing the measurement times of 1997, 1998, 1999, and 2000 (Column 3), respectively. In this case, the intercept coefficient (β_{00}) in Equation 11.1 represents the average family routine score across all participants in 1997. Alternatively, the age variable can be used as an indicator of the discrete measurement time. In this case, the age ranges between 12 and 15, depending on the year of measurement (Column 5). Under this coding scheme, β_{00} is less meaningful because it represents the average family routine score when the child is 0 years old. We note that the V–CV matrix Σ is a [4 by 4] matrix in all conditions when the data are

TABLE 11.1 Balanced on Time and Unbalanced on Time in Longitudinal Data

		Balanced on Time			Unbalanced on Time	
(1) ID	(2) Family Routine Score	(3) Year	(4) Year Centered on 1997	(5) Age	(6) Age	(7) Age Centered on 12
1	15	1997	0	12	12.6	0.6
1	10	1998	1	13	13.2	1.2
1	8	1999	2	14	14.4	2.4
1	6	2000	3	15	15.7	3.7
2	11	1997	0	12	12.3	0.3
2	7	1998	1	13	13.5	1.5
2	7	1999	2	14	14.3	2.3
2	6	2000	3	15	15.2	3.2
3	9	1997	0	12	12.8	0.8
3	7	1998	1	13	13.6	1.6
3	2	1999	2	14	14.7	2.7
3	1	2000	3	15	15.6	3.6

balanced on time. In our illustrative example, we treat the time matrix of the repeated measures as balanced on time because we are more interested in the growth of family routines by different age groups and assume that the family routines do not sensitively change by days or months.

Being *unbalanced on time* means that the repeated measures are not equally spaced in time across all individuals. Instead, the spacing of the repeated measures varies across people. For example, when the age of the adolescent is not roughly computed by the year information provided by subjects, but rather by year and month information, the ages at the four measurement time points vary across different subjects. For instance, the subject with ID = 1 is measured at the ages of 12.6, 13.2, 14.4, and 15.7, while the subject with ID = 2 is measured at the ages of 12.3, 13.5, 14.3, and 15.2 (see Column 6 in Table 11.1). A time variable created by centering this age variable on 12 for the unbalanced time scale appears in Column 7. In this case, the data become sparse, and there may be only one observation at a certain time point, so no variance exists under the extreme scenarios. For this reason, the within-subject V–CV structure tends not to be further considered in model specifications but is simplified to be a single estimate of error variance. Developmental research usually treats the time matrix for repeated measures as unbalanced on time when the observed outcomes are subject to change by days or months. For further discussion, we refer readers to Raudenbush and Bryk (2002, pp. 188–199).

Within-Subject Residual V–CV Structures

When the data structure is set to be balanced on time, the within-subject residual V–CV structures can be flexibly altered depending on the available options embedded in the statistical software. Table 11.2 presents several alternative within-subject residual V–CV structures for cases that have four repeated measures in the analysis. The program commands for two popular MLM software packages, the SPSS MIXED and SAS PROC MIXED procedures, are also provided in Table 11.2. The alternative structures in Table 11.2 are special types of the matrix Σ in Equation 11.5. The identity V–CV at the top of Table 11.2 is the simplest error variance structure, the unstructured V–CV is the most complex structure (Wolfinger, 1996). We present three additional V–CV structures lying between these two structures: *first order autoregressive, Toplitz,* and *banded main diagonal*. To interpret the results of data analyses, we employ the basic rationale of "parsimony means power" from Wolfinger (1996), meaning that researchers should specify the most parsimonious covariance structure possible that still reasonably fits the data.

Identity

The identity (ID) structure, the simplest error variance structure, contains only one parameter estimate (σ^2) on the within-subject residual V–CV structure. The ID structure assumes a homogeneous within-subject residual variance across all repeated measures; the covariance between any pair of repeated measures is assumed to be equal to 0 (see Table 11.2). When the random intercept-only model is specified (i.e., only τ_{00} in Equation 11.5), this structure is equivalent to the compound symmetry (CS) structure in the univariate repeated-measures analysis of variance (UANOVA) model. Under the sphericity assumption, equal variances (σ^2) and equal covariances (τ_{00}) are estimated as follows:

$$\begin{bmatrix} \tau_{00}+\sigma^2 & & & \\ \tau_{00} & \tau_{00}+\sigma^2 & & \\ \tau_{00} & \tau_{00} & \tau_{00}+\sigma^2 & \\ \tau_{00} & \tau_{00} & \tau_{00} & \tau_{00}+\sigma^2 \end{bmatrix} \qquad (11.7)$$

Although adoption of the ID structure for the within-subject V–CV has been criticized for its strong assumption, it is still the most frequently implied variance structure in applied studies. Notably, it is the *default* setting

TABLE 11.2 Alternative Within-Subject Variance–Covariance Structures for Four Repeated Measures of Data

Structure	Example	Description	Number of parameters	SPSS Mixed COVTYPE( )	SAS Proc Mixed TYPE=
Identity	$\sigma^2\begin{bmatrix}1&0&0&0\\0&1&0&0\\0&0&1&0\\0&0&0&1\end{bmatrix}$	constant variance; zero covariance	1	ID	VC
First-order-autoregressive	$\sigma^2\begin{bmatrix}1&\rho&\rho^2&\rho^3\\\rho&1&\rho&\rho^2\\\rho^2&\rho&1&\rho\\\rho^3&\rho^2&\rho&1\end{bmatrix}$	constant variance; exponentially decling covariance over time	2	AR1	AR(1)
Toeplitz	$\begin{bmatrix}\sigma^2&\sigma_1&\sigma_2&\sigma_3\\\sigma_1&\sigma^2&\sigma_1&\sigma_2\\\sigma_2&\sigma_1&\sigma^2&\sigma_1\\\sigma_3&\sigma_2&\sigma_1&\sigma^2\end{bmatrix}$	constant variance; constant covariance at the same lag	t	TP	TOEP
Banded-main-diagonal	$\begin{bmatrix}\sigma_1^2&0&0&0\\0&\sigma_2^2&0&0\\0&0&\sigma_3^2&0\\0&0&0&\sigma_4^2\end{bmatrix}$	heterogeneous variance; zero covariance	t	DIAG	UN(1)
Unstructured	$\begin{bmatrix}\sigma_1^2&\sigma_{12}&\sigma_{13}&\sigma_{14}\\\sigma_{21}&\sigma_2^2&\sigma_{23}&\sigma_{24}\\\sigma_{31}&\sigma_{32}&\sigma_3^2&\sigma_{34}\\\sigma_{41}&\sigma_{42}&\sigma_{43}&\sigma_4^2\end{bmatrix}$	freely estimated variance and covariance	$t(t+1)/2$	UN	UN

for the within-subject V–CV structure in many MLM software packages under the *standard* 2-level MLM specification, including SPSS MIXED and SAS PROC MIXED, if the program subcommands (e.g., REPEATED) are not designated to identify data as longitudinal.

First-Order Autoregressive

The first-order autoregressive structure [AR(1)] assumes a constant error variance (σ^2) and a systematic correlation between the pairs of error variances. More specifically, the correlation between any pair of time points is a function of ρ^d, where d is the distance between the two time points and ρ is the autoregressive parameter. For example, the correlation between the first and second time points is ρ (i.e., $\rho^1 = \rho$), the first and third time points are ρ^2, and so on. The AR(1) structure is generally used when the measures are equally spaced and assumed to have test-retest effects. The AR(1) structure allows a stronger correlation between adjacent time points than those that are farther apart.

Toeplitz

The Toeplitz structure (TOEP) is another more general type of V–CV structure than the ID structure. TOEP can specify any value between 0 and 1 for the correlation between any pair of time points laying on the same band. The first band of TOEP refers to the diagonal in the V–CV matrix, which is the variance, and the second band (TOEP2) refers to the covariance between two adjacent time points (e.g., Time 1 and Time 2, Time 2 and Time 3, etc.). Like AR(1), the TOEP structure is generally applied to longitudinal data with equally spaced timepoints. In our example we employ TOEP(2), which estimates the constant variance (σ^2) and covariance on the second band (σ_1) but constrains all other covariances to be zero.

Banded Main Diagonal

The banded main diagonal structure [UN(1)] is another commonly used within-subject V–CV structure, especially when applying latent growth models in the SEM framework.[2] UN(1) specifies a completely general (unstructured) covariance matrix for the first band of the matrix, which comprises the variance components on the main diagonal. Thus, the number of free parameters equals the number of repeated measures (t). No covariance is assumed under UN(1); this is done by setting all of the covariance components on the higher bands to zero.

Unstructured

In the most general V–CV structure, unstructured (UN), each pair of time points can have a unique covariance. All the V–CV matrices mentioned above are special cases of the UN structure. Because the other matrices can be reproduced from the UN structure by constraining some parts of the variances and covariances, they are nested within the UN structure. The UN structure produces the best model fit with the smallest deviance because it is the most parameterized (freed) model. Although specifying the UN structure may improve model fit, there is a downside: UN has a saturated V–CV structure, leaving zero degrees of freedom left to estimate the between subject variances. In other words, specifying the UN structure does not allow researchers to decompose the between- and within-subject variability when a researcher's interests are often in understanding differences in the growth trajectories across people. Thus, the UN structure is usually used as a starting point to diagnose the pattern of the residual variances to search for the optimal structure (Kim et al., 2016).

Table 11.3 presents the example of the combined V–CV structures of both between- and within-subject variance components for the alternative level-1 error variance structures with four repeated measures. In this example, the *standard* MLM with the saturated between-subject V–CV structure (Equation 11.5) is used for illustrating the combination of level-1 and level-2 V–CV structures.

ILLUSTRATIVE EXAMPLE SPECIFYING THE ALTERNATIVE LEVEL-1 ERROR VARIANCE STRUCTURES

Data Source

We illustrate how to specify the level-1 error variance structure in a two-level growth model using data from the NLSY97, a nationally representative sample of approximately 9,000 youths who were 12 to 16 years old as of December 31, 1996. The NLSY97 was the first nationally representative data set that collected family process measures and perspectives from multiple family members, allowing researchers to investigate the interactions between family members (Jones-Sanpei, Day, Holmes, & van Langeveld , 2009). The NLSY97 conducted 18 rounds of interviews (1997–2015), and the analytical data for our illustrative example were drawn from the first 4 years of survey data (1997–2000). We studied the trajectory of adolescents' engagement in family routines across 4 years. Specifically, participants were asked to report the number of days per week they spent on each of three routine activities with their family members: (a) doing housework, (b) recreational activities, and (c) religious activities. The family routines index, which is a three-item summative index ranging from 0 (no time spent) to 21 (engaged in three activities every day of the week), was used for growth modeling. Although this index can be considered as a frequency index, we treat the variables as continuous for further growth analysis, given that

TABLE 11.3 Combined V–CV Structures of the Saturated Between- and the Alternative Within-Subject Variance Components

Level-1 Structure	Combined V–CV structure
Identity	$\begin{bmatrix} \tau_{00}+\sigma^2 & & & \\ \tau_{00}+\tau_{01} & \tau_{00}+2\tau_{01}+\tau_{11}+\sigma^2 & & \\ \tau_{00}+2\tau_{01} & \tau_{00}+3\tau_{01}+2\tau_{11} & \tau_{00}+4\tau_{01}+4\tau_{11}+\sigma^2 & \\ \tau_{00}+3\tau_{01} & \tau_{00}+4\tau_{01}+3\tau_{11} & \tau_{00}+5\tau_{01}+6\tau_{11} & \tau_{00}+6\tau_{01}+9\tau_{11}+\sigma^2 \end{bmatrix}$
First-order-autoregressive	$\begin{bmatrix} \tau_{00}+\sigma^2 & & & \\ \tau_{00}+\tau_{01}+\rho\sigma^2 & \tau_{00}+2\tau_{01}+\tau_{11}+\sigma^2 & & \\ \tau_{00}+2\tau_{01}+\rho^2\sigma^2 & \tau_{00}+3\tau_{01}+2\tau_{11}+\rho\sigma^2 & \tau_{00}+4\tau_{01}+4\tau_{11}+\sigma^2 & \\ \tau_{00}+3\tau_{01}+\rho^3\sigma^2 & \tau_{00}+4\tau_{01}+3\tau_{11}+\rho^2\sigma^2 & \tau_{00}+5\tau_{01}+6\tau_{11}++\rho\sigma^2 & \tau_{00}+6\tau_{01}+9\tau_{11}+\sigma^2 \end{bmatrix}$
Toeplitz	$\begin{bmatrix} \tau_{00}+\sigma^2 & & & \\ \tau_{00}+\tau_{01}+\sigma_1 & \tau_{00}+2\tau_{01}+\tau_{11}+\sigma^2 & & \\ \tau_{00}+2\tau_{01}+\sigma_2 & \tau_{00}+3\tau_{01}+2\tau_{11}+\sigma_1 & \tau_{00}+4\tau_{01}+4\tau_{11}+\sigma^2 & \\ \tau_{00}+3\tau_{01}+\sigma_3 & \tau_{00}+4\tau_{01}+3\tau_{11}+\sigma_2 & \tau_{00}+5\tau_{01}+6\tau_{11}+\sigma_1 & \tau_{00}+6\tau_{01}+9\tau_{11}+\sigma^2 \end{bmatrix}$
Banded-main-diagonal	$\begin{bmatrix} \tau_{00}+\sigma_1^2 & & & \\ \tau_{00}+\tau_{01} & \tau_{00}+2\tau_{01}+\tau_{11}+\sigma_2^2 & & \\ \tau_{00}+2\tau_{01} & \tau_{00}+3\tau_{01}+2\tau_{11} & \tau_{00}+4\tau_{01}+4\tau_{11}+\sigma_3^2 & \\ \tau_{00}+3\tau_{01} & \tau_{00}+4\tau_{01}+3\tau_{11} & \tau_{00}+5\tau_{01}+6\tau_{11} & \tau_{00}+6\tau_{01}+9\tau_{11}+\sigma_4^2 \end{bmatrix}$
Unstructured	$\begin{bmatrix} \sigma_1^2 & \sigma_{12} & \sigma_{13} & \sigma_{14} \\ \sigma_{21} & \sigma_2^2 & \sigma_{22} & \sigma_{24} \\ \sigma_{31} & \sigma_{32} & \sigma_3^2 & \sigma_{34} \\ \sigma_{41} & \sigma_{42} & \sigma_{43} & \sigma_4^2 \end{bmatrix}$

the distributions are relatively normal with skewness < 3 and kurtosis < 7 (Kline, 2015).

We further investigated the trajectory of family routines once after accounting for the effect of maternal monitoring measured at each concurrent time point. Specifically, we used the adolescents' report on their perspective regarding how well their mothers knew the adolescents' friends, friends' parents, and teachers and how the adolescents spent time when they were not home. The responses to four questionnaires with a five-point scale (0–4) were summed to create a maternal monitoring score for in our analysis. The score ranges between 0 and 16, with a higher number indicating greater mother familiarity with the adolescent's friends and activities (Jones-Sanpei, Day, & Holmes, 2009). Table 11.4 presents descriptive statistics for the family routine scores and maternal monitoring scores.

For demonstration purposes, we restricted the analytical sample to adolescents who were 12 years old in 1997, who had at least one time point for the family routines index information, and who were White, African American, or Latinx ($n = 1{,}751$). In this sample, 51.5% were male. Regarding race/ethnicity, the majority were White (52.8%), followed by African American (25.0%), and Latinx (22.2%). Because our example focuses on the variance and covariance components of the time variable, we regard the data to be *balanced on time with data missing at random* (Raudenbush & Bryk, 2002) by restricting the age variable to be discrete (i.e., 12, 13, 14, or 15) rather than continuous (e.g., 12.14, 12.59, 13.02).

TABLE 11.4 Unweighted Descriptive Statistics of Variables Derived from NLSY97

Variable	Descriptive statistics	
Gender		
Male	51.46%	
Female	48.54%	
Race/Ethnicity		
White	52.77%	
Black	25.01%	
Hispanic	22.22%	
Family Routines Scores	**Mean (SD)**	**Skewness/Kurtosis**
Year 1997 ($n = 1{,}728$)	9.95 (4.60)	.08 / 2.98
Year 1998 ($n = 1{,}578$)	8.81 (4.36)	.25 / 3.13
Year 1999 ($n = 1{,}604$)	8.40 (4.27)	.40 / 3.23
Year 2000 ($n = 1{,}564$)	7.42 (4.20)	.49 / 3.32

Research Questions

Family routines are a means of establishing family connection and cohesion (Day et al., 2009; Fiese et al., 2002). Previous research has suggested that family routines are a protective factor against adolescent substance use, health risks (Manlove et al., 2012), and problem behavior at school (Taylor & Lopez, 2005). It is important to understand whether family routines can be maintained over a longer period of time as a protective factor. Our illustrative example therefore addresses the following research questions: As children move from early to later adolescence, do they tend to engage in fewer family routines, after accounting for the effect of maternal monitoring measured at each concurrent time point? What is the effect of maternal monitoring on family routines over time, considering that family routines change? Is the effect of maternal monitoring on family routines consistent over time?

Data Analysis

We analyzed the data for our example using the SAS PROC MIXED procedure. We used a restricted maximum likelihood estimation (REML) method to analyze the 2-level growth models with five alternative level-1 error variance structures (ID, AR(1), TOEP(2), UN(1), and UN). Example syntax code for SAS PROC MIXED is provided in the Appendix.

Preliminary Analysis

First, we examine and report descriptive statistics in Table 11.4. The mean values of engagement in family routines for four times (i.e., T1 = 9.95, T2 = 8.81, T3 = 8.40, & T4 = 7.42) indicate that engagement in family routine activities decreases over time. The intraclass correlation coefficient (ICC) of the repeated measures nested within individuals is .42, indicating that 42% of the variance is between people and 58% of the variance is within people, and there is sufficient amount of between-subject variance to use multilevel modeling. Next, we visually inspect the individual growth in engagement in family routines by plotting the reported scores for a subset of the total sample. Figure 11.1 presents a scatter plot for the four waves of family routine scores from a random sample of 16 subjects, and Figure 11.2 displays a lined plot of family routine scores from 100 randomly selected subjects. Both figures support the preliminary finding that the amount of time spent on family routines decreases over time.

Mean Structure

Based on the mean values and the visual inspection, we analyze both linear- and quadratic-growth models to examine whether there is a non-linearly declining pattern on the outcome variable across time. For the model

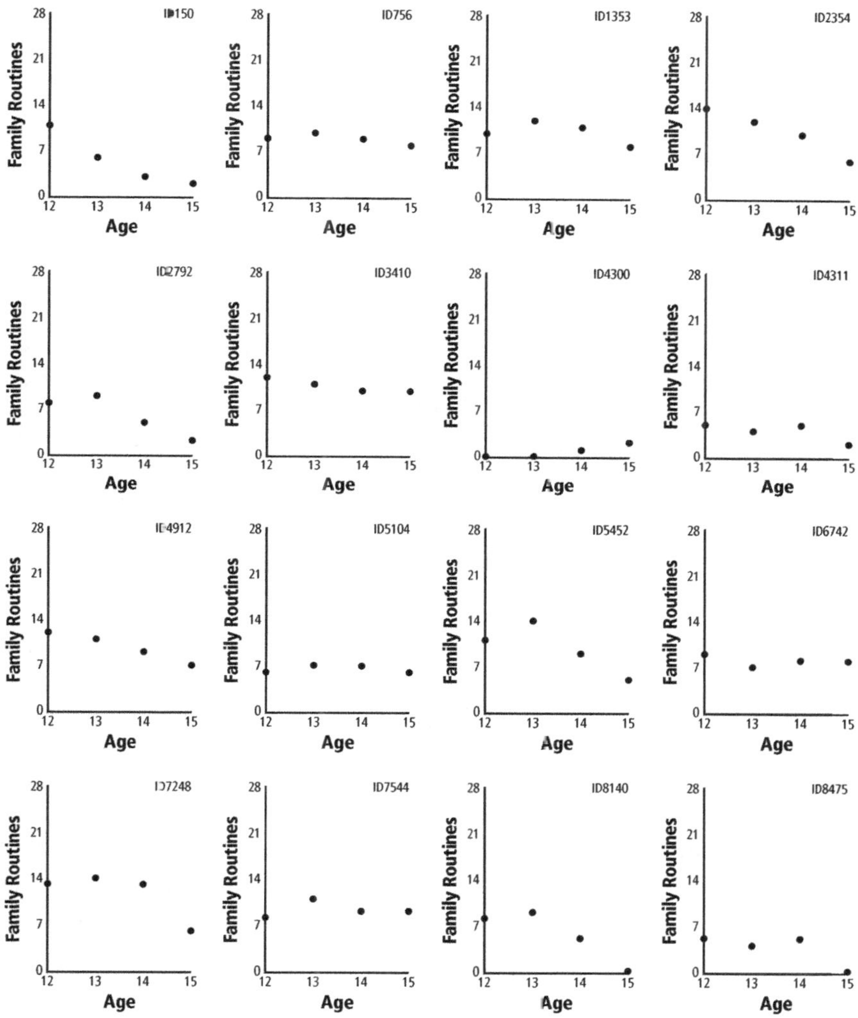

Figure 11.1 Scatter plot for the four waves of family routine scores from a random sample of 16 subjects.

comparison between the two different growth trajectories (i.e., linear vs. quadratic), we employed a search strategy suggested by Kim et al. (2016), which specifies the unstructured (or saturated) pattern for the within-subject V–CV structure. On the other hand, we constrained the between-subject V–CV structure to be null for model identification purposes (Kwok et al., 2007; Wu & West, 2010). The time variable is coded as 0, 1, 2, and 3 for the ages 12, 13, 14, and 15, respectively, so that the intercept value represents the mean values of family routines at age 12.

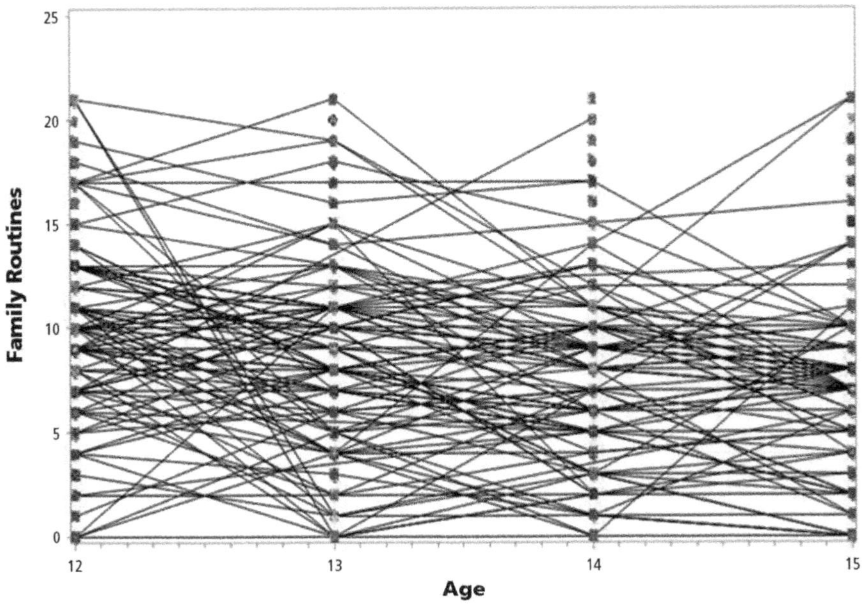

Figure 11.2 Lined plot of family routine scores for 100 randomly selected subjects.

Table 11.5 displays the model fit indices from the linear- and quadratic-growth models with UN V–CV structures, respectively. The fit statistics show that the linear-growth model fits the data better than the quadratic model, given the smaller AIC and BIC.

After choosing the linear model for the mean trajectory, we respecified the model by switching the within-subject V–CV to the alternative structures in Table 11.2. The linear growth model of engagement in family routines can be written as follows:

Level 1:

$$FamilyRoutines_{ti} = \pi_{0i} + \pi_{1i}Time_{ti} + e_{ti}$$

Level 2:

$$\pi_{0i} = \beta_{00} + u_{0i} \quad \pi_{1i} = \beta_{10} + u_{1i} \tag{11.8}$$

Mixed Model:

$$FamilyRoutines_{ti} = \beta_{00} + \beta_{10}Time_{ti} + u_{0i} + u_{1i}Time_{ti} + e_{ti}$$

where $FamilyRoutines_{ti}$ is the family routine activity score for individual i at time t, $Time_{ti}$ is the linear growth term (T1 = 0, T2 = 1, T3 = 2, T3 = 3), e_{ti} is the

TABLE 11.5 Model Fit of Two Nested Growth Models: Linear and Quadratic Models

Model	AIC	BIC	–2LL
Linear model	35799.7	35854.4	35779.7
Quadratic model	35804.2	35858.8	35784.2

within-subject residual, β_{00} represents the grand mean of the family routine scores across all subjects at age 12, β_{10} is the annual rate of decrease in family routines across the four time points, u_{0i} is the deviation from the grand mean β_{00} at age 12 for individual *i*, and u_{1i} is the deviation from the grand slope β_{10} for individual *i*.

Variance Structure

As presented in Equation 11.4, there are two components of the variance structure in 2-level growth models: the between-subject V–CV (T) and the within-subject V–CV (Σ). In this example, we specified the saturated T structure with freely estimating all three components (i.e., τ_{00}, τ_{10}, and τ_{11}), which is the *standard* MLM specification. The residual variance and covariance of the time-specific outcomes are modeled by the five alternative within-subject V–CV structures: (a) identity, (b) first-order autoregressive, (c) Toeplitz, (d) banded main diagonal, and (e) unstructured. When the UN structure is employed for the level-1 error variance structure, the components in the between-subject V–CV structure are constrained to be null given the saturated level-1 error variance structure with zero degree of freedom (Kwok et al., 2007; Wu & West, 2010).

Results

Table 11.6 displays the results of the series of linear growth models with five different specifications in the V–CV structures. The SAS PROC MIXED output provides four model fit statistics values: the –2 residual log likelihood value (–2LL), AIC, AIC corrected for small sample size (AICC), and BIC. Note that AIC, AICC, and BIC are a function of –2LL; smaller values indicate a better model fit to the data. We use the AIC and BIC for the model comparison in this example. We also note that the more parsimonious (simpler) model is selected when the model fit statistics are reasonably comparable.

In the results, Model 4 with the UN(1) structure appears to be the best-fitting model: It has the smallest BIC (35845.4) among all five models. Although Model 5 shows the lowest AIC (35799.7), we follow the rule of parsimony for choosing Model 4 over Model 5. We use Model 5 as a diagnostic tool to observe the pattern of the total residual variance and covariance

TABLE 11.6 A Comparison of Five Models with Different Level-1 Error Variance Structures

	Model 1: ID		Model 2: AR(1)		Model 3: Toeplitz(2)		Model 4: UN(1)		Model 5: UN	
	Coeff.	*SE*	*Coeff.*	*SE*	*Coeff.*	*SE*	*Coeff.*	*SE*	*Coeff.*	*SE*
Level-2 Coefficients										
Average intercept, $\hat{\beta}_{00}$	9.87*	(0.10)	9.88*	(0.10)	9.87*	(0.10)	9.86*	(0.10)	9.87*	(0.10)
Average linear growth rate (Time), $\hat{\beta}_{10}$	–0.80*	(0.04)	–0.81*	(0.04)	–0.81*	(0.04)	–0.80*	(0.04)	–0.80*	(0.04)
Variance-Covariance Component										
Level 2	$\hat{\tau}_{00} = 11.34$ $\hat{\tau}_{11} = 0.95$ $\hat{\tau}_{10} = -1.44$	(0.63)* (0.11)* (0.21)*	$\hat{\tau}_{00} = 10.04$ $\hat{\tau}_{11} = 0.65$ $\hat{\tau}_{10} = -1.03$	(0.86)* (0.17)* (0.28)*	$\hat{\tau}_{00} = 10.33$ $\hat{\tau}_{11} = 0.70$ $\hat{\tau}_{10} = -1.10$	(0.75)* (0.15)* (0.25)*	$\hat{\tau}_{00} = 9.70$ $\hat{\tau}_{11} = 0.95$ $\hat{\tau}_{10} = -0.90$	(0.67)* (0.13)* (0.24)*	—	
Level 1	$\hat{\sigma}^2 = 8.81$	(0.23)*	$\hat{\sigma}^2 = 9.90$ $\hat{\rho} = 0.11$	(0.56)* (0.04)*	$\hat{\sigma}^2 = 9.65$ $\hat{\sigma}_1 = 0.85$	(0.42)* (0.33)*	$\hat{\sigma}_1^2 = 12.15$ $\hat{\sigma}_2^2 = 9.53$ $\hat{\sigma}_3^2 = 8.12$ $\hat{\sigma}_4^2 = 5.21$	(0.69)* (0.44)* (0.39)* (0.50)*	$\begin{bmatrix} 21.22 & & & \\ 8.80 & 19.22 & & \\ 7.23 & 9.99 & 18.21 & \\ 6.51 & 9.27 & 10.80 & 17.78 \end{bmatrix}$ *All elements are statistically significant.	
Model Fit										
AIC	35899.4		35894.4		35894.9		35807.2		35799.7	
AAIC	35899.4		35894.4		35894.9		35807.2		35799.7	
BIC	35921.4		35921.7		35922.2		35845.4		35854.4	
–2LL	35891.4		35884.4		35884.9		35793.2		35779.7	

* $p < .05$.

across repeated measures. The estimated parameters for the random effects show that there certainly is a decreasing pattern over time for the variances on the diagonal (ranging from 17.78 to 21.12) and a significant covariance on the off-diagonal. Since Model 5 assumes no particular pattern or structure for the variance and covariance, it often is referred to as the starting model to further examine the alternatives to better understand the data. On the other hand, Model 1 (standard MLM), which is the most popularly adopted model in practice shows poor fit compared to other models. Although the fixed effects are similar between the models, the time-specific variances in Model 4 clearly show that variances are not equal across repeated measures. We provide more details about the estimated parameters in the next section.

Fixed Effects

The fixed effects of interest in Model 4 with the UN(1) structure are the intercept coefficient ($\hat{\beta}_{00} = 9.86$, SE = 0.10) and linear-growth coefficient ($\hat{\beta}_{10} = -0.80$, SE = 0.04), which are both statistically significant ($p < .05$). On average, the participants scored 9.86 (out of 21) for doing housework, recreational activities, and religious activities with their family at age 12. This score decreases over time, with a rate of change of –.80 per year. In other words, the average adolescent engaged in about 10 activities per week with his or her family at age 12, and this was reduced by almost one activity per year. Regarding our first research question, the results suggest that as children move from early (age 12) to later adolescence (age 15), they tend to engage in fewer family routines.

Random Effects at Level 2

Table 11.6 presents the estimated parameters for the variance components for all five level-1 error structures. Given that we selected Model 4 with the UN(1) structure as the final model, our interpretations for the V-CV components are also based on Model 4. First, the intercept variance ($\hat{\tau}_{00}$) from the between-subject V–CV (T) structure is statistically significant ($\hat{\tau}_{00} = 9.70$, SE = 0.67; $p < .05$), meaning that there is a substantial amount of variation among subjects responding to the family routine activities at age 12. The 95% plausible value range for $\hat{\tau}_{00}$ of family routine scores at age 12 falls between 3.76 and 15.96 [i.e., $\hat{\beta}_{00} \pm 1.96(\hat{\tau}_{00})^{1/2} = 9.86 \pm 1.96(9.70)^{1/2}$]. Second, the slope variance is also statistically significant ($\hat{\tau}_{11} = 0.95$, SE = 0.13; $p < .05$), indicating that the linear growth trajectories of engagement in family routines across age 12 to 15 vary among adolescents. The 95% plausible value range for the growth slopes falls between –2.71 and 1.11 [i.e., $\hat{\beta}_{10} \pm 1.96(\hat{\tau}_{11})^{1/2} = -0.80 \pm 1.96(0.95)^{1/2}$], meaning some adolescents have decreasing trajectories of family routines scores, while others have increasing trajectories. Third, the covariance between intercepts and slopes ($\hat{\tau}_{10}$ or $\hat{\tau}_{01}$) is

–0.90 (SE = 0.24), which is statistically significant ($p < .05$), suggesting that adolescents with family routines scores at age 12 (intercepts) exhibit more negative family routine growth from age 12 to 15 (i.e., slopes).

Random Effects at Level 1

Next, the within-subject variance components are all statistically significant at $p < .05$, indicating that there is a significant amount of residual variance after accounting for between-subject variance and covariance. Under the UN(1) structure in Model 4, the four variance components on the diagonal of the Σ matrix are freely estimated, while the covariance components on the off-diagonal are constrained to be zero. Specifically, time-specific residual variance is the average squared deviation of the observed score from the predicted score for each individual across all subjects. As shown in Table 11.6, the overall pattern for the within-subject residual variance is gradually decreasing across four time points (ages 12 to 15), indicating that, in general, the unexplained variability of the family routine score is decreasing over time. This is evident in Figure 11.2, which contains the plots of 100 randomly selected individuals across four waves of family routine scores.

Recall that the fixed effect $\hat{\beta}_{10}$ suggests the general trajectory of family routine activities decreases across the ages of 12 to 15. The between-subject (level-2) random effects indicate significant variatiability in intercepts ($\hat{\tau}_{00}$) and slopes ($\hat{\tau}_{11}$) among adolescents. Further, the within-subject (level-1) random effects show that the level of participation in family routine activities tends to be less spread out between the ages of 12 and 15. This implies that youths behave more similarly to each other in terms of their involvement in family routines as they move from early to later adolescence. Given that family routines are a protective factor against adolescent substance use, health risks, and problem behavior, researchers might want to study the factors that are related to greater participation in family routines.

It is interesting to compare the results of Model 4, with the UN(1) structure, to the model with the simplest (ID) structure, Model 1. In Table 11.6, the single variance estimate on the diagonal of the within-subject V–CV structure for Model 1 is 8.81 (SE = 0.23; $p < .05$), which is constant across all four time points. In other words, it assumes that the standard deviation of the predicted score accounting for the time effect is around 2.97 (i.e., $\sqrt{\sigma^2} = \sqrt{8.81}$) and is equal across time. Model 1 does not capture the fact that at later time points the pattern of the responses is more homogeneous.

Conditional Model

In the following analysis, we further investigate whether the decreasing pattern of adolescents' involvement in family routines over time is associated with maternal monitoring of their child's friends, their school activities, and their whereabouts when not at home. In the initial model, we include the main effect of maternal monitoring as well as the interaction effect between the monitoring variable and the time variable to test whether the decreasing pattern of engagement in family routines is dependent on the level of maternal monitoring. The preliminary results show that there is no significant interaction effect between monitoring and time. Thus, we do not estimate the interaction between monitoring and time in the final conditional model. In addition, to facilitate the interpretation of the intercept (β_{00}) in the final conditional model, we center the maternal monitoring value at its grand mean (10.19), so that the intercept of the conditional model indicates the average engagement in family routines at age 12 (i.e., $Time_{ti} = 0$) for those who have an average level of maternal monitoring (i.e., centered $Monitoring_{ti} = 0$). Thus, our conditional model is specified as follows:

Level 1:

$$FamilyRoutines_{ti} = \pi_{0i} + \pi_{1i}Time_{ti} + (Monitoring_{ti} - \overline{Monitoring_{..}}) + e_{ti}$$

Level 2:

$$\pi_{0i} = \beta_{00} + u_{0i}, \quad \pi_{1i} = \beta_{10} + u_{1i}, \quad \pi_{2i} = \beta_{20} \tag{11.9}$$

Mixed model

$$FamilyRoutnes_{ti} = \beta_{00} + \beta_{10}Time_{ti} + \beta_{20} + (Monitoring_{ti} - \overline{Monitoring_{..}})$$
$$+ u_{0i} + u_{1i}Time_{ti} + e_{ti}$$

where β_{00} is the grand mean of the family routine scores across all subjects at age 12 for those who have average maternal monitoring, β_{10} is the annual rate of change in family routines across the four years holding maternal monitoring constant, β_{20} is the effect of maternal monitoring on family routines holding the time effect constant, e_{ti} is the within-subject V–CV with a freely estimated variance and zero covariance [UN(1)], u_{0i} is the deviation from the grand mean β_{00} for individual i, and u_{1i} is the deviation from the grand slope β_{10} for individual i.

Table 11.7 presents the results of the final conditional model—a linear growth model with a time-varying predictor of maternal monitoring and a level-1 error variance structure equal to UN(1). Corresponding to our questions, we first interpret the growth slope after controlling for the effect of maternal monitoring. The results show that the conditional effect of time is statistically significant ($\hat{\beta}_{10} = -0.73$, SE = 0.04), indicating that the youth's family routine score decreases by .72 per year from ages 12 to 15, holding

TABLE 11.7 Conditional Models with a UN(1) or ID Level-1 Error Variance Structure

Model parameters	Conditional Model With UN(1) structure		Conditional Model With ID structure	
	Coeff.	*SE*	*Coeff.*	*SE*
Fixed Effect				
Intercept, $\hat{\beta}_{00}$	9.74*	(0.10)	9.74*	(0.10)
Linear growth rate (Time), $\hat{\beta}_{10}$	–0.73*	(0.04)	–0.73*	(0.04)
Mother monitoring, $\hat{\beta}_{20}$	0.29*	(0.02)	0.29*	(0.02)
Variance-Covariance Component				
Level 2	$\hat{\tau}_{00}$ = 8.40	(0.64)*	$\hat{\tau}_{00}$ = 10.31	(0.60)*
	$\hat{\tau}_{11}$ = 0.70	(0.12)*	$\hat{\tau}_{11}$ = 0.75	(0.11)*
	$\hat{\tau}_{10}$ = –0.55	(0.22)*	$\hat{\tau}_{10}$ = –1.21	(0.20)*
Level 1	$\hat{\sigma}_1^2$ = 11.91	(0.66)*		
	$\hat{\sigma}_2^2$ = 8.88	(0.41)*		
	$\hat{\sigma}_3^2$ = 7.18	(0.37)*		
	$\hat{\sigma}_4^2$ = 4.45	(0.47)*	$\hat{\sigma}^2$ = 8.27	(0.22)*
Model Fit				
AIC	32419.7		32536.1	
AAIC	32419.7		32536.1	
BIC	32457.9		32557.9	
–2LL	32405.7		32528.1	

* $p < .05$.

the maternal monitoring constant across time. Compared to the results of Model 4 in Table 11.6, the magnitude of $\hat{\beta}_{10}$ in the conditional model (–0.73) is smaller than in the unconditional model (–0.80). After taking the effect of maternal monitoring into account, the decrease in family routines over time is less intense (less negative).

We also examine the impact of the maternal monitoring. Maternal monitoring statistically significantly positively affects the youths' engagement in family routines ($\hat{\beta}_{20}$ = 0.29, SE = 0.02). This suggests that adolescents who are more willing to share information about their friends and school with their mother tend to be more involved in family routine activities.

The statistically significant effect of maternal monitoring on family routines also affects the estimated parameters for the level-1 variance components. As shown in Tables 11.6 and 11.7, the estimates on the within-subject V–CV structure are updated from the unconditional linear growth model (Model 4 in Table 11.6) to the conditional growth model (left side of Table 11.7):

$$\underset{\text{Unconditional model}}{\begin{bmatrix} 12.15 & & & \\ 0 & 9.53 & & \\ 0 & 0 & 8.12 & \\ 0 & 0 & 0 & 5.21 \end{bmatrix}} \rightarrow \underset{\text{Conditional model}}{\begin{bmatrix} 11.91 & & & \\ 0 & 8.88 & & \\ 0 & 0 & 7.18 & \\ 0 & 0 & 0 & 4.45 \end{bmatrix}}.$$

The residual variances at all four time points are reduced when the effect of time-specific maternal monitoring is taken into account. For example, the residual variance of the initial response at time 1 ($\sigma^2_{1\text{uncond}} = 12.15$) is reduced to 11.91, meaning that 2% of the variance is explained by the maternal monitoring at age 12 ($\frac{12.15-11.91}{12.15} = 0.02$). At age 15 (time 4), about 15% of the residual variance is explained by the maternal monitoring (from $\sigma^2_{4\text{uncond}} = 5.21$ to $\sigma^2_{4\text{cond}} = 4.45$), which might have important implications for educational practice.

We also notice the changes in the between-subject variance components where all three components ($\hat{\tau}_{00}$, $\hat{\tau}_{11}$, and $\hat{\tau}_{10}$) are reduced after maternal monitoring is included in the model:

$$\underset{\substack{\text{Unconditional model}\\ \text{(Model 4 in Table 11.6)}}}{\begin{bmatrix} 9.70 & \\ -0.90 & 0.95 \end{bmatrix}} \rightarrow \underset{\substack{\text{Conditional model}\\ \text{(left side in Table 11.7)}}}{\begin{bmatrix} 8.40 & \\ -0.55 & 0.70 \end{bmatrix}}.$$

It turns out that the maternal monitoring also explains the between-person variation in intercept ($\hat{\tau}_{00}$), slope ($\hat{\tau}_{11}$), and covariance between intercept and slope ($\hat{\tau}_{10}$).

In Table 11.7, we again compare the model with the UN(1) structure and the model with the simplest ID structure to clearly illustrate the additional information provided by alternative within-subject V–CV structure. As shown on the right side of Table 11.7, when the simplest ID structure is used, the estimated variance of e_{ti} (σ^2) is 8.27, meaning that the average deviation from the predicted scores to the observed scores is about 2.88 (i.e., $\sqrt{\sigma^2} = \sqrt{8.27}$) across all four time points, after partialling out the variability due to the subjects. When adopting the UN(1) structure, we discover there is actually less within-person variability in the youth's responses at age 15 (i.e., $\sqrt{\sigma^2} = \sqrt{4.45} = 2.11$) compared to age 12 (i.e., $\sqrt{\sigma^2} = \sqrt{11.91} = 3.45$).

CONCLUDING REMARKS

As discussed earlier, misspecifying the structure of level-1 error (co)variances can result in inaccurate and inefficient inferences of fixed effects as well as biased variance parameters (Ferron et al., 2002; Kwok et al., 2007). More importantly, the within-subject V–CV contains unique and valuable information for longitudinal data: the change in the variation of the average trend over time. From this perspective, determining the form of the level-1 error (co)variance structure that best fits the model given multiwave longitudinal data should not be overlooked. We encourage practitioners to consider an intermediary form for the level-1 error (co)variance structure that is between extremely restricted (e.g., CS) and extremely liberal structures (Wolfinger, 1996). In practice, overly parsimonious structures (such as the ID structure) are often contradictory to the features of multiwave longitudinal data and are therefore usually not favored by the AIC or BIC (as in our illustrative example). On the other hand, adopting an overly liberal structure is not also a good direction because this requires (exponentially) increasing the number of parameters that need to be estimated, resulting in inefficient inferences about the fixed effects (Liu et al., 2012; Wolfinger, 1996). Therefore, selecting an intermediary form for the level-1 error (co)variance structure is recommended as a useful strategy. We encourage practitioners to take advantage of the features of level-1 error (co) variance modeling offered by statistical packages and compare the model fit of several intermediary structures (e.g., AR(1), TOEP, and UN(1)).

How to decide on an optimal structure among many intermediary structures is still challenging for practitioners. Currently, the AIC and BIC are probably the most frequently used indices for selecting the best-fitting V–CV structure, and they are provided in many MLM statistical packages. However, prior research has raised concerns that these two indices may fail to select the best structure under certain conditions. For example, in their simulation study, Keselman et al. (1998) found that the percentages of correct identification of four time-point level-1 error (co)variance structures by the AIC and BIC were 47% and 35%, respectively. In a more recent study, Liu et al. (2012) provided a cautionary note regarding using the AIC and BIC to select the covariance structure because the estimation of the growth trajectory (fixed effects) is confounded with the covariance structure. They also noted that the AIC and BIC may not perform adequately when the sample size is small (e.g., less than 20 individuals in their simulation study). For this reason, we encourage practitioners to use the AIC and BIC thoughtfully. For instance, Ferron et al. (2002) found that, in general, the performance of the AIC and BIC can be improved by increasing series lengths (the number of time points) or sample sizes. In other words, practitioners should be aware that both indices are less informative about the structure of level-1 error (co)variances when the longitudinal data have a short series length or a small sample size.

Practitioners may feel trapped when two or more intermediary structures have similar AIC and BIC values. In this scenario, should the most parsimonious model be adopted? The answer to this question depends on what researchers are most interested in. Based on Ferron et al.'s (2002) study, we suggest that if the estimates or tests of the fixed effects are the main focus, researchers can select a more parsimonious structure, but if the estimates of the random effects are as important as the growth shape, researchers might want to select a less parsimonious structure to gain further understanding of the variation in the average growth trend. Liu et al. (2012) suggest selecting the more parsimonious model if it is not substantially worse than the more complex model. Chapter 3 of this volume, "Evaluation of Model Fit and Adequacy," provides additional details on model fit and model selection.

In summary, specifying the optimal within-subject V–CV model can be a challenging task for the substantive researchers who use MLM for analyzing longitudinal data. Nevertheless, testing and comparing the model with alternative level-1 error variance structures can be useful as a diagnostic tool or for guidance to better understand their data and should not be overlooked when modeling changes over time.

Recommended Readings

Kim, M., Kwok, O.-M., Yoon, M., Willson, V., & Lai, M. H. (2016). Specification search for identifying the correct mean trajectory in polynomial latent growth models. *The Journal of Experimental Education, 84*(2), 307–329.

Kwok, O.-M., Underhill, A. T., Berry, J. W., Luo, W., Elliott, T. R., & Yoon, M. (2008). Analyzing longitudinal data with multilevel models: An example with individuals living with lower extremity intra-articular fractures. *Rehabilitation Psychology, 53*, 370–386.

Liu, S., Rovine, M. J., & Molenaar, P. C. (2012). Selecting a linear mixed model for longitudinal data: Repeated measures analysis of variance, covariance pattern model, and growth curve approaches. Psychological Methods, *17*(1), 15–30.

TRY THIS!

For this exercise, we created a subset dataset named "NLSY97_exe" by drawing approximately 50% of samples from the analytical NLSY97 data for this chapter's illustrative example. The dataset comprises of individual id (*id*), score of family routines index (*fam_rou*), time variable (*time*), score of adolescents' substance use index (*substance*) as well as substance use score centered to its grand mean (*substance_c*). Note *substance* has scores ranging from 0 to 3, where higher scores indicate more instances of substance use (e.g., smoke a cigarette, drink an alcoholic beverage, use marijuana).

1. Fit the data to a series of linear growth models (see Equation 11.17) with five different specifications of alternative level-1 error variance structures, including identity (ID), first-order autoregressive structure [AR(1)], Toeplitz structure 2 (TOEP2), banded main diagonal structure [UN(1)], and unstructured structure (UN).
2. Compare model fit statistics values (e.g., BIC, AIC) across five growth models and determine the best-fitting model.
3. Interpret the parameter estimates in the best-fitting model, including fixed effects ($\hat{\beta}_{00}$ and $\hat{\beta}_{10}$), random effects at level 2 ($\hat{\tau}_{00}$, $\hat{\tau}_{11}$, and $\hat{\tau}_{10}$), and random effect at level 1.
4. Further investigate whether the adolescents' involvement in family routines is associated with their substance use at different time points. To do so, include the substance use variable (*substance_c*) as a predictor in the best-fitting model (see Equation 11.9).
5. Interpret the parameter estimate of adolescents' substance use score and compare the fixed effects and random effects between the model with and without the substance variable.

APPENDIX
Example syntax code for SAS PROC MIXED

```
proc mixed data=NLSY covtest noclprint noinfo noitprint method=reml;
  title3 'Unconditional linear growth model with Identity';
  class id ;
  model fam_rou=time / solution notest ddfm=kr;
  random intercept time/ type=UN subject=id;
  repeated / type=VC subject=ID r;
run;

proc mixed data=NLSY covtest noclprint noinfo noitprint method=reml;
  title3 'Unconditional linear growth model with AR(1)';
  class id ;
  model fam_rou=time / solution notest ddfm=kr;
  random intercept time/ type=UN subject=id;
  repeated / type=AR(1) subject=ID r;
run;

proc mixed data=NLSY covtest noclprint noinfo noitprint method=reml;
  title3 'Unconditional linear growth model with TOEP(2)';
  class id ;
  model fam_rou=time / solution notest ddfm=kr;
  random intercept time/ type=UN subject=id;
  repeated / type=TOEP(2) subject=ID r;
run;

proc mixed data=NLSY covtest noclprint noinfo noitprint method=reml;
  title3 'Unconditional linear growth model with UN(1)';
  class id ;
  model fam_rou=time / solution notest ddfm=kr;
  random intercept time/ type=UN subject=id;
  repeated / type=UN(1) subject=ID r;
run;

proc mixed data=NLSY covtest noclprint noinfo noitprint method=reml;
  title3 'Unconditional linear growth model with UN';
  class id ;
  model fam_rou=time / solution notest ddfm=kr;
  repeated / type=UN subject=ID r;
run;
```

NOTES

1. A list of the studies included in this review is available upon request. In one study, authors allowed the within-subject variance to be group-specific. In another study, the autoregressive residual [AR(1)] was adopted for the within-subject V–CV structure.
2. UN(1) is the default structure for the growth modeling embedded in M*plus* 8.2, which is a popular SEM software package.

REFERENCES

Day, R. D., Gavazzi, S. M., Miller, R., & van Langeveld, A. (2009). Compelling Family Processes. *Marriage & Family Review, 45*(2–3), 116–128. https://doi.org/10.1080/01494920902733260

Ferron, J., Dailey, R., & Yi, Q. (2002). Effects of misspecifying the first-level error structure in two-level models of change. *Multivariate Behavioral Research, 37*(3), 379–403.

Fiese, B. H., Tomcho, T. J., Douglas, M., Josephs, K., Poltrock, S., & Baker, T. (2002). A review of 50 years of research on naturally occurring family routines and rituals: Cause for celebration? *Journal of Family Psychology, 16*(4), 381–390. https://doi.org/10.1037//0893-3200.16.4.381

Jones-Sanpei, H. A., Day, R. D., & Holmes, E. K. (2009). Core family process measures in the NLSY97: Variation by gender, race, income, and family structure. *Marriage & Family Review, 45*(2–3), 140–167. https://doi.org/10.1080/01494920902733468

Jones-Sanpei, H. A., Day, R. D., Holmes, E., & van Langeveld, A. (2009). Family process variables in the NLSY97: A primer. *Marriage & Family Review, 45*(2–3), 129–139.

Keselman, H. J., Algina, J., Kowalchuk, R. K., & Wolfinger, R. D. (1998). A comparison of two approaches for selecting covariance structures in the analysis of repeated measurements. *Communications in statistics: Simulation, 27,* 591–604.

Kim, M., Kwok, O.-M., Yoon, M., Willson, V., & Lai, M. H. (2016). Specification search for identifying the correct mean trajectory in polynomial latent growth models. *The Journal of Experimental Education, 84*(2), 307–329.

Kline, R. B. (2015). *Principles and practice of structural equation modeling* (4th ed.). The Guilford Press.

Kwok, O.-M., West, S. G., & Green, S. B. (2007). The impact of misspecifying the within-subject covariance structure in multiwave longitudinal multilevel models: A Monte Carlo study. *Multivariate Behavioral Research, 42*(3), 557–592.

Laird, N. M., & Ware, J. H. (1982). Random-effects models for longitudinal data. *Biometrics,* 963–974.

Liu, S., Rovine, M. J., & Molenaar, P. C. (2012). Selecting a linear mixed model for longitudinal data: Repeated measures analysis of variance, covariance pattern model, and growth curve approaches. *Psychological Methods, 17*(1), 15–30.

Manlove, J., Wildsmith, E., Ikramullah, E., Terry-Humen, E., & Schelar, E. (2012). Family environments and the relationship context of first adolescent sex:

Correlates of first sex in a casual versus steady relationship. *Social Science Research, 41*(4), 861–875. https://doi.org/10.1016/j.ssresearch.2012.02.003

Raudenbush, S. W., & Bryk, A. S. (2002). *Hierarchical linear models: Applications and data analysis methods* (Vol. 1). SAGE.

Sclove, S. L. (1987). Application of model-selection criteria to some problems in multivariate analysis. *Psychometrika, 52*(3), 333–343. https://doi.org/10.1007/BF02294360

Singer, J. D. (1998). Using SAS PROC MIXED to fit multilevel models, hierarchical models, and individual growth models. *Journal of Educational and Behavioral Statistics, 24(4)*, 323–355.

Singer, J. D., & Willett, J. B. (2003). *Applied longitudinal data analysis: Modeling change and event occurrence.* Oxford University Press.

Taylor, R. D., & Lopez, E. I. (2005). Family management practice, school achievement, and problem behavior in African American adolescents: Mediating processes. *Journal of Applied Developmental Psychology, 26*(1), 39–49. https://doi.org/10.1016/j.appdev.2004.10.003

Weiss, R. E. (2005). *Modeling longitudinal data.* Springer Science & Business Media.

Wolfinger, R. (1996). Heterogeneous variance: Covariance structures for repeated measures. *Journal of Agricultural, Biological, and Environmental Statistics, 1,* 205–230.

Wu, W., & West, S. G. (2010). Sensitivity of fit indices to misspecification in growth curve models. *Multivariate Behavioral Research, 45*(3), 420–452.

CHAPTER 12

MODELING VARIATION IN INTENSIVE LONGITUDINAL DATA

Donald Hedeker
The University of Chicago

Robin J. Mermelstein
University of Illinois at Chicago

Modern data collection procedures, such as ecological momentary assessments (EMA) (Shiffman et al., 2008; Smyth & Stone, 2003; Stone & Shiffman, 1994), experience sampling (de Vries, 1992; Feldman Barrett & Barrett, 2001; Scollon et al., 2003), and diary methods (Bolger et al., 2003), have been developed to record the momentary events and experiences of subjects in daily life. These procedures yield relatively large numbers of subjects and observations per subject, and data from these designs are sometimes referred to as intensive longitudinal data (Walls & Schafer, 2006). Such designs follow the "bursts of measurement" approach described by Nesselroade and McCollam (2000), who called for such an approach in order to assess intra-individual variability. In this approach, a large number of

Multilevel Modeling Methods with Introductory and Advanced Applications, pages 421–456

measurements are obtained over a relatively short time span (e.g., a week). As noted by Nesselroade and McCollam (2000), this increases the research burden in several ways; however, it is important for studying intra-individual variation and to explain why subjects differ in variability rather than solely in mean level (Bolger et al., 2003). In this chapter, we describe data from an EMA study of adolescents, where interest was on determinants of the variation in the adolescents' moods.

In mental health research, EMA procedures have been used in studying pediatric affective disorders (Axelson et al., 2003), eating disorders (Boseck et al., 2007; le Grange et al., 2002), drug abuse (Epstein et al., 2009), schizophrenia (Granholm et al., 2008; Kimhy et al., 2006), borderline personality disorder (Trull et al., 2008), stress and anxiety (de Vries et al., 2001; Yoshiuchi et al., 2008), and sexual abuse (Simonich et al., 2004). Similarly, in smoking research, EMA studies include those studying relapse in people who are quitting smoking (Shiffman, 2005), relapse among adolescent smokers (Gwaltney et al., 2008), examining the urge to smoke (O'Connell et al., 1998), and our own EMA studies on adolescents (Mermelstein et al., 2002, 2007). Recently, a number of review articles on EMA studies have been published in several diverse research areas that indicate the wide range of studies using EMA methods (aan het Rot et al., 2012; Armey et al., 2015; Heron et al., 2017; Liao et al., 2016; May et al., 2018; Rodriguez-Blanco et al., 2018; Serre et al., 2015; Walz et al., 2014; Wen et al., 2017).

Data from EMA studies are inherently multilevel with, for example, (level-1) observations nested within (level-2) subjects. Thus, linear mixed models (LMMs, aka multilevel or hierarchical linear models) are often used for EMA data analysis, and several books and/or book chapters describe mixed model analysis of EMA data (Bolger & Laurenceau, 2013; Schwartz & Stone, 2007; Walls & Schafer, 2006). A basic characteristic of these models is the inclusion of random subject effects into regression models in order to account for the influence of subjects on their repeated observations. The variance of these random effects indicates the degree of variation that exists in the population of subjects, or the between-subjects variance. Analogously, the error variance characterizes how much variation exists within a subject, or the within-subjects variance. These variances are usually treated as being homogeneous across subject groups or levels of covariates.

In EMA studies, it is common to have up to 30 or 40 observations per subject, and this allows greater modeling opportunities than what conventional LMMs allow. In particular, one very promising extended approach is the modeling of both between-subject (BS) and within-subject (WS) variances as a function of covariates, in addition to their effect on overall mean levels. For example, if a person's mood is the outcome, then one can consider the effect of covariates on their mood level (e.g., how happy/sad they are on average), how similar/different they are in their mood levels to others (e.g., how homogeneous/heterogeneous are particular groups of subjects), as well as on their own variation in mood (e.g., how consistent/erratic their mood is).

Momentary mood may be influenced by both stable trait factors and situational or momentary influences and contexts. A persistent debate among researchers interested in personality and psychopathology, for example, has been whether mood variability is a more stable trait or a more situationally specific state; parsing out the between-subject and within-subject variances helps to better address this research question. Of interest to researchers has been whether mood variability is related to a host of standard personality traits (e.g., introversion, extraversion; e.g., Hepburn & Eysenck [1989]) or how much it may be influenced by being with others, such as contagion effects on mood (e.g., Neumann & Strack [2000]). Examining whether these influences on mood are more personality (e.g., extraversion) or situational (e.g., influenced by others) becomes possible by examining the effects of specific covariates on the BS and WS variances.

Expanding on the work of Cleveland et al. (2000), Hedeker et al. (2008) describe an extended LMM for variance modeling of EMA data, dubbed the mixed-effects location scale (MELS) model. Like all LMMs, this model allows covariates and a random subject effect to influence the mean response of a subject. However, this model also includes a log-linear structure for both the WS and BS variance, allowing covariates to influence both sources of variation. Finally, a random subject effect is included in the WS variance specification. This permits the WS variance to vary at the subject level, above and beyond the influence of covariates on this variance. In this chapter we more fully describe the MELS model, and show how it can be used to model changes in mood levels and mood variation as a function of covariates. We will also describe some of the software programs that can be used to estimate the parameters of the MELS model towards the end of the chapter. A word of caution about the notation used in this chapter should be made. Because of the modeling of the variances, the notation of the MELS model is perhaps more involved than standard multilevel and/or hierarchical linear models (HLMs). For example, it is customary to use τs to represent variances in HLMs, however in what follows τs will be used to denote fixed-effects in the WS variance submodel. Similarly, we will use αs to represent fixed-effects in the BS variance submodel. We apologize for any confusion that our choices for notation creates.

MELS MODEL

Consider the following mixed-effects regression model (aka hierarchical linear or multilevel model) for the measurement y of subject i ($i = 1,2,\ldots,N$ subjects) on occasion j ($j = 1,2,\ldots,n_i$ occasions):

$$y_{ij} = \boldsymbol{x}'_{ij}\boldsymbol{\beta} + \upsilon_i + \varepsilon_{ij}, \tag{12.1}$$

where $\boldsymbol{x}_{ij}$ is the $p \times 1$ vector of regressors (typically including a "1" for the intercept as the first element) and $\boldsymbol{\beta}$ is the corresponding $p \times 1$ vector of regression coefficients. The regressors can either be at the subject level, vary across occasions, or be interactions of subject-level and occasion-level variables. In the multilevel terminology, subjects are at level 2, while the repeated observations are at level 1. Thus, the level-2 random subject effect υ_i indicates the influence of individual i on his/her repeated level-1 measurements. The population distribution of these random effects is assumed to be a normal distribution with zero mean and variance σ_υ^2. The errors ε_{ij} are also assumed to be normally distributed in the population with zero mean and variance, σ_ε^2 and independent of the random effects. Here, σ_υ^2 represents the BS variance and σ_ε^2 is the WS variance.

For simplicity, consider the null model with no covariates, namely, $y_{ij} = \beta_0 + \upsilon_i + \varepsilon_{ij}$. Here, υ_i represents a subject's mean deviation from the population intercept β_0, the latter representing the population mean of the outcome variable in this model with no covariates. A subject's mean is therefore $\beta_0 + \upsilon_i$. If subjects are very similar to each other, then $\upsilon_i \approx 0$ and σ_υ^2 will approach 0. Conversely, as subjects differ, $\upsilon_i \neq 0$ and σ_υ^2 will increase from 0. Thus, the magnitude of the BS variance σ_υ^2 indicates how different subjects are from each other in terms of their means. We refer to this as the degree of homogeneity/heterogeneity across subjects.

Analogously, ε_{ij} is subject i's error at time j, which represent deviations from their mean. If the observations from all subjects are all close to their means, $\varepsilon_{ij} \approx 0$ and σ_ε^2 will approach 0. Alternatively, as the observations from subjects deviate from their means, $\varepsilon_{ij} \neq 0$ and σ_ε^2 will increase from 0. The magnitude of the WS variance σ_ε^2 indicates how data vary within subjects, which we refer to as the degree of consistency/erraticism within subjects.

To allow covariates (i.e., regressors) to influence the BS and WS variances, we can utilize a log-linear representation, as has been described in the context of heteroscedastic (fixed-effects) regression models (Aitkin, 1987; Harvey, 1976); namely,

$$\sigma_{\upsilon_{ij}}^2 = \exp(\boldsymbol{u}'_{ij}\boldsymbol{\alpha}), \tag{12.2}$$

$$\sigma^2_{\varepsilon_{ij}} = \exp(\omega'_{ij}\tau). \tag{12.3}$$

The variances are subscripted by i and j to indicate that their values change depending on the values of the covariates $\boldsymbol{u}_{ij}$ and $\boldsymbol{\omega}_{ij}$ (and their coefficients). Both $\boldsymbol{u}_{ij}$ and $\boldsymbol{\omega}_{ij}$ would usually include a (first) column of ones for the reference BS and WS variances (α_0 and τ_0), respectively. Thus, the BS variance equals exp α_0 when the covariates $\boldsymbol{u}_{ij}$ equal 0, and is increased or decreased as a function of these covariates and their coefficients $\boldsymbol{\alpha}$. Specifically, for a particular covariate u^*, if $\alpha^* > 0$, then the BS variance increases as u^* increases (and vice versa if $\alpha^* < 0$). Note that the exponential function ensures a positive multiplicative factor for any finite value of α, and so the resulting variance is guaranteed to be positive. The WS variance is modeled in the same way. The coefficients in $\boldsymbol{\alpha}$ and $\boldsymbol{\tau}$ indicate the degree of influence on the BS and WS variances, respectively, and the ordinary random intercept model is obtained as a special case if $\boldsymbol{\alpha} = \boldsymbol{\tau} = 0$ for all covariates in $\boldsymbol{u}_{ij}$ and $\boldsymbol{\omega}_{ij}$ (i.e., excluding the reference variances α_0 and τ_0).

We can further allow the WS variance to vary across subjects, above and beyond the contribution of covariates; namely,

$$\sigma^2_{\varepsilon_{ij}} = \exp(\omega'_{ij}\tau + \omega_i), \tag{12.4}$$

where the random subject (scale) effects ω_i are distributed in the population of subjects with mean 0 and variance σ^2_ω. The idea for this is akin to the inclusion of the random (location) effect in Equation 12.1; namely, covariates do not account for all of the reasons that subjects differ from each other. In this model, υ_i is a random effect which characterizes a subject's mean, or location, and ω_i is a random effect which characterizes a subject's variance, or (square of the) scale. These two random effects are correlated with covariance parameter $\sigma_{\upsilon\omega}$, which indicates the degree to which the random location and scale effects are associated with each other.

A BETTER UNDERSTANDING OF THE MELS MODEL PARAMETERS

Here, we try to provide a more concrete description of the parameters of the mean and variance submodels with some simple illustrations. Consider, first, the modeling of the BS variance in Equation 12.2. It might seem odd that the BS variance can change depending on a time-varying variable, say u_{ij}. In fact, in Hedeker et al. (2008) we did not consider this possibility. However, this is clearly possible as we will now explain. Suppose that at each prompt a subject indicates whether they are alone or with others. This variable, denoted as `Others`, is therefore a time-varying (level-1) variable which could be coded 0 when a subject is alone and 1 if the subject is with others for that prompt. Suppose that the BS variance is decreased when subjects are with others. In other words, subjects are more heterogeneous in terms of their mood responses when those mood responses are obtained while they are alone, and less heterogeneous when their mood responses are obtained while they are with others. Figure 12.1 depicts such a situation.

In Figure 12.1, the "alone" responses are grouped together on the left-hand side and the "with others" responses are on the right-hand side. Consider the solid lines first, which represent the mean and the effect on the mean attributable to being with others. Notice that the mean level is lower for alone (approximately 0) than with others (approximately 1). Thus, on average, being with others raises the mean level of mood by 1 unit. To represent subject heterogeneity, the figure presents the hypothetical data of two subjects, one below and one above the mean levels. These subject lines are depicted as dotted lines, and, as mentioned, their alone/with others responses are separated and grouped together on the left/right-hand side of the figure. For simplicity, only two subjects are presented, but for a real dataset there would be as many subject lines as there are subjects in the sample. Also, for each subject, the dots represent the prompt-level (level-1) responses. In Figure 12.1, each subject has many responses when alone and many responses with others, however in a real dataset the numbers of

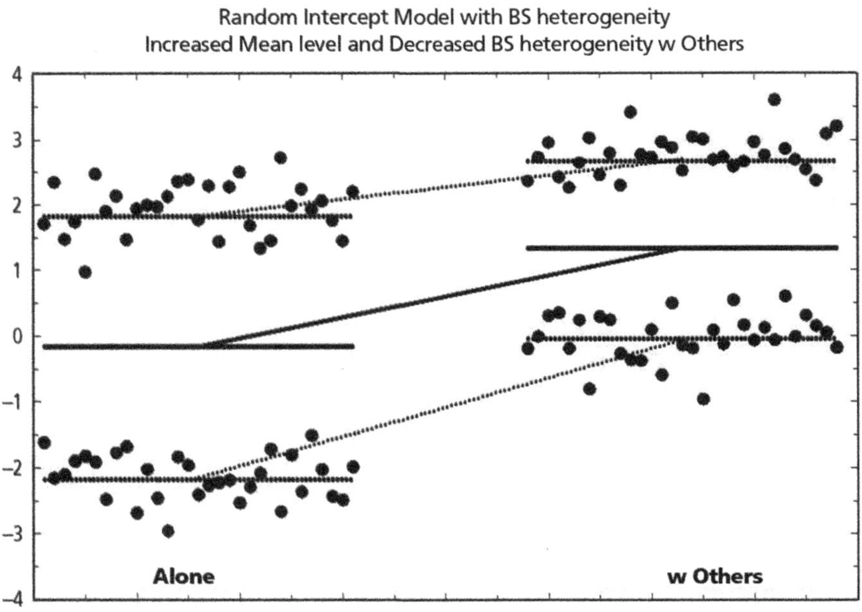

Figure 12.1 Visual representation of the mean and BS variance submodels.

observations could vary considerably, both across subjects and the alone/with others situations. Each subject's dotted lines represent their mean mood level when alone and with others, respectively. Thus, the subject in the upper part of the figure (above the mean) has higher relative mood levels than the subject in the lower part of the figure (below the mean), both when alone and with others. Notice that the distance between the dotted lines of the two subjects is greater for the alone responses (a difference of about 4 units) than when the two subjects are with others (a difference of about 2.5 units). In other words, subjects are more similar to each other in their mood responses when those mood responses are obtained with others, relative to when they are alone. Or, the level of subject heterogeneity varies across the different values of the Others variable. This is precisely how a level-1 variable (Others) can influence the level-2 BS variance.

In terms of a model of the BS variance, we might posit

$$\sigma^2_{\upsilon_{ij}} = \exp(\alpha_0 + \alpha_1 \texttt{Others}_{ij}), \tag{12.5}$$

and according to our figure, $\alpha_1 < 0$, since the BS variance is reduced for the "with others" responses ("with others" responses coded as 1, relative to the "alone" responses coded as 0). Notice that even though the α coefficients can be negative, the exponential function ensures that the resulting variance is a positive value. Also, for simplicity, here we simply have a single (level-1) covariate in our BS variance model, but more generally the model could include level-2 variables (e.g., a subject's gender), other level-1 variables (e.g., day of the prompt), and cross-level interactions.

As depicted in Figure 12.1, the WS (level-1) variance σ^2_ε is constant. Thus, the dispersion of the dots around the subjects' mean levels is the same across subjects and both levels of the `Others` variable. This implies that subjects are equally consistent in their mood reports, and likewise that the mood responses obtained when alone or with others are also equally consistent. Suppose, however, that the mood responses obtained when alone vs with others are not equally consistent. This would imply that the level-1 WS variance could be influenced by covariates, say,

$$\sigma^2_{\varepsilon_{ij}} = \exp(\tau_0 + \tau_1 \texttt{Others}_{ij}), \tag{12.6}$$

in a similar manner as the BS variance model. More generally, the WS variance could be influenced by covariates at level-1, level-2, or cross-level interactions. Additionally, subjects themselves could vary in their response consistency/erraticism, over and above the effects of covariates. Certainly, in terms of mood, there is likely to be a large unique subject component to the consistency/erraticism of their responses. To extend the WS variance model, consider

$$\sigma^2_{\varepsilon_{ij}} = \exp(\tau_0 + \tau_1 \texttt{Others}_{ij} + \omega_i) \qquad (12.7)$$

where ω_i is the random effect of subject i on the WS (level-1) variance. More consistent subjects would have negative values of ω_i, while more erratic subjects would have positive ω_i values. Similar to the random subject effects υ_i on the mean (or location), these random subject effects ω_i on variance (or scale) are assumed to be normally distributed in the population of subjects with mean 0 and variance σ^2_ω.

Figure 12.1, with constant error variance, would assume that $\tau_1 = \sigma^2_\omega = 0$. Namely, there is no effect of `Others` on the dispersion of points, and all subjects have the same degree of dispersion (i.e., no subject heterogeneity in the WS variance). Alternatively, Figure 12.2 illustrates the effects of `Others` and subjects on the dispersion of points (i.e., the WS variance). Notice that the subject in the upper part of the plot has more dispersed points than the subject in the lower part of the plot, for both "alone" and "with others" observations. Thus, the top subject is more erratic and would have a larger value of ω_i than the bottom subject. Also, for each subject, the dispersion of points is greater when subjects are alone than with others, which would imply that $\tau_1 < 0$. To summarize the effects of the level-1 Others variable that is illustrated in Figure 12.2: it raises the mean level ($\beta_1 > 0$), while reducing

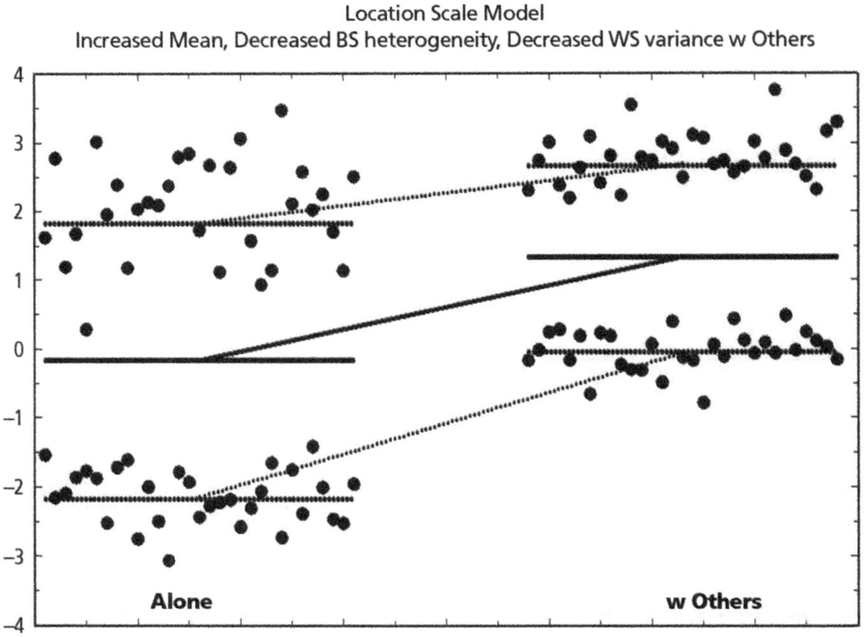

Figure 12.2 Visual representation of the mean, BW and WS variance submodels.

both the level-1 and level-2 variances ($\alpha_1 < 0$ and $\tau_1 < 0$, respectively). Thus, when subjects are with others, their mood levels are elevated, more similar to each other, and more consistent.

A final parameter in the MELS model is the association or covariance between the random location and scale parameters $\sigma_{\upsilon\omega}$. Based on Figure 12.2, the subject in the top part has both a higher mean level and a greater dispersion of points, than the subject in the lower part of the figure. This would imply a positive covariation between these two random effects, or namely, $\sigma_{\upsilon\omega} > 0$.

Further Interpretation of Variance Effects

Suppose now that the BS variance submodel includes a subject-level (level-2) variable $\texttt{Male}_i$ (coded 0 for females and 1 for males), in addition to the time-varying (level-1) variable $\texttt{Others}_{ij}$ (coded 0 for alone and 1 for with others).

$$\sigma^2_{\upsilon_{ij}} = \exp(\alpha_0 + \alpha_1 \texttt{Male}_i + \alpha_2 \texttt{Others}_{ij}) \tag{12.8}$$

Holding the variable `Others` constant, the BS variance equals $\exp(\alpha_0)$ for females, and $\exp(\alpha_0 + \alpha_1)$ for males. The ratio of the BS variances (male to female) equals $\exp(\alpha_0 + \alpha_1)/\exp(\alpha_0) = \exp(\alpha_1)$. Thus, exponentiating the slope parameters in the BS variance model yields variance ratio interpretations: the ratio of the BS variances for a unit change in the regressor. This is akin to the interpretation in a Poisson regression model (see Chapter 8 for example), which also has a log link function, in which the exponentiated slopes have rate ratio interpretations. Similarly, the exponentiated slopes in the WS variance submodel represent ratios of the WS variances for a unit change of the regressor. Thus, for the variance submodels, reporting of the variance ratios (exp α for BS variance, and exp τ for WS variance), along with 95% confidence intervals, can be useful. Notice that a variance ratio of 1 would imply no effect of the variable on the variance, so if the 95% confidence interval includes 1 then the variable does not have a statistically significant effect at the 0.05 level.

Coding of Regressors

The mean, BS variance, and WS variance submodels are all regression models, and so the coding of the regressors is an important aspect to consider. For example, in the BS variance submodel above, if both regressors `Male` and `Others` are dummy-coded variables (0 or 1), then α_0 represents the BS variance when females are alone (i.e., when both regressors equal 0). Alternatively, if both variables were coded 1 or 2, then the slopes would be the same, but the intercept α_0 would represent the BS variance when both variables equal 0 (which does not occur).

For continuous regressors, the coding of the variables is perhaps of greater consideration. Suppose that instead of a subject's sex, we included their age in the model:

$$\sigma^2_{\upsilon_{ij}} = \exp(\alpha_0 + \alpha_1 \texttt{Age}_i + \alpha_2 \texttt{Others}_{ij}). \tag{12.9}$$

In this case, α_0 would represent the BS variance when a subject aged 0 is alone. Such an interpretation is often an extrapolation outside of the data range (unless, perhaps, one was studying infants). Thus, for a continuous variable like Age it often makes sense to subtract off the lowest level in the sample, or to center the variable around its mean. This is especially important if one includes interactions. For example, suppose we included an interaction of age by being with others:

$$\sigma^2_{\upsilon_{ij}} = \exp(\alpha_0 + \alpha_1 \texttt{Age}_i + \alpha_2 \texttt{Others}_{ij} + \alpha_3 \texttt{Age}_i \times \texttt{Others}_{ij}). \tag{12.10}$$

Here, with the interaction in the model, the main effect of Others (α_2) represents the effect of being with others versus being alone for subjects of zero age. Again, in most studies this would be an extrapolation away from the data range, and essentially uninterpretable. Subtracting off the sample's low value or its mean from $\texttt{Age}_i$ greatly helps here, so that α_2 would represent the effect of being with others versus being alone for subjects at the lowest or mean age level. Also, the scaling of continuous regressors is worth considering. For example, one might want to express the age variable in deciles, say $\texttt{Age}_i/10$, so that the coefficients associated with age pertain to a 10-unit change in age (rather than a unit change in age). These considerations can lead to more interpretable results, as well as ease the computational complexity inherent in estimation of the model parameters.

Between-Subject and Within-Subject Effects

As is well-known in the multilevel literature, the effects of time-varying (level-1) variables can be decomposed into BS and WS effects (Hedeker & Gibbons, 2006; Neuhaus & Kalbfleisch, 1998; van de Pol & Wright, 2009). For example, suppose we are considering the following mean submodel for our modeling of mood (y_{ij}):

$$y_{ij} = \beta_0 + \beta_1 \texttt{Age}_i + \beta_2 \texttt{Others}_{ij} + \upsilon_i + \varepsilon_{ij}. \tag{12.11}$$

Suppose that we obtain a positive effect for $\texttt{Others}_{ij}$ (i.e., $\hat{\beta}_2 > 0$). Does this indicate that (a) mood is elevated for a subject when they are with others, or does it indicate that (b) the average mood is elevated for subjects that tend to be with others to a greater extent? This latter interpretation (b) is the BS effect, namely the association between a subject's average mood ($\bar{y}_i$) and their average of being with others ($\overline{\texttt{Others}_i}$). In the case of a dummy variable, this average would be the proportion of prompts that a subject reports being with others. The former interpretation (a) is the WS effect, or the relationship for a given subject momentary mood ($y_{ij} - \bar{y}_i$) and being with others ($\texttt{Others}_{ij} - \overline{\texttt{Others}_i}$), relative to their averages on both variables. In Equation 12.11 above, it is assumed that the BS and WS effects (of being with others) are equal. This may not be the case, and in EMA studies it is often the momentary or WS effect that is of greatest interest. To allow for separate BS and WS effects, we simply augment the model in the following way:

$$y_{ij} = \beta_0 + \beta_1 \texttt{Age}_i + \beta_2 \overline{\texttt{Others}_i} + \beta_3(\texttt{Others}_{ij} - \overline{\texttt{Others}_i}) + \upsilon_i + \varepsilon_{ij}. \quad (12.12)$$

Here, β_2 represents the BS effect and β_3 is the WS effect of the time-varying $\texttt{Others}_{ij}$ variable. Because the model of Equation 12.11 is nested within the model of Equation 12.12, under the assumption that the BW and WS effects are equal, one can use a likelihood-ratio test to assess this assumption. However, because of the primary interest in the WS effect in EMA studies, the model of Equation 12.12 would generally be preferred, regardless of the results of such a test.

This decomposition can also be applied to the variance submodels, for example:

$$\sigma^2_{\upsilon_{ij}} = \exp\left(\alpha_0 + \alpha_1 \texttt{Age}_i + \alpha_2 \overline{\texttt{Others}_i} + \alpha_3(\texttt{Others}_{ij} - \overline{\texttt{Others}_i})\right), \quad (12.13)$$

for the BS variance. Here, the BS effect α_2 characterizes heterogeneity/homogeneity for subjects that are more/less with others. The WS effect α_3 is what is depicted in Figure 12.2; namely, the degree to which subject heterogeneity (dispersion of the dotted lines) is affected when a subject is with others, relative to being alone. Similarly, the WS variance submodel can include WS and BS effects of time-varying (level-1) variables:

$$\sigma^2_{\varepsilon_{ij}} = \exp\left(\tau_0 + \tau_1 \texttt{Age}_i + \tau_2 \overline{\texttt{Others}_i} + \tau_3(\texttt{Others}_{ij} - \overline{\texttt{Others}_i}) + \omega_i\right). \quad (12.14)$$

The BS effect τ_2 represents the degree of consistency/erraticism comparing subjects with different levels of being with others, while the WS effect τ_3 reflects consistency/erraticism differences when a subject is with others, relative to being alone. Again, the WS effect is depicted in Figure 12.2, in which subjects display more consistency in their responses when they are with others.

VARIANCE-COVARIANCE AND STANDARDIZATION OF THE RANDOM EFFECTS

The random location υ_i and scale ω_i effects follow a bivariate normal distribution (in the population of subjects) with means equal to 0 and variance covariance matrix $\boldsymbol{\Sigma}$ given by:

$$\boldsymbol{\Sigma} = \begin{bmatrix} \sigma_\upsilon^2 & \sigma_{\upsilon\omega} \\ \sigma_{\upsilon\omega} & \sigma_\omega^2 \end{bmatrix}. \tag{12.15}$$

Here, σ_υ^2 is the variance of the location effects, σ_ω^2 is the variance of the scale effects, and $\sigma_{\upsilon\omega}$ is the covariance of the two. For estimation of the model parameters, it is beneficial to standardize the random effects (i.e., as standard normals). The reason for this is that the maximum likelihood solution requires integration over the bivariate normal distribution of the random effects, and this is facilitated if the random effects are always in standardized form, rather than taking on a different unstandardized form for each dataset. To standardize the random effects, we can use the Cholesky factorization (Bock, 1975), or matrix square-root, of the variance covariance matrix, namely $\boldsymbol{\Sigma} = \boldsymbol{SS}'$, where $\boldsymbol{S}$ is the lower triangular Cholesky factor (or matrix square-root).

$$\begin{bmatrix} \upsilon_i \\ \omega_i \end{bmatrix} = \begin{bmatrix} s_{1ij} & 0 \\ s_{2ij} & s_{3ij} \end{bmatrix} \begin{bmatrix} \theta_{1i} \\ \theta_{2i} \end{bmatrix} = \begin{bmatrix} \sigma_{\upsilon_{ij}} & 0 \\ \sigma_{\upsilon\omega} / \sigma_{\upsilon_{ij}} & \sqrt{\sigma_\omega^2 - \sigma_{\upsilon\omega}^2 / \sigma_{\upsilon_{ij}}^2} \end{bmatrix} \begin{bmatrix} \theta_{1i} \\ \theta_{2i} \end{bmatrix} \tag{12.16}$$

Here, θ_{1i} and θ_{2i} are the standardized location and scale random effects, respectively, with means of 0 and variances of 1. In this representation, the Cholesky elements s_{1ij}, s_{2ij}, and s_{3ij} would be estimated, and subscripts i and j are included on these elements because the BS variance $\sigma_{\upsilon_{ij}}^2$ can vary across subjects and/or occasions. The model can now be written as

$$y_{ij} = \boldsymbol{x}_{ij}'\beta + s_{1ij}\theta_{1i} + \varepsilon_{ij} \tag{12.17}$$

where $s_{1_{ij}=\sigma_{\upsilon ij}} = \sqrt{\exp(\boldsymbol{u}_{ij}'\alpha)}$, and the errors ε_{ij} have variance given by

$$\sigma_{\varepsilon_{ij}}^2 = \exp(\boldsymbol{\omega}_{ij}'\tau + s_{2ij}\theta_{1i} + s_{3ij}\theta_{2i}) \tag{12.18}$$

In this representation, the covariance of the random effects $(\sigma_{\upsilon\omega})$ is obtained as the product of the Cholesky elements $s_{1ij} \times s_{2ij}$, and the variance of the random scale (σ_ω^2) equals $s_{2ij}^2 + s_{3ij}^2$.

Alternative Formulation for Association of Location and Scale

Suppose that instead of allowing the location and scale random effects to be correlated, we assume that they are independent (i.e., $\sigma_{\upsilon\omega} = 0$, and therefore $s_{2ij} = 0$), but that the location random effect θ_{1i} explicitly influences the WS variance. In this case, the WS variance could be expressed as

$$\sigma^2_{\varepsilon_{ij}} = \exp(\omega'_{ij}\tau + \tau_l\theta_{1i} + \sigma_\omega\theta_{2i}). \tag{12.19}$$

where the regression coefficient τ_l represents the (linear) influence of the location random effect θ_{1i} on the (log of the) WS variance. These two models of the WS variance are essentially the same, although in Equation 12.18 the parameter s_{2ij} is indicative of the covariance between the random location and scale effects, whereas in Equation 12.19 the parameter τ_l represents the effect of the random location effect on the WS variance. Also, the Cholesky element s_{3ij} in Equation 12.18 is replaced by the (simpler) square root of the random scale σ_ω in Equation 12.19. We have merely shifted from a correlation-like association between the mean and variance to a regression setting in which the mean influences the variance.

Although equivalent in the present case, the latter representation can be more easily generalized to represent various forms of the relationship between the random location effect and the WS variance. For example, one can easily extend the model to allow for a quadratic relationship; namely,

$$\sigma^2_{\varepsilon_{ij}} = \exp(\omega'_{ij}\tau + \tau_l\theta_{1i} + \tau_q\theta^2_{1i} + \sigma_\omega\theta_{2i}). \tag{12.20}$$

Here, τ_l and τ_q represent the linear and quadratic effect, respectively, of the random location effect θ_{1i} on the (log of the) WS variance. A quadratic relationship between the mean and variance would seem to be useful for rating scale data with ceiling and/or floor effects, where subjects that have mean levels at either the maximum or minimum value of the rating scale also have near-zero variance. For example, if the rating scale goes from 1 to 10, then any subject with a mean level near either 1 or 10 would almost certainly have a very small variance, giving rise to the potential for a quadratic relationship between the mean and variance. In this regard, the program MIXREGLS (Hedeker & Nordgren, 2013) allows for three possibilities: (a) no association ($\tau_l = \tau_q = 0$); (b) linear association only ($\tau_l \neq \tau_q = 0$); and (c) linear and quadratic association ($\tau_l \neq \tau_q \neq 0$).

Intraclass Correlation

The expectation of y_{ij}, $E(y_{ij})$, is simply $\boldsymbol{x}'_{ij}\beta$, as in an ordinary multiple regression model. Given that the MELS model allows extensive modeling of the BS and WS variances, the variance of y_{ij} is a bit more involved. For example, for the situation in which the random location θ_{1i} has a quadratic effect on the WS variance (i.e., $\tau_q \neq 0$), the variance of y_{ij} varies as a function of the random location effect. However, for the simpler situation in which $\tau_q = 0$ (i.e., the random location θ_{1i} has only a linear effect on the WS variance), the variance of y_{ij} is given by

$$V(y_{ij}) = \exp(\boldsymbol{u}'_{ij}\alpha) + \exp\left(\omega'_{ij}\tau + \frac{1}{2}\left[\tau_l^2 + \sigma_\omega^2\right]\right). \tag{12.21}$$

This is the sum of the contributions of the BS and WS variance submodels. Within the latter, the factor is $\frac{1}{2}\left[\tau_l^2 + \sigma_\omega^2\right]$ based on the expectation of log-normally distributed variables (Skrondal & Rabe-Hesketh, 2004). The covariance for any two observations j and j' that are nested within the same subject i (e.g., two different observations made on the same subject) equals

$$C(y_{ij}, y_{ij'}) = \sigma^2_{\upsilon_{ij}} = \exp(\boldsymbol{u}'_{ij}\alpha) \quad \text{for } j \neq j'. \tag{12.22}$$

As in an ordinary random-intercept multilevel model, this is simply the BS variance. However, here, because the BS variance is modeled in terms of covariates it can vary across values of these covariates $\boldsymbol{u}_{ij}$. Expressed as a correlation, this yields the intraclass correlation (ICC), denoted as r_{ij},

$$r_{ij} = \frac{\exp(\boldsymbol{u}'_{ij}\alpha)}{\exp(\boldsymbol{u}'_{ij}\alpha) + \exp\left(\omega'_{ij}\tau + \frac{1}{2}\left[\tau_l^2 + \sigma_\omega^2\right]\right)} \tag{12.23}$$

Note that the ICC, which is equal to the BS variance divided by the sum of the BS and WS variance, represents the proportion of total *unexplained* variation that is at the subject level. Here, the word *unexplained* refers to the residual variation of the dependent variable not explained by the mean submodel covariates $\boldsymbol{x}$. In the MELS model, the ICC can be obtained for specific values of the covariates $\boldsymbol{u}_{ij}$ and ω_{ij}, which can include both time-invariant and time-varying covariates. This is why r carries both i and j subscripts in Equation 12.23. Thus, another use of the MELS model can be to examine the degree to which the ICC varies across particular covariate values, or sets of covariate values.

EXAMPLE

Data for the analyses reported here come from a longitudinal, natural history study of adolescent smoking (Mermelstein et al., 2002). Students included in the study were either in 8th or 10th grade at baseline, and self-reported on a screening questionnaire 6–8 weeks prior to baseline that they either had never smoked, but indicated a probability of future smoking, or had smoked in the past 90 days, but had not smoked more than 100 cigarettes in their lifetime. Written parental consent and student assent were required for participation. The data collection modalities included self-report questionnaires, a week-long time/event sampling method via palmtop computers (EMA), and in-depth interviews. Data for the current analyses came from the EMA portion. Adolescents carried the handheld computers with them at all times during the seven consecutive day data collection period and were trained to both respond to random prompts from the computers and to event record (initiate a data collection interview) smoking episodes. Questions included ones about place, activity, companionship, mood, and other subjective items. The handheld computers date and time-stamped each entry. For the analyses reported, we treated the responses obtained from the random prompts at baseline. In all, there were 17,402 random prompts obtained from 510 students with an approximate average of 34 prompts per student (range = 3 to 58).

The dependent variable considered is a measure of the subject's negative affect (`NegAff`) at each random prompt. This measure consists of the average of several individual mood items that were identified via factor analysis. Each item was rated from 1 to 10 with higher values indicating higher levels of negative mood. Over all prompts, and ignoring the clustering of the data, the marginal mean of `NegAff` was 2.41 ($sd = 1.49$). Of interest is the degree of heterogeneity in this mood measure in terms of both WS and BS variation. To get a sense of this, Figures 12.3–12.4 provide histograms of the subject-level means and variances, calculated for each subject based on their negative affect responses. The variances are expressed on the natural log scale to reflect the metric in which they will be modeled. Notice that both means and variances vary rather considerably across subjects. Modeling of the BS variance will attempt to relate covariates to the variability in the distribution of the subject means of negative affect depicted in Figure 12.3. In other words, what might be related to the homogeneity/heterogeneity in subject mean levels of negative affect. Similarly, modeling of the WS variance will examine if there are covariates that are related to the variability levels in the distribution of the subject variances of negative affect depicted in Figure 12.4. This might help us better understand the factors that explain why subjects are more/less consistent/erratic in negative affect.

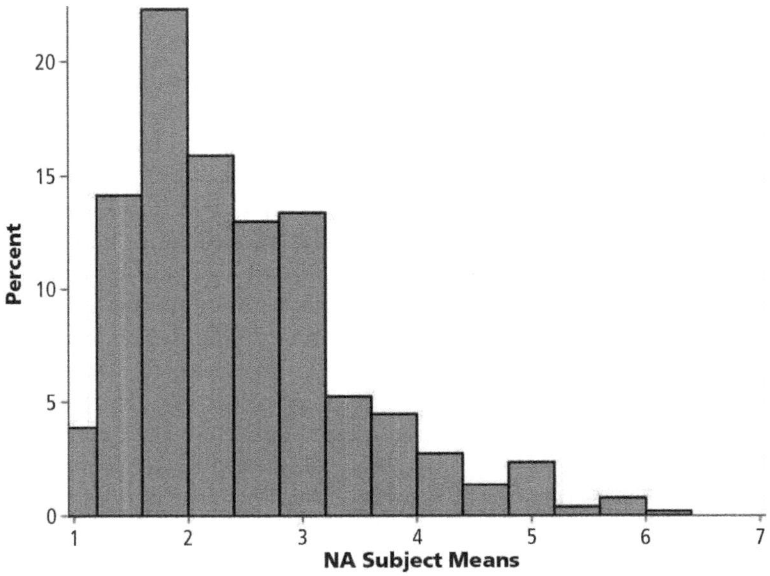

Figure 12.3 Histogram of subject-level means of negative affect.

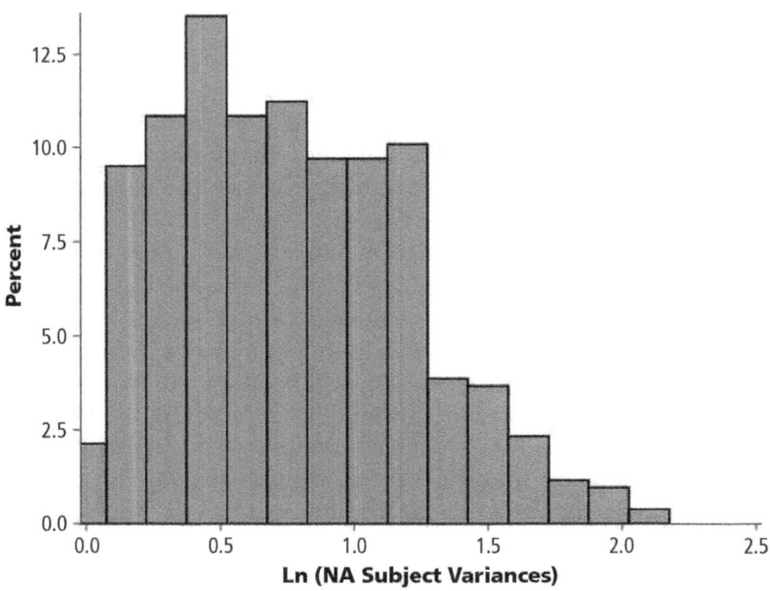

Figure 12.4 Histogram of subject-level ln(variances) of negative affect.

To begin, we will estimate a null model with no covariates, and no association between the random location and scale effects. This can be written as:

$$y_{ij} = \beta_0 + \upsilon_{1i} + \varepsilon_{ij}, \tag{12.24}$$

with

$$\sigma^2_{\upsilon_{ij}} = \exp(\alpha_0), \tag{12.25}$$

and

$$\sigma^2_{\varepsilon_{ij}} = \exp(\tau_0 + \omega_i). \tag{12.26}$$

For consistency we include the subject and occasion subscripts i and j, respectively, on the variances in the previous equations, though without covariates, the BS variance does not vary with i or j and the WS variance only varies with i (subjects). The estimates from this model are $\hat{\beta}_0 = 2.407(se = 0.034)$, $\hat{\alpha}_0 = -0.164(se = 0.049)$, $\hat{\tau}_0 = -0.074(se = 0.049)$, and $\hat{\sigma}_\omega = 1.061(se = 0.038)$. In contrast to a null multilevel model, here the BS and WS variance parameters are estimated on the log scale ($\hat{\alpha}_0$ and $\hat{\tau}_0$, respectively), and this model additionally includes the random subject scale effect (whose contribution is estimated as the standard deviation $\hat{\sigma}_\omega$). Here, the mean negative affect is estimated as 2.407, and following Equation 12.21, the BS variance is estimated as $\exp(-0.164) = 0.849$, and the WS variance is estimated as $\exp(-0.074 + 0.5 \times 1.061^2) = 1.630$.[1]

Using Equation 12.23, an estimate of the intraclass correlation is obtained as:

$$r_{ij} = 0.849 / (0.849 + 1.630) = 0.342, \tag{12.27}$$

which indicates that approximately one-third of the variation in negative affect is at the subject level. Notice that, without covariates in the variance submodels, the ICC does not vary with subjects or occasions, and so the subscripts i and j are, strictly speaking, not necessary here.

From this null model, it is interesting to examine the estimates of the random scale effects, since these random scale effects are a distinguishing feature of the MELS model relative to a standard multilevel model. As in multilevel models, the random effects are estimated using empirical Bayes estimation. Figures 12.5–12.7 provide histograms of the observed data for selected subjects with low scale, average scale, and high scale, respectively.

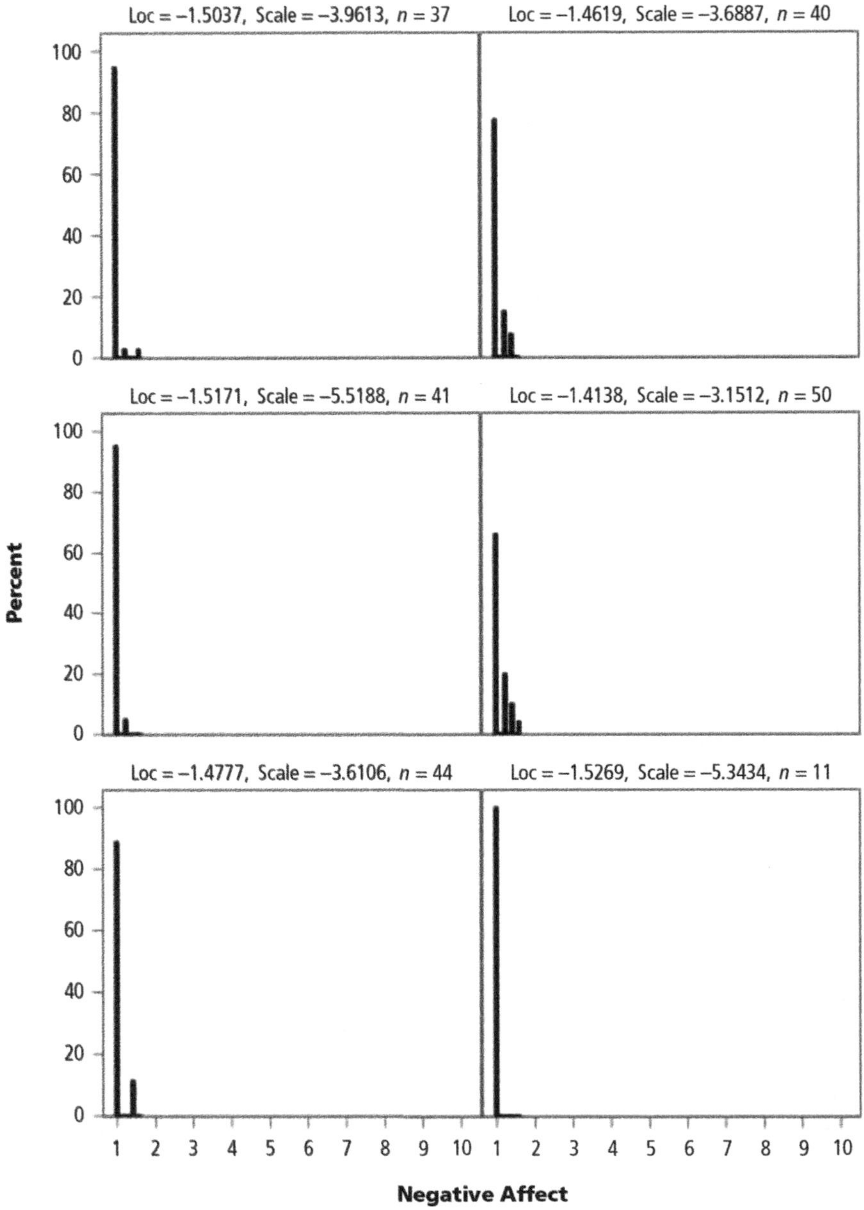

Figure 12.5 Null Model of Negative Affect: Data histograms from subjects with low scale estimates.

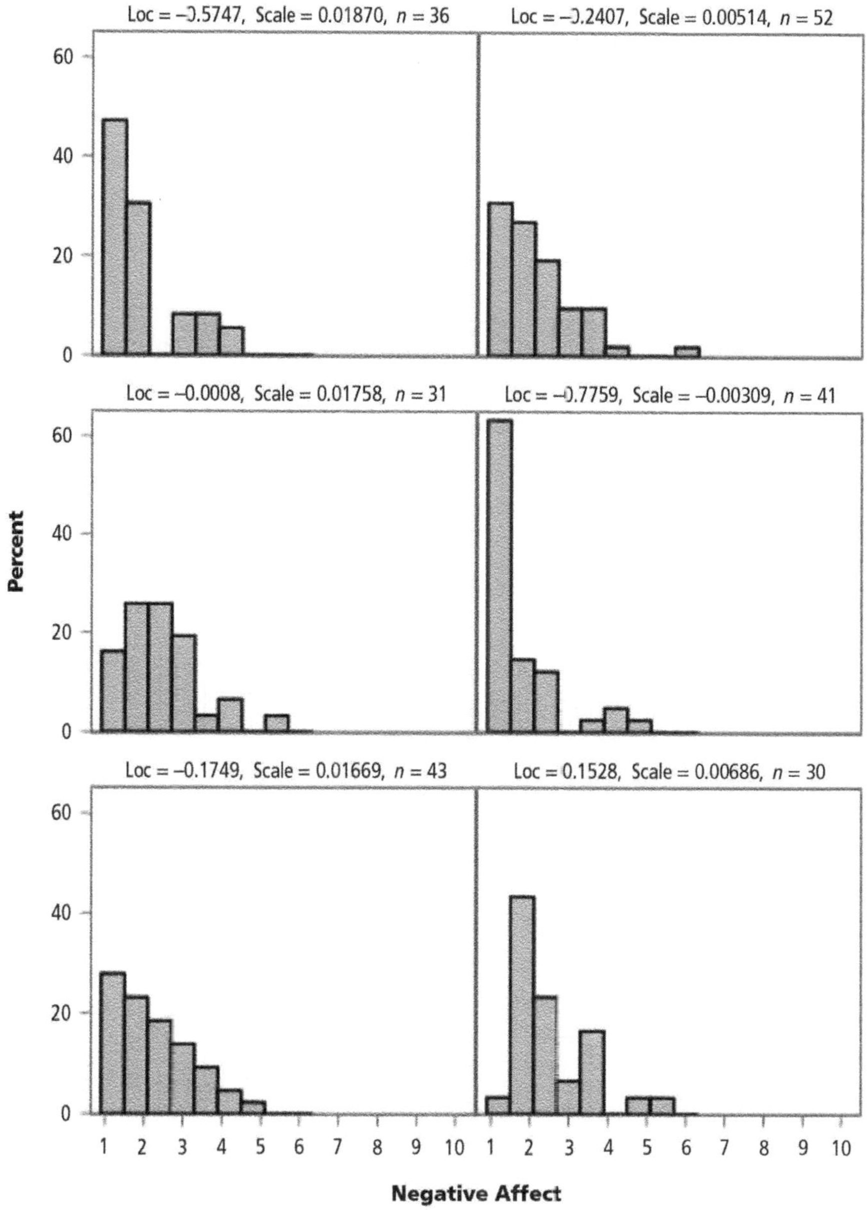

Figure 12.6 Null Model of Negative Affect: Data histograms from subjects with near-zero scale estimates.

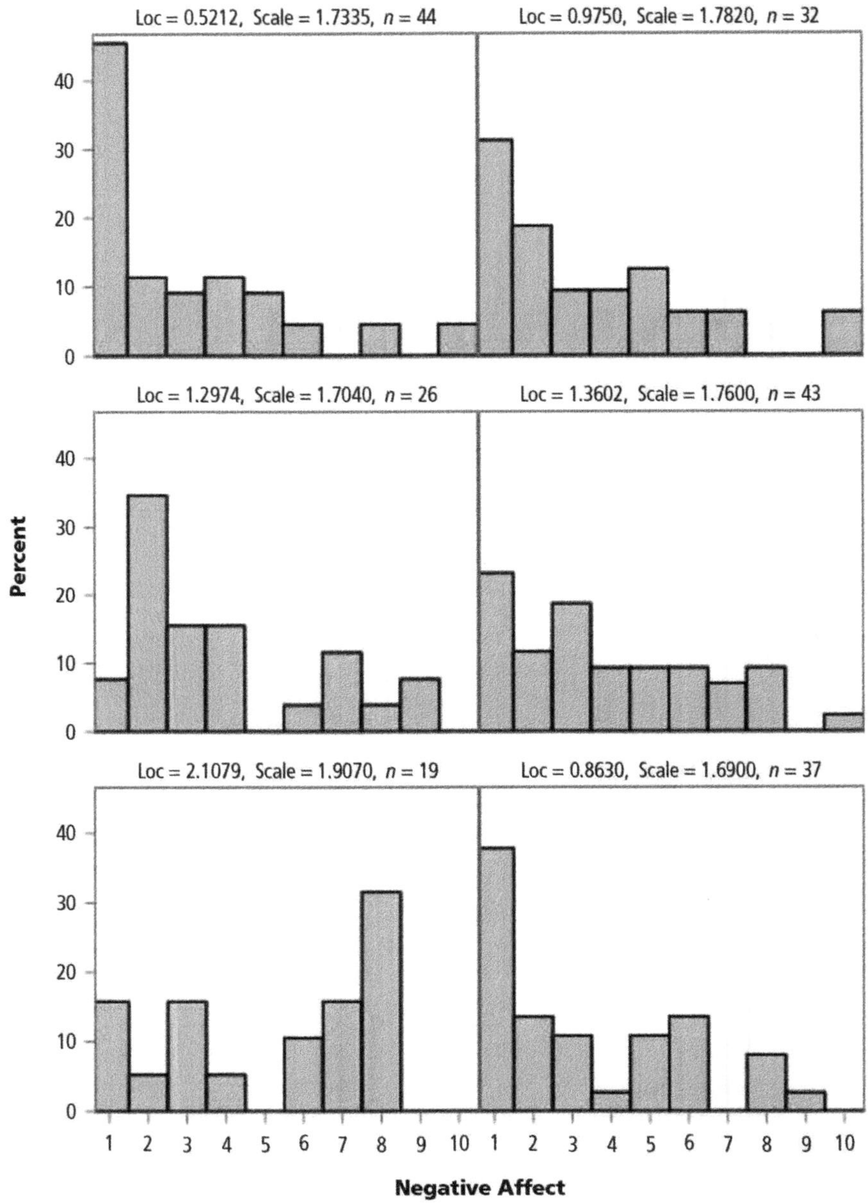

Figure 12.7 Null Model of Negative Affect: Data histograms from subjects with large scale estimates.

In each of these histograms, estimated values of the location and scale random effects are listed as "Loc" and "Scale," respectively, and the number of observations for each subject is listed as "*n*." Since the random effects are expressed as standard normals, these estimates are akin to *z*-values. The y-axis in these histograms represents percentages of each subject's responses, with maximums of 100%, 60%, and 40% for the three figures, respectively. Figure 12.5 provides histograms of six subjects with very low scale estimates (–3 to –5.5), who also have very low location estimates (approximately –1.5). Notice that these are subjects with the bulk of their responses in the lowest level of negative affect. Thus, they are very consistently low in their negative affect responses. Figure 12.6 includes subjects with near-zero estimates of scale, and with various levels of location. These subjects provide average levels of variability in their negative affect responses. Finally, Figure 12.7 includes subjects with high scale estimates (approximately 1.7 to 1.9), and with various positive location estimates. These are subjects who are more erratic in their negative affect responses, largely providing responses across the entire range of negative affect. Note that a standard multilevel model would assume that all subjects have the same level of dispersion across the response categories, but these plots clearly indicate that subjects do differ in terms of their consistency/erraticism in negative affect.

Building from the null model, in terms of covariates, we will examine `GenderF`, which is a subject (level-2) covariate coded 0 for males and 1 for females, `Others`, which is a time-varying (level-1) covariate coded 0 if the subject was alone or 1 if with others, `Age15` which is the subject's age at baseline minus 15 (since the average age in the sample was 14.93), and `COPEc`, which is a grand-mean centered version of a coping scale measurement at baseline. For the prompt-level variable `Others`, we created both a BS and WS version (`Others_BS` and `Others_WS`) as described in the preceding section, namely $X_{ij} = \overline{X}_i + (X_{ij} - \overline{X}_i)$. Here, `Others_BS`, the first term on the right-hand side, equals the proportion of random prompts in which a subject was with others, and `Others_WS`, the latter term on the right-hand side, is the prompt-specific deviation relative to a subject's proportion. Thus, it equals $0 - \overline{X}_i$ or $1 - \overline{X}_i$ if the subject was alone or was with others, respectively, for the given random prompt. Finally, the time of the prompt was recorded and the following indicators were created and treated as covariates: 9am–2pm, 2pm–6pm, 6pm–1pm, and 10pm–3am; leaving 3am–9am as the reference time indicator. All covariates were included in the mean, BS and WS variance submodels.

Table 12.1 lists the maximum likelihood estimates, standard errors, *z*-, and *p*-values for these regressors in terms of the (a) mean submodel, (b) BS

TABLE 12.1 Mixed-Effects Location Scale Model of Negative Affect, $N = 510$, $\Sigma n_i = 17{,}402$, Maximum Likelihood Estimates, Standard Errors, *z*-Values, and *p*-Values

Variable	Estimate	Std Error	*z*-value	*p*-value
Beta (regression coefficients)				
Intercept	2.869	0.175	16.355	0.001
9am-2pm	0.054	0.024	2.252	0.024
2pm-6pm	0.063	0.025	2.562	0.010
6pm-10pm	0.048	0.025	1.901	0.057
10pm-3am	0.024	0.041	0.582	0.560
Others_BS	–0.701	0.227	–3.083	0.002
Others_WS	–0.229	0.021	–10.831	0.001
genderf	0.016	0.067	0.234	0.815
age15	0.020	0.030	0.669	0.504
COPEc	–0.523	0.065	–8.032	0.001
Alpha (BS variance parameters: log-linear model)				
Intercept	–0.070	0.296	–0.236	0.814
9am-2pm	0.063	0.055	1.150	0.250
2pm-6pm	0.092	0.056	1.647	0.100
6pm-10pm	0.006	0.057	0.109	0.913
10pm-3am	0.047	0.090	0.526	0.599
Others_BS	–0.572	0.401	–1.426	0.154
Others_WS	–0.299	0.044	–6.833	0.001
genderf	0.054	0.132	0.409	0.682
age15	0.054	0.059	0.914	0.361
COPEc	–0.872	0.126	–6.898	0.001
Tau (WS variance parameters: log-linear model)				
Intercept	0.227	0.229	0.992	0.321
9am-2pm	0.174	0.042	4.110	0.001
2pm-6pm	0.248	0.043	5.797	0.001
6pm-10pm	0.297	0.043	6.918	0.001
10pm-3am	0.355	0.064	5.528	0.001
Others_BS	–0.930	0.297	–3.133	0.002
Others_WS	–0.220	0.030	–7.254	0.001
genderf	0.296	0.089	3.321	0.001
age15	–0.038	0.040	–0.938	0.348
COPEc	–0.391	0.080	–4.884	0.001
Random scale standard deviation				
Std Dev	0.719	0.028	25.64	0.001
Random location (mean) effect on WS variance				
Loc Eff	0.676	0.042	16.217	0.001

variance submodel, and (c) WS variance submodel. For the mean submodel, all of the time interval effects are positive, but only the first two are statistically significant. The estimated effects for these two intervals equal 0.054 ($z = 2.25$, $p < .024$) and 0.063 ($z = 2.56$, $p < .01$), respectively. Thus, though these effects are not large, negative affect is significantly elevated during 9am–2pm and 2pm–6pm, relative to the early morning period of 3am–9am. For the Others variable, both the BS and WS versions indicate diminished negative affect ($z = -3.08$, $p < 0.002$, and $z = -10.83$, $p < 0.001$ respectively). Thus, being with others to a greater extent (higher on `Others_BS`) as well as the momentary effect of being with others (`Others_WS`) decrease negative affect. Neither age nor gender have significant effects on negative affect, but the coping variable does ($z = -8.03$, $p < 0.001$), with higher coping scores leading to lower negative affect.

Turning to the effects on the BS variance, most of the variables do not have statistically significant effects. There are two notable exceptions: being with others (`Others_WS`) and higher scores on the coping variable decrease subject heterogeneity ($z = -6.83$, $p < 0.001$, and $z = -6.90$, $p < 0.001$, respectively).

For the effects on WS variance, nearly all variables have significant effects. All time interval effects are positive indicating increased WS variability in the mood responses relative to the early morning reference period of 3am–9am (all have $p < 0.001$. Being with others to a greater extent (higher on `Others_BS`) as well as the momentary effect of being with others (`Others_WS`) decrease WS variation on negative affect ($z = -3.13$, $p < 0.002$, and $z = -7.25$, $p < 0.001$, respectively). Females exhibit greater WS variance on negative affect ($z = 3.32$, $p < 0.001$), while higher values of coping lead to diminished WS variance ($z = -4.88$, $p < 0.001$). Finally, in terms of the random scale effect, it is highly significant ($z = 25.64$, $p < 0.001$) indicating that subjects exhibit different degrees of consistency/erraticism on negative affect, over and above the covariate effects on the WS variance of this outcome. Also, the location effect on the WS variance is positive and highly significant ($z = 16.22$, $p < 0.001$), indicating that subjects with higher levels of negative affect are more erratic, and that subjects with lower levels of negative affect are more consistent in their responses. The latter is suggestive of a floor effect of measurement in that subjects with below-average means on negative affect (say means of 1 or 2) must be consistent in their responses in order to have such a low mean.

Table 12.2 lists the exponentiated estimates, and 95% confidence limits, of the effects in the BS and WS variance submodels. For the intercepts in these submodels, the exponentiated values represent the reference BS and WS variance, respectively. Namely, the variances when all of the regressors equal zero. In the present case, for the BS variance, that would correspond to the 3am–9am time interval for males aged 15 with average coping values

TABLE 12.2 Variance Ratios and 95% Confidence Intervals (lower, upper)

Variable	Ratio	Lower	Upper
Alpha (BS variance parameters)			
Intercept	0.933	0.523	1.665
9am-2pm	1.065	0.957	1.185
2pm-6pm	1.097	0.983	1.224
6pm-10pm	1.006	0.900	1.126
10pm-3am	1.048	0.879	1.250
Others_BS	0.564	0.257	1.239
Others_WS	0.742	0.681	0.808
genderf	1.055	0.815	1.366
age15	1.055	0.940	1.184
COPEc	0.418	0.327	0.536
Tau (WS variance parameters)			
Intercept	1.255	0.802	1.964
9am-2pm	1.190	1.095	1.294
2pm-6pm	1.282	1.179	1.394
6pm-10pm	1.346	1.237	1.464
10pm-3am	1.426	1.257	1.617
Others_BS	0.394	0.220	0.706
Others_WS	0.803	0.756	0.852
genderf	1.344	1.129	1.600
age15	0.963	0.890	1.042
COPEc	0.677	0.578	0.791
Random scale standard deviation			
Location Effect	1.966	1.817	1.085
Random location (mean) effect on WS variance			
Std Dev	2.052	1.942	2.168

who are always alone. For the WS variance, it would also be for subjects with average levels of both the random location and scale effects. For the regressors, as mentioned above, the exponentiated estimates correspond to variance ratio estimates which indicate the ratio of variance per unit increase in the regressor. For example, for `genderf` it is the (estimated) BS (or WS) variance for females divided by the same quantity for males. In this metric, if the 95% confidence limit includes a ratio of one, then the variable's effect is not statistically significant at the $\alpha = 0.05$ level. For the effects on the BS variance, only `Others_WS` and `COPEc` have intervals that do not include one. Thus, when subjects are with others the BS variance is reduced by a factor of 0.74. In percentage terms, this would represent a 100–74 = 26% reduction in subject heterogeneity (in the mean levels of negative affect). Conversely, a unit increase in coping leads to a 58% (100–42) reduction in the BS variance. The standard deviation for the coping variable equals 0.54, so a unit change on coping represents a large change (i.e., nearly a two sigma change on coping). As noted, nearly all variables had significant effects on the WS variance, and so the corresponding CIs do not include unity. For the time intervals, with 3am–9am as the reference, we see increases in subject erraticism of 19%, 28%, 35%, and 43% for the four time bins, respectively (the variance ratios are multiplied by 100 and then 100 is subtracted off to yield the percentage increases in WS variance). Subjects who are always with others are 61% (100–39) less erratic than subjects that are always alone. When a subject is with others, they are 20% (100–80) less erratic than when they are alone. Females are 34% (134–100) more erratic than males, and a unit increase in coping leads to a 32% (100–68) reduction in WS variance.

Figure 12.8 presents coefficient plots for the covariate effects on the mean, BS, and WS variance submodels, including 95% confidence intervals

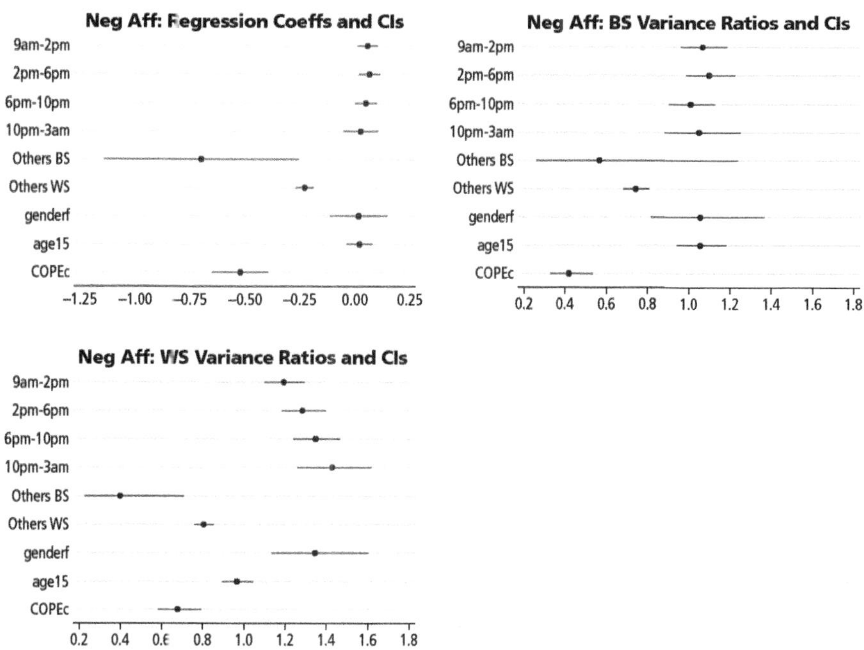

Figure 12.8 Modeling of Negative Affect: (a) regression coefficients and 95% confidence intervals, (b) BS variance ratios and 95% confidence intervals, (c) WS variance ratios and 95% confidence intervals.

(CIs). These plots are useful as a visual representation of the regressor effects. For the effects on the mean submodel, the figure indicates the estimated regression coefficients, 95% CIs, and a reference line of zero. If the CI does not include zero for a variable, then that variable's effect is statistically significant at the two-tailed 0.05 level. For the BS and WS variance submodels, the plots provide estimates of the variance ratios associated with each variable, and their corresponding 95% confidence intervals. As mentioned, if the CI for a variance ratio includes one, then the effect of that variable is not statistically significant at the two-tailed $\alpha = 0.05$ level. As we have already interpreted the effects of this analysis, we will not repeat that here. However, there are a few things that the plots help to reveal. For example, it is apparent that the CI for `Others_BS` is quite large in all submodels. The scaling of this variable plays a role here, as the effect is for a unit change on this variable. For `Others_BS`, this represents a comparison of a mean of 1 on `Others_BS` versus a mean of 0 on `Others_BS`. In other words, comparing subjects who are always with others to subjects that are always alone. Clearly, a different scaling of this variable might yield a more interpretable effect for this variable (e.g., a change of say .1 on this variable, rather than a unit change). The plots also reveal how a variable's effect on the three submodels might behave similarly (or not). Thus, both coping and being with others have a consistent negative effect on the mean, BS variance, and WS variance.

SOFTWARE

In terms of the major statistical packages, Hedeker and Mermelstein (2012) and the supplemental materials of Hedeker et al. (2008) and Li and Hedeker (2012) provide examples of SAS PROC NLMIXED code. Unlike many software procedures, SAS PROC NLMIXED includes programming features and requires starting values for all model parameters. For this, it is advisable to estimate somewhat simpler models to get reasonable starting values for most of the model parameters. For example, using SAS PROC MIXED (or some other standard software program for multilevel models) can provide good starting values for the regression coefficients and some of the variance parameters. Another software option for the MELS model is the free stand-alone MIXREGLS software program (Hedeker & Nordgren, 2013).[2] It runs on the Windows operating system and the manual describes program usage, including how it can be accessed via R. Also, Leckie (2014) provides software to run the MIXREGLS program from within Stata. The analyses presented in this chapter were done using MIXREGLS, while the coefficient plots in Figure 12.8 were obtained using the coefplot module developed in Stata (Jann, 2014). An extension of MIXREGLS, that includes a graphical user interface (GUI), has recently been developed and is also accessible at the aforementioned website. This program, named MixWILD (Dzubar et al., 2020), allows for multiple random location effects (i.e., random slopes), and is available for both the Windows and Mac OS X operating systems. SAS PROC NLMIXED, MIXREGLS, and MixWILD all use full-likelihood estimation; details are provided in the Appendix of Hedeker and Nordgren (2013).

Another option for estimation is the use of Bayesian software. For this, Rast et al. (2012) describe estimation using BUGS/JAGS software. Similarly, Leckie et al. (2014) discuss Bayesian estimation using the Stat-JR statistics package (Charlton et al., 2013) for an example with clustered data (students in neighborhoods). Lin, Hedeker, and Mermelstein (2018) provide code and details using Stan in the supplemental materials. These articles using Bayesian estimation also provide various extensions to the MELS model presented in this chapter. Similarly, the most recent version of M*plus* (Muthén & Muthén, 1998–2017) also uses Bayesian estimation methods and has capability to estimate MELS models and extensions.

DISCUSSION

This chapter has illustrated how the MELS model can be used to model differences in variances, and not just means, across subject and time-varying covariates. As such, the MELS model can help to identify predictors of both within-subjects and between-subjects variation, and to test hypotheses about these variances. Additionally, by including a random subject effect on the WS variance, this model can examine the degree to which subjects are heterogeneous in terms of their variation on the outcome variable. Our example with negative affect clearly shows that subjects are quite heterogeneous in terms of their mood variation, as one might expect.

More applications of this class of models clearly exist. For example, many questions of both normal development and the development of psychopathology address the issue of variability or stability in emotional responses to various situations and/or contexts. Often, a concern is with the range of responses an individual gives to a variety of stimuli or situations, and not just with the overall mean level of responsivity. Intraindividual variability in mood reflects a different aspect of one's emotional life than overall mean levels, and is important in itself in predicting future behavior and psychological states (Eid & Diener, 1999; Kuppens et al., 2007). Emotion dysregulation is at the core of several psychological disorders (e.g., borderline personality disorder, bipolar disorder), and understanding more about what factors or covariates may influence affective instability may help in guiding treatment and predictability of future behavior. EMA provides a window into the affective lives of individuals and provides opportunity to both monitor and intervene, in the moment, in real-time, to help correct or prevent disruptive mood dysregulation (Ebner-Priemer & Trull, 2009). Having the analytic ability to tease apart controlling variables on individual level mood variability may help enhance the understanding and treatment of psychological disorders. The MELS model also allows us to examine hypotheses about cross-situational consistency.

Modern data collection procedures, such as EMA and/or real-time data captures, usually provide a fair amount of both WS and BS data, and so give rise to the opportunity for modeling of both WS and BS variances as a function of covariates. One might wonder about how much WS and BS data are necessary for estimation and variance modeling purposes. For random coefficient models, Longford (1993) noted the difficulty with providing general guidelines about the degree of complexity, for the variation part of a model, that a given dataset could support. This would also seem to be true here. Nonetheless, carrying out some simulations with relatively small sample sizes (e.g., 20 subjects with 5 observations each) gives the general impression that the primary issue is that the estimation algorithm does not often converge, but instead has estimation difficulties of one sort or another, in small sample situations. This improves dramatically as the number of subjects and observations increases to even modest sizes of 100 subjects and 10 observations.

Various extensions of the MELS model presented in this chapter have been developed. Li and Hedeker (2012) extends the model to three levels to allow for the nesting of observations within days within subjects. This permits one to model the within-days within-subjects variance, within-subjects between-days variance, and the between-subjects variance as functions of covariates. Pugach et al. (2014) describes a bivariate MELS model that can be used to additionally model the association between the bivariate outcomes. Kapur et al. (2015) builds on this work to the multivariate outcome setting using Bayesian estimation methods. Ferrer and Rast (2017) extend the MELS model for dyadic data, including multiple random location and scale effects. Leckie et al. (2014) describe MELS models with multiple random location and scale effects applied to clustered data where students are nested within schools. Lin, Hedeker, and Mermelstein (2018) extends the model to multiple waves of EMA data, and allows multiple random location and scale effects. Brunton-Smith et al. (2016) develops a MELS model for cross-classified random effects. For longitudinal data with skewness, detection limits, and measurement errors, Lu (2017) develops a Tobit MELS model. For longitudinal human stature data, Goldstein et al. (2017) develop a non-linear growth MELS model that includes multiple location and scale random effects. Scherer et al. (2016) and Courvoisier et al. (2019) describe approaches for MELS modeling of time-to-event data. For missing EMA data, Cursio et al. (2019) and Lin, Mermelstein, and Hedeker (2018) develop methods that jointly model the missingness and the outcomes. Hedeker, Demirtas, and Mermelstein (2009) describes an extension of the model for ordinal EMA outcome data, and Hedeker et al. (2016) further extends this approach for cross-sectional ordinal questionnaire data. Finally, Walters et al. (2018) comprehensively evaluate and develop methods for power analysis using the MELS model.

As this is a relatively new modeling technique, certain limitations and cautions should be mentioned. The MELS model assumes that the random location effects are normally distributed and that the random scale effects are log-normally distributed. It is unclear how robust this model is to violations of these assumptions. To some extent, this can be examined empirically using the approach of Liu and Yu (2008) for estimating models with non-normal random effects. Attention should also be paid to outliers and influential observations, as these might have undue effects on estimation of the model parameters, especially the variance parameters.

RECOMMENDED READINGS

Recommended Technical Readings

In terms of technical articles, Hedeker et al. (2008) describes development and application of the MELS model to an EMA dataset including many covariates. However, this initial paper did not allow for the possibility that time-varying (level-1) covariates could influence the BS variance. As detailed in this chapter, and illustrated in Figures 12.1 and 12.2, this is clearly possible. Hedeker, Mermelstein, and Demirtas (2012) further describes application of the MELS model, including piecewise linear effects of covariates on the mean, BS variance, and WS variance submodels. As mentioned, the software program MIXREGLS is described in Hedeker and Nordgren (2013), which includes estimation details in the Appendix. Two examples are included with the program and described in the manual. In the context of educational data consisting of students nested within schools, Leckie et al. (2014) comprehensively describe MELS models with multiple random location and scale effects. They also include results of a simulation study that shows why inclusion of the random scale is necessary for correct inference for the covariates of the level-1 variance submodel.

Recommended Applied Readings

Recently, several applications of the MELS model have been published in the substantive literatures. Ong et al. (2016) applied the MELS model to examine sleep variability in a sample of subjects with chronic insomnia. Piasecki et al. (2016) examined mood variability and nicotine dependence in adolescent smokers. From the same adolescent study, Gorka et al. (2017) used the MELS model to test for the unique and interactive effects of alcohol coping motives and internalizing symptoms on mood changes during drinking. Gerhart et al. (2018) examined relationships between negative mood mean and variability with pain levels, comparing a sample of chronic low back pain subjects and their pain-free spouses. Geukes et al. (2017) applied the MELS model to analyze self-esteem level and variability in narcissists. Sandhu and Leckie (2016) examined orthodontic pain trajectories in adolescents. Maher, Huh, et al. (2018) examined variability in physical activity and its association with mental health outcomes among obese adults, and Maher, Dzubur, et al. (2018) further examined mood variability and its effect on physical activity and sedentary behavior. Clearly, use of the MELS model is increasing as researchers learn about its use and capabilities.

TRY THIS!

The dataset used in this chapter (named VarModel.dat) is available from the online appendix and includes the variables in the following order: `id`, `PosAff`, `NegAff`, `t1`, `t2`, `t3`, `t4`, `Others_BS`, `Others_WS`, `genderf`, `age15`, `COPEc`. In this chapter, we have presented analyses of the negative affect outcome `NegAff`. Readers are encouraged to replicate the analyses presented, as well as running similar models on the positive affect outcome (i.e., the variable `PosAff`).

NOTES

1. Estimates from a null multilevel are similar: $\hat{\beta}_0 = 2.435$, $\hat{\sigma}_\upsilon^2 = 0.869$, and $\hat{\sigma}_\varepsilon^2 = 1.379$.
2. This program is available at the website: https://voices.uchicago.edu/hedeker/mixwild_mixregls

REFERENCES

aan het Rot, M., Hogenelst, K., & Schoevers, R. A. (2012). Mood disorders in everyday life: A systematic review of experience sampling and ecological momentary assessment studies. *Clinical Psychology Review, 32*(6), 510–523.

Aitkin, M. (1987). Modelling variance heterogeneity in normal regression using GLIM. *Applied Statistics, 36,* 332–339.

Armey, M. F., Schatten, H. T., Haradhvala, N., & Miller I. W. (2015). Ecological momentary assessment (ema) of depression-related phenomena. *Current Opinion in Psychology, 4,* 21–25.

Axelson, D., Bertocci, M., Lewin, D., Trubnick, L., Birmaher, B., Williamson, D.,... Dahl, R. (2003). Measuring mood and complex behavior in natural environments: Use of ecological momentary assessment in pediatric affective disorders. *J Child Adolescent Psychopharmacology, 13,* 253–266.

Bock, R. D. (1975). *Multivariate statistical methods in behavioral research.* McGraw-Hill.

Bolger, N., Davis, A., & Rafaeli, E. (2003). Diary methods: Capturing life as it is lived. *Annual Review of Psychology, 54,* 579–616.

Bolger, N., & Laurenceau, J.-P. (2013). *Intensive longitudinal methods: An introduction to diary and experience sampling research.* Guilford Press.

Boseck, J., Engel, S., Allison, K., Crosby, R., MitchellL, J., & de Zwaan, M. (2007). The application of ecological momentary assessment to the study of night eating. *The International Journal of Eating Disorders, 40,* 271–276.

Brunton-Smith, I., Sturgis, P., & Leckie, G. (2016). Detecting and understanding interviewer effects on survey data by using a cross-classified mixed effects location-scale model. *Journal of the Royal Statistical Society: Series A (Statistics in Society), 180*(2), 551–568.

Charlton, C. M. J., Michaelides, D., Parker, R., Cameron, B., Szmaragd, C., Yang, H., Zhang, Z., Frazer, A. J., Goldstein, H., Jones, K., Leckie, G. Moreau, L., & Browne, W. (2013). *Stat-JR version 1.0.* Centre for Multilevel Modelling, University of Bristol & Electronics and Computer Science, University of Southampton.

Cleveland, W., Denby, L., & Liu, C. (2000). *Random location and scale effects: Model building methods for a general class of models. Computing Science and Statistics,* 32. https://www.researchgate.net/publication/2436589_Random_Location_and_Scale_Effects_Model_Building_Methods_for_a_General_Class_of_Models

Courvoisier, D., Walls, T. A., Cheval, B., & Hedeker, D. (2019). A mixed-effects location scale model for time-to-event data: A smoking behavior application. *Addictive Behaviors,* 94, 42–49.

Cursio, J. F., Mermelstein, R. J., & Hedeker, D. (2019). Latent trait shared-parameter mixed models for missing ecological momentary assessment data. *Statistics in Medicine,* 38, 660–673.

de Vries, M. (1992). *The experience of psychopathology: Investigating mental disorders in their natural settings.* Cambridge University Press.

de Vries, M., Caes, C., & Delespaul, P. (2001). The experience sampling method in stress and anxiety research. *Anxiety Disorders,* 289–306.

Dzubur, E., Ponnada, A., Nordgren, R., Yang, C.-H., Intille, S., Dunton, G., & Hedeker, D. (2020, epub). MixWILD: A program for examining the effects of variance and slope of time-varying variables in intensive longitudinal data. *Behavior Research Methods, 52,* 1403–1427. https://doi.org/10.3758/s13428-019-01322-1

Ebner-Priemer, U., & Trull, T. J. (2009). Ecological momentary assessment of mood disorders and mood dysregulation. *Psychological Assessment, 21*(4), 463–475.

Eid, M., & Diener, E. (1999). Intraindividual variability in affect: Reliability, validity, and personality correlates. *Journal of Personality and Social Psychology, 76*(4), 662–676.

Epstein, D., Willner-Reid, J., Vahabzadeh, M., Mezghanni, M., Lin, J.-L., & Preston, K. (2009). Real-time electronic diary reports of cue exposure and mood in the hours before cocaine and heroin craving and use. *Arch Gen Psychiatry, 66,* 88–94.

Feldman Barrett, L., & Barrett, D. (2001). An introduction to computerized experience sampling in psychology. *Social Science Computer Review, 19,* 175–185.

Ferrer, E., & Rast, P. (2017). Partitioning the variability of daily emotion dynamics in dyadic interactions with a mixed-effects location scale model. *Current Opinion in Behavioral Sciences, 15,* 10–15.

Gerhart, J. I., Burns, J. W., Bruehl, S., Smith, D., Post, K., Porter, L., Schuster, E., Buvanendran, A., Fras, A. M., & Keefe, F. J. (2018). Variability in negative emotions among individuals with chronic low back pain: Relationships with pain and function. *Pain, 159,* 342–350.

Geukes, K., Nestler, S., Hutteman, R., Dufner, M., Küfner, A. C. P., Egloff, B., Denissen, J. A., & Back, M. D. (2017). Puffed-up but shaky selves: State self-esteem level and variability in narcissists. *Journal of Personality and Social Psychology, 112*(5), 769–786. https://doi.org/10.1037/pspp0000093

Goldstein, H., Leckie, G., Charlton, C., Tilling, K., & Browne, W. J. (2017). Multilevel growth curve models that incorporate a random coefficient model for the level 1 variance function. *Statistical Methods in Medical Research, 27*(11). https://doi.org/10.1177/0962280217706728

Gorka, S. M., Hedeker, D., Piasecki, T. M., & Mermelstein, R. (2017). Impact of alcohol use motives and internalizing symptoms on mood changes in response to drinking: An ecological momentary assessment investigation. *Drug and Alcohol Dependence, 173,* 31–38.

Granholm, E., Loh, C., & Swendsen, J. (2008). Feasibility and validity of computerized ecological momentary assessment in schizophrenia. *Schizophr Bull, 34,* 507–514.

Gwaltney, C. J., Bartolomei, R., Colby, S. M., & Kahler, C. (2008). Ecological momentary assessment of adolescent smoking cessation: A feasibility study. *Nicotine & Tobacco Research, 10,* 1185–1190.

Harvey, A. C. (1976). Estimating regression models with multiplicative heteroscedasticity. *Econometrica, 44,* 461–465.

Hedeker, D., Demirtas, H., & Mermelstein, R. J. (2009). A mixed ordinal location scale model for analysis of ecological momentary assessment (EMA) data. *Statistics and Its Interface, 2,* 391–402.

Hedeker, D., & Gibbons, R. D. (2006). *Longitudinal data analysis.* Wiley.

Hedeker, D., & Mermelstein, R. J. (2012). Mood changes associated with smoking in adolescents: An application of a mixed-effects location scale model for longitudinal EMA data. In G. R. Hancock & H. J. (Eds.), *Advances in longitudinal methods in the social and behavioral sciences* (p. 59–79). Information Age Publishing.

Hedeker, D., Mermelstein, R. J., & Demirtas, H. (2008). An application of a mixed-effects location scale model for analysis of ecological momentary assessment (EMA) data. *Biometrics, 64*, 627–634.

Hedeker, D., Mermelstein, R. J., & Demirtas, H. (2012). Modeling between-subject and within-subject variances in ecological momentary assessment data using mixed-effects location scale models. *Statistics in Medicine, 31*, 3328–3336.

Hedeker, D., Mermelstein, R. J., Demirtas, H., & Berbaum, M. L. (2016). A mixed-effects location-scale model for ordinal questionnaire data. *Health Services and Outcomes Research Methodology, 16*(3), 117–131.

Hedeker, D., & Nordgren, R. (2013). Mixregls: A program for mixed-effects location scale analysis. *Journal of Statistical Software, Articles, 52*(12), 1–38.

Hepburn, L., & Eysenck, M. W. (1989). Personality, average mood, and mood variability. *Personality and Individual Differences, 10*(9), 975–983.

Heron, K. E., Everhart, R. S., McHale, S. M., & Smyth, J. M. (2017). Using mobile-technology-based ecological momentary assessment (ema) methods with youth: A systematic review and recommendations. *Journal of Pediatric Psychology, 42*(10), 1087–1107.

Jann, B. (2014). Plotting regression coefficients and other estimates. *The Stata Journal* (4), 708–737.

Kapur, K., Li, X., Blood, E. A., & Hedeker, D. (2015). Bayesian mixed-effects location and scale models for multivariate longitudinal outcomes: An application to ecological momentary assessment data. *Statistics in Medicine, 34*(4), 630–651.

Kimhy, D., Delespaul, P., Corcoran, C., Ahn, H., Yale, S., & Malaspina, D. (2006). Computerized experience sampling method (ESMc): Assessing feasibility and validity among individuals with schizophrenia. *Journal of Psychiatric Research, 40*, 221–230.

Kuppens, P., Van Mechelen, I., Nezlek, J. B., Dossche, D., & Timmermans, T. (2007). Individual differences in core affect variability and their relationship to personality and psychological adjustment. *Emotion*, 7(2), 262–274.

Leckie, G. (2014). Runmixregls: A program to run the mixregls mixed-effects location scale software from within stata. *Journal of Statistical Software, Code Snippets, 59*(2). https://doi.org/10.18637/jss.v059.c02

Leckie, G., French, R., Charlton, C., & Browne, W. (2014). Modeling heterogeneous variance–covariance components in two-level models. *Journal of Educational and Behavioral Statistics, 39*(5), 307–332.

le Grange, D., Gorin, A., Dymek, M., & Stone, A. (2002). Does ecological momentary assessment improve cognitive behavioural therapy for binge eating disorder? A pilot study. *European Eating Disorders Review, 10*, 316–328.

Li, X., & Hedeker, D. (2012). A three-level mixed-effects location scale model with an application to ecological momentary assessment data. *Statistics in Medicine, 31*(26), 3192–3210.

Liao, Y., Skelton, K., Dunton, G., & Bruening, M. (2016). A systematic review of methods and procedures used in ecological momentary assessments of diet and physical activity research in youth: An adapted strobe checklist for reporting ema studies (cremas). *Journal of Medical Internet Research, 18*(6), e151.

Lin, X., Hedeker, D., & Mermelstein, R. J. (2018). A three level Bayesian mixed effects location scale model with an application to ecological momentary assessment data. *Statistics in Medicine, 37*(13), 2108–2119.

Lin, X., Mermelstein, R., & Hedeker, D. (2018). A shared parameter location scale mixed effect model for EMA data subject to informative missing. *Health Services and Outcomes Research Methodology, 18*(4), 227–243.

Liu, L., & Yu, Z. (2008). A likelihood reformulation method in non-normal random effects models. *Statistics in Medicine, 27,* 3105–3124.

Longford, N. T. (1993). *Random coefficient models.* Oxford University Press.

Lu, T. (2017). Mixed-effects location and scale Tobit joint models for heterogeneous longitudinal data with skewness, detection limits, and measurement errors. *Statistical Methods in Medical Research, 27*(12), 3525–3543.

Maher, J. P., Dzubur, E., Nordgren, R., Huh, J., Chou, C.-P., Hedeker, D., & Dunton, G. F. (2018). Do fluctuations in positive affective and physical feeling states predict physical activity and sedentary time? *Psychology of Sport and Exercise. 41,* 153–161.

Maher, J. P., Huh, J., Intille, S., Hedeker, D., & Dunton, G. F. (2018). Greater variability in daily physical activity is associated with poorer mental health profiles among obese adults. *Mental Health and Physical Activity, 14* , 74–81.

May, M., Junghaenel, D. U., Ono, M., Stone, A. A., & Schneider, S. (2018). Ecological momentary assessment methodology in chronic pain research: A systematic review. *The Journal of Pain, 19*(7), 699–716.

Mermelstein, R., Hedeker, D., Flay, B., & Shiffman, S. (2002). *Situational versus intraindividual contributions to adolescents' subjective mood experience of smoking.* Annual Meeting for the Society for Research on Nicotine and Tobacco.

Mermelstein, R., Hedeker, D., Flay, B., & Shiffman, S. (2007). Real-time data capture and adolescent cigarette smoking: Moods and smoking. In A. Stone, S. Shiffman, A. Atienza, & L. Nebeling (Eds.), *The science of real-time data capture: Self-report in health research* (p. 117–135). Oxford University Press.

Muthén, L., & Muthén, B. (1998–2017). *Mplus user's guide, eighth edition.* Muthén & Muthén.

Nesselroade, J. R., & McCollam, K. M. S. (2000). Putting the process in developmental processes. *International Journal of Behavioral Development, 24,* 295–300.

Neuhaus, J. M., & Kalbfleisch, J. D. (1998). Between- and within- cluster covariate effects in the analysis of clustered data. *Biometrics, 54,* 638–645.

Neumann, R., & Strack, F. (2000). "Mood contagion": The automatic transfer of mood between persons. *Journal of Personality and Social Psychology, 79*(2), 211–223.

O'Connell, K. A., Gerkovich, M. M., Cook, M. R., Shiffman, S., Hickcox, M., & Kakolewski, K. E. (1998). Coping in real time: Using ecological momentary assessment techniques to assess coping with the urge to smoke. *Research in Nursing & Health, 21,* 487–497.

Ong, J. C., Hedeker, D., Wyatt, J. K., & Manber, R. (2016). Examining the variability of sleep patterns during treatment for chronic insomnia: Application of a location-scale mixed model. *Journal of Clinical Sleep Medicine, 12*(6), 797–804.

Piasecki, T. M., Hedeker, D., Dierker, L. C., & Mermelstein, R. J. (2016). Progression of nicotine dependence, mood level, and mood variability in adolescent smokers. *Psychology of Addictive Behaviors, 30*(4), 484–493.

Pugach, O., Hedeker, D., & Mermelstein, R. J. (2014). A bivariate mixed-effects location-scale model with application to ecological momentary assessment (EMA) data. *Health Services and Outcomes Research Methodology, 14,* 194–212.

Rast, P., Hofer, S., & Sparks, C. (2012). Modeling individual differences in within-person variation of negative and positive affect in a mixed effects location scale model using bugs/jags. *Multivariate Behavioral Research, 47*(2), 177–200.

Rodriguez-Blanco, L., Carballo, J. J., & Baca-Garcia, E. (2018). Use of ecological momentary assessment (EMA) in non-suicidal self-injury (NSSI): A systematic review. *Psychiatry Research, 263,* 212–219.

Sandhu, S. S., & Leckie, G. (2016). Orthodontic pain trajectories in adolescents: Between-subject and within-subject variability in pain perception. *American Journal of Orthodontics and Dentofacial Orthopedics, 149*(4), 491–500.e4.

Scherer, E. A., Huang, L., & Shrier, L. A. (2016). Application of correlated time-to-event models to ecological momentary assessment data. *Psychometrika, 82*(1), 233–244.

Schwartz, J. E., & Stone, A. (2007). The analysis of real-time momentary data: A practical guide. In A. A. Stone, S. S. Shiffman, A. Atienza, & L. Nebeling (Eds.), *The science of real-time data capture: Self-report in health research* (pp. 76–114). Oxford University Press.

Scollon, C. N., Kim-Prieto, C., & Diener, E. (2003). Experience sampling: Promises and pitfalls, strengths and weaknesses. *Journal of Happiness Studies, 4,* 5–34.

Serre, F., Fatseas, M., Swendsen, J., & Auriacombe, M. (2015). Ecological momentary assessment in the investigation of craving and substance use in daily life: A systematic review. *Drug and Alcohol Dependence, 148,* 1–20.

Shiffman, S. (2005). Dynamic influences on smoking relapse process. *Journal of Personality, 73,* 1715–1748.

Shiffman, S., Stone, A. A., & Hufford, M. R. (2008). Ecological momentary assessment. *Annual Review of Clinical Psychology, 4,* 1–32.

Simonich, H., Wonderlich, S., Crosby, R., Smyth, J., Thompson, K., Redlin, J., Mitchell, J., & Haseltine. (2004). The use of ecological momentary assessment approaches in the study of sexually abused children. *Child Abuse and Neglect, 28,* 803–809. https://doi.org/10.1016/j.chiabu.2004.01.005

Skrondal, A., & Rabe-Hesketh, S. (2004). *Generalized latent variable modeling.* Chapman & Hall/CRC.

Smyth, J. M., & Stone, A. A. (2003). Ecological momentary assessment research in behavioral medicine. *Journal of Happiness Studies, 4,* 35–52.

Stone, A., & Shiffman, S. (1994). Ecological momentary assessment (EMA) in behavioral medicine. *Annals of Behavioral Medicine, 16,* 199–202.

Trull, T., Solhan, M., Tragesser, S., Jahng, S., Wood, P., Piasecki, T., & Watson, D. (2008). Affective instability: Measuring a core feature of borderline personality disorder with ecological momentary assessment. *Journal of Abnormal Psychology, 117,* 647–661.

van de Pol, M., & Wright, J. (2009). A simple method for distinguishing within- versus between-subject effects using mixed models. *Animal Behaviour, 77,* 753–758.

Walls, T. A., & Schafer, J. L. (2006). *Models for intensive longitudinal data.* Oxford University Press.

Walters, R. W., Hoffman, L., & Templin, J. (2018). The power to detect and predict individual differences in intra-individual variability using the mixed-effects location-scale model. *Multivariate Behavioral Research,* 1–15.

Walz, L. C., Nauta, M. H., & aan het Rot, M. (2014). Experience sampling and ecological momentary assessment for studying the daily lives of patients with anxiety disorders: A systematic review. *Journal of Anxiety Disorders, 28*(8), 925–937.

Wen, C. K. F., Schneider, S., Stone, A. A., & Spruijt-Metz, D. (2017). Compliance with mobile ecological momentary assessment protocols in children and adolescents: A systematic review and meta-analysis. *Journal of Medical Internet Research, 19*(4), e132.

Yoshiuchi, K., Yamamoto, Y., & Akabayashi, A. (2008). Application of ecological momentary assessment in stress-related diseases. *BioPsychoSocial Medicine, 2*(1), 13.

SECTION III

DESIGN AND SPECIAL ISSUES

CHAPTER 13

USING LARGE-SCALE COMPLEX SAMPLE DATASETS IN MULTILEVEL MODELING

Laura M. Stapleton
University of Maryland

Scott L. Thomas
University of Wyoming

Researchers recently exposed to instruction in multilevel modeling techniques often hunt for data with which to wield their newfound knowledge and skills. This chapter can be used as a starting point in a search for appropriate data for example analyses and to support research utilizing large-scale complex samples. We focus on data available from the National Center for Education Statistics ([NCES], 2018). Because most of the data collected by NCES is obtained through complex sampling procedures, we first introduce the reader to basic issues in sampling theory, including the concepts of multistage sampling, stratification, and disproportionate selection. Second, we provide detailed descriptions of two popular publicly available datasets that might be useful to build examples of multilevel models.

Multilevel Modeling Methods with Introductory and Advanced Applications, pages 459–494

For these datasets, we provide basic information, including the sampling plan, the content covered, and how the data and documentation can be obtained. Next, we discuss important issues to consider when working with some of these datasets, including the incorporation of sampling weights, the handling of missing data, and a variety of other technical issues. In the chapter's final section, we provide a few recommendations for readers undertaking research using complex sample survey datasets. We share suggested applied readings, as well as pose some challenges for the reader in terms of downloading data and producing analyses. Readers should note that working with these datasets is not an easy task, and there is no way that we can provide all the detailed knowledge necessary to address the complexities in the data in this short chapter. We strongly recommend that readers read the accompanying user's guides for their dataset of interest.

BASIC ISSUES IN SAMPLING THEORY

Before introducing readers to sources of available data for multilevel modeling examples, we need to lay out some terms that we will use regarding the sampling process used to collect data. There are two basic data collection problems encountered when collecting large-scale sample survey data. First, with many large-scale surveys, no simple sampling frame for the target population exists. Second, even if a sampling frame existed, we most likely would want to ensure that our sample has a sufficient number of respondents with certain characteristics (e.g., certain racial/ethnic groups, from different types of schools or colleges, from different regions of the country). Many researchers are interested in research focusing on these smaller segments of a population; thus, a simple random sample (SRS), in which one forms the sample by selecting a pre-specified number of respondents randomly from a sampling frame, might not yield adequate numbers of observations in the subpopulations of interest.

Both of these data collection problems—lack of a sampling frame and need for adequately sized subsamples of interest—are addressed by stratified multistage cluster sampling strategies (Lohr, 1999). Such strategies usually involve sampling in several stages to overcome the problem of not having a single list of the entire population as well as oversampling individuals with certain characteristics. The process of oversampling results in certain elements or participants being sampled at a higher probability of selection than is the case for others in the sample. Sampling in stages as well as oversampling are design issues that need to be accommodated during the analysis phase.

Most NCES survey samples involve several stages of selection and this selection may be stratified on many different variables at each level of sampling. For example, selection of schools might be stratified by sector to include both public and private schools, or selection of students might be stratified by both English learner (EL) status and grade to include specifically EL students in third grade, EL students in fourth grade, and non-EL students in both third and fourth grade.

One implication of complex sampling designs plays out in analyses during estimation of standard errors for parameter estimates, such as a mean or a regression coefficient. For complex sampling designs (be they multistage, stratified, etc.), a measure of the difference in standard errors that would result between the complex sample data and simple random sample data is called the design effect (Kalton, 1983). Design effects (DEFF) greater than 1.0 indicate that the complex sampling design is not as efficient (yields greater standard errors) as compared to a simple random sample of the same size. Specifically, the design effect is the ratio of an estimate's sampling variance (its squared standard error) under the actual complex sample design to the estimate's variance that would have resulted under simple random sampling assumptions (see Rust, 1985; Thomas & Heck, 2001; Wolter, 1985; Woodruff, 1971):

$$DEFF = \frac{SE^2_{COMPLEXSAMPLE}}{SE^2_{SRS}}. \tag{13.1}$$

In the following sections, we provide an overview of the details associated with these particular types of sampling strategies (multistage, stratified, and disproportionate selection) and consider the implications of common sample characteristics for multilevel analyses. Heeringa et al. (2017) provide a particularly helpful and accessible text that addresses issues related to the analysis of data from complex sampling strategies.

Multistage Sampling

Multistage sampling is one of the more common strategies for sample selection, often used when groups are sampled more easily than are individual units. Some example situations in which multistage sampling might be beneficial include sampling children who are clustered or nested within classrooms or schools, sampling teachers who are nested within schools, sampling faculty who are nested within universities, and sampling households nested in census tracts. Generally, each nesting level defines a discrete stage of the sampling process, and each stage has its own set of probabilities defining population members' likelihood for inclusion. In the first stage of selection, the primary sampling units (PSUs) are selected. For example, this might involve obtaining a random sample of schools or universities. In the second or subsequent stages, participants are selected from within those sampled clusters; for example, children or teachers from the selected schools would be included in the final sample. Technically, the term "cluster sample" denotes the situation in which all of the elements in a selected PSU are included in the sample (Kalton, 1983; Kish, 1965). The sampling design for the Trends in International Math and Science Study (TIMSS) is an example of such a sampling strategy where intact classrooms are sampled from selected schools (Martin et al., 2016). When only some of the elements from a selected PSU are included, the sampling design is termed "multistage sampling" (Kalton, 1983).

As is well known to sampling theorists and multilevel modelers, when a sample has been collected via multistage or cluster sampling, the estimates of single-level parameters (such as regression coefficients, means, or correlations) typically will be less precise than had a sample of the same size been collected through simple random sampling (Lee et al., 1989; Skinner et al., 1989). Moreover, if clustered data are analyzed assuming independence of observations, that is, ignoring the clustering, the standard errors for estimates will be underestimated and the degree of this bias will depend on the homogeneity of the response variable(s) across clusters (Lee et al., 1989).

Stratification

In the literature, less attention has been paid to the issue of stratification in complex sample data. Many sampling designs used in the collection of national and international complex sample data sets include a process by which the elements in the population are stratified into mutually exclusive categories and then sampled from these strata in order to ensure representation of each category in the sample. For example, a sample of schools might be stratified by school type (private and public) and/or by U.S. Census region (Northeast, South, Midwest, and West) and then schools are sampled explicitly or implicitly from each of the stratification categories. Explicit sampling involves selecting a specified number or percentage of elements within each stratum (Kalton, 1983). Implicit sampling is accomplished by sorting all of the elements in the sampling frame by stratum and then choosing every ith element on the list, where i is a predetermined interval defined as the ratio of the number of elements in the sampling frame over the number of elements to be sampled (Kalton, 1983). This systematic movement by a specified interval through an entire sorted list typically will result in each of the strata being represented proportionately to the distribution in the population. When database documentation manuals refer to sorting the elements in the sampling frame by a specific set of variables and then using systematic sampling, it is referring to an implicit stratification process.

Stratification can be used at any or all levels of a multistage sampling design. For example, suppose that to obtain a sample of fourth-grade students in U.S. elementary schools, first counties are sampled, then schools in those selected counties are chosen, and then the children in those selected schools are sampled. At the first stage of selection, the counties might be stratified by U.S. Census region and the resulting sample of counties will include some counties from each region of the country. At the second stage of selection, the schools in the selected counties might be stratified by public/private status and sorted by school size, resulting in the use of explicit stratification on school type and implicit stratification on size. At the final stage of selection, the students might be stratified by English learner status.

In general, any time stratification is used as part of the sampling design and the response variable is homogenous within strata, the estimates from the sample will be more precise than had a sample of the same size been obtained through simple random sampling (Kalton, 1983; Kish, 1965; Lee et al., 1989). This increase in precision is related to the guarantee that different types of people (or units) will be included in the sample—it will be nearly impossible to obtain a "bad" (or non-representative) sample with an extreme parameter estimate. Related to this increase in precision of estimates, analyses ignoring the fact that stratification was used tend to result in overestimates of the standard errors associated with parameter estimates and a subsequent Type I error rate lower than the nominal rate. In essence, the result is the converse of that found with multistage or cluster sampling. The use of stratification alone would result in a design effect less than 1.0 (Kalton, 1983). However, most sampling designs include both stratification and multistage sampling and the increase in precision of estimates resulting from stratification is smaller than the decrease in precision found with multistage sampling (Kalton, 1983). Consequently, the issue of stratification has received less attention than the issue of multistage sampling in recent modeling literature.

Disproportionate Selection

If not properly accounted for in the analysis, disproportionate selection of elements into the sample can adversely affect the resulting analysis estimates (e.g., regression coefficients). This brief section introduces the concepts surrounding disproportionate selection and sampling weights; methods to accommodate these sampling weights into an analysis will be discussed in the third section of this chapter. At various stages of the sampling design, the elements (such as schools and students) typically are selected at different sampling rates. We discuss three types of disproportionate sampling here: probability proportionate to size (PPS) sampling, over-sampling by stratum, and post-stratification adjustments. First, consider the example of a simple two-stage sampling design. Suppose we want to obtain a sample of 10,000 seniors from public high schools in the United States. In Fall 2015, there were about 3,500,000 12th-grade students in the United States in about 20,000 public schools (U.S. Department of Education, 2018). One way to obtain a sample of 10,000 students would be to select students randomly out of the total pool of 3,500,000 students; for a given individual i who is in school j we would be using a selection probability of π_{ij} = 10,000/3,500,000 = .0028 (or 28 out of every 10,000 students). We could, for example, take a list of 3,500,000 students, select every 350th name on the list, and we would obtain a sample of 10,000 names; each of the names had a .0028 chance of being selected into the sample. The sampling interval of 350 was obtained by dividing the total population by the number of desired sample elements. Note that this sampling interval is also the reciprocal of the selection probability: 1/.0028 = 350 which is also referred to as the "sampling weight." Each person in our hypothetical sample represents 350 people from the original population.

There is a complication in our example, however, in that there is no single complete list or existing sampling frame of all students at every school, so it is not logistically feasible for us to conduct a simple random sample of students. Multistage sampling solves this problem: first sample a group of schools, asking those selected schools for a list of their students, and then take a selection of those listed students. The difficulty with this approach, however, is that two selection probabilities must be determined, one for the schools (π_j) and one for the students within the selected schools ($\pi_{i|j}$). To continue the example from above, to obtain a sample of 10,000 students with a multistage approach, the product of these two probabilities must equal the overall selection probability of .0028. A value for one of these probabilities can be selected, and the other value will, therefore, be determined. For example, suppose it was decided to sample 5% (or .05) of the students within each selected school ($\pi_{i|j}$ = .05). That would mean that we would need to select 5.6% of the schools because .056 × .05 = .0028. So, if we sample 5.6% of the schools (π_j = .056) and sample 5% of the students at each selected school ($\pi_{i|j}$ = .05), we would obtain a sample with an expected size of 10,000 and each element in the population would have an overall selection rate of .0028 ($\pi_{ij} = \pi_j \times \pi_{i|j}$ = .056 × .05 = .0028).

However, the number of enrolled students typically varies across schools and that with this proposed process we might sample a relatively large number of students in very large schools and a small number in relatively small schools. It is usually more administratively efficient to conduct surveys and assessments in a standardized manner across schools. Looking across the sample designs for the publicly available data from the National Center for Education Statistics (NCES), in recent years it is apparent that this number is often around 20 to 25 participants per site. Therefore, an alternate plan might be to sample schools at a fixed selection rate and then to sample a prespecified number of students at each school, for example, 20 students. With this two-stage sampling design and taking a specified number of participants at each site, the students in small schools have a very high probability of selection into the sample if their school is chosen. For a school with only 20 12th graders, the selection rate would be $\pi_{i|j} = 20/20 = 1.00$. Conversely, students who are in very large schools have a relatively small chance of being selected for the sample if their school is chosen. Students in a school with 1,000 12th graders would have a selection rate of $\pi_{i|j} = 20/1000 = .02$. If the schools are sampled with equal probabilities (say, $\pi_j = .056$), then students from these two different schools would have very different overall rates of selection: $\pi_{ij} = \pi_j \times \pi_{i|j} = .056 \times 1.00 = .056$ for students in the small school and $\pi_{ij} = \pi_j \times \pi_{i|j} = .056 \times .02 = .00112$ for students in the large school. Consequently, students in small schools would be overrepresented in the sample.

To avoid having unequal numbers of selected students per school or unequal selection probabilities for individual students across schools, schools can be selected with a method that samples large schools at higher rates and smaller schools at lower rates, rather than sample schools with equal probability. This method, called probability proportionate to size (PPS) sampling, is commonly used in national data collection strategies. With PPS sampling, the overall selection probabilities for students (π_{ij}) will tend to be equivalent, but their conditional probabilities ($\pi_{i|j}$) within their respective schools will differ. For example, suppose with our previous example that we wanted to select about 20 students at each school no matter the size of the school. A school with 20 12th graders would need to have a .0028 chance of selection into the sample to result in an overall probability of inclusion for a student in that school of .0028 given that the conditional selection probability is 1.0 (20/20): ($\pi_{ij} = \pi_j \times \pi_{i|j} = .0028 \times 1.0 = .0028$). Alternately, a school with 1,000 12th graders (and thus a conditional probability of .02 (20/1,000) would need to have a chance of selection of .14 to result in an overall selection probability for the students in those schools of .0028 ($\pi_{ij} = \pi_j \times \pi_{i|j} = .14 \times .02 = .0028$).

In addition to accounting for differing sizes of initial sampling units, disproportionate selection also may be used to obtain sufficient numbers of elements to undertake subgroup reporting. For example, special interest might lie in reporting estimates for Asian American students or teachers (as is the case in the Early Childhood Longitudinal Study of Kindergarten students and the Schools and Staffing Survey involving teachers; Tourkin et al., 2004; Tourangeau et al., 2006). When the desire is to report estimates by subgroup, sample designers might employ a higher rate of sampling in certain strata than the rate used in other strata. Thus, with NCES datasets, it is not unusual to find sampling weights that differ for teachers or students within the same school.

Finally, similar to the approach of over-sampling, non-response weighting adjustments might be employed by the survey designers to adjust for those who opt not to participate or respond to the survey. Non-response to the survey might result in differing proportions of participants in the respondent sample as compared to the desired sample. Although equal selection probabilities might have been used for all elements in the initial sampling plan, some groups typically respond at lower rates. For example, women tend to respond to surveys at higher rates than men (Dillman, 2000), and in a survey of faculty, although men and women might have been sampled at an equal rate, women might have responded at a higher rate. After survey data are collected, sampling weights for responses from men would be adjusted higher to reflect their true proportion in the population (if it is known or can be approximated), and women's responses would be associated with relatively smaller adjustments to the weights.

Statistically, the use of sampling weights to address non-response reduces bias in parameter estimates resulting from differential selection or non-response but does so at the expense of precision (Kalton, 1983; Skinner et al., 1989). If the differential selection or non-response is unrelated to the response variable, then no bias will exist in the parameter estimate, thus rendering the use of weights unnecessary. By including weights (either developed through post-stratification adjustment or due to initial disproportionate selection probabilities), estimated standard errors will be larger than if they had been estimated from an equal probability of selection sample. Therefore, the weighted analysis becomes less powerful. A primary (and familiar) challenge for the analyst is to find the appropriate trade-off between biased standard error and biased coefficient estimates.

We will return to these issues of sampling and disproportionate selection rates in the third section of this chapter when we discuss the use of sampling weights in multilevel analyses. Given these brief descriptions of issues in the commonly used sampling procedures by NCES, we hope data users will be more comfortable when they encounter discussions of strata, primary sampling units, and weights in technical manuals or journal articles or the variables themselves in the data set.

PUBLICLY AVAILABLE DATASETS FOR USE IN MULTILEVEL MODELING

In this section, we provide descriptions of two education-related datasets from NCES that are available to the public and that can be used for research as well as in multilevel courses for demonstration purposes. We present one cross-sectional database (Schools and Staffing Survey [SASS]) and one longitudinal database (Early Childhood Longitudinal Study, Kindergarten Class of 1998–1999 [ECLS-K]). For each dataset, we provide information on the important aspects of the sampling design used for data collection, a review of the content of the collected data, how to obtain the data, and special concerns with the dataset structure or required analysis approach of which we are aware. In an appendix, we provide a brief list of additional datasets that may be of interest to multilevel modelers. We cannot stress enough that the few paragraphs that we provide here are not a replacement for a careful reading of the user's guides that are associated with each of these data sets. Our goal here is simply to put those user's guides in a more accessible and helpful context. Documentation on the NCES website should be consulted for further information on the surveys discussed below.

In NCES datasets, there are three general types of data available: sample design data, questionnaire responses or assessment results for participants, and derived variables created by the study administrators. The sample design information includes data about the sample elements or higher-level units that are available prior to data collection, such as U.S. Census region, public or private status of schools, and urban/rural location of the schools. The great bulk of the variables on any dataset are specific responses to questionnaires that were administered to the participants; copies of these questionnaires are usually available on the survey program's website and in technical manuals written for the specific dataset. For participants who are too young to complete written questionnaires (such as the youngest children who participate in the ECLS-K), results from interviewer-administered standardized assessments typically are provided. Finally, on any dataset, the study developers usually include derived variables, which can be extremely useful for research and analysis purposes. An example might be socioeconomic status (SES); this variable often is created from participant and/or parent/guardian responses to questions about their income, their education level, and their occupation as reported in database user guides.

Before we present our discussion of the specific databases, we note that two different databases are usually available for any given dataset: publicly available and restricted-use datasets. There is concern that if all available data for participants were provided to the public, it might be possible to identify particular individuals. For example, within the SASS dataset, there might be an Asian American male teacher from North Dakota who responded to the set of questionnaires. If data were available about the state and level of the school, coupled with the school enrollment size, the discipline in which he teaches, and his age, we very well might be able to pinpoint the exact individual who must have answered the questionnaire, compromising the confidentiality of his responses. Therefore, on its publicly-available datasets,

NCES masks some of the data by either removing it from the database or by categorizing continuous variables. In this example, perhaps the variable "discipline in which the respondent teaches" would not be available on the public-use dataset. Additionally, the variable "age" might be recorded as an ordinal variable with categories of "younger than 30," "30 to 44," "45 to 55," and "56 and over." As a precaution, researchers should not assume that all data collected with the questionnaires will be available to the end-user on the public-use data files. If a particular piece of data is crucial to an analysis, a researcher has the option to apply for a restricted-use license. To obtain such a license, the researcher must document the security procedures that will be in place to keep the data confidential. More information is available at the NCES website: http://nces.ed.gov/statprog/rudman/

The datasets described below are available free to the public via the NCES website https://nces.ed.gov/OnlineCodebook/. Note that the Online Codebook replaced the former EDAT system in 2018. On this website, as of the writing of this chapter, datasets for seven survey programs are available. The researcher or analyst then can use the application on the website to identify variables of interest. Once all variables needed for analysis are identified, the user can then request the creation of SPSS, SAS, S-Plus, Stata or R syntax for extraction of the selected data elements. The user can then download both the full datafile as well as the syntax file that will extract the selected data elements (the user may need to edit this syntax slightly to refer to local path names). It should be noted that some variables are automatically included in the extract syntax; these are typically accompanied by selected sampling design variables that the database designers at NCES deem important for all analysts to have (e.g., sample design information or sampling weights, etc.).

Example Dataset 1: Schools and Staffing Survey

Sampling Design

The Schools and Staffing Survey (SASS) is a cross-sectional survey that involves the collection of data at two levels: At each selected school, responses are obtained from a sample of teachers (level 1) and responses are provided by the principal and by an administrative contact (level 2; Tourkin et al., 2004). This survey is conducted periodically, and the most recent data available to the public are from the 1999–2000 school year (released April, 2005). Two additional administrations of the SASS survey occurred in subsequent years but those data are available on restricted release only. After 2010–2011, NCES redesigned SASS and renamed it. It is now referred to as the National Teacher and Principal Survey (NTPS); the redesigned study's focus is on the teacher and principal labor market as well as K–12 school staff. For the 1999–2000 SASS survey, two separate sampling processes were undertaken to obtain the sample, one for public schools and one for private schools, thus utilizing explicit stratification on school type. For the public school sample, the goals of sampling included the desire to report both national and state estimates of school and teacher characteristics at both the elementary and secondary level and to over-sample schools with a higher percentage of Native American enrollments so that national estimates for these schools might be produced. Some schools, such as public charter schools, were selected with certainty (they automatically were included in the sample). For the remainder of the schools in the public school sampling frame, several levels of stratification were used: Schools were divided into the categories of Native American schools (19.5% or more of the enrolled students were Native American), schools in Nevada, West Virginia, or Delaware (which required a different sampling strategy given the school system structure), and all other schools (note that all Alaskan schools were placed in this third group regardless of Native American enrollment). At the second level of stratification, in general, the schools were stratified by state or state and district. At the third level of stratification, within states, the schools were grouped into categories of elementary, secondary, or combined (spanning at least Grades 6 and 9) schools. Within each of these strata, the public schools were sorted by region, highest grade, percent minority, and enrollment. Systematic PPS sampling then was used to select schools. Disproportionate rates of sampling of schools were used across the explicit strata, and as an example, the sampling rates for states ranged from .051 in Illinois to .429 in the District of Columbia. Using this sampling process, a total of 11,139 public schools were selected.

To develop the private school sample, a list of schools was developed as the sampling occurred; 120 PSUs (counties) were chosen across the United States in pairs within 60 strata defined by region. Because no inclusive list already existed for all private schools in the United States, all private schools within these PSUs were attempted to be identified using a variety of methods. The goals of the sample allocation included the ability to provide estimates for each of 20 school affiliations and provide national comparisons of private versus public schools. Schools were therefore stratified explicitly by affiliation and school grade level (elementary, secondary, and combined). Within PSUs, schools were sorted by affiliation, highest grade in school, location type and enrollment. Systematic PPS sampling then was used to select schools. In all, 3,560 private schools were selected for the sample.

For each sampled public and private school, a list of current teachers was obtained and stratified into one of five categories: teachers who are Asian or Pacific Islander (API), teachers who are American Indian or Alaskan Native (AIAN), teachers who are assigned to teach limited English proficient (LEP) classes, teachers with fewer than 4 years of experience, and teachers who have 4 or more years of experience. Teachers falling into more than one category were placed in the first as listed above. The sampling rates of teachers within a school depended on the selection rates for the school, resulting in approximately equal overall rates of selection for all teachers in the sampling frame. A maximum of 20 teachers were intended to be selected from a school, with oversampling of API, AIAN, and LEP teachers. Within private schools, new teachers were sampled at higher rates than experienced teachers. Across all categories of schools, a total of 72,058 teachers were selected for the sample. Approximately 80% of teachers responded to the survey; the final dataset contains data for 42,086 public school teachers from 8,712 schools, and 7,098 private school teachers from 2,504 schools.

Content

The publicly available dataset contains hundreds of variables on school, principal, and teacher characteristics. The dataset contains teacher respondent information, such as satisfaction with class size, job security, and salary (measured on Likert-type scales), classes assigned and grade levels, income and benefits, licensure and academic preparation, perceived professional development support and availability, demographic information, the teacher's career path, and instructional practices, including hours spent on specific subjects in class and computer use. Principal respondent information includes the principal's career path and demographic information, goals for the school and an assessment of how close to the goals the school is perceived to be, responses to questions pertaining to levels of influence of various in-school and out-of-school forces on school policies and practices, problems with student behavior and teacher performance, activities undertaken as principal in the last month, and whether the school had or had not met standards and, if not, the required action. School-level data, which were provided by a contact at the school (who may or may not have been the principal), contain information on grade levels at the school, enrollments by race/ethnicity and gender, absentee information, admission requirements, program availability (such as Advanced Placement and technical preparatory programs), student scheduling policies (use of tracking, block schedules, team teaching, etc.), school calendar, student outcomes after graduation, parental involvement activities and levels of participation, security and violence prevention policies, staffing numbers, and vacancies.

Summary of the Data Structure

The available data lend themselves to many 2-level modeling examples, using principal and school data as level-2 variables and teacher data at level-1. The number of teachers per school is somewhat low, however, ranging from 1 to 19 with an average of about 4.8. The distribution of schools by number of sampled teachers is shown in Table 13.1.

TABLE 13.1 Number and Percent of Schools by Size of Teacher Sample from the 1999–2000 SASS Dataset

Size of Teacher Sample	Number of Schools	Percent of Schools
1	504	5.8
2	1,089	12.5
3	1,473	16.9
4	1,543	17.7
5	1,217	14.0
6	929	10.7
7	661	7.6
8	455	5.2
9	298	3.4
10 or more	543	6.2

In general, this dataset would be useful for 2-level contextual analyses, given the simple structure of teachers within schools. However the number of teachers within a school may have implications for some multilevel analyses. Most immediately, the number of within-cluster observations (teachers in this case) has an impact on the reliability with which the population means for each cluster, and relations within each cluster, are estimated. Smaller within-cluster samples yield relatively less reliable estimates of these population parameters than do larger within-cluster samples. Because sample sizes within each level-2 unit (schools in this case) are likely to differ, reliability will vary across level-2 units, also. Some multilevel modeling programs (e.g., HLM) provide a summary measure of these reliabilities that takes the form of the average reliability across all level-2 units. Whereas the ICC provides a sense of the degree to which variability exists between level-2 units, the reliability of these within-school estimates determines the degree to which observed differences between-groups are representative of real population differences rather than sampling error. It should be clear that the ICC and associated reliabilities are related directly to the ability to detect between-unit differences at level 2 (i.e., power; Heck & Thomas, 2020).

A user should note that when data are downloaded from the Online Codebook, several datasets may be downloaded after variable tagging, one for each type of survey respondent (public school, public principal, public teacher, private school, private principal, and private teacher). The public and private datasets need to be appended (vertically concatenated into one file); these files already contain the same variable names, and thus concatenating the files is generally a simple process. Depending on the multilevel software to be used (see McCoach et al., 2018), the principal and/or school information can be merged with the teacher data by SCHCNTL, a school identification code. Individuals within schools are identified by a personal identification number, CNTLNUM. For more investigation of this data set, see exercises 1–4 in the "Try This!" section at the end of the chapter.

On each of the SASS datasets, there is a unique sampling weight variable, and thus separate weights exist for the school, principal, and teacher levels. These weights represent the reciprocal of the overall selection rates of the element in the sample, adjusted for non-response. School and principal rates are not necessarily the same, due to principal non-response.

Where to Obtain More Information

More information about the SASS dataset is available at: http://nces.ed.gov/surveys/sass/. At the website, a user can obtain copies of the questionnaires, along with more information about the data collection process, summary data tables, and a listing of all the reports that the NCES has produced using this dataset. In the documentation, a user's guide provides the estimated design effects for key variables across the sample design (see Appendix G of the 1999–2000 SASS User's Guide). A review of these design effects may help to identify variables that will likely show larger intraclass correlations. Note that under the tab at the top called "Data Projects," a user can see all data available for the SASS survey program, through either a restricted license process or public release.

Example Dataset II: Early Childhood Longitudinal Study, Kindergarten Class of 1998–1999

Sample Design

In 1998, the U.S. Department of Education began a longitudinal assessment of cognitive and noncognitive abilities of children, along with a collection of information about their home and school environments. This study began with a sample of kindergarten children and assessed these children repeatedly at different times over the first 9 years of their school careers (kindergarten through approximately eighth grade). Briefly, the ECLS-K (1998–1999) sampling design includes three stages of sampling: PSUs of single counties or groups of counties, schools within those counties, and students within the selected schools. First-stage strata were defined by a combination of census metropolitan statistical area, census region, proportion of the population of a specific race/ethnicity, size of the PSU and average per capita income of the PSU. One hundred PSUs were selected within these strata and within each of the PSUs, schools were stratified by public/private status and then were sampled with further implicit stratification on size of the school and proportion of Asian/Pacific Islander (API) students. Finally, within 1,280 sampled schools, students were selected using two strata: API students and non-API students. API students were selected at a rate three times greater than non-API students (when population numbers allowed), for a target size of 24 students per school.

Six follow-up data collection waves occurred for the kindergarten cohort. After the kindergarten students were assessed initially in the Fall of 1998, the students were again assessed in the Spring of 1999. In the subsequent year, a 30% subsample was selected in the fall of the first grade to facilitate the assessment of "summer loss" in achievement. In the Spring of 2000, a "freshened" sample of the original cohort of students, now mostly in first grade, was assessed. A freshened sample was required to capture students who were now in first grade but had not been enrolled in kindergarten the prior year (and thus did not have a chance to be selected in the original cohort). This freshened sample allows analysts to make inferences to the population of first grade students in the United States. If the freshening had not occurred, the inference population is limited to first grade students who had been in a U.S. kindergarten the prior year. Two years later, in the Spring of 2002, the freshened sample again was assessed. NCES refers to this data collection period as "third-grade collection" but not all students were in the third grade; 89% of the respondents were actually in third grade, 8% were in second grade, and the remaining students were in other grades or not enrolled. Data from the wave of data collection in Spring 2004 is referred to as the fifth-grade follow-up, although just 87% are actually in fifth grade. In Spring of 2007, data were collected on these students; NCES refers to this data collection period as the eighth-grade follow-up, although, again, not all students actually were in eighth grade.

Content

During each of the data collection stages, data were collected from the child, the child's parent or guardian, a teacher of the child, and the principal or school administrator. In the data collection that took place in kindergarten and first grade, children were assessed as they took part in cognitive and non-cognitive tasks. In the third, fifth, and eighth grade collections, in addition to the assessments, students also were asked to complete questionnaires about their perceptions of their abilities and their interest and attitudes toward school subjects and learning. Using item-response theory, summary scores were created from the assessment information for the cognitive and non-cognitive tasks and are available on the data set.

Parents were interviewed to obtain data, including information about the home, past day care arrangements, family structure, and child behaviors and activities, as well as information to assess the parent's level of depression and family functioning. Teachers provided information using a self-administered questionnaire on their own teaching practices and background as well as an evaluation of the child's cognitive and noncognitive development.

School staff were asked to provide information about student attendance, transfer, and individual educational plans, if applicable.

Summary of Data Structure

Out of the 21,409 records on the kindergarten to eighth-grade longitudinal dataset, there are 19,173 observations that have sampling weights greater than 0 in the first wave of data collection (the remaining records are part of the freshened samples for spring of kindergarten and spring of first grade or non-responders in the first wave). These students attended 949 separate schools. Students of multilevel modeling may be interested in either a cross-sectional analysis of students within schools, a longitudinal analysis of students within occasions, or a cross of the two with a 3-level model of occasions nested within students nested within schools. Note, however, the difficulty of modeling students within schools over time. Students change schools (especially between elementary and middle school, or by entering or leaving school districts) and thus a pure nesting does not exist. Multiple membership and cross-classified models are likely needed in these situations (see Beretvas, 2011, for more discussion of these models as well as Chapter 6: "Cross-Classified Random-Effects Models" in this volume); note that it is likely that these models will be extraordinarily complex given that school-level sampling weights are not defined for schools over time—only at initial sample selection.

For those interested in cross-sectional data analyses, Table 13.2 shares information on the number of students per school in the two cohorts that allow for generalization to the population (the initial cohort in fall kindergarten and the freshened spring first grade). In the fall of kindergarten, data were available for 19,173 students across 949 schools, resulting in an average of

TABLE 13.2 Number of Students within Schools at Fall of Kindergarten and Spring of 1st Grade in the ECLS-K 1998–1999 Dataset

	Wave 1 (Fall of Kindergarten)		Wave 4 (Spring of 1st Grade)	
Size of Student Sample	Number of Schools	Percent of Schools	Number of Schools	Percent of Schools
1–5	32	3%	1,060	53%
6–10	42	4%	78	4%
11–15	52	5%	263	13%
16–20	187	20%	440	22%
21–25	624	66%	138	7%
26 or more	12	1%	7	<1%
Total Number of Schools	949	100%	1,986	100%

about 20 students per school, and this number ranged from 1 to 27 students per school. As shown in Table 13.2, the majority of schools had data available for between 21 and 25 students. Over time, as students transferred to new schools, the number of schools with data available for only one student rises dramatically across the data collection waves. For the spring of first grade data collection, over 800 schools included only one student. By design, students were followed, even if they changed schools. In the ECLS–K, when a student changed schools, he or she typically entered a school that was not already in the study and, therefore, created a new school in the sample with only one student. The reliabilities of schools represented by a single respondent are, of course, quite low, and algorithms in most multilevel modeling programs place very little weight on the contributions of these schools to level-1 estimates. The decision to include or exclude level-2 units with small sample sizes should depend on the goals of the analysis.

Of the initial Fall 1998 kindergarten cohort of students, data for 75% are available up to the third-grade data collection period. Moreover, about 62% of the initial cohort were assessed at waves 1, 2, 4, 5, and 6, making for a fairly large set of complete longitudinal data. Note that some students were not available to be assessed and dropped from the study at various time points. For most of the children who participated in all data collection periods, their parent and teacher data are also available, roughly coinciding with the time frame during which the children were assessed. Table 13.3 provides data on the drop-out/stop-out pattern for those who are interested in longitudinal modeling of occasions nested within students.

Note that users must be cognizant of their population of interest, whether it be kindergarten students, first graders, and so forth. Analyses using the kindergarten data and any longitudinal analyses that start with students in

TABLE 13.3 Selected Patterns of Data Collection for the *N* = 19,173 Kindergarten Students Who Participated in the First Wave of Data Collection of ECLS-K 1998–1999

	Wave1 Fall K	Wave2 Spring K	Wave4 Spring 1	Wave5 Spring 3	Wave6 Spring 5	Wave7 Spring 8	*N*
Pattern 1	X	X	X	X	X	X	7,803
Pattern 2	X	X	X	X	X		1,993
Pattern 3	X	X	X	X			2,594
Pattern 4	X	X	X				2,162
Pattern 5	X	X					2,784
Pattern 6	X						747
Pattern 7	X	*at least one in between*				X	667
Pattern 8	X	*other pattern*					423

Note: X indicates that data are observed.

kindergarten will be representative of all kindergarten students in the United States in 1998; teacher data from such analyses are nationally representative of teachers of kindergarten students for the 1998–1999 academic year, and the school-level data represent all schools with kindergartens during the 1998–1999 academic year. Data for first graders (from the Spring 2000 follow-up) are representative of all first-grade students in the United States (because of the use of a freshened sample), but the teacher and school-level data are not representative of all U.S. first-grade teachers and schools. The Spring 2002 and 2004 and 2007 follow-up data are *neither* representative of all students in the third, fifth, or eighth grade, nor all teachers of third, fifth, or eighth graders, nor all schools with third, fifth, or eighth grades. These data are intended to be used for longitudinal purposes and not for cross-sectional analyses.

Another possible complication in the use of the ECLS-K data is that not all students are assessed at the same time point. It is not feasible to have assessment personnel visit all 16,000 students at the same time. In fact, the assessments are spread out across a few months. For example, in the Spring 1999 data collection, students were assessed anytime between March and July. Most students (92%) were assessed in April and May, but others were outside that time window. With the cognitive growth that is occurring in children at these ages, the difference between March and July assessments might be quite large. There is a section of the fifth-grade combined user's manual (Tourangeau et al., 2006) that discusses measuring gains (Section 3.1.6) when students are assessed at differing time points across data collection waves, and the NCES cautions users to consider child age and assessment date when comparing gains within relatively short periods of time (such as within-grade or across a 1-year time point).

Unlike the SASS dataset that contains just one sampling probability weight variable for each observation, the kindergarten to eighth-grade longitudinal ECLS-K dataset contains 71 sampling probability weights for each child record in the dataset, not including the hundreds of replicate weights. Because there are several sources of information about each child (child, parent, teacher, school) and because there are several data collection waves, the appropriate weight to use will depend on whether the analysis model includes certain kinds of variables. In Section 4.8 of the user's manual, cross-sectional weights are described, and Section 9.3 of the user's manual contains a discussion on longitudinal analysis strategies including weighting options (Tourangeau et al., 2009). To gain experience with this data set, attempt exercises 5 through 10 in the "Try This!" section at the end of the chapter.

Where to Obtain More Information

More information about the dataset is available at: https://nces.ed.gov/ecls/kindergarten.asp. At the website, a user can obtain copies of the questionnaires, along with more detailed information about the data collection process, summary data tables, and a listing of all the reports that the NCES has produced using this dataset. There was a new data collection program started in the 2010–2011 school year for a second kindergarten cohort. This cohort has now been followed through fourth grade. The public release file is not on the Online Codebook and is available instead through a tool called the electronic codebook (ECB) which must first be installed on a user's computer. For more information, see: https://nces.ed.gov/ecls/dataproducts.asp#K-2.

ADDITIONAL INFORMATION: DATASETS AND TRAINING

In the appendix, we have provided a list of additional datasets that may be of interest to readers. For those readers who are interested in more targeted training in learning how to work with NCES data sets, NCES provides training opportunities and these opportunities typically are announced at the following website: http://nces.ed.gov/whatsnew/conferences/ in the "Conferences, Workshop/Training & Technical Assistance" section.

ISSUES WHEN WORKING WITH LARGE-SCALE COMPLEX SAMPLE DATASETS

In this section, we focus on several important analytic issues for researchers working with datasets such as the surveys we have described above. These issues include the use of sampling weights, the accommodation of missing data, the careful consideration of the level of analysis, and attention to the sampling design.

Sampling Weights

Probably one of the most challenging aspects of working with these datasets is determining the appropriate way to accommodate sampling weights into the analysis. First, we review a simple single-level research situation, and then we extend the use of weights to multilevel designs. Suppose that data were collected using a simple stratified sampling design. Specifically, 100 children were sampled from a school of 1,000 children. In the population, there were 800 non-Hispanic children and 200 Hispanic children. For the sample, however, explicit stratification by ethnicity was used, and 50 non-Hispanic children and 50 Hispanic children were sampled. The probability of selection for a non-Hispanic child was 50/800 = .0625 and the probability of selection for a Hispanic child was 50/200 = .25. After data were collected, suppose interest lies in using the data to estimate the mean level of student achievement. We hope that it is clear from our discussion to this point in the chapter that the disproportionate selection in the sampling design demands attention.

To obtain unbiased parameter estimates, the observations in the sample dataset most likely should be weighted to bring the intentionally uneven sample proportions back in correspondence with the known population proportions. Failure to apply sample weights will result in an artificial influence of groups that were oversampled. In this case, Hispanic children were oversampled at a rate of 4:1 and failure to weight the sample back to the correct proportion will result in any estimates being biased toward the characteristics of the Hispanic children in the sample—that is, Hispanic children will be accorded disproportionate weight in any estimates made directly from the unweighted sample. If Hispanic children perform higher on the achievement test of interest, then the overall estimated mean will be biased upward by the effective quadruple-counting in the sample, and thus the estimate will not be reflective of the true overall population mean on the test. The weights used to "balance" such quadruple-counting resulting from the sample design simply reflect the inverse of any sample element's overall probability of their inclusion in the sample. Thus, after applying the sampling weights in this simple single-level example, a response from a Hispanic student will be accorded a weight of $w_i = 1/.25 = 4$, and a response from a non-Hispanic student will be accorded a weight of $w_i = 1/.0625 = 16$. These weights are used in determining the parameter estimate. For any dataset, summing the raw weights across all observations yields an estimate of the population N:

$$\sum_{i=1}^{n} w_i = N. \tag{13.2}$$

While statistical packages vary in the way they use weights to calculate certain statistics, most calculate the weighted mean as follows:

$$\bar{x} = \sum_{i=1}^{n} w_i x_i \Big/ \sum_{i=1}^{n} w_i, \tag{13.3}$$

which is the sum of the products of each observation's raw weight and value for x, divided by the sum of the raw weights, N. Notice that the sum of the raw weights (Σw_i), or the size of the target population, now becomes what is referred to as the "effective sample size" in this calculation. A consequence of using the raw weights supplied with most complex survey data is that, when calculating SE estimates, some statistical packages that treat weights as "frequency weights" (IBM SPSS included when you select "weight cases" under the "Data" menu) are fooled into believing that the sample size is much larger than it really is. While use of the raw weights yields an unbiased point estimate for the mean in all software packages, in some software packages, like SPSS, analyses using the raw weights result in an effective sample size that is the same as the population *N*. The use of this inflated effective sample size can compromise some calculations, such as standard errors and sampling variances as well as variances and covariances. Most software now correct for this problem with special survey analysis procedures that rescale the weights so that the effective sample size resembles the actual sample size and not the population size. In particular, SPSS now has a "Complex Samples" function under the "Analysis" menu, through which the weights can be specifically identified for a particular analysis.

The user is encouraged to review the most recent technical manuals that are available for the software to be sure the weight is being handled in an expected manner. For most types of analyses, multilevel modeling software offers researchers the opportunity to provide sampling weights at each of the levels of their model. The ability to weight at different levels of the model requires a consideration of the ways in which disproportionate sampling occurs at different levels and in how combining weights at these different levels can affect parameter estimates. To this point in the chapter, we have considered weighting issues in the context of a single level of analysis only. As we hope to make clear in this section, multilevel analyses require a more complex way of thinking about the purpose and behavior of sample weights. On which level(s) should one apply weights? Which weights are appropriate—weights that reflect the sampling rate of the school, weights that reflect the conditional sampling rate of the student, or weights that combine those sampling rates? These are crucial issues to consider when using data from complex samples in multilevel analyses. Considerable debate exists about the correct choice of weights to use for a multilevel analysis. The central features of that debate include the appropriate scaling of the weight and the best estimation method for the parameter and variance components with models incorporating sampling weights. Simulation studies have been conducted, and no method is clearly the most appropriate under all circumstances; success of the methods depend on conditions such as whether the response variable is correlated with the sampling weights (referred to as informative sampling weights), whether the model includes continuous or dichotomous outcomes, and the sample sizes at level 1 and level 2 (Asparouhov, 2005; Pfeffermann et al., 1998; Rabe-Hesketh & Skrondal, 2006; Snijders & Bosker, 2012).

To clarify some of the decisions that a researcher must address, let us examine a very simple example. Suppose that we draw a stratified sample from a school district. In the population there are five public schools with 300 students enrolled in each of them and 25 private schools, each with just 20 students enrolled, for a total population of 2,000 students (1,500 public and 500 private). We select two schools, using explicit stratified sampling and select one public school at a probability of $\pi_1 = 1/5 = .2$ and one private school at a probability of $\pi_2 = 1/25 = .04$. The respective raw sampling weights at the school level for these two schools are $w_1 = 1/.2 = 5$ and $w_2 = 1/.04 = 25$. The public school represents five schools, and the private school represents 25 schools. Once the schools are selected, 10 students are selected from each. Students in the public school are selected with a conditional probability of $\pi_{i|1} = 10\ /\ 300 = .033$, and students at the private school are selected at a conditional probability of $\pi_{i|2} = 10/20 = .5$. The conditional raw sampling weight for students in the public school is therefore $w_{i|1} = 1/.033 = 30$; thus, each public-school student in the sample represents 30 students in the population from that specific school. In the private school the conditional raw sampling weight is $w_{i|2} = 1/.5 = 2$; thus, each private-school student in the sample represents just two students from that specific school in the population. Overall sampling weights can be calculated by taking the product of the school-level and conditional student-level weights. These overall weights indicate the number of students in the overall population represented by each student selected for the sample, after adjusting for strata (public or private school). Thus, the selected students from the public schools each represent $w_{i1} = w_1 \times w_{i|1} = 5 \times 30 = 150$ students from the original school district population, and the selected students from the private schools each represent $w_{i2} = w_2 \times w_{i|2} = 25 \times 2 = 50$ students from the original school district population. Given this scenario, the typical weight provided on an NCES database would be 150 for public school students and 50 for private school students (NCES usually provides overall weights on databases and not within-school conditional sampling weights). For more investigation of weights, see exercises 11 and 12 in the "Try This!" section at the end of the chapter.

Assuming we had access to the overall sampling weight and each of the conditional weights, on which level of a multilevel analysis does one specify the weight? Most multilevel software now will allow users some options: to provide a weight at level 1 only, a weight at level 2 only, or weights at both levels of the analysis. Using the HLM software as an example, if we provide our overall raw sampling weights at level 1 ($w_{i1} = 150$ for public school students and $w_{i2} = 50$ for private school students) and no weight information at level 2, the software accurately will assume that the level-1 weights provided are overall sampling weights. In the HLM software, these weights then will be scaled as relative weights by dividing each overall weight by the average of the weights:

$$w_{relative_{ij}} = w_{ij}\ /\ \bar{w}. \tag{13.4}$$

In this example, the average weight is 100, and thus, the new, scaled relative weight for students from public schools becomes 1.5, and the new, scaled relative weight for private school students becomes .5. Note that these new weights will sum to the actual sample size, thus allowing for more appropriate standard error estimation than if unscaled raw weights were used. However, appropriate scaling is still under investigation in a multilevel context (Asparouhov, 2005; Rabe-Hesketh & Skrondal, 2006). In obtaining level-1 parameter estimates, data from public-school students in our example will weigh more heavily in the estimation than data from private-school students since their relative weight is larger. However, at level 2, because we provided no weight information (it was incorporated into the overall relative weight), school-level relations would be determined considering the information from each of the schools as equally informative.

Alternatively, in this example we could choose to provide weights at both level 1 and level 2. With this approach, the HLM software will make the assumption that the weights at level 1 are conditional weights (i.e., within level 2). Unfortunately, this assumption would be incorrect for most analyses using publicly-available data from NCES; recall that NCES typically only provides overall sampling weights. However, the violation of this assumption is relatively benign. For example, if we provide the overall sampling weights to HLM (w_{ij} = 150 for public school students and w_{ij} = 50 for private school students), HLM automatically will rescale the weights within each level-2 unit such that the sum of the weights within a level-2 unit equals the sample size within that level-2 unit. The HLM software does this by dividing each raw weight by the average of the weights within each level-2 unit, similar to the previously-described scaling procedure. For public-school students in our example, their raw weights would be divided by 150, resulting in a relative weight of 1 (150/150 = 1), and for private-school students, this scaling would result in each raw weight being divided by 50, resulting in a relative weight of 1 (50/50 = 1). Note that if we had access to the conditional sampling weights (of 30 and 2 respectively) to provide to HLM, the relative weights still would have been equal to 1 The distinction between marginal and conditional weights is meaningless for this rescaling because the overall sampling probability weight differs from the conditional probability weight only by a constant for all elements within a level-2 unit.

With weighting at both level 1 and level 2 in the analysis, the level-1 relations will be estimated based on equal weighting of the data on students from public and private schools; however, the level-1 estimation is moderated by the weights applied at level 2. When raw weights are provided at level 2 (in this example, 5 and 25 for public and private schools, respectively), HLM will rescale these weights by dividing each school-level weight by the mean of the school weights (i.e., yielding relative weights). In our simple case, each raw school weight would be divided by 15 to result in new, scaled relative weights of .33 for public schools and 1.67 for private schools. These new school-level relative weights will sum to the actual sample size at level 2.

To obtain estimates of school-level relations that can be generalized to the population of schools in this district, the data for the private school will be weighted more heavily than the public school. Thus, the level-2 estimates will play a role in level-2 estimation.

Finally, a third option for weighting is to provide sampling weights at level 2 only. In HLM, the weights would be rescaled automatically at level 2 as described above (i.e., relative weights) and responses from individuals within the level-2 units are assumed to contribute to the estimation equally. This method may appear to result in the same approach as providing both level-1 and level-2 weights as described above; however, this similarity is a result of a simplified example. If unequal weighting is applied *within*-schools (some students selected at higher rates than others), then the two approaches would yield different estimates, assuming the responses differed across strata of students within schools (such as males versus females).

Estimation methods and accommodations for addressing weights in multilevel software appear to be changing with each new version of software, so users are encouraged to consult the manuals for the appropriate version of their software before deciding on which weighting approach to use. For example, the weighting options and assumptions in the MLwiN software are similar to those provided with HLM; however, MLwiN appears a little more flexible. MLwiN allows the user to model with raw weights (instead of rescaling automatically to use relative weights, although the rescaling is available) and will estimate level-2 weights from overall inclusion weights at level 1. M*plus* (Muthén & Muthén, 1998–2017) also allows a similar flexibility in the choice of scaling of weights.

The examples presented above reveal that the estimates a user obtains for level-1 and level-2 regression coefficients can differ depending on the weighting approach chosen. In light of these differences, perhaps the safest approach is to analyze the model with and without weights and for the various level-weighting combinations that can be attempted (Aitkin & Sheih, 2002). If the interpretations of model estimates do not differ across these various analyses, then the researcher can be confident in reporting either set of estimates, at least to the extent that the model is assumed to be specified correctly. If the interpretations do differ, then the researcher should clarify the weighting approach chosen and might opt to present and discuss the differential results in the manuscript or research report (Aitkin & Sheih, 2002).

Regression estimates that are dependent on the weighting scheme selected provide evidence that the stratification variables (which led to the disproportionate selection probabilities) are related to the response variable, and thus the sampling weights are informative. Including salient stratification variables in the analysis may help to remove the apparent discrepancy between weighted and unweighted results. For example, a researcher might choose to include a public/private indicator variable in the level-2 equations.

The idea here is that one can control for variation on the stratification elements by including these in the model itself. In most complex sample surveys, the incorporation of all such sample design variables is unrealistic. Nonetheless, the thoughtful inclusion of design elements most central to the analysis (such as public/private school stratum indicators or demographic stratum indicators at the student-level) can prove helpful in understanding potential biases and in obtaining a relevant, interpretable model.

Depending on the type of database (cross-sectional or longitudinal) and the types of respondents (student, parent, teacher, etc.), the sampling probability weights can be easy or quite complex to select. NCES attempts to provide weights to handle the representation of any particular observation that may be in an analysis. An assortment of different types of weights for different kinds of analyses is included in NCES databases and are explained in all NCES user's guides. For example, a user might be analyzing data from the ECLS-K database to examine children's cognitive growth from the spring of kindergarten to the spring of first grade and only intend to include child-provided information, such as assessment data in the analysis. Thus, the researcher needs to use weights specifically designed to represent the sampling probability and non-response adjustment for child data that were collected in spring of kindergarten and first grade. The suitable weight in this case is contained in the variable called C24CW0, and children with non-zero values for this weight will yield a nationally representative sample suitable for studying children's growth from spring of kindergarten to spring of first grade. However, suppose that the analyst also wanted to include data that the parent and teacher had provided. For some of these students, the teacher and/or parent may not have provided information; thus, using the C24CW0 weight may not yield the most representative accounting of the relations for hypotheses involving national estimates. In this case, a more suitable weight would contain values of 0 for children who have no responses from teachers and parents. In the user's manuals, NCES typically explains the set of weights available to the researcher and provides example analyses that would be appropriate for each type of weight. In some cases, however, an analysis may not fit with any of the pre-specified weights, and a choice must be made among those available. This process of choosing a weighting strategy based on the availability of data (i.e., from teachers, from parents, etc.) must be used with caution because it assumes that the researcher does not want to model incomplete observations or wants to exclude people who may have left the study in later years.

Theoretically, as alluded to earlier, it is possible to avoid using the sampling weights entirely by including the components of the sampling design into the analysis. For example, suppose you were working with a dataset that had been collected by stratified sampling on gender and that male and female teachers were sampled at disproportionate rates. Further, suppose that you desired to calculate the (overall) mean income from this sample. An analysis not using sampling weights (disregarding the disproportionate rates of selection) would lead to a biased estimate of the mean if income of teachers differed across gender in the population. That is, the mean would be more reflective of the group that was oversampled. However, if appropriately derived sampling weights were used in the calculation of the sample mean, then this estimate of the population mean would be unbiased. Using weights in a statistical analysis is referred to as a "design-based" analysis (Kalton, 1983). An alternative to using the sampling weights is to include the stratum information in the statistical model. Such an approach would be considered a "model-based" analysis (i.e., including the sampling design into the analytic model). For example, one could regress income on gender without using sampling weights and thus obtain two appropriate estimates; assuming gender was effect-coded (i.e., using codes of 1 and –1 for each gender), this regression would yield an intercept "baseline" income estimate as well as an estimate of the incremental income associated with being male (or female). Note that we no longer would have one estimate of overall average income but two estimates (the intercept and an incremental gender coefficient), therefore answering a somewhat different research question than originally posed. Instead of asking, "What is the average income?" we would be asking two questions: "What is the average income, and does it differ between men and women?" Given a population difference in income between genders, perhaps this statistical model leads to more appropriate inference about average income. However, the descriptions of the sampling designs provided in the first section of this chapter suggest that it would be extremely difficult to include all of the many elements of typical NCES sampling designs into the statistical model without losing some clarity in interpretation of the model. In summary, the correct application of carefully constructed weights will obviate the need for the incorporation of additional variables controlling for elements of the sampling design—elements that may have little or no part in the substantive analysis of the question at hand.

Non-Response and Non-Response Coding

Missing data can occur at both the respondent/occasion level and at the item level. On its datasets, NCES provides flags on records to indicate whether a certain respondent provided data for a particular wave of the data collection. Missing data can arise in various ways: a respondent may miss an entire wave of data collection (unit non-response); a respondent might have missed, ignored, or refused to answer a particular item (item non-response); or a respondent may have been re-routed through a section of the survey based on known non-applicability of a particular item or items—known as a "legitimate skip." On most NCES datasets, these missing item-level data are given codes that are negative numbers. It is extremely important that data users understand this coding and properly recode the data or identify the values as missing values to the analysis software being used. Without such action, erroneous estimates would be obtained. As an example, there is a variable in the ECLS-K (1998–1999) database from the parent survey that holds the response to, "In a typical week, how often does [child] use the computer? Would you say..." and the parent is offered the response options of *never, once or twice a week, three to six times a week,* or *everyday.* This question was asked in the spring of kindergarten (1999) as well as in later waves; P2COMPWK is the variable holding the response from the Spring 1999 parent interview. In the dataset, there were 18,950 observations with C2PW0 weights that were greater than 0, and for these observations, the frequency distribution for the responses is shown in Table 13.4.

Note that the only way to know the context for a question and whether or not the non-response categories represent legitimate skips or otherwise is to review the questionnaire and item wording carefully. For this question, a sizable portion of the respondents, more than 8,000, are in the "not applicable" category because they actually have no computer in the home, as determined by their answer to an earlier item during the interview. Because the question was in reference to how often the child used the home computer, it was not asked of respondents who indicated they did not have a home computer, and thus, these responses are legitimate skips. According to the data, no parents refused to answer the question on home computer usage, 12 parents were not sure how many days a week their child used the computer, and an answer was not obtained for 15 parents, possibly because

TABLE 13.4 Distribution of Parent Responses to Frequency With Which Child Uses Computer in the Home, ECLS-K 1998–99

Response	Numeric Value in Datafile	Number of Parent Responses	Percent of Parent Responses
Not Ascertained	–9	15	<1%
Don't Know	–8	12	<1%
Refused	–7	0	0%
Not Applicable	–1	8,545	45%
Never	1	322	2%
Once or Twice a Week	2	4,622	24%
3–6 Times a Week	3	3,819	20%
Every Day	4	1,615	9%
Total Responses		18,950	100%

the interview was ended prematurely. The analyst must decide how to treat each of these responses for each of the variables used in any analysis with great care. For this example, the researcher might opt to recode the response of "not applicable" to be 1 (*never* use the computer in the home) and consider the "don't know" and "not ascertained" responses to be truly missing values. The recoding will depend on the intended analysis to be undertaken, but analysis results must be interpreted according to how any recoding was conducted.

A full discussion on how to accommodate missing data in multilevel analyses is beyond the scope of this chapter. For those who need a basic introduction to the topic of missing data, we highly recommend the coverage provided by Graham and Hofer (2000) and Enders (2010). Regarding choices for treatment of missing data during multilevel modeling, researchers should investigate the options available in the software program being used (and these procedures may change as software versions are updated). Alternatively, researchers can consider approaches to managing missing data that are handled outside of existing multilevel software packages or procedures, such as the development and use of multiply imputed datasets. For example, a powerful new package called Blimp allows for the imputation of missing multilevel data (http://www.appliedmissingdata.com/multilevel-imputation.html; see also Enders & Hayes, this volume).

Accommodating Other Aspects of the Sample Design

When researchers choose to use multilevel modeling to answer a specific research question, they usually have to make choices about the levels to include in the analysis and the predictor variables to include at each level. Based on the research question, the analyst may or may not incorporate all of the important features of the sample design (such as school strata indicators and student demographic strata indicators) into the analysis (see earlier discussion on model-based versus design-based analysis). When the analysis model does not include all sampling design information, researchers should be careful to temper their interpretation of the results according to the complexities of the sample data. For example, the SASS dataset was collected using a two-stage sampling design: schools were selected, and then teachers within those schools were sampled. Given the two stages, a 2-level analysis appropriately would account for clustering effects. On the other hand, the ECLS-K sample is a product of a three-stage sampling design: counties were selected, then schools, then students. A 2-level analysis of students within schools would disregard the dependence of schools within counties; therefore, it is possible that standard errors of parameter estimates might be downwardly biased. Additionally, although the use of sampling weights can account for disproportionate selection across strata in a sampling design, it does not address the effect of stratification on standard error estimation. For the SASS data, schools were selected from within categories of public/private status, region, and location type (urban/suburban/rural); thus, one can assume that if the response variable is homogenous within these categorizations, efficiencies would be gained by this stratified sampling design (Kalton, 1983). However, an analysis not including these stratum indicators likely would overestimate standard errors. Stapleton and Kang (2018) provide a discussion of strategies to use when not modeling all elements of a complex sampling design. We hope from this discussion that the challenges of modeling all parts of the sampling design are apparent and that the reader recognizes that not every analysis can accommodate all of the potentially available information. This difficulty, however, does not relieve the researcher from the responsibility of reporting the possible problems associated with these sampling issues.

SUMMARY WITH RECOMMENDATIONS FOR PRACTICE

Our intent in this chapter was to provide readers with the basic knowledge necessary to begin working confidently with large-scale complex survey data, freely available through national, public-use databases. Our discussion was designed to help readers locate data suitable to develop and answer research questions requiring multilevel modeling, either for a contextual/organizational analysis or a longitudinal analysis. We would like to conclude this chapter with some suggestions for practice.

First, for any given database, researchers carefully must consider the substantive nature of the variables on the data file before jumping into data management and analysis tasks. These tasks often consume a fair amount of time and energy and will tend to yield only disappointment if variable definition and identification is not given careful and thoughtful attention beforehand. We have noted the temptation that well-designed multilevel datasets present to some analysts. Once the power of the multilevel model is understood, exuberance sometimes blinds researchers from the hard realities defined by variables available on a given dataset. No matter how well the sample design may lend itself to multilevel exploitation, the lack of proper variables for the substantive research question at hand quickly will grind this exuberance to a halt.

Second, researchers should always review the user's guide that comes with the database. A quick read will provide a general overview of the sampling design and help the user identify whether there are any design variables, other than the clustering, that will need to be accommodated in the analysis. In addition, the user's guide will provide information regarding whether the sample size is sufficient given the levels included in the intended analysis. A detailed reading of the guide is not necessary at this initial stage (it will be later), but reviewing this information early will save valuable time—compared to finding out after extensive work involving database manipulation and programming that the data do not support the intended analysis.

Third, before starting to use the Online Codebook data extraction tool to identify specific variables for analysis, researchers should read through the questionnaires or interview protocols used during data collection. The variables on these datasets number in the hundreds or thousands, and the variables in the Online Codebook tool can be alluring. As you look through the list of variables to determine which to tag for extraction to an outside software package, suddenly every one of them will look interesting even though they may be ancillary to the research question. In addition, the Online Codebook sometimes contains only partial wording for an interview or questionnaire item, and the items may not be in the same order as they appear in the questionnaires or interviews. Before extracting items to address a substantive research question, it is important to understand exactly how the items were worded, what the response alternatives were, and the order in which each item was asked of the respondent in order to fully appreciate the measurement issues or the possibility of order effects.

Fourth, after gaining an understanding of the general sampling design, data collection procedures, and actual question responses with which they would like to work, researchers should then go to the Online Codebook website and find the variables to tag. It takes time to become oriented to the program's functions and the steps necessary to identify, examine, and download the data and program code. We encourage the reader not to underestimate the time that such orientation may require. With that caveat in mind, most of the data from NCES allow for a simple click on a variable name to obtain basic aggregate descriptive information, such as the mean, range, and/or frequency distribution. Reviewing this descriptive information for each variable is useful in identifying potential problems with missing data that might be overwhelming and preclude further analysis using that variable. Also, at this time, users should identify the variables needed to accommodate the sampling design appropriately, including weights. Detailed sections in the user's guides cover selection of sample weights and variance estimation; in general, it is better to extract more sample-design or weighting-related variables than you may use eventually for any specific analysis.

Fifth, before extracting data and beginning to model, researchers should examine the frequency distributions or other descriptive statistical information for the extracted dataset. In order to identify any possible problems in creating the dataset, determine whether the distributions and basic descriptive statistics for the variables in your dataset match what was reported on the Online Codebook, and verify the adequacy of your sample size. Next, be sure to note the value labels. It can be dangerous to assume that you know what each value of a variable represents on a public-use database. For example, in the ECLS-K dataset, the variable T5GLVL represents the grade level that the student was in at the fifth data collection wave. The values of this variable are –9, 1, 2, 3, 4, 5, 6, and 7. Unfamiliar users may assume that the value of "4" represents fourth grade. In fact, it (non-intuitively) represents third grade. Without carefully examining your data and the documentation provided with it, such mistakes will be easy to make.

Finally, researchers should carefully evaluate the sufficiency of the data in terms of the issues we have laid out in this chapter. What is the level of ICC for the dependent variable of interest? Are sufficient data available to ensure desired power to detect effects (which requires consideration of the ICC and the number of observations at each level)? What are the missing data codes for the variables of interest, and how will these be recoded or imputed appropriately? Researchers have a responsibility to remain aware of the lack of firm resolution about weighting in the multilevel context and about the availability and appropriateness of weights themselves. Moreover, the estimation procedures being used by the various software programs can and do differ, and there is likely to remain some controversy about which procedures are most appropriate under specific circumstances. Careful thought to questions and issues such as these will better enable profitable use of large-scale datasets, such as those we have outlined in this chapter. Helpful resources to aid researchers in sharing the information in manuscripts about how their analyses were conducted are provided by Bell et al. (2012) and Stapleton (2018).

RECOMMENDED READINGS

Readings That Use the SASS Data With a Multilevel Analysis

Fitchett, P. G., Heafner, T. L., & Lambert, R. G. (2014). Examining elementary social studies marginalization: A multilevel model. *Educational Policy, 28*(1), 40–68.

Liu, X. S. (2007). The effect of teacher influence at school on first-year teacher attrition: A multilevel analysis of the Schools and Staffing Survey for 1999–2000. *Educational Research and Evaluation, 13*(1), 1–16.

National Center for Education Statistics. (1997). *Teacher professionalization and teacher commitment: A multilevel analysis* (NCES 97-069). U.S. Department of Education.

Readings That Use the ECLS-K Data With a Multilevel Analysis

Finn, J. D., & Pannozzo, G. M. (2004). Classroom organization and student behavior in kindergarten. *Journal of Educational Research, 98,* 79–92.

Leroux, A. J. (2018). Student mobility in multilevel growth modeling: A multiple membership piecewise growth model. *The Journal of Experimental Education,* 1–19.

Li, A., & Fischer, M. J. (2017). Advantaged/disadvantaged school neighborhoods, parental networks, and parental involvement at elementary school. *Sociology of Education, 90*(4), 355–377.

TRY THIS!

1. From the OnlineCodebook, download the 1999–2000 SASS public teacher datafile and a syntax file, after tagging at least one variable.
2. Run the downloaded syntax to create a file containing a subset of the variables that you tagged.
3. Replicate the data from Table 13.1 by using an aggregate function in your selected software. SCHCNTL is the School ID variable.
4. What additional variables were you required to download (that automatically appeared in your datafile?)
5. From the OnlineCodebook, download the ECLS-K 1998–1999 longitudinal data file and a syntax file, after tagging the seven school ID variables (Sx_ID) and the seven cross sectional weight variables (CxCW0) where x is the wave number.
6. Run the downloaded syntax to create a file containing a subset of the variables that you tagged.
7. How many records are on the datafile?
8. How many student records have a C1CW0 weight that is greater than 0?
9. Of those students, how many different schools did they attend?
10. Replicate the remaining data from Table 13.2 using your selected software.
11. Using the ECLS-K dataset that you downloaded, examine the C1CW0 weights for students who attended the same school (S1_ID=0001). Do students have equal or unequal weights?
12. Now, compare those weights with those of students from a different school (S1_ID = 0002). Are they similar or different?

APPENDIX
Other Publicly Available Datasets (some will require application for a restricted license)

The following is a short list of other datasets that readers may find useful for undertaking multilevel modeling. All but two of these studies are conducted by NCES, and more information for these NCES data sets can be found at http://nces.ed.gov/surveys/

The *Baccalaureate and Beyond Longitudinal Study* (B&B) is a survey of about 11 000 students who completed their bachelor's degree in 1992–1993 and were surveyed in 1994, 1997, and 2003. Information collected in 1994 included immediate post-graduate plans and transcript information from the undergraduate years. In 1997, information about graduate school attendance and workforce outcomes were collected.

Beginning Postsecondary Student Longitudinal Study (BPS) follows students from their first attendance in college (in 1997), to 1 year later, to a final follow up in 2001. About 9,000 students attending 800 institutions are included in the study, and interest lies in examining college persistence, financial aid, college transfer behavior, and work status.

High School and Beyond (HS&B) followed two cohorts, one of which were high school sophomores in 1980. They were followed up in 1982, 1984, 1986, and for a final time in 1992. Topics include the educational attainment, employment outcomes, and family formation of these young adults during the 12 years after their sophomore year in high school.

The IEA *Civic Education Study* (CivEd) is a 28-country data set that contains information about the level of civic knowledge and skills as well as attitudes toward social issues of 14-year-old students. The sampling included whole-class cluster sampling within selected schools. Across all countries, over 90,000 students participated in the study. For the United States, there are 2,811 students from 124 schools.

The National Longitudinal Survey of Youth 1997 (NLSY97) is a longitudinal survey program of 12 to 17 year olds in 1997; they are surveyed twice a year (https://www.nlsinfo.org/content/cohorts/nlsy97). Data are available for 18 waves of data collection regarding a range of topics including education, health, employment, attitudes, and behaviors.

The *National Study of Postsecondary Faculty* (NSOPF) is a cross-sectional survey of faculty at colleges and universities in the United States. The study has been repeated several times, with the most recent being in 2003–2004. The sample contains 35,000 teaching and research faculty from 1,080 postsecondary institutions. Questions pertaining to job satisfaction, workload, scholarly productivity, and career plans are covered. Data are available at the institution level as well as the faculty level.

The *National Longitudinal Study of Adolescent to Adult Health* (Add Health), housed at the University of North Carolina Population Research Center, is a longitudinal data set that examines the correlates of health-related behaviors of adolescents in Grades 7 through 12 (https://www.cpc.unc.edu/projects/addhealth). Data at the individual, family, school, and community levels have been collected for four waves, in 1994, 1996, 2001–2002, 2008–2009. A fifth wave of data collection is underway in 2018–2019.

REFERENCES

Aitkin, M., & Shieh, Y.-Y. (2002, April). *Longitudinal modeling of incomplete response data from clustered and stratified survey designs.* Paper presented at the annual meeting of the American Educational Research Association, New Orleans, Louisiana.

Asparouhov, T. (2005). Sampling weights in latent variable modeling. *Structural Equation Modeling, 12,* 411–434.

Bell, B. A., Onwuegbuzie, A. J., Ferron, J. M., Jiao, Q. G., Hibbard, S. T., & Kromrey, J. D. (2012). Use of design effects and sample weights in complex health survey data: A review of published articles using data from 3 commonly used adolescent health surveys. *American Journal of Public Health, 102*(7), 1399–1405.

Beretvas, S. N. (2011). Cross-classified and multiple-membership models. *Handbook of advanced multilevel analysis,* 313–334.

Dillman, D. A. (2000). *Mail and Internet surveys: The tailored design method.* John Wiley & Sons.

Enders, C. K. (2010). *Applied missing data analysis.* Guilford Press.

Graham, J. W., & Hofer, S. M. (2000). Multiple imputation in multivariate research. In T. D. Little, K. U. Schnabel, & J. Baumert (Eds.), *Modeling longitudinal and multilevel data* (pp. 201–218). Lawrence Erlbaum Associates.

Heck, R. H., & Thomas, S. L. (2020). *An introduction to multilevel modeling techniques: MLM and SEM approaches.* Routledge.

Heeringa, S. G., West, B. T., & Berglund, P. A. (2017). *Applied survey data analysis.* Chapman and Hall/CRC.

Kalton, G. (1983). *Introduction to survey sampling.* SAGE Publications.

Kish, L. (1965). *Survey sampling.* John Wiley & Sons.

Lee, E. S., Forthofer, R. N., & Lorimor, R. J. (1989). *Analyzing complex survey data.* SAGE Publications.

Lohr, S. (1999). *Sampling: Design and analysis.* Duxbury.

Martin, M. O., Mullis, I. V. S., & Hooper, M. (Eds.). (2016). *Methods and procedures in TIMSS Advanced 2015.* Boston College, TIMSS & PIRLS International Study Center. http://timssandpirls.bc.edu/publications/timss/2015-a-methods/chapter-15.html

McCoach, D. B., Rifenbark, G. G., Newton, S. D., Li, X., Kooken, J., Yomtov, D., Gambino, A. J., & Bellara, A. (2018). Does the package matter? A comparison of five common multilevel modeling software packages. *Journal of Educational and Behavioral Statistics, 43*, 594–627.

Muthén, L. K., & Muthén, B. O. (1998–2017). *Mplus user's guide* (8th ed.). Muthén & Muthén. https://www.statmodel.com/download/usersguide/MplusUserGuideVer_8.pdf

National Center for Education Statistics. (2018). *Survey and program areas.* http://nces.ed.gov/surveys/

Pfeffermann, D., Skinner, C. J., Holmes, D. J., Goldstein, H., & Rasbash, J. (1998). Weighting for unequal selection probabilities in multilevel models. *Journal of the Royal Statistical Society, Series B, 60*, 23–40.

Rabe-Hesketh, S., & Skrondal, A. (2006). Multilevel modeling of complex survey data. *Journal of the Royal Statistical Society.* https://doi.org/10.1111/j.1467-985X.2006.00426.x

Rust, K. (1985). Variance estimation for complex estimators in sample surveys. *Journal of Official Statistics, 4*, 381–397.

Skinner, C. J., Holt, D., & Smith, T. M. F. (1989). *Analysis of complex surveys.* John Wiley & Sons.

Snijders, T. A. B. & Bosker, R. J. (2012). Discrete dependent variables. In *Multilevel analysis: An introduction to basic and advanced multilevel modeling* (pp. 304–307). SAGE Publications.

Stapleton, L. M. (2018). Survey sampling, administration, and analysis. In G. R. Hancock, L. M. Stapleton, & R. Mueller (Eds.), *The reviewer's guide to quantitative methods in the social sciences* (2nd ed.). Taylor & Francis.

Stapleton, L. M., & Kang, Y. (2018). Design effects of multilevel estimates from national probability samples. *Sociological Methods & Research, 47*(3), 430–457.

Thomas, S. L., & Heck, R. H. (2001). Analysis of large-scale secondary data in higher education Potential perils associated with complex sample designs. *Research in Higher Education, 42*, 517–540.

Tourangeau, K., Nord, C., Lê, T., Sorongon, A. G., & Najarian, M. (2009). *Early childhood longitudinal study, Kindergarten Class of 1998–99 (ECLS-K): Combined user's manual for the ECLS-K eighth-grade and K–8 full sample data files and electronic codebooks* (NCES 2009–004). National Center for Education Statistics, Institute of Education Sciences, U.S. Department of Education. https://nces.ed.gov/ecls/data/ECLSK_K8_Manual_part1.pdf

Tourangeau, K., Pollack, J. M., Atkins-Burnett, S., & Hausken, E. G. (2006). *Early childhood longitudinal study, kindergarten class of 1998–99 (ECLS-K): Combined user's manual for the ECLS-K fifth-grade data files and electronic codebooks* (NCES 2006-032). https://nces.ed.gov/pubs2006/2006032.pdf

Tourkin, S. C., Pugh, K. W., Fondelier, S. E., Parmer, R. J., Cole, C., Jackson, B., Warner, T., Weant, G., & Walter, E. (2004). *1999–2000 schools and staffing survey (SASS) data file user's manual* (NCES 2004-303). U.S. Department of Education, National Center for Education Statistics.

U.S. Department of Education. (2018). *Digest of Education Statistics.* https://nces.ed.gov/programs/digest/d17/tables/dt17_203.40.asp

U.S. Department of Education, National Center for Education Statistics, Statistics of Public Elementary and Secondary School Systems, 1980–81; Common Core of Data (CCD), "Table 203.10. Enrollment in public elementary and secondary schools, by level and grade: Selected years, fall 1980 through fall 2026. https://nces.ed.gov/programs/digest/d16/tables/dt16_203.10.asp?current=yes

Wolter, K. M. (1985). *An introduction to variance estimation.* Springer.

Woodruff, R. S. (1971). A simple method for approximating the variance of a complicated estimate. *Journal of the American Statistical Association, 66*, 411–414.

CHAPTER 14

COMMON MEASUREMENT ISSUES IN A MULTILEVEL FRAMEWORK

Brian F. French
Washington State University

W. Holmes Finch
Ball State University

Thao Vo
Washington State University

In this chapter, we provide a general introduction to issues in measurement and psychometrics that are particularly salient within a multilevel modeling context. It is important to understand potential problems with ignoring nested data, particularly in the examination of validity evidence. Specifically, in this chapter, we examine the impact of multilevel data structure on common psychometric questions concerning internal structure, internal consistency, and fairness. These ideas are situated in a conversation about what to consider in developing a measure intended to be used

Multilevel Modeling Methods with Introductory and Advanced Applications, pages 495–533

www.infoagepub.com

in a multilevel framework, as well as examining a measure that is used in such a framework but was not originally designed with multilevel data in mind. Our primary focus is on multilevel factor analysis, and multilevel differential item functioning (MDIF). We also touch on multilevel reliability estimates. Throughout the chapter, we refer to an example instrument that is used in a higher education context. It was originally designed for individual assessment, but we discuss how it is used at an aggregate level and whether a multilevel construct could be defensible from a multilevel theoretical perspective.

GENERAL ISSUE

Common in many educational and psychological data collection situations is a hierarchical or multilevel data structure. These structures are recognized in measurement and psychometric work at a greater frequency than a decade ago. However, this structure is often ignored when analyses are conducted on large-scale programs. Not accounting for this nested structure can result in problems for statistical analyses commonly used in gathering validity evidence, such as factor analysis and differential item functioning. These problems can include inaccurate estimation of parameters (Muthén, 1994), bias in standard errors used to form statistical tests, and resulting inflation of the Type I error, even in studies of the measurement properties of assessments (e.g., Finch & French, 2010; French & Finch, 2010). For the estimates of the standard errors of interest to be accurate (Bloom, 2005) the simple random sampling variance needs to be adjusted to account for this clustering (Kish, 1965). Briefly, when adjusting the sampling variance to account for clustering, certain parameters have to be considered, including the intraclass correlation coefficient (ρ), which is equal to the proportion of total population variance ($\tau^2 + \sigma^2$) across the clusters versus variance within clusters, σ^2 (Bloom, 2005). To obtain the operating sampling variance, the cluster effect multiplier (CEM = $\sqrt{1+(n-1)\rho}$; Bloom, 2005) is applied to the simple random sampling variance, where n is the size of the cluster. Larger values of the CEM result in larger values of the sampling variance. Only when the cluster size is 1 for all clusters or the ρ is 0.0 are the two standard errors the same. In all other cases, as cluster size and/or ρ increases so does the adjustment to the standard errors. Consider an example with $J = 50$ clusters to be randomly selected, each containing $n = 50$ people. When the ICC is .05 this design produces standard errors $\sqrt{1+(50-1)(.05)} = 1.86$ times larger as those produced by separately randomizing $50J$ individuals. With an ICC or .10 and $n = 50$, standard errors are 2.43 times larger than if individuals were randomly sampled (e.g., Bloom, 2005).

There are two overarching approaches for addressing the issue of nested data. The first is a design-based analysis strategy (e.g., Wu, 2010), which involves statistical adjustments of the standard errors (SEs) and parameter estimates to account for the nested structure. The second is a model-based analysis strategy (e.g., Wu, 2010). This approach involves analyses where the nested structure is explicitly included as a component of the statistical model used. That is, a form of multilevel model is estimated to account for this structure. For example, both within (e.g., student) and between (e.g., school) level components are included in the model, and variance is estimated for each level. We suggest that the researcher select the strategy (i.e., design or model-based) that (a) accurately reflects the underlying data structure, and (b) perhaps most importantly, matches the research question and theoretical framework. Let us turn our attention to measurement issues related to building validity arguments for the use of scores to make decisions about individuals and groups. We begin by reviewing the basic concepts of reliability and validity with a focus on developing an argument with evidence to support inferences about score use.

OVERVIEW OF RELIABILITY AND VALIDITY

A brief overview of key reliability and validity concepts will ensure that we have a common language as we work through the chapter concepts. A central concept in the area of psychometrics is reliability, which from a classical test theory (CTT) lens, can be described by the fundamental equation of $X = T + E$. In other words, the observed test score X, can be decomposed into a portion associated with an individual's true score on that assessment (T), and random error (E). The true score on a psychological variable can be roughly thought of as the person's average score on that assessment if that assessment were taken an infinite number of times. A primary assumption underlying CTT is that E is random, encompasses every factor that might influence X other than an individual's true score on the trait measured, and is uncorrelated with either T or X. Therefore, if an examinee were taking a reading test, T would reflect only their knowledge of the areas of reading assessed, and E would include all other factors influencing performance (e.g., fatigue, external distractions). Thus, reliability can be defined as the proportion of variance in an observed test score that is accounted for by the construct being measured by the instrument.

With the true score idea in mind, the variance in an observed score, X, can be expressed as:

$$\sigma_X^2 = \sigma_T^2 + \sigma_E^2 + 2COV(T, E) \tag{14.1}$$

where σ_T^2 = variance of the true scores, σ_E^2 = error variance, and $COV(T, E)$ = covariance of the true score and error. As E is assumed to be uncorrelated with T (i.e., $COV(T, E) = 0$), Equation 14.1 reduces to:

$$\sigma_X^2 = \sigma_T^2 + \sigma_E^2 \tag{14.2}$$

With reliability, users of an assessment are focused on the amount of variation in the observed score that can be attributed to the true score. Given Equation 14.2, this quantity can be expressed as the ratio of the true score and observed score variances, or:

$$\rho_{xx} = \frac{\sigma_T^2}{\sigma_X^2} = \frac{\sigma_T^2}{\sigma_T^2 + \sigma_E^2} \tag{14.3}$$

This quantity is called reliability and it represents the proportion of observed score variance accounted for by the true score. This can also be thought of as the R^2 value from a regression model where X is the dependent variable, and T is the independent variable. As ρ_{xx} increases in value, the observed score is more predicted by the latent trait, whereas smaller values indicate that the observed score is more a function of other factors not assessed (i.e., error). We can also think of reliability as a measure of instrument consistency in the scores it yields. That is, if we asked a person to repeatedly complete a measure, without recalling the time before, the quantity of ρ_{xx} would indicate how consistent the scores were from these repeated assessments. This, of course, is not possible but does convey the concept of reliability. Another key idea is that reliability of scores is sample dependent. That is, the estimates vary depending on the sample assessed and the resulting scores from the sample. Thus, we talk about the reliability of scores, not the instrument or scale (e.g., Thompson, 2003). There is a rich literature on issues related to reliability (e.g., Feldt & Brennan, 1989; Raykov & Marcoulides, 2011) that can be consulted for a deep dive into the topic.

In addition to reliability, the other major measurement concept associated with CTT is score validity. One of the primary modern views of validity was first articulated by Messick (1989) and has been expanded by a number of authors. This definition asserts that validity statements cannot be made about an instrument, but rather must refer to inferences that are made using scores from the scale. Furthermore, in this view of scale validity, the various types of validity evidence are not compartmentalized (i.e., criterion validity, content validity, construct validity) in contrast to what is sometimes presented in the literature or textbooks. Instead, a variety of evidence is used together to build an argument for the extent to which valid inferences and uses can be made with scores derived from the instrument (e.g., Kane, 2013). Given that psychometrics is focused on producing scores to be used for informing decisions about individuals, we should remain focused on validity to support such inferences, claims, and decisions about persons. However, this is not the only way to think about validity. For example, work by Borsboom and Markus (2013) and others describe validity in terms of the instrument itself and offer a thoughtful argument about this topic. This conception relies on a much simpler formulation: The construct that the instrument is meant to measure must exist, and it must have a causal relationship with responses to the items on the scale. We encourage exploration of these views. For this chapter, we focus on inferences and follow the *Standards for Educational and Psychological Testing* (American Educational Research Association [AERA] et al., 2014).

In accordance with the *Standards* (AERA et al., 2014), we focus our validation conversation on the interpretations and uses of an instrument's scores. This approach is identical whether scores are derived at different levels such as a level-1 (e.g., individual student) or level-2 (e.g., classroom) in a nested context, or at a single level (e.g., students only). We acknowledge instrument validation can revolve around five types of inferences—domain description, scoring, generalization, extrapolation, and implications (Kane, 2013). However, for this chapter, we focus on three of these, including (a) scoring, (b) generalization, and (c) extrapolation. Within each, we focus on key aspects of evidence that are either typically popular, or evidence that is not sought out when it should be. This evidence includes differential item functioning (DIF) and confirmatory factor analysis (CFA) for scoring and extrapolation inferences, respectively. We also include a brief discussion of reliability estimation as evidence for a generalization inference in this context.

Briefly, scoring inferences relate the individual's observed performance on a task (i.e., item response) to an observed score, dependent upon assumptions that "scoring procedures are appropriate, are applied as intended, and are free of overt bias" (Kane, 2013, p. 25). The extrapolation inference examines evidence that scores appear to be associated with constructs in a manner consistent with the theory and provide an accurate level of the abilities measured. The generalization inference concerns the conceptual notion that tasks or indicators represent a sample of tasks that could be selected to assess the factors and do so consistently as a representation of a universe of items. We describe these inferences in relation to the modeling strategy that accounts for multilevel data and contrast this with a single level modeling strategy. First, we set the measurement context that we use for the examples.

MEASUREMENT CONTEXT

For this chapter, our example is grounded in a dataset of responses to the Institutional Integration Scale (IIS), an instrument that has been used commonly in the higher education environment to study the connections students have to their university. Such student-level factors are of interest as they relate to academic success and persistence at the college level. Students with a high sense of institutional integration or connection may feel that they are a part of, and identify with, the university, thereby leading to positive academic and social outcomes. Integration has been conceptualized as two factors: faculty and students (e.g., a student's perception of their relationships with faculty and peers (Bean, 1980; Berger & Milem, 1999). This idea fits with long-standing and revised versions of student persistence (e.g., Tinto, 1975, 1987). We do not spend time in conceptual detail here, but many articles and books on the topic are easy to locate if of interest. Rather, for this chapter we focus specifically on the IIS (French & Oakes, 2004; Pascarella & Terenzini, 1980). The IIS measures five facets of institutional integration using a 5-point scale with response options ranging from 1 (*strongly disagree*) to 5 (*strongly agree*). The revised 34-item IIS (French & Oakes, 2004) is comprised of the following subscales (a) Peer-Group Interactions (10 items), (b) Interactions With Faculty (5 items), (c) Faculty Concern for Student Development and Teaching (5 items), (d) Academic and Intellectual Development (8 items), and (e) Institutional and Goal Commitment (6 items). The scale possesses desirable properties that are appealing for use with college students (e.g., short, less than 10 minutes to complete), has been supported for a two-factor model (French & Oakes, 2004), is predictive of student outcomes, demonstrates item invariance (e.g., Baker et al., 2007), and had been used in other methodological studies (e.g., Levy, 2011). That said, this measurement work has supported score use at the individual level (i.e., student).

Measurement Beyond the Individual Level

In education, as well as other domains (e.g., organizational leadership), there is interest in aggregating data for use in multilevel situations consistent with theories that are multilevel by nature (e.g., ecological system models). The aggregation of data allows for scores at a higher level to provide information about that higher level of the data structure. For example, individual student feedback on teacher quality can be aggregated at the classroom level so that scores represent a construct at that higher classroom level. The critical assumption to consider here is that the level-1 scores and the construct meaning are also retained at the next level in the hierarchy, for example, that the aggregated satisfaction scores for individuals provide an overall satisfaction score of the teacher (Huang et al., 2015; Toland & De Ayala, 2005). However, this assumption of direct correspondence between the individual and aggregate levels of measurement may not hold across all situations (Bliese, 2000; Kozlowski & Klein, 2000). That is, the construct meaning and structure at one level may not be the same at another level. In our example, the construct of teacher satisfaction may be different for an individual student than it is for the entire class. In other words, the internal structure of a measure may not be the same at level-1 and level-2 (Huang & Cornell, 2016). This raises questions about validity inferences for score use at each level. We focus our discussion on these situations. If these scales are to be informative for researchers and practitioners, then there must be evidence supporting their use at each level (i.e., student and school).

A scale may have been designed originally for use at the individual student level, as was the IIS in our example. However, there may also be interest in using institutional integration information at a higher level to account for between-group variability. For example, there could be interest in accounting for variability in student retention rates across different classes or majors by understanding something about how unique groups of students within classes or majors (i.e., class-level or major-level score) integrate as a group into an academic environment. Thus, validation efforts need to consider both levels of use; a critical component that has lacked attention for more than 20 years (Chan, 1998) and likely still lacks sufficient attention today. That is changing, with excellent work discussing these issues in depth (e.g., Stapleton et al., 2016a). In either case, the researcher will have to specify the level(s) of interest (i.e., individual, cluster) for score use, guided by a theory of the underlying constructs, to determine the appropriate modeling strategy (i.e., design or model-based) to account for the nested data structure. We begin with confirmatory factor analysis to show how this can be applied, and provide a sample of models through the design and model-based approach.

Internal Structure of the IIS via Confirmatory Factor Analysis: Multilevel Confirmatory Factor Analysis

Factor analysis is a major source of evidence that is used in the investigation of validity. Indeed, this technique is one of the most widely used statistical methods in the social and behavioral sciences and is used increasingly in other disciplines. We focus specifically on confirmatory factor analysis (CFA). As a demonstration of its popularity and widespread use, using the search terms "confirmatory factor analysis and higher education" in Google Scholar for 2010 to 2020, as one example, resulted in nearly 46,000 results. As we mentioned before, CFA can be used to support a scoring inference (Kane, 2013). The evidence allows us to obtain empirical information about how responses to items relate to the underlying constructs we are attempting to measure. The items most strongly related to the constructs generally inform which items are used to create the scores for those constructs. Before discussing how CFA can be used in developing validity evidence, we first need to briefly review the model.

The standard factor model is defined as:

$$x = \Lambda\xi + \varepsilon \tag{14.4}$$

where x = observed indicator variables (e.g., item responses), ξ = factors, Λ = factor pattern coefficients (i.e., loadings) linking observed indicators with factors and ε = unique variances for the indicators. Each indicator has a loading (i.e., parameter estimate, λ) associated with a factor. These loadings reflect the relationships between the factors and the indicators, with larger values indicating a closer association between the latent and observed variables. This factor model is used to estimate the covariance matrix of the indicator variables as expressed below.

$$\Sigma = \Lambda\Psi\Lambda' + \Theta \tag{14.5}$$

Here, Σ = the model predicted covariance (correlation) matrix of the indicators, correlation matrix for the factors, and Θ = diagonal matrix of unique error variances.

Figure 14.1 displays the factor model in matrix form, such that each indicator is only associated with a single latent trait, as reflected by the inclusion of 0's in the loading matrix. To clarify, Indicators 1, 2, and 3 are assumed to be associated with Factor 1, and Indicators 4 and 5 are assumed to be associated with Factor 2, representing a 2-factor model in this example.

In Figure 14.2, we turn the matrix implied by Equation 14.5 into a typical path diagram of the 2-factor model.

$$\begin{bmatrix} X_1 \\ X_2 \\ X_3 \\ X_4 \\ X_5 \end{bmatrix} = \begin{bmatrix} \lambda_{11} & 0 \\ \lambda_{21} & 0 \\ \lambda_{31} & 0 \\ 0 & \lambda_{42} \\ 0 & \lambda_{52} \end{bmatrix} \begin{bmatrix} \phi_{11} & \\ \phi_{21} & \phi_{22} \end{bmatrix} + \begin{bmatrix} \theta_{11} & & & & \\ 0 & \theta_{22} & & & \\ 0 & 0 & \theta_{33} & & \\ 0 & 0 & 0 & \theta_{44} & \\ 0 & 0 & 0 & 0 & \theta_{55} \end{bmatrix}$$

Figure 14.1 Matrix representation of a CFA model.

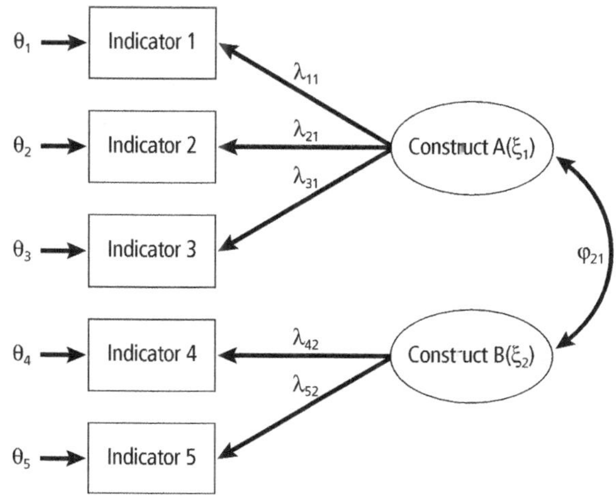

Figure 14.2 Two-Factor Model Specified in Figure 14.1.

This CFA model describes relationships of indicators and factors at the individual (e.g., student) level. However, in a multilevel world, we need a model that considers both the within and between-level variance. That is, the measurement model is hypothesized to have a within-level covariance matrix and a between cluster covariance matrix. Thus, we modify Equation 14.4 to account for both levels of the data, which now becomes:

$$x = \Lambda_w \xi_w + \varepsilon_w + \Lambda_b \xi_b + \varepsilon_b \qquad (14.6)$$

where the indicators (x's) have factor loadings and residuals at both the within and between levels of the data structure; for example, for individual students, and for schools that the students attend. This is an extension of Figure 14.2 as shown in Figure 14.3. The indicators now have a residual variance estimate for both the between (b) and within (w) components, as indicated in Equation 14.6. For simplicity, the figures do not include the residuals. This is a conceptual figure, because the five indicators are actually treated as observed variables at the within-level (e.g., students) but latent variables at the between level (e.g., classes; Muthén, 1994). The between-level latent variables represent five random intercepts, and the two factors at the between level are specified to account for variation among the random intercepts. This figure represents one visualization; others represent the within- and between-levels differently (e.g., Muthén, 1994; Stapleton et al., 2016b).

Another way to think about the multilevel factor model is with respect to how variance in indicator or item responses at the within (individual) and between (cluster) levels are understood. We can ask, for example, if the

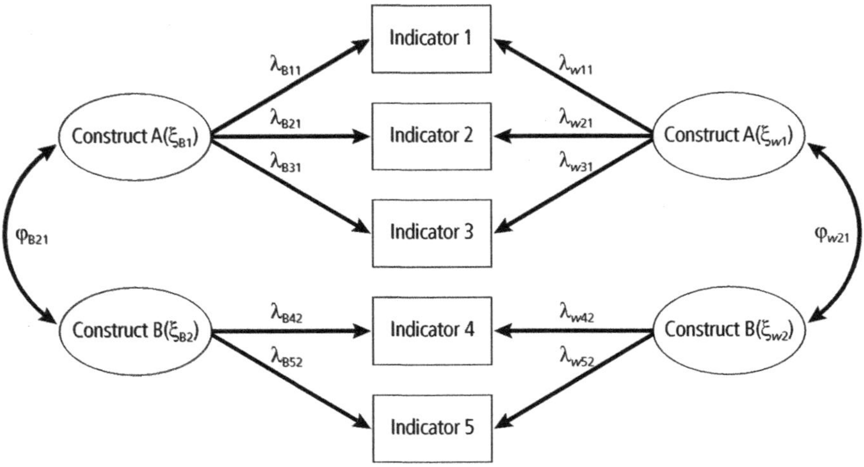

Figure 14.3 Sample of a Multilevel Confirmatory Factor Model where B and W indicate between and within levels.

factor structure at the individual level is the same at the cluster level, and if not, why not? Returning to our example scale, we might be interested in assessing not only a student's level of faculty and student integration to their campus, but also levels of integration across majors or programs, perhaps as a way to assess integration for certain majors where retention is an issue. Modeling the data in this way may help account for differences among majors in outcomes such as retention/persistence behavior. In the following section of the chapter, we cover two approaches in which we might consider such a scenario, beyond the standard individual-level approach. Recall from the beginning of the chapter we discussed a model-based approach, in contrast to a design-based approach, to account for multilevel data. We continue with this idea by discussing how to account for the nested data structure simply by adjusting the standard errors and fit indices in a design-based approach, or model the variance at the between and within levels using a model-based approach (Stapleton et al., 2016b; Wu, 2010). Wu (2010) provides a thorough mathematical explanation of these approaches, and Stapleton and colleagues build from these by emphasizing the importance of matching theory to such analyses by showing several modeling possibilities, many more than we can cover here. Muthén (1994) also provides excellent coverage of foundation ideas for multilevel CFA. We discuss the basic models in this chapter to provide a foundation from which one can build more advanced models.

One way to estimate a factor model is to ignore the clustering of the data and conduct a CFA by treating the data as a single level (i.e., clustering ignored). Alternatively, we can estimate a factor model using the design-based approach where the standard errors and model fit indices are adjusted to account for the nested data. These adjustments are mostly automatic in many software programs with properly specified options. A third way to handle the nested data structure is to use a model-based approach, which requires considering and modeling the multilevel nature of the data. In the examples, we demonstrate three different models within the model-based approach, describe their specification, and compare the results to the clustering-ignored and design-based approaches.

Model-Based Approaches

To begin thinking about the model-based approach to understanding multilevel constructs, we can consider a *within*-cluster factor model, where the interest in the construct is at the within-cluster level, and not at the between-cluster level. Scores on the construct for the *within* model represent individual standing on the construct within a cluster, thus, the use of scores would be tied to comparing individuals within a cluster. To understand this model, Panel A of Figure 14.4 contains the model to be specified in the example that follows. This model has been altered compared to the model in Figure 14.3. Specifically, Figure 14.4 contains three conceptual figures in the panels, with the five indicators as observed variables at the within-level (e.g., students), but with the latent variables at the within-level (Panel A), between-level (Panel B), or both (Panel C). We use these visual representations to aid the understanding of the M*plus* code examples we provide, which are also consistent with other examples in the literature.

For the model in Panel A, the within component contains the two constructs or factors of interest. However, the between-level constructs are no longer specified as constructs. Since our primary interest is in the within-cluster, the between-level is specified as correlations between the indicators.

Alternatively, construct assessed might be conceptualized as a characteristic of the cluster. These cluster-level constructs are often referred to as *shared* (Stapleton et al., 2016a) or *climate* (Marsh et al., 2012) constructs. This shared/climate construct represents a characteristic of the cluster (e.g., college major), where responses to items are from individuals but are an indication or reflection of the cluster-level construct, thereby implying that the indicator scores are useful for investigating cluster-level differences. That is, interest in the shared/climate construct is at the between-cluster level. In a shared construct, it is expected that individuals from a given cluster would respond similarly to each other. Thus, there is little interest in the within cluster-level information and focus is on the between-level information in the CFA. Cluster-level scores would be of interest for use in studying differences across majors, classrooms, or schools (i.e., clusters) but not

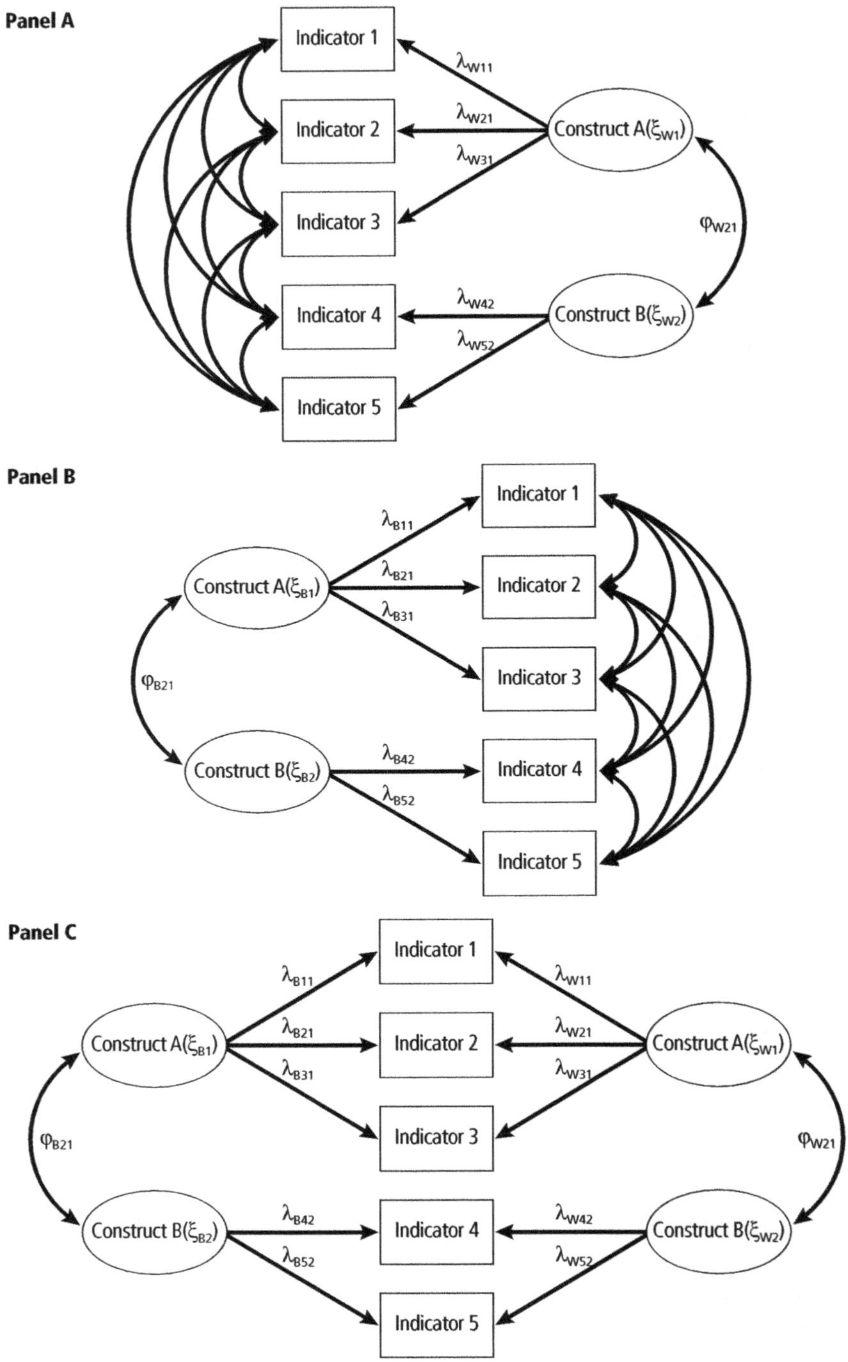

Figure 14.4 Model-based Confirmatory Factor Analysis for a Within (Panel A), Shared (Panel B), and Configural model (Panel C).

focusing on individual (i.e., level-1) differences. As an example with the IIS scale, perhaps communications majors feel more integrated as a group due to the welcoming climate unique to their department compared to engineering majors, such that individual item responses are driven largely by this group-level variability. In Figure 14.4, Panel B, the two between-cluster factors are specified, and the within-level is captured or specified as correlations among the indicators. The expectation in this model is that there is very little variability or covariation of responses at the within-level (e.g., Bliese, 2000), although work is needed to establish empirical criteria for determining guidelines to inform decisions for low variation at the within-level.

Another way to think about multilevel factor analysis is in terms of a contextual or *configural* construct, in which context refers to the aggregation of the individual responses (e.g., Stapleton et al., 2016a). That is, an aggregation of the data from individuals in a given cluster is formed, referred to as cluster aggregates of the level-1 constructs. An example of this approach may be that people who choose sales and marketing as a major in college are by nature more extroverted, and thus more likely to become engaged in the university. Thus, their item responses reflecting greater integration will also be reflected in higher aggregate integration scores at level 2. There is interest in the construct at both the within- and between-level for the clusters. Unlike shared constructs, here there is no expectation that individuals in a particular cluster would respond similarly to items as a function of any specific trait at level-2. Rather, the tendency to be affiliative is reflected in greater integration scores for individuals in this particular major. One version of this model, under certain assumptions, is seen in Panel C of Figure 14.4, where it is assumed that the between and within loadings are equivalent. This implies the measurement model is the same at each level.

The most appropriate model across these options would be selected on more criteria than comparing model fit. We argue that validation efforts and their associated inferences and claims (Kane, 2013) would need to be made at the cluster-level as well as at the individual-level, and depends on the underlying theory supporting the use of the scores and the models examined. In other words, there should be evidence for the validity of a given scale at both the individual student level, as well as at the classroom, school, or major level.

DEMONSTRATION

For our example, we use a subset of our IIS data for the analyses. Additional data are available for the "Try This!" exercises. The dataset contains 670 students from 33 classrooms with an average of 20 students per class. As a first step, the ICCs should be computed. The ICCs for the five subscale scores are (a) .082 for Peer-Group Interactions (PGI), (b) .026 for

Academic and Intellectual Development (AID), (c) .058 for Institutional and Goal Commitment (IGC), (d) .073 for Interactions with Faculty (IWF), and (e) .043 for Faculty Concern for Student Development and Teaching (FCWSD). These ICCs straddle non-trivial values (< .05) for construct validation work (Geldhof et al., 2014). We estimate a total of five models including a (a) standard individual model assuming no clustering or ignoring clustering (i.e., factor structure at the student-level), (b) design-based model correcting for clustering (i.e., factor structure at the student-level with adjusted standard errors to acknowledge the multilevel nature of the data), (c) *within-cluster* model accounting for between variability via correlated between-level indicators, (d) *shared* (between-cluster construct) model accounting for within variability with correlated within level indicators, and (e) *configural* (within and between) model with between- and within-level constructs specified in the model.

The latter three models have a different level of focus for the construct interpretation including the within-cluster, the between-cluster, and both the between- and within-cluster, which we refer to as the *within-cluster, shared,* and *configural* models, respectively. Thus, the interpretation of scores across these models differs as well. The within-cluster model represents individual standing on the construct within the cluster. The shared model represents scores only at the cluster-level, and the construct is a characteristic of the cluster. The configural model, assuming invariant model parameters for the within and between components, represents a construct that is the aggregate of the dissimilar characteristics of respondents within a cluster (e.g., Stapleton et al., 2016a; Wu, 2010). Of course, the invariance assumption can be tested via a chi-square difference test between models with and without this invariance constraint.

Given that the IIS was designed for use at the individual-level and there is no explicit theory to suggest a need to model within- and between-level constructs, it may be expected that the first two models fit better than the latter three. That is, the items were developed for making decisions about individual students from a finite population from which a sample is drawn, assuming individual and independent sampling, and with item content focused on the individual. Given this fact, any cluster dependency could be considered spurious and a design-based approach could be recommended for analyzing the data (e.g., Stapleton et al., 2016b). However, researchers may want to explore whether or not this assumption holds for this population, particularly if there is interest in using the instrument within a cluster (e.g., university, major) or for university climate research to ask questions about different experiences at the major-level. If so, a different structural validation model is needed. For example, a theory might suggest that there may be a cluster-level influence on responses (e.g., students of a certain major respond similarly—i.e., shared model), and thus a multi-level model may actually fit better to capture non-spurious dependency. Thus, the models discussed are viable options that can support a validity argument and can be tested empirically, both of which depend on researchers' underlying theory and research question of interest.

Model Comparison

To evaluate model fit, and thereby its suitability for a particular set of data, several statistics are used in conjunction with common guidelines for their interpretation (e.g., Fan & Sivo, 2005; Hu & Bentler, 1999). We do acknowledge that the use of such fit criteria is an ongoing area of research, especially for multilevel models. Though we are comfortable relying on them in the following examples, keep in mind that future work may help to identify better or more accurate guidelines for interpretation. The complication of fit in multilevel CFA, and multilevel SEM, in general, is that fit indices apply to the complete model (i.e., within and between components). Recall that the chi-square is sensitive to sample size and is used in the calculations of the fit indices. In the MLM framework, the within component of the model will have the largest sample size and will have more influence on fit indices. Methods have been proposed to obtain these indices (e.g., Ryu, 2014) but they are not easily obtained in all software packages at this time. In addition, between-level fit indices may not be accurate when the ICCs are low (e.g., .09; Hsu et al., 2017). This is an area of work to watch over the next few years to understand the current recommendations. In our example, with ICCs less than .09, we rely on the Comparative Fit Index (CFI), Root Mean Squared Error of Approximation (RMSEA) and respective 90% confidence intervals, Standardized Root Mean Square Residual (SRMR within and between), and chi-square (i.e., maximum likelihood ML χ^2). Model-data fit was considered adequate when CFI > .95, RMSEA with CIs < .06, and SRMR < .08 criteria were applied to both within and between estimates. The $SRMR_{between}$ gives an indication of misfit for the within, shared, and configural models for the between components. Additionally, parameter estimates were inspected for appropriate values.

Results

To begin our investigation as to which conceptualization (individual-level only, or multilevel) might be optimal given the data, we estimated both the individual-level model and the design-based model using M*plus* software. The design-based model employs the TYPE = COMPLEX in M*plus* to adjust the standard errors for the nested data. We provide sample code at

TABLE 14.1 Model Fit Results for the Individual and Multilevel Confirmatory Factor Analysis of the IIS-Two Factor Model

	Clustering Ignored	Design-Based Adjusted	Within-Cluster	Between Cluster (Shared)	Configural
Model χ^2	23.48 (df = 4)	19.62 (df = 4)	16.60 (df = 4)	17.55 (df = 6)	83.65 (df = 14)
RMSEA (90% CI)	.085 (.05 – .12)	.077 (.04 – .11)	.069	.054	.086
SRMR (between)	.024	.024	.024 (.020)	.003 (.190)	.029 (.426)
CFI	.982	.985	.988	.989	.936

Note: CI = Confidence Interval; SRMR = Standardized Root Mean Square Residual; CFI = Comparative Fit Index; RMSEA = Root Mean Square Error of Approximation; *df* = degrees of freedom. For Within-Cluster, Shared, and Configural the values in parentheses are for SRMR between. All models account for nested data except for the Cluster Ignored.

the book's website www.modeling.unconn.edu. Tables 14.1 and 14.2 contain model fit and factor pattern coefficients for both models, respectively. Both the clustering-ignored and design-based models exhibited adequate fit and had strong and statistically significant factor loadings. The design-based adjusted model had a slightly lower chi-square value and RMSEA, although both chi-square values were statistically significant ($p < .01$) indicating possible misfit. However, it is likely that given the strong factor loadings and acceptable fit collectively, most would agree that these scales do measure the two aspects of student and faculty integration at the university level. Thus, this evidence would support a claim about the underlying factor structure of the instrument, supporting a score inference at the individual-level.

Next, we address the question as to whether there is support for a factor structure beyond the individual level. We first estimate a within-cluster model with a saturated between level model, as seen in Figure 14.4, Panel A. In our example, this model includes two factors at the within-level. At the between level, there are no factors, but all indicators at the between level are correlated. This model does not assume a cluster-level construct, but instead, it accounts for cluster-level variability on the indicators. As seen in Table 14.2, the loadings are also very similar to the loadings for the individual-level analyses. As one might expect, and due to the relatively low values of the ICCs across the indicators in our data, the overall model fit is similar to that of the individual level only analysis: χ^2 (4) = 16.60, CFI = .988, $SRMR_{between}$ = .020, $SRMR_{within}$ = .024, and RMSEA = .069. In other words, because the cluster-level effect on the item responses is relatively low, models accounting for and ignoring the cluster effect yield similar results. Given the results of these three models, we can support a statement that a two-factor within-cluster model that allows for cluster variability may be viable.

TABLE 14.2 Standardized Factor Pattern Coefficients for the Estimated Models

	Clustering Ignored	Design-Based Adjusted	Within-Cluster	Between Cluster (Shared)	Configural
PGI	.683	.683	.666	.938	.685
AID	.796	.796	.789	1.00	.763
IGC	.631	.631	.639	.465	.623
IWF	.738	.738	.731	.959	.704
FCWSD	.841	.841	.841	.865	.846

Note: Student Integration Factor = PGI, AID, IGC; Faculty Integration Factor = IWF, FCWSD. PGI = Peer Group Interactions, AID = Academic and Intellectual Development, IGC = Institutional Goal Commitment, IWF = Interactions with Faculty, FCWSD = Faculty Concern with Student Development.

We reach this conclusion because the fit indices (Table 14.2) met standard criteria and were nearly identical across the models; large and statistically significant parameter estimates were within bounds; and we had no convergence issues or negative variance estimates. This conclusion also is consistent with the ICC values of the subscale scores, which were low and ranged from .02 to .08. That said, we do not have evidence of cluster-level factors at this point. Additional models require further examination to support such a validity claim.

Next, we continue the modeling process with the consideration of the level-2 *shared* factor model, Figure 14.4, Panel B. From a theoretical point of view, integration to faculty and peers or students may very well be a cluster-level construct. That is, responses to the scales about the integration into a university or school might be influenced by the cluster of students with whom an individual spends most of their time, especially during the first year at a university. Another way to think about this is that we might expect students in certain environments (e.g., courses, majors, learning communities) to respond similarly to one another when compared to students from different environments. Responses are related within a cluster and reflect cluster-level constructs. Thus, theory is driving the modeling process, as it should. However, the scale's use and content of items were not originally constructed with such a purpose in mind. Rather, the scale developers were interested in understanding individual students' integration into campus life, not the collective integration of students from a common grouping or cluster. Nonetheless, if modeled correctly, such questions of collective integration may be considered, even if they were not intended by the test developer. For a shared construct or factor structure, we would expect that the indicators of this example scale would be highly among individuals within a cluster, and the focus would be on the factors at the between-level. Recall that the ICCs for the subscales ranged from .02 to .08. With these low and straddling values (>.05), some might argue that multilevel factor analysis is needed; however, it could be difficult to justify the use of the integration scores as a shared construct among students in the same classroom. We note that the ICC > .05 is a somewhat arbitrary criterion for such a conclusion. That is, each scale is comprised of items asking each student about their personal experiences, and not a single experience shared by their peers or students in a given class. There are no set guidelines for how large an ICC should be to justify a multilevel CFA; however, many MCFA studies report values above .10 as being important (Dedrick & Greenbaum, 2011).

As shown in Figure 14.4, Panel B, we could model level-2 constructs with the 5 indicators by estimating two factors at the between level. At the within-level, we could allow the 5 indicators to be correlated with one another to allow for covariation among the indicators at the within level. This model does not specify within-level factors, but allows for correlations among the indicators. Thus, this model specification is similar to the within-model described previously, but now we have between-level loadings and each within-level path is replaced with co-varying level-1 indicators. Initially, this model yielded a negative residual variance at the between-level for the IWF and PGI factors. All between-level residuals were not statistically significant and near 0.01. Although not ideal and in need of theoretical justification but for the purpose of this exercise, we set the negative variances to the average value of .005 (or 0.0) to achieve fit (Muthén & Muthén, 1998–2017). In fact, in other areas (e.g., generalizability theory) when such estimates occur, it is also recommended that negative variance estimates be set to 0.0 (e.g., Brennan, 2001, p. 85) without influencing practical conclusions. Methodologically, this can assist with such convergence issues and is not uncommon (Hox, 2002) when the sample sizes at level 2 are small and between-level residual variability is nearly zero. Substantively, having a between-level residual variance near 0 may indicate that there is very little between-level residual variability. Such a situation does appear to be the case in this example, given the low ICC values, indicating that a shared construct may not be viable. As seen in Table 14.1, fit indices were not horrible, except for the SRMR with the between-level construct (i.e., .19 shared model), which likely signals the aforementioned issues with the residuals. The between-level factor loadings were all large and statistically significant. However, they differ noticeably from the other solutions. This is likely the result of such low variability at the between level. The correlations between the within-level subscales ranged from .156 to .382 with an average of .234.

The ISS scale was originally constructed for individual student measurement. After conducting more diagnostic digging (e.g., examining localized strain and residuals after we fit the model; Brown, 2014) given the SRMR and residual issues, we would be hesitant to support a *shared* construct without replicating the structure with an independent sample and having additional theoretical and practical use to support this level. In short, with the low ICC values, model fit issues, and a lack of theory, it could be difficult to justify a shared factor structure. If such a construct were of interest, item content would likely need to be reconsidered and revised to capture greater between cluster variance. In addition, other models could be considered where the factor structure may be different at the between-level (e.g., one factor) compared to the within-level. We leave these issues for the interested reader to examine in the "Try This!" section.

The last model of the five that we considered is the configural model. In contrast to the shared model, the configural model does not include the expectation that students within a certain cluster, such as a classroom, to respond similarly to other members of that cluster. This model can be estimated in multiple ways, depending on the assumptions made about the nature of the constructs (e.g., Forer & Zumbo, 2011; Geldhof et al., 2014). The two factors in this example can be modeled across the within- and between-levels such that the two factors are estimated at both levels with separate factor loadings. However, we agree with Stapleton and colleagues (2016a, 2016b), who recommend modeling the factors at both levels and constraining the loadings at the between cluster-level to be equal or invariant to those at the within-cluster level (i.e., the between factors are configural factors, Figure 14.4, Panel C). This method is similar to modeling contextual effects in standard MLM analyses with a level-1, group mean-centered predictor. In such a case, the correlation at the between-level reflects the within- and between-level relationships. In this model specification, the configural factor, a cluster-level factor, is the mean of the latent factor based on persons within a given cluster. Thus, the cluster-level factor is a representation of the within factor, given model constraints of equality between levels, and indicates differences in means on the factors across the clusters. Another way to conceptualize this is that the structure is the same at the between and within levels. That is, we impose cross-level measurement invariance in this model, given (a) the low ICCs, (b) the classroom or cluster is not considered as a source of variability of the individual construct. Of course, other non-constrained models can be specified testing this invariance constraint. However, if this model does not fit well, then it can become complicated to sort out if there is cluster bias where measurement is not the same at the within- and between-levels (Jak et al., 2013), or if a shared construct also exists (Stapleton et al., 2016a).

As in the previous examples, this model also yielded a negative residual variance at the between-level, possibly due to the low amount of variance at the between level, as seen in the ICCs. All between-level residuals were non-statistically significant and near 0.01. As in the shared model, we set the negative variance for AID to the average value of .014 to achieve convergence (Muthén & Muthén, 1998–2017). This issue may be signaling that the residual variability at the between-level is very low, and/or the small between-level sample size may be partially responsible for the negative variance estimates. As seen in Table 14.1, the large SRMR of .426 at the between-level and the large increase in the model chi-square value for the configural model indicate the fit for this model is substantially worse than that of the other models, including the baseline within-cluster model. At this stage, we do not have evidence to support a configural or contextual factor with a factor structure at the between-level. That is, no evidence suggests that the cluster level constructs reflect the cluster aggregation of the individual constructs at level 1. In terms of our example, this means that we cannot support the claim of an average level of the faculty integration construct in a given classroom with internal structure evidence from a measurement perspective. Additional model could consider relaxing constraints across levels, examining cluster measurement invariance assumption, or examining a mixture of shared and configural model, as proposed by Stapleton and colleagues (2016a). Again, we encourage the reader to explore such models with the data accompanying this chapter.

Internal Consistency of the IIS Scores

The next concept we introduce is internal consistency reliability of the latent factor scores, used to gather evidence to support a generalization inference. In practice, we would address this issue if we have identified a well-fitting model justified by theory and empirical evidence, thereby suggesting that the internal structure is sound. Reliability can be defined as the squared correlation between true and observed scores (e.g., Lord & Novick, 1968) or as the ratio of true score variance to total variance within a scale (e.g., McDonald, 1999). In the multilevel framework, we are interested in assessing whether the scores we wish to use at levels 1 and 2 have sufficiently high reliability to justify score use in making decisions. Single-level reliability estimates do not differentiate reliability across different levels, even when accounting for nested structures (e.g., Raykov & du Toit, 2005), leaving the user with a reliability estimate value does not apply to the different levels of use. Multilevel CFA does allow us to partition the variance in scores into within- and between-level components. Other methods that account for nesting do not separate within and between error variances, leaving the researcher with a less precise estimate. When a multilevel factor model fits the data, the reliability of the scores for the factors can be obtained at each level, as we describe below. There are several forms of reliability for multilevel situations that have been discussed in the literature including multilevel Cronbach's alpha, multilevel composite reliability (multilevel omega), and multilevel maximal reliability, to name a few (e.g., Forer & Zumbo,

2011; Geldhof et al., 2014; Raykov & du Toit, 2005; Raykov & Marcoulides, 2006; Zinbarg et al., 2005).

Cronbach's Alpha coefficient has been the most widely used index of reliability in the social and behavioral sciences. However, it is recognized that it is not a perfect index and other indices can be more precise. See McNeish (2018) for a review of issues associated with Alpha. Thus, we chose to use omega to estimate reliability to show some examples because we have worked with multilevel factor analysis in this chapter. McDonald (1999) described how the true score idea can be re-conceptualized through the factor analysis model. To understand this, recall that variance in the observed score is a function of variance in the true score and error variance. The observed score variance is equal to the sum of variance due to the latent trait (true score) that is being measured by the indicators, and the sum of unique (error) variance that is idiosyncratic for each indicator. Using factor analysis, we can generate estimates of error variance for the indicators through the unique variance associated with each indicator. Furthermore, under the assumption that the individual indicator errors are uncorrelated, the unique variances can be summed to obtain an estimate of overall error variance for the scale. Similarly, the factor loadings provide a direct measure of the relationship between each item and the latent trait. McDonald (1999) showed that the square of the sum of the factor loadings is an estimate of the true score variance, the variance in the indicators that is due to the latent trait. Even though alpha is widely used, composite reliability via omega calculated from factor loading coefficients can produce more precise reliability estimates compared to alpha (e.g., Geldhof et al., 2014). An estimate of reliability via omega based upon parameters from the factor model can be written as:

$$\omega = \frac{\left(\sum_{i=1}^{I} a_i\right)^2}{\sigma_X^2} = \frac{\left(\sum_{i=1}^{I} a_i\right)^2}{\left[\left(\sum_{i=1}^{I} a_i\right)^2 + \sum_{i=1}^{I} \psi_i^2\right]} \tag{14.7}$$

where

a_i = Factor loading for item i

ψ_i^2 = Error variance for item i obtained from the factor analysis model.

Across reliability estimates mentioned above, a simulation study by Geldhof and colleagues supported the use of the index omega, as it performed well in many settings and was similar to the performance of alpha. The estimates of maximal reliability tended to be more biased in comparison to omega. The results also provide evidence that the single-level reliability estimate cannot capture either within or between score reliability unless reliability is nearly the same at each level of the nested structure. In practice, we would obtain the estimate for by fitting the factor analysis model to indicators and extracting the estimated loadings and error variances for each indicator. We can then generate an estimate of reliability using Equation 14.7. Given that we have parameter estimates presented in Table 14.2, it is straightforward to show how to obtain estimates of omega based on the various models. Applying Equation 14.7 to the estimates from the model where clustering was ignored produces omega reliabilities for student integration ($\omega = .75$), and faculty integration ($\omega = .77$). Recall from Table 14.2, the estimates for the design-based approach where standard errors were corrected for the nested structure were identical to the naïve model (ignoring clustered data). Thus, omega estimates for reliability are identical.

There has not been much discussion in the literature directly addressing internal consistency reliability standards for omega or other indices in the multilevel context. Thus, we rely on what has been suggested in the literature for standard situations. Internal consistency reliability estimates should be greater than 0.95 for making decisions about individuals, but 0.80 is acceptable for other purposes such as inferences at the group-level (e.g., identifying low or high aggression traits; Nunnally & Bernstein, 1994). Thus, the values here are lower than we would like. For the within-model, the omega reliability estimates for the faculty integration factor are similar ($\omega = .77$). The student integration omega estimate is slightly lower ($\omega = .74$), given the slightly lower coefficients for the three indicators.

Given that we do not have evidence in our example for configural or a contextual factor that supports a factor structure for the between-subjects level, we do not need to calculate separate reliability estimates at the between- and within-levels. However, in the "Try This!" section, we encourage you to calculate these to observe what occurs. Note that to do so, we cannot constrain between-level residual estimates to zero (e.g., Gottfredson et al., 2009) because this assumes perfect between-level reliability. We must choose between freeing the between-level reliability so that it can be estimated or constraining between-level variance estimates to zero and thereby assuming reliability at that level to be 1. If we choose the former option, we must ensure that model estimation without the between-level residual variance constraints is possible. This is not surprising, given the instrument was originally constructed at the individual level.

As we mentioned previously, to measure faculty and student integration at the classroom or group level would likely require the development of a new set of indicators to gather cluster-specific data. If we did have such evidence, it would be a simple matter to obtain a within- and between-level omega estimates. That is, Equation 14.7 can produce a within-level or a between-level omega estimate. The within-level estimate is a function of the within-level factor loadings and residual variance estimates. The between-level omega estimate is a function of the between-level factor loadings and residual variance estimates (Geldhof et al., 2014).

In relation to the IIS scale and reliability claims, we can reach several conclusions. First, the reliability estimates obtained when accounting for the nested structure are similar to the reliability estimates that ignore the multilevel structure. Second, the model fit indicators do not meet generally used standards for denoting good fit, meaning that making decisions about individuals in this sample using the factors may not be warranted. Finally, sufficiently high-reliability levels were not achieved at either the within- or between-levels of the data. We recommend estimating the between and within CFA model parameters and omega at these levels. Even if estimating even single-level reliability, we would argue for using a design-based approach for the CFA so the modeling strategy still accounts for the fact that the data are nested. In addition, we encourage learning more about reliability in the multilevel formwork by following the work in this area over the past 10 years (e.g., Raykov et al., 2016; Raykov & Penev, 2010; Raykof et al., 2015).

Reliability estimates can accurately be obtained for the within- and between-level factors (e.g., Geldhof et al., 2014; Raykov et al., 2016). These estimates are sample dependent, as are traditional estimates of reliability. Bias in the between-level estimates occurs in the presence of low ICCs or a small number of observations per cluster (e.g., Geldhof et al., 2014). Such conditions can also be accompanied by model convergence issues and/or negative variance estimates, as we saw in our example. That said, with IIC values above .05 and theory supporting a between structure, compared to single-level estimates, level-specific reliability estimates may be preferred and are generally more accurate. The most limiting factor in this area is the accessibility of implementation to obtain such estimates. Applied researchers are accustomed to automatic functions that produce reliability, but software for MCFA may not produce multilevel reliability estimates automatically.

DIFFERENTIAL ITEM FUNCTIONING AND MULTILEVEL DIFFERENTIAL ITEM FUNCTIONING

Measurement invariance (MI) is a critical component in constructing a validity argument to support test score use and interpretation in the context of fairness. At the item level, MI indicates that the statistical properties characterizing an item (e.g., difficulty) are equivalent across diverse examinee groups (e.g., native language). Thus, it represents a critical aspect of validity, particularly for ensuring the comparability of item and total scores to guide decisions (e.g., placement) across groups, both at the individual and group level. Differential item functioning (DIF) is present when item parameters differ across equal ability (e.g., trait, behavior) groups, resulting in the differential likelihood of a particular item response (Raju et al., 2002).

One of the most well-known DIF procedures is the Mantel-Haenszel (MH), which has been applied to both dichotomous and polytomous items (Holland & Thayer, 1988; Narayanan & Swaminathan, 1994; Penfield, 2001). In fact, in Google Scholar, there are over 13,000 references (2020) to the term *Mantel-Haenszel DIF* procedure. It is a straightforward extension of the chi-square test of association. The MH compares item responses between groups across multiple levels of a matching score.

The MHχ^2 statistic (Holland & Thayer, 1988) is used to assess the presence of DIF with the following:

$$\frac{\left\{\left|\sum_{j=1}^{S}\left[A_j - E(A_j)\right]\right| - .5\right\}^2}{\sum_{j=1}^{S} Var(A_j)}, \tag{14.8}$$

where

$$Var(A_j) = \frac{n_{Rj} n_{Fj} m_{1j} m_{0j}}{T_j^2(T_j - 1)}, \tag{14.9}$$

In Equations 14.8 and 14.9, $A_j - E(A_j)$ is the difference between the observed number of correct responses for the reference group on the item being studied for DIF (A) and the expected correct number ($E(A_j)$; n_{Rj} and n_{Fj} are the sample sizes for the reference and focal group respectively at score j of the matching subtest; m_{1j} and m_{0j} represent the number of correct and incorrect responses, respectively matching subtest score j; and T represents the total number of examinees at matching subtest score j. This statistic is distributed as a chi-square with 1 degree of freedom and tests the null hypothesis of no uniform DIF. This statistic can be readily extended to accommodate items with more than two categories using the generalized MH procedure (Penfield, 2001).

One limitation of using MH in the standard way is that it does not account for data that may have been collected in a clustered fashion, such as students within schools. Not accounting for such structure can increase Type I errors in DIF detection (e.g., Finch & French, 2010; French & Finch, 2010, 2013). The MH for DIF has been adjusted to account for clustered data using the Begg Mantel-Haenszel (MH_B) method proposed by Begg (1999). French and Finch (2013) provide details for analyzing dichotomous data and codes for analysis. Extensions of the model have been extensively evaluated (e.g., Beretvas et al., 2012; French & Finch, 2015; Jin et al., 2014). Details also appear in French, Finch, and Immekus (2019) for an application with ordinal (e.g., rating scale) data, which we use here, given the IIS items are scored on a scale from 1 to 5.

The MH_B technique is based on the observation that the score statistic obtained from logistic regression is equivalent to the MH test statistic when the ICC is equal to 0. The variance associated with the logistic regression score statistic is proportional to the variance of the MH test statistic in DIF detection. MH_3 adjusts the MH test statistic by the ratio of the score statistic variance determined through a generalized estimating equation (GEE) approach to estimating the logistic regression (LR) model, which accounts for the nested data structures, to a naïve score statistic variance that does not account for the nested data. The naïve and GEE-based logistic regression models both take the form:

$$\ln\left(\frac{P_{ki}}{1-P_{ki}}\right)=\beta_0+\beta_1 X_i+\beta_2 Y_i \tag{14.10}$$

where

P_{kj} = probability of a correct response to item k
β_0 = intercept
X_i = group membership for subject i
Y_i = *matching* subtest score for subject i
β_1 = coefficient for group variable
β_2 = coefficient for matching subtest variable

For the standard LR model, the covariance matrix for the dependent variable with respect to clusters is the identity matrix, in which the off-diagonal elements are 0, reflecting no clustering effects on the outcome. The GEE model estimates the off-diagonal elements of the covariance matrix and accounts for clustering in the data. With results from both naïve LR and GEE, the variance of the score statistic is used to calculate an adjustment factor, f, defined as:

$$f=\frac{\sigma^2_{GEE}}{\sigma^2_{Naïve}} \tag{14.11}$$

Here, σ^2_{GEE} is the GEE adjusted variance of the score statistic accounting for clustering, and $\sigma^2_{Naïve}$ is the naïve variance of the score statistic ignoring clustering, and is proportional to the variance of the MH test statistic. The adjustment factor will increase as the ICC moves away from 0.0. Given the proportionality, f can be used to adjust the MH test statistic and is defined as:

$$MH_B=\frac{MH}{f} \tag{14.12}$$

where MH is the standard MH chi-square test statistic. As the ICC increases, the use of the adjustment factor will widen the difference between MH_B and MH. The MH_B is smaller in value compared to the MH as the MH_B is corrected for the presence of the multilevel structure. Most simply, this adjustment helps control Type I error. Because our example scale has items that are ordinal, a generalized MH adjusted method is needed and made possible using the MH_B framework. That said, the process is similar in terms of how the software carries out the procedure. Details can be found for both cases in the literature (French & Finch, 2013; French et al., 2019) and now available in the R package DIFplus (https://cran.r-project.org/web/packages/DIFplus/).

DEMONSTRATION: MDIF

The IIS data for this example were collected from 78 classes consisting of 1,544 students at a large public university. The intraclass correlations for the six scores that comprise the two factors were PGI (.037), AID (.013), IGC (.034), IWF (.041), and FCWSD (.029). For each subscale, the ICC values indicate that less than 5% of the variance in scores occurs between the 78 university classes spread across disciplines. In other words, this is a minor amount of variance between classes. Notice that these do differ a bit from our MCFA analysis, but all remain somewhat low. However, recalling our discussion about the cluster effect, standard errors (SE) would be 1.7 to 2.2 times higher given the number of clusters and range of ICCs. This means if standard statistical approaches are used then the SEs can be biased downward, thus inflating Type I error. We examine the IWF scale using both the standard MH, and the MH_B to be able to compare differences in their DIF detection performance. Scale purification or scale refinement is important in the DIF detection process, where generally an iterative search procedure is employed to establish a matching score or ability estimate free from DIF items so as not to contaminate the matching process (French & Maller, 2007). This is similar to establishing appropriate invariant referents in multigroup confirmatory factor analysis (French & Finch, 2008; Meade & Wright, 2012). These are issues well-documented in the literature (e.g., Dorans & Holland, 1992, French & Maller, 2007; Navas-Ara & Gómez-Benito, 2002; Zumbo, 1999, Zwick, 2012). We did not conduct a full DIF analysis (e.g., scale purification, effect size estimation, and classification of DIF type), due to space limitations.

RESULTS

DIF falls within the scoring inference aspect of a validation argument. In part, this inference depends on the assumption that the scoring procedures are applied as intended to all members of the sample, and specific to DIF, are free of bias (Kane, 2013). For the IIS items, we might make the claim that IIS scores from the individuals' item responses, *accurately represent differential endorsement of the items relevant to the domain, regardless of a respondent's group membership.* Thus, support for the inference would be gained by showing that there is little to no DIF in the items across male and female students, for example. Given our discussion of MLM and knowledge of the ICC for IWF, we might expect that more items would be identified with DIF under the standard MH procedure, as compared to the adjusted procedure using the Begg adjustment discussed above, which is indeed the case. In Table 14.3, we see that two items would be identified using the standard generalized MH method with a criteria of $p < .01$, but only one item is identified using the generalized MH_B method where the multilevel structure is accounted for in the analysis. As this simple example shows, 40% vs. 20% of the items were identified as DIF items on the initial analysis. We provide sample code in our online appendix materials.

In reviewing these results, it would seem that for the IWF subscale, we could use this DIF evidence to show support for the scoring inference in terms of having minimal differential item functioning across males and female respondents. Of course, to build more evidence we would need to analyze all of the subscales. See our "Try This!" for practice at doing this initial step of DIF analysis. Additionally, logistic regression models can be used for DIF detection as well and have advantages over the MH procedure (e.g., French & Finch, 2010; Zumbo, 1999). In general, all methods that have been adjusted for multilevel data show the same trend with good control of Type I error rates when accounting for the nested structure compared to when not accounted for in the analysis (e.g., French & Finch, 2010, 2013; French et al., 2019; Jin et al., 2014; Jin & Eason, 2016).

TABLE 14.3 Generalized MH (GMH) and Generalized MH_B (GMH_B) DIF Results for IWF Items

	$GMH\chi^2$	*p*-value	GMH_B	*p*-value
IWF 21	8.71	.003	7.69	.010
IWF 23	3.95	.046	2.67	.102
IWF 24	0.36	.540	0.30	.585
IWF 25	12.35	<.001	10.24	.001
IWF 26	6.07	.013	5.19	.022

DISCUSSION

This chapter provides an introduction to components of multilevel measurement issues in the context of building a validity argument based on the notion of inferences and claims (Kane, 2013). We introduced analyses that utilize both design-based and model-based approaches to addressing the presence of a multilevel data structure. In addition, we provided evidence to support a scoring inference via our design-based DIF analysis, as well as to support a scoring and extrapolation inference with our primarily model-based multilevel confirmatory factor analysis. Remember that no single analysis is sufficient to support a validity inference. It requires many components of the argument to do so. However, the general processes and issues to consider could be applied to any measurement scenario in a multilevel framework (e.g., classification, associations with other variables, reliability). Indeed, a single chapter cannot cover all of these issues. However, the reading of this chapter should provide a better understanding of multilevel measurement and the exciting challenges it presents to applied researchers, measurement professionals and psychometricians.

Key Issues to be Certain to Report in Such Analyses

When documenting validity evidence for a multilevel environment, there are several critical components that we suggest need to be clearly reported. These recommendations are in addition to the standard components that should be reported in a single-level analysis. See Hancock and Mueller (2010) for an excellent guide to reporting such results. First, it is essential that the theory underlying the need for a multilevel model be clearly articulated. The reader should understand the logical argument and data structure underlying the multilevel model. For a DIF analysis, as an example, it may be sufficient to simply select a method that corrects or adjusts the standard errors to correctly account for the nested data structure. Such a design-based approach may fit the research questions at hand and be sufficient to support the validity inference(s) under investigation. In addition, the design-based approach may be appropriate when there is simply not enough data or clusters at a higher level to estimate a model beyond the individual level. However, if there are questions relating to the explanation of differences in item functioning that revolve around contextual factors (e.g., Zumbo et al., 2015), a model-based approach is justified. Such a model may involve including school, teacher, or neighborhood level information in accounting for DIF. Doing so in a confirmatory framework would follow recommendations (e.g., Roussos & Stout, 2004) that have received insufficient attention, in our opinion, in the last 15 years.

Second, the ICC for each outcome should be reported. These statistics are important because they provide an understanding of the amount of variance that lies at the between-clusters level compared to the within-clusters level. In addition, ICCs are likely to differ by outcome and context (e.g., Hedges & Hedberg, 2007), as was the case in the current study. For future planning of studies in the measurement field, this documentation can allow for empirical information to inform our assumptions and justify our projected conditions (e.g., statistical power, sample sizes) which can be critical to the work in which we and our colleagues in other areas engage. Third, we suggest that clear statements be offered around the inferences and claims that are supported by the multilevel measurement work. We do have a preference for focusing on validity inferences (e.g., Kane, 2013). Regardless of the validity lens that is used for validity work, it should be clear how results connect to conclusions that can be made based on the evidence (e.g., Gotch & French, 2020). Finally, with all of the design and analysis strategies employed here, the role of theory is central. It is not sufficient to argue for the appropriateness of a given model based solely on statistical grounds (i.e., model fit indices, factor loadings, *p*-values). Such arguments must always be couched in a theoretical framework about the constructs themselves, and the role that multilevel structures may play in their manifestation.

Topics to Consider for Future Work

The field of psychometrics and measurement, multilevel modeling, and quantitative methods, in general, are expanding at a rate perhaps never witnessed before. The amount of new information that is released each year from unique applications to new methodological approaches, keeps researchers busy reading and digesting the material. Key ideas in the measurement realm that weave through topics of reliability and validity issues include (a) adjusting standard errors in our models to account for multilevel structures through design-based strategies, (b) specifying models to not only adjust for the nested structure but also to model the structure through model-based strategies, and (c) determining practical aspects of our MLM measurement work (i.e., sample size, power analysis). However, much needs to be clarified and expanded upon as the field moves forward with the use of multilevel models in measurement research and practice. Here, we suggest several topics that span both theoretical and methodological work, with a focus on leveraging the power of MLM to gain a deeper understanding of contextual factors' influence on the measurement issue at hand, and to inform future test development and use of scores to make decisions.

First, we do not know how incongruence between theory and modeling strategy influences test design, validity evidence, and score use. A focus on the theory underlying the modeling strategy during the test construction phase is needed. Specifically, clarity in the domain mapping and score use stages of instrument development would specify whether scores are to be used at the individual and or aggregate level for decisions. In turn, this would determine the type of validity evidence needed to support score use as well as bring coherence to the type of modeling strategy required to match theory (e.g., design vs. model-based). For example, a state using a kindergarten readiness measure designed to assess and make individual-level decisions may also hope to use it as an indicator of collective readiness at the school or classroom level. Multilevel CFA could be used to assess the structure at that level to provide support for scoring and perhaps extrapolation inferences. If there is a desire to have decisions made at a particular level (e.g., classroom) other than the level for which a measure was originally designed (e.g., student), then it is imperative to develop the measure from the beginning with this purpose as the goal. Attempting to address unintended score use post hoc with complex statistical models may be insufficient, and possibly even harmful depending on the context of score use, particularly if it is for individual decision-making. Thus, we suggest that this backward-looking gathering of validity information is not the best practice and could influence or alter the meaning of the construct (e.g., Stapleton et al., 2016a). The establishment of a partnership between the user and test developer encourages a proactive role in ensuring evidence is aligned with the intended use, in helping to guide and monitor change (Gotch & French, 2020), and in producing high-level research focused on solving problems of practice (Coburn & Penuel, 2016).

Second, there remains a lack of clarity in our understanding of commonalities across modeling strategies. Such commonalities have been advanced in other areas of measurement. For example, over the past 10 years, there has been an increase in work demonstrating the connection between factor analysis and item response theory. Such research has filtered into instruction, understanding about model fit and limitations to a greater extent, methodological advancements, as well as user-friendly software for the modeling of either framework. The same attention is needed between item response theory in the multilevel context and multilevel structural equation models to continue to advance the use, understanding, and application of these latent variable modeling strategies (e.g., Cai et al., 2016; Huang, 2015, 2017). Given the proliferation of computing power and available models and modeling strategies, this area of research is ripe for many advances. Such work could help inform the development of effect sizes and guidelines for conducting multilevel measurement invariance and application of scale reliability to the multilevel data context (e.g., Geldhof et al., 2014).

Third, and perhaps most importantly, we suggest that, as a field, we do not completely understand how to fully leverage MLM to gain a deeper understanding of contextual factors that influence measurement accuracy. In the next section, we provide an example of the types of questions that can be asked using MLM measurement models to understand contextual influences through an ecological and sociocultural theoretical lens in a measurement invariance context. We consider this to be an important area of future work, as it sits snugly at the interaction of methodological and applied measurement work.

Example: Multilevel Measurement Models and Context Factors

As our chapter has highlighted, DIF falls within the scoring inference component of a validation argument, critical for maintaining fairness across populations, languages, and cultures. For our brief example, we use the Washington Assessment of the Risk and Needs of Students (WARNS; George et al., 2015), where total scores are used to gauge maladaptive, truancy-related behaviors among adolescents to help guide decisions about student interventions. Thus, DIF detection is needed to support equal item-functioning across various populations—especially for groups disproportionately represented as "high-risk" in schools (e.g., underrepresented students; Skiba et al., 2014). Although differences at the individual-level (i.e., sex, age, race/ethnicity) are important characteristics to consider for DIF detection, perhaps more importantly for measurement accuracy and test equity is to understand not only *whether* group differences exist, but, *why* these differences are observed.

In the context of the WARNS, which measures areas of adolescent development including depression, substance abuse, family engagement, and other truancy-related domains, model-based analyses allow us to propose theoretically driven questions concerning measurement invariance. Specifically, we ask how ecological, contextual, or sociocultural variables can be used to account for the observed differences in item-responding across groups, through a sociocultural theory lens (SC; John-Steiner & Mahn, 1996). SC theory proposes that the act of learning, or the formal and informal participation in the shared language, norms, and practices of an individual's environmental context, can occur on two levels: the individual and the societal level. In considering data structures, it is clear to see how this theory can map onto the logic of a multilevel modeling of DIF.

In our example, high school students (Black = 337, White = 512, Latinx = 593) nested in districts completed the WARNS as a standard practice in their schools. Students responded to 40 items on a scale from *never, hardly never to always, almost always*. DIF detection was conducted with multilevel ordinal logistic regression between two ethnic group comparisons at a time, using the student's self-reported "race/ethnicity" variable as a level-1 predictor. Random-intercept (RI-DIF; French & Finch, 2010) and random-coefficient (RC; Shear, 2018) models were then used to identify DIF items, where RI-DIF models allowed the intercepts to vary across school districts, and RC-DIF allowed for item coefficients to vary. The latter model can be interpreted as the DIF effect for the grouping variable ethnicity having varying magnitudes of DIF across districts. Next, level-2 predictors were obtained from state-level administrative data through the Office of the Superintendent for Public Instruction. Specifically, school-level discipline and chronic absenteeism rates were obtained as stand-in estimates or "proxy" variables representing the students' *school climate,* while family income and SAT scores were used as indicators of the students' *family climate.* To explore the ecological models, items with statistically significant level-1 predictors using either RI- or RC-DIF models were retained for level-2 modeling; the school and family proxies were then added to the models to account for the variance associated with the DIF.

Overall, the models identified 11 unique DIF items between at least two ethnic groups, accounting for 1.0 to 7.4% of the between district variance. When level-2 proxies were added to the models, family and school contextual variables accounted for 0.30 to 74.6% of the district variance associated with the DIF items. Although the percentage of district variance was small and the proxies were non-significant in the models, two possible takeaways for this example are that (a) WARNS items may be functioning differently for particular student groups and (b) the sociocultural variables examined in the study could be associated with the district variance. Indeed, much work is needed to examine the relevance, sensitivity, and salience of our sociocultural and contextual variables, as well as the conditions in which multilevel DIF models can be adequately estimated (e.g., free of model convergence errors). Nevertheless, by leveraging the nested nature of the data structure for student characteristics (level-1) as well as district-level characteristics (level-2), this example allows for a deeper, theoretically-driven, and confirmatory approach to account for the observed differences in item responding.

This example should push psychometricians and measurement professionals to think beyond the conventional ways that data are currently being modeled in order to address key issues in measurement revolving around the underlying mechanisms that may be driving the observed item response patterns. Specifically, this brief example highlights how data from an assessment program (individual-level) and a state educational office (district-level) can be leveraged in a multilevel model to study a social phenomenon. This combination allows both theory and methodology to drive different research questions in an appropriate model-based approach in contrast to a design-based approach where methodological adjustments are post hoc.

As with any new methodological approach, framework, or even process, there are a plethora of opportunities to address the challenges that come with these advancements. For this WARNS example, readers may be left wondering to what degree do the DIF items impact the groups in the MLM model? Under what conditions are these complex models successful or unsuccessful in computing? What additional variables in community, state, or country can be used as sociocultural indicators in the model? With much left to explore, we note that multilevel measurement modeling in building validity arguments deserves continued attention to understand both the meaning of constructs at the different levels and how best to model the properties of such data to provide empirical evidence to support score use. It is this type of modeling that attempts to account for, or explain, differences for such groups that allow student, examinee, and participant context behind the data to be brought back into the story we tell with our multilevel models in the educational and psychological measurement realm.

RECOMMENDED READINGS

Technical Readings

French, B. F., & Finch, W. H. (2015). SIBTEST in a multilevel data environment. *Journal of Educational Measurement,* 52, 159–180.

Geldhof, G. J., Preacher, K. J., & Zyphur, M. J. (2014). Reliability estimation in a multilevel confirmatory factor analysis framework. *Psychological Methods, 19,* 72–91.

Stapleton, L. M., Yang, J. S., & Hancock, G. R. (2016). Construct meaning in multilevel settings. *Journal of Educational and Behavioral Statistics, 41,* 481–520.

Applied Readings

French, B. F., Finch, W. H., & Valdivia Vazquez, J. A. (2016). Predicting differential item functioning on mathematics items using multilevel SIBTEST. *Psychological Test and Assessment Modeling, 58,* 471–483.

Huang, F. L., & Cornell, D. G. (2016). Using multilevel factor analysis with clustered data: Investigating the factor structure of the Positive Values Scale. *Journal of Psychoeducational Assessment, 34,* 3–14.

Konold, T., Cornell, D., Huang, F., Meyer, P., Lacey, A., Nekvasil, E., Heilbrun, A., Shukla, K. (2014). Multilevel multi-informant structure of the Authoritative School Climate Survey. *School Psychology Quarterly, 29,* 238.

TRY THIS!

The data file contains scores for the five subscales from the IIS used in the example. You are wondering whether the items on the subscales that we did not examine in the chapter function the same across male and female students. You also wonder if a different factor structure could exist. That is, there is theory that suggests a two-factor model may still be best but that the factors are an academic factor with indicators AID, FCWSD, and IGC and a social factor with indicators of PGI and IWF. Your job is to estimate (a) a new two factor model and compare fit with the model shown in the chapter, (b) estimate reliability, and (c) conduct DIF analysis for at least one of the subscales besides IWF. The items for each are as follows:

PGI: is11-is20
IWF: is21 is23-is26
AID: is1-is6 is8 is10
FCWSD: is22 is27-is30
IGC: is7 is9 is31-is34

1. Estimate the new two-factor model both for assuming an individual level of measurement and at the aggregated level.
   a. Do both fit well? Can you support a validity inference with either one? Which validity inference does this evidence fall under?
   b. Compare fit of these models to the two factor model in the chapter. Write a few sentences describing which model you would select based on the evidence.
2. Estimate reliability for the new two-factor model from #1 both at the individual level and at the aggregate level.
   a. Write a few sentences about how this compares to the estimates in the chapter. What evidence does this support in a validity argument?
3. Select at least one subscale other than IWF. Conduct a DIF analysis on the items following what was conducted for IWF in the chapter.
   a. What do you conclude about how these items function across male and female students?
   b. What evidence does this support for score use in terms of an inference around validity?

REFERENCES

AERA, APA, and NCME. [American Educational Research Association, American Psychological Association, & National Council on Measurement in Education] (2014). *Standards for educational and psychological testing.*

Baker, B. A., Caison, A. L., & Meade, A. W. (2007). Assessing gender-related differential item functioning and predictive validity with the institutional integration scale. *Educational and psychological measurement, 67*(3), 545–559. https://doi.org/10.1177/0013164406292088

Bean, J. P. (1980) Dropouts and turnover: The synthesis and test of a causal model of student attrition. *Research in Higher Education, 12*(2), 155–187.

Beretvas, S. N., Cawthon, S. W., Lockhart, L. L., & Kaye, A. D. (2012). Assessing impact, DIF, and DFF in accommodated item scores: a comparison of multilevel measurement model parameterizations. *Educational and Psychological Measurement, 72*(5) 754–773. https://doi.org/10.1177/0013164412440998

Berger, J. B., & Milem, J. F. (1999). The role of student involvement and perceptions of integration in a causal model of student persistence. *Research in Higher Education, 40*(6), 641–664.

Bliese, P. D. (2000). Within-group agreement, non-independence, and reliability: Implications for data aggregation and analysis. In K. J. Klein & S. W. J. Kozlowski (Eds.), *Multilevel theory, research, and methods in organizations: Foundations, extensions, and new directions* (pp. 349–381). Jossey-Bass.

Bloom, H. S. (Ed.). (2005). *Learning more from social experiments: Evolving analytic approaches.* Russell Sage Foundation.

Borsboom, D., & Markus, K. A. (2013). Truth and evidence in validity theory. *Journal of Educational Measurement, 50*(1), 110–114.

Brennan, R. L. (2001). *Generalizability theory.* Springer-Verlag.

Brown, T. A. (2014). *Confirmatory factor analysis for applied research.* Guilford Publications.

Cai, L., Choi, K., Hansen, M., & Harrell, L. (2016). Item response theory. *Annual Review of Statistics and Its Application, 3,* 297–321. https://doi.org/10.1146/annurev-statistics-041715-033702

Chan, D. (1998). Functional relations among constructs in the same content domain at different levels of analysis: A typology of composition models. *Journal of Applied Psychology, 83*(2), 234.

Coburn, C., & Penuel, W.R. (2016). Research–practice partnerships in education: Outcomes, dynamics and open questions. *Educational Researcher, 45*(1), 48–54. https://doi.org/10.3102/0013189X16631750

Dedrick, R. F., & Greenbaum, P. E. (2011). Multilevel confirmatory factor analysis of a scale measuring interagency collaboration of children's mental health agencies. *Journal of Emotional and Behavioral Disorders, 19*(1), 27–40. https://doi.org/10.1177/1063426610365879

Dorans, N. J., & Holland, P. W. (1992). DIF detection and description: Mantel-Haenszel and standardization. *ETS Research Report Series, 1992,* 1. https://doi.org/10.1002/j.2333-8504.1992.tb01440.x

Fan, X., & Sivo, S. A. (2005). Sensitivity of fit indexes to misspecified structural or measurement model components: Rationale of two-index strategy revisited. *Structural Equation Modeling, 12*(3), 343–367.

Feldt, L. S., & Brennan, R. L. (1989). Reliability. In R. L. Linn (Ed.), *Educational measurement* (3rd ed). Macmillan.

Finch, W. H., & French, B. F. (2010). Detecting differential item functioning of a course satisfaction instrument in the presence of multilevel data. *The Journal of the First-Year Experience and Students in Transition, 22,* 27–48.

French, B. F., & Finch, W. H. (2008). Multigroup confirmatory factor analysis: Locating the invariant referent sets. *Structural equation modeling: A multidisciplinary journal, 15*, 96–113. https://doi.org/10.1080/10705510701758349

French, B. F., & Finch, W. H. (2010). Hierarchical logistic regression: Accounting for multilevel data in DIF detection. *Journal of Educational Measurement, 47*, 299–317. https://doi.org/10.1111/j.1745-3984.2010.00115.x

French, B. F., & Finch, W. H. (2013). Extensions of Mantel-Haenszel for multilevel DIF detection. *Educational and Psychological Measurement, 73*(4), 648–671. https://doi.org/10.1177/0013164412472341

French, B. F., & Finch, W. H. (2015). SIBTEST in a multilevel data environment. *Journal of Educational Measurement, 52*, 159–180.

French, B. F., Finch, W. H., & Immekus, J. C. (2019). Multilevel generalized Mantel-Haenszel for differential item functioning detection. *Frontiers in Education: Assessment, Testing, and Applied Measurement, 4*. https://doi.org/10.3389/feduc.2019.00047

French, B. F., & Maller, S. J. (2007). Iterative purification and effect size use with logistic regression for differential item functioning detection. *Educational and Psychological Measurement, 67*(3), 373–393.

French, B. F., & Oakes, W. (2004). Reliability and validity evidence for the institutional integration scale. Educational and Psychological Measurement, 64(1), 88-98.

Geldhof, G. J., Preacher, K. J., & Zyphur, M. J. (2014). Reliability estimation in a multilevel confirmatory factor analysis framework. *Psychological methods, 19*(1), 72. https://doi.org/10.1037/a0032138

George, T., Coker, E., French, B., Strand, P., Gotch, C., McBride, C., & McCurley, C. (2015). *Washington assessment of the risks and needs of students (WARNS)* [User manual]. Center for Court Research, Administrative Office of the Courts.

Gotch, C. M., & French, B. F. (2020). A validation trajectory for the Washington Assessment of Risks and Needs of Students. *Educational Assessment*, 65–82. https://doi.org/10.1080/10627197.2019.1702462

Gottfredson, N. C., Panter, A. T., Daye, C. E., Allen, W. F., & Wightman, L. F. (2009). The effects of educational diversity in a national sample of law students: Fitting multilevel latent variable models in data with categorical indicators. *Multivariate Behavioral Research, 44*(3), 305–331. https://doi.org/10.1080/00273170902949719

Hancock, G. R., & Mueller, R. O. (Eds.). (2010). *The reviewer's guide to quantitative methods in the social sciences*. Routledge.

Hedges, L. V., & Hedberg, E. C. (2007). Intraclass correlation values for planning group-randomized trials in education. *Educational Evaluation and Policy Analysis, 29*(1), 60–87. https://doi.org/10.3102/0162373707299706

Holland, P. W., & Thayer, D. T. (1988). Differential item performance and the Mantel-Haenszel procedure. In H. Wainer & H. I. Braun (Eds.), *Test validity* (pp. 129–145). Lawrence Erlbaum Associates Publishers.

Hox, J. (2002). *Multilevel analysis: Techniques and applications*. https://doi.org/10.4324/9780203852279

Hsu, H. Y., Lin, J. H., Kwok, O. M., Acosta, S., & Willson, V. (2017). The impact of intraclass correlation on the effectiveness of level-specific fit indices in multilevel structural equation modeling: A Monte Carlo study. *Educational and psychological measurement,* 77(1), 5–31. https://doi.org/10.1177/0013164416642823

Hu, L., & Bentler, P. M. (1999). Cutoff criteria for fit indexes in covariance structure analysis: Conventional criteria versus new alternatives. *Structural Equation Modeling,* 6(1), 1–55. https://doi.org/10.1080/10705519909540118

Huang, H. Y. (2015). A multilevel higher order item response theory model for measuring latent growth in longitudinal data. *Applied psychological measurement, 39*(5), 362–372. https://doi.org/10.1177/0146621614568112

Huang, H. Y. (2017). Mixture IRT model with a higher-order structure for latent traits. *Educational and psychological measurement,* 77(2), 275–304. https://doi.org/10.1177/0013164416640327

Huang, F. L., & Cornell, D. G. (2016). Multilevel factor structure, concurrent validity, and test–retest reliability of the high school teacher version of the Authoritative School Climate Survey. *Journal of Psychoeducational Assessment, 34*(6), 536–549. https://doi.org/10.1177/0734282915621439

Huang, F. L., Cornell, D. G., Konold, T., Meyer, J. P., Lacey, A., Nekvasil, E. K., Heilbrun, A., & Shukla, K. D. (2015). Multilevel factor structure and concurrent validity of the teacher version of the Authoritative School Climate Survey. *Journal of School Health, 85*(12), 843–851. https://doi.org/10.1111/josh.12340

Jak, S., Oort, F. J., & Dolan, C. V. (2013). A test for cluster bias: Detecting violations of measurement invariance across clusters in multilevel data. *Structural Equation Modeling: A Multidisciplinary Journal, 20*(2), 265–282. https://doi.org/10.1080/10705511.2013.769392

Jin, Y., & Eason, H. (2016). DIF analysis with multilevel data: A simulation study using the latent variable approach. *Journal of Educational Issues, 2*(2), 290–306.

Jin, Y., Myers, N. D., & Ahn, S. (2014). Complex versus simple modeling for DIF detection: when the intraclass correlation coefficient (ρ) of the studied item is less than the ρ of the total score. *Educational and Psychological Measurement, 74*(1), 163–190. https://doi.org/10.1177/0013164413497572

John-Steiner, V., & Mahn, H. (1996). Sociocultural approaches to learning and development: A Vygotskian framework. *Educational Psychologist, 31*(3–4), 191–206.

Kane, M. T. (2013). Validating the interpretations and uses of test scores. *Journal of Educational Measurement, 50,* 1–73. https://doi.org/10.1111/jedm.12000

Kish, L. (1965). *Survey Sampling.* John Wiley.

Kozlowski, S. W., & Klein, K. J. (2000). A multilevel approach to theory and research in organizations: Contextual, temporal, and emergent processes. In K. J. Klien & S. W. J. Kozlowski (Eds.), *Multilevel theory, research and methods in organizations: Foundations, extensions, and new directions.* (pp. 3–90). Jossey-Bass.

Levy, R. (2011). Bayesian data-model fit assessment for structural equation modeling. *Structural Equation Modeling: A Multidisciplinary Journal, 18*(4), 663–685. https://doi.org/10.1080/10705511.2011.607723

Lord, F. M., & Novick, M. R. (1968). *Statistical theories of mental test scores.*

Marsh, H. W., Lüdtke, O., Nagengast, B., Trautwein, U., Morin, A. J., Abduljabbar, A. S., & Köller, O. (2012). Classroom climate and contextual effects: Conceptual and methodological issues in the evaluation of group-level effects. *Educational Psychologist, 47*(2), 106–124.

McDonald, R. P. (2013). *Test theory: A unified treatment.* Psychology Press.

McNeish, D. (2018). Thanks coefficient alpha, we'll take it from here. *Psychological Methods, 23*(3), 412. https://doi.org/10.1037/met0000144

Meade, A. W., & Wright, N. A. (2012). Solving the measurement invariance anchor item problem in item response theory. *Journal of Applied Psychology, 97*(5), 1016. https://doi.org/10.1037/a0027934

Messick, S. (1989). Meaning and values in test validation: The science and ethics of assessment. *Educational Researcher, 18*(2), 5–11.

Muthén, B. O. (1994). Multilevel covariance structure analysis. *Sociological Methods & Research, 22*(3), 376–398. https://doi.org/10.1177/0049124194022003006

Muthén, L. K., & Muthén, B. O. (1998–2017). *Mplus user's guide* (8th ed.).

Narayanan, P., & Swaminathan, H. (1994). Performance of the Mantel-Haenszel and simultaneous item bias procedures for detecting differential item functioning. *Applied Psychological Measurement, 18*(4), 315–328.

Navas-Ara, M. J., & Gómez-Benito, J. (2002). Effects of ability scale purification on the identification of dif. *European Journal of Psychological Assessment, 18*(1), 9–15. https://doi.org/10.1027/1015-5759.18.1.9

Nunnally, I. H., & Bernstein, I. H. (1994). *Psychometric theory.* McGraw-Hill.

Pascarella, E. T., & Terenzini, P. T. (1980). Predicting freshman persistence and voluntary dropout decisions from a theoretical model. *The Journal of Higher Education, 51,* 60–75.

Penfield, R. D. (2001). Assessing differential item functioning among multiple groups: A comparison of three Mantel-Haenszel procedures. *Applied Measurement in Education, 14*(3), 235–259. https://doi.org/10.1207/S15324818AME1403_3

Raju, N. S., Laffitte, L. J., & Byrne, B. M. (2002). Measurement equivalence: A comparison of methods based on confirmatory factor analysis and item response theory. *Journal of Applied Psychology, 87*(3), 517.

Raykov, T., & du Toit, S. H. C. (2005). Estimation of reliability for multiple-component measuring instruments in hierarchical designs. *Structural Equation Modeling: A Multidisciplinary Journal, 12*(4), 536–550. https://doi.org/10.1207/s15328007sem1204_2

Raykov, T., & Marcoulides, G. A. (2006). On multilevel model reliability estimation from the perspective of structural equation modeling, *Structural Equation Modeling: A Multidisciplinary Journal, 13*(1), 130–141. https://doi.org/10.1207/s15328007sem1301_7

Raykov, T., & Marcoulides, G. A. (2011). *Introduction to psychometric theory.* Routledge.

Raykov, T., & Marcoulides, G. A. (2016). Scale reliability evaluation under multiple assumption violations. *Structural Equation Modeling: A Multidisciplinary Journal, 23*(2), 302–313.

Raykov, T., & Penev, S. (2010). Estimation of reliability coefficients in two-level designs via latent variable modeling. *Structural Equation Modeling, 17*(4), 629–641. https://doi.org/10.1080/10705511.2010.510052

Raykov, T., West, B. T., & Traynor, A. (2015). Evaluation of coefficient alpha for multiple component measuring instruments in complex sample designs. *Structural Equation Modeling, 22,* 429–438. https://doi.org/10.1080/10705511.2014.936081

Roussos, L. A., & Stout, W. (2004). Differential item functioning analysis. In D. Kaplan (Ed.), *The SAGE handbook of quantitative methodology for the social sciences* (pp. 107–116). SAGE.

Ryu, E. (2014). Model fit evaluation in multilevel structural equation models. *Frontiers in psychology, 5,* 81. https://doi.org/10.3389/fpsyg.2014.00081

Shear, B. R. (2018). Using hierarchical logistic regression to study DIF and DIF variance in multilevel data. *Journal of Educational Measurement, 55*(4), 513–542.

Skiba, R. J., Chung, C. G., Trachok, M., Baker, T. L., Sheya, A., & Hughes, R. L. (2014). Parsing disciplinary disproportionality: Contributions of infraction, student, and school characteristics to out-of-school suspension and expulsion. *American Educational Research Journal, 51*(4), 640–670.

Stapleton, L. M., McNeish, D. M., & Yang, J. S. (2016). Multilevel and single-level models for measured and latent variables when data are clustered. *Educational Psychologist, 51*(3–4), 317–330.

Stapleton, L. M., Yang, J. S., & Hancock, G. R. (2016). Construct meaning in multilevel settings. *Journal of Educational and Behavioral Statistics, 41*(5), 481–520.

Thompson, B. (2003). *Score reliability: Contemporary thinking on reliability issues.* Sage Publications.

Tinto, V. (1975). Dropout from higher education: A theoretical synthesis of recent research. *Review of Educational Research, 45*(1), 89–125.

Tinto, V. (1987). *Leaving college: Rethinking the causes and cures of student attrition.* University of Chicago Press.

Toland, M. D., & De Ayala, R. J. (2005). A multilevel factor analysis of students' evaluations of teaching. *Educational and Psychological Measurement, 65*(2), 272–296. https://doi.org/10.1177/0013164404268667

Wu, J. Y. (2010). *Comparing model-based and design-based structural equation modeling approaches in analyzing complex survey data* [Doctoral dissertation, Texas A&M].

Zinbarg, R. E., Revelle, W., Yovel, I., & Li, W. (2005). Cronbach's α, Revelle's β, and McDonald's ω_H: Their relations with each other and two alternative conceptualizations of reliability. *Psychometrika, 70*(1), 123–133. https://link.springer.com/content/pdf/10.1007/s11336-003-0974-7.pdf

Zumbo, B. D. (1999). *A handbook on the theory and methods of differential item functioning (DIF).* Ottawa: National Defense Headquarters, 160.

Zumbo, B. D., & Forer, B. (2011). Testing and measurement from a multilevel view: Psychometrics and validation. In J. A. Bovaird, K. F. Geisinger, & C. W. Buckendahl (Eds.), *High-stakes testing in education: Science and practice in K–12 settings* (pp. 177–190). American Psychological Association. https://doi.org/10.1037/12330-011

Zumbo, B. D., Liu, Y., Wu, A. D., Shear, B. R., Olvera Astivia, O. L., & Ark, T. K. (2015). A methodology for Zumbo's third generation DIF analyses and the ecology of item responding. *Language Assessment Quarterly, 12*(1), 136–151. https://doi.org/10.1080/15434303.2014.972559

Zwick, R. (2012). A review of ETS differential item functioning assessment procedures: Flagging rules, minimum sample size requirements, and criterion refinement. *ETS Research Report Series, 2012*(1). https://doi.org/10.1002/j.2333-8504.2012.tb02290.x

CHAPTER 15

MISSING DATA HANDLING FOR MULTILEVEL DATA

Craig K. Enders
University of California Los Angeles

Timothy Hayes
Florida International University

A substantial body of methodological research supports techniques that assume a missing at random (MAR) mechanism, whereby the missing data indicator for an incomplete variable (e.g., 0 = complete, 1 = missing) is unrelated to that variable's scores after conditioning on observed data (Little & Rubin, 2002; Rubin, 1976). Maximum likelihood estimation and multiple imputation are two such methods that enjoy widespread use, and Bayesian estimation is a third approach gaining in popularity (van de Shoot et al., 2017). Despite the literature's preference for MAR-based techniques, multilevel software packages often force the user to remove cases with missing data, thereby assuming a missing completely at random (MCAR) mechanism where missingness is unrelated to the analysis variables (i.e., no systematic predictors of nonresponse). Deletion has been characterized as "among the worst methods available for practical applications" (Wilkinson

Multilevel Modeling Methods with Introductory and Advanced Applications, pages 535–566

& Taskforce on Statistical Inference, 1999, p. 598), and its effects can be particularly devastating with incomplete level-2 variables because the entire cluster is excluded from analysis, even if the corresponding level-1 variables are complete.

Sophisticated missing data handling methods for single-level analyses were developed more than 2 decades ago (Arbuckle, 1996; Little & Rubin, 1987; Rubin, 1987; Schafer, 1997), and analogous procedures for multilevel data are now widely available. We focus on multiple imputation in this chapter because it is currently the most flexible and mature technology for dealing with the complexities of multilevel models. Briefly, multiple imputation creates many copies of a data set (e.g., 20 or more is a recent rule of thumb; Graham et al., 2007), each containing different estimates of the missing values. This procedure employs an iterative Bayesian estimation algorithm that generates multilevel regression model parameters and random effects, and these quantities in turn define a distribution of plausible replacement values for each observation (e.g., a predicted score and within-cluster residual variance define the center and spread of each distribution). The algorithm draws imputations from these distributions, and it uses the filled-in data to update the parameter values and random effects at the next iteration. After generating the desired number of filled-in data sets, the researcher fits the analysis model to each imputed data set and subsequently uses Rubin's pooling rules (Little & Rubin, 2002; Rubin, 1987) to combine the estimates and standard errors into a single set of results. Multilevel analyses do not require specialized pooling procedures, so we focus on the mechanics of imputation and refer readers to other resources for an overview of Rubin's rules (Enders, 2010; Little & Rubin, 2002; Schafer & Olsen, 1998; Sinharay et al., 2001; van Buuren, 2012).

Readers may wonder why we focus on multiple imputation, given that maximum likelihood is the estimator of choice in multilevel software programs. Maximum likelihood can handle a wide range of single-level missing data problems, but software platforms are currently less adept with multilevel missing data. Most multilevel programs allow for incomplete outcomes (e.g., a longitudinal model where measurements are unbalanced across respondents), but they necessarily exclude cases with missing explanatory variables because the conventional mixed modeling framework treats these variables as fixed constants. Certain structural equation modeling packages (e.g., M*plus*; Muthén & Muthén, 1998–2012) can accommodate missingness by treating predictors as normally distributed random variables, but this solution is not ideal for models that include mixtures of categorical and continuous variables, and it does not appear to work well with random coefficient models (Enders, Hayes et al., 2018). The development of flexible maximum likelihood estimation routines is an important area of ongoing methodological research (Raudenbush et al., 2013; Shin & Raudenbush, 2007, 2010, 2013), but imputation is arguably more flexible, at least for now.

Illustrative Analysis Models

Throughout the chapter, we consider imputation for three common multilevel analyses: a random intercept model, a random coefficient model, and a model with a cross-level interaction effect, as follows.

$$y_{ij} = \beta_0 + \beta_1(x_{ij}) + \beta_2(w_j) + b_{0j} + \varepsilon_{ij} \tag{15.1}$$

$$y_{ij} = \beta_0 + \beta_1(x_{ij}) + \beta_2(w_j) + b_{0j} + b_{1j}(x_{ij}) + \varepsilon_{ij} \tag{15.2}$$

$$y_{ij} = \beta_0 + \beta_1(x_{ij}) + \beta_2(w_j) + \beta_3(x_{ij})(w_j) + b_{0j} + b_{1j}(x_{ij}) + \varepsilon_{ij} \tag{15.3}$$

Each model features an explanatory variable at level-1 and level-2, with b_{0j} and b_{1j} denoting random intercepts and slopes, respectively, for level-2 unit j, and ε_{ij} representing the within-cluster residual for observation i in group j. In line with conventional mixed models, level-2 residuals are multivariate normal with zero means a covariance matrix Σ_b, and within-cluster residuals are normal with constant variance σ^2_ε.

Models such as these abound in multilevel textbooks and in published applications. For example, in an education context with students nested in schools, X and Y could represent socioeconomic status and academic achievement, respectively, and W could be a school-level predictor such as school size (Raudenbush & Bryk, 2002). In a diary study with repeated measurements nested in individuals, X and Y could represent daily measurements of stress and mood, respectively, and W could be an individual-level personality variable such as neuroticism (Nezlek, 2012). We chose these models as examples because they are exceedingly common in the applied literature, yet they make different propositions about the population covariance structure that are important to consider when selecting an imputation procedure.

BAYESIAN ESTIMATION FOR MULTILEVEL MODELS

Multiple imputation creates several copies of the data set, each with different estimates of the missing values. Understanding how the procedure creates sets of exchangeable imputations requires a brief discussion of Bayesian estimation for multilevel models, as the Bayesian paradigm provides the mathematical machinery for multiple imputation. Returning to the example analyses, the three models require the following quantities: a vector of regression coefficients, $\boldsymbol{\beta}$, a vector of random effects for each cluster, $\boldsymbol{b}_j$, a level-2 covariance matrix for these residuals, $\boldsymbol{\Sigma}_b$, and a within-cluster residual variance, σ_ε^2. The Bayesian paradigm views the parameters and level-2 residuals as random variables that follow a probability distribution. Because the joint distribution that describes the cooccurrence of the variables is very complicated, an iterative procedure—a Markov chain Monte Carlo (MCMC) algorithm—estimates one quantity at a time, treating all other variables as known (e.g., the algorithm estimates coefficients, conditional on the current values of the level-2 residuals and variance estimates).

Each MCMC estimation step uses computer simulation to "sample" parameter values or random effects from a probability distribution that conditions on the current values of all other quantities. With complete data, the estimation steps for a single MCMC iteration *t* are

$$\begin{aligned}
&\text{1. Sample fixed coefficients from } p\left(\dot{\boldsymbol{\beta}}^{(t)} \middle| \dot{\boldsymbol{b}}_j^{(t-1)}, \dot{\sigma}_\varepsilon^{2(t-1)}, \dot{\boldsymbol{\Sigma}}_b^{(t-1)}, \text{data, priors}\right)\\
&\text{2. Sample level-2 residuals from } p\left(\dot{\boldsymbol{b}}_j^{(t)} \middle| \dot{\boldsymbol{\beta}}^{(t)}, \dot{\sigma}_\varepsilon^{2(t-1)}, \dot{\boldsymbol{\Sigma}}_b^{(t-1)}, \text{data}\right)\\
&\text{3. Sample residual variance from } p\left(\dot{\sigma}_\varepsilon^{2(t)} \middle| \dot{\boldsymbol{\beta}}^{(t)}, \dot{\boldsymbol{b}}_j^{(t)}, \dot{\boldsymbol{\Sigma}}_b^{(t-1)}, \text{data, priors}\right)\\
&\text{4. Sample level-2 covariance matrix from } p\left(\dot{\boldsymbol{\Sigma}}_b^{(t)} \middle| \dot{\boldsymbol{\beta}}^{(t)}, \dot{\boldsymbol{b}}_j^{(t)}, \dot{\sigma}_\varepsilon^{2(t)}, \text{data, priors}\right)
\end{aligned} \tag{15.4}$$

where *p* represents a probability distribution (e.g., a normal distribution), the variable to the left of a vertical pipe is the target of a particular estimation step, and the variables to the right of each vertical pipe function as known constants that define the shape of the distribution. Following van Buuren (2012), we use a dot accent to denote synthetic estimates drawn at random from these distributions. Additional technical details about these distributions are widely available in the literature (Browne & Draper, 2000; Enders, Keller et al., 2018; Goldstein et al., 2007; Goldstein et al., 2009; Kasim & Raudenbush, 1998; Schafer & Yucel, 2002; van Buuren, 2012).

Repeating the MCMC steps many times (e.g., for hundreds or thousands of iterations) produces empirical distributions of the parameter values (i.e., posterior distributions). Although the Bayesian paradigm does not rely on the concept of repeated sampling, each posterior is conceptually analogous to a frequentist sampling distribution. For example, the distribution of $\dot{\beta}_1^{(t)}$ across a large number of MCMC cycles should approximate a normal curve with roughly the same center and spread as the theoretical sampling distribution of β_1 from a frequentist analysis. The goal of a Bayesian analysis is to summarize the empirical distributions of the synthetic parameter values, and the mean and standard deviation function roughly like point estimates and standard errors from the frequentist framework (e.g., $\bar{\dot{\beta}}_1$ is conceptually and numerically similar to $\hat{\beta}_1$, and $s_{\dot{\beta}_1}$ is comparable to $SE_{\hat{\beta}_1}$).

UNIVARIATE MISSINGNESS ON THE OUTCOME

When missing values are restricted to the outcome variable, each iteration of the MCMC algorithm can be viewed as a complete-data Bayesian analysis followed by an additional step that updates imputations, conditional on the current parameter estimates and residual terms. To illustrate, consider the random intercept model from Equation 15.1. After applying the estimation sequence from Equation 15.4 to the filled-in data from the previous iteration, the resulting parameter values and residual terms define a distribution of plausible replacement scores from which the algorithm uses computer simulation to draw new imputations.

$$\begin{aligned} \mu_{Y|X,W} &= \dot{\beta}_0 + \dot{\beta}_1(x_{ij}) + \dot{\beta}_2(w_j) + \dot{b}_{0j} \\ y_{ij(mis)} &\sim N(\mu_{Y|X,W}, \dot{\sigma}_\varepsilon^2) \end{aligned} \tag{15.5}$$

In words, Equation 15.5 says to draw an imputation for observation i in group j from a normal distribution with center and spread equal to a predicted value and within-cluster residual variance, respectively. After updating all imputations, the filled-in data carry forward to the next iteration's estimation sequence.

As explained previously, the MCMC algorithm repeats the estimation and imputation sequence for many (usually several thousand) iterations. In the context of multiple imputation, the synthetic parameter values are not the primary focus as they are in a Bayesian analysis. Rather, these quantities are just a means to an end, which is to characterize the distribution of missing values. To create multiple imputed data sets, researchers typically employ a "thinning" procedure that culls filled-in data sets at regular intervals in a long MCMC sequence. The first data set is saved after a burn-in period that allows the MCMC algorithm to stabilize or converge, and subsequent data sets are saved at regular intervals thereafter (e.g., after every 1,000th iteration). In our experience, multilevel imputation algorithms can take much longer to converge than their single-level counterparts, so it is important to examine graphical or numeric diagnostics such as trace plots and potential scale reduction factors from a preliminary diagnostic run (Gelman et al., 2014; Gelman & Rubin, 1992; Schafer & Olsen, 1998).

MULTIVARIATE IMPUTATION FRAMEWORKS

Imputation is straightforward when missing values are restricted to the outcome because conventional mixed models specify a distribution for this variable. For example, the three analysis models assume that *Y* is conditionally normal given the fixed and random effects. We occasionally use the generic notation $p(Y|X,W)$ (i.e., the distribution *Y* given particular values of *X* and *W*) to reference this distribution. Equation 15.5 shows the model-implied distribution of *Y* for a random intercept analysis, and random coefficients and interactive or non-linear effects would simply add to the linear predictor that defines the mean parameter (e.g., the random coefficient model has a predicted value equal to $\mu_{Y|X,W} = \dot{\beta}_0 + \dot{\beta}_1(x_{ij}) + \dot{\beta}_2(w_j) + \dot{b}_{0j} + b_{1j}(x_{ij})$). The situation changes considerably when predictors are missing because these variables are usually treated as fixed constants during estimation. In order to do any type of missing data handling, we must specify models and distributions for *all* variables in a given analysis, not just the outcome. As we will see, the composition of the analysis model—particularly whether or not it includes random coefficients or nonlinear (e.g., interactive) effects—has important implications for modeling incomplete explanatory variables.

We describe three imputation frameworks in this chapter: joint model imputation, fully conditional specification, and fully Bayesian (model-based) multiple imputation. The strategies are similar in the respect that they use multilevel model parameters and random effects to characterize the distributions of the incomplete variables, but they differ in how they specify these distributions. In particular, joint model imputation specifies a multivariate distribution for the incomplete variables, whereas fully conditional specification uses a sequence of univariate conditional distributions to characterize their cooccurrence. To illustrate this difference, Figure 15.1 displays a contour plot (i.e., overhead view) of a bivariate normal distribution where *X* and *Y* are positively correlated. The joint modeling approach draws imputations directly from the bivariate distribution, whereas fully conditional specification carves the surface into two sets of univariate distributions: the slices on the vertical hash marks represent normal distributions of *Y* at different values of *X*, and the slices on the horizontal hash marks are normal distributions of *X* at different values of *Y*.

Returning to our analysis examples, Table 15.1 uses generic notation to describe the probability distributions invoked by each imputation procedure. Because we are restricting our attention to continuous variables, $p(Y,X,W)$ can be viewed as a trivariate normal distribution. Fully conditional specification uses a round robin algorithm to carve this distribution into three sets of univariate normal distributions, where the variable to the left of the vertical pipe is the target of imputation, and the remaining variables to the right of the pipe (the variables being conditioned on) are imputed

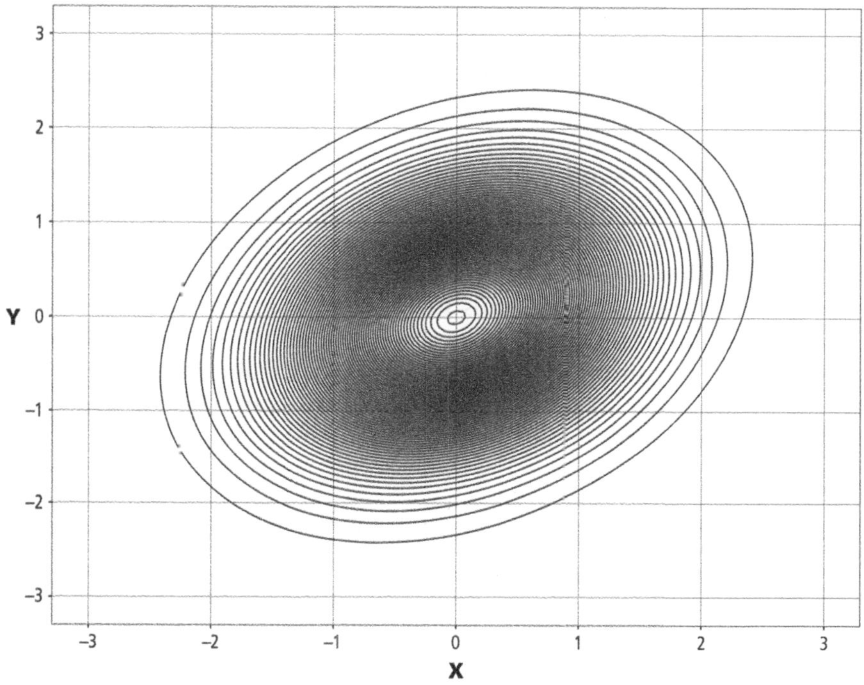

Figure 15.1 Bivariate normal contour plot with positively correlated variables. The vertical slices represent the conditional distributions of Y at different values of X, and the horizontal slices represent the conditional distributions of X at different values of Y.

at a previous step. Finally, a fully Bayesian imputation scheme can be viewed as a two-part specification where the researcher's analysis model defines a normal distribution for Y, and a bivariate normal distribution is applied to the predictor variables. Because the fully Bayesian approach is built around a particular analysis, we henceforth refer to it simply as model-based imputation, and variants of this procedure are also known as substantive model-compatible imputation in the literature (Bartlett et al., 2015; Quartagno & Carpenter, 2018).

TABLE 15.1 Model-Implied Distributions of the Major Imputation Frameworks

Imputation Framework	Model-Implied Distributions
Joint Model Imputation	$p(Y, X, W)$
Fully Conditional Specification	$p(Y \mid X, W)$, $p(X \mid Y, W)$, and $p(W \mid Y, X)$
Model-Based Imputation	$p(Y \mid X, W)\,p(X, W)$

Imputation models are characterized by van Buuren (2012) as ranging from broad in scope to narrow in scope. In a broad-scope application, the researcher specifies a model that creates a single set of imputations to be used for a variety of different analyses. A narrow-scope application, on the other hand, creates imputations for one specific analysis. Joint model imputation and fully conditional specification can be used for broad or narrow problems, whereas model-based imputation is inherently narrowly focused. In our view, the complexities of multilevel data make it difficult to apply broad imputation models that are compatible with a wide range of substantive hypotheses. Although this strategy might be feasible for analyses with random intercepts, random coefficients and interactive or non-linear effects typically require a narrow imputation model tailored to a particular analysis. The imputation models and data analysis examples that we present in this chapter reflect this narrow focus.

IMPUTATION FOR RANDOM INTERCEPT ANALYSES

This section considers multiple imputation for the random intercept model from Equation 15.1. Returning to the education example, the β_1 coefficient gives the constant influence of socioeconomic status across schools, β_2 is the net influence of a school-level predictor such as school size, and b_{0j} and ε_{ij} capture between- and within-school residual variation in achievement scores, respectively. In the daily diary example, β_1 reflects the constant influence of stress on mood, β_2 is the net influence of neuroticism on person-average mood scores, and b_{0j} and ε_{ij} are between- and within-person residuals. When the analysis variables are continuous and normally distributed, the three imputation frameworks and their model-implied distributions from Table 15.1 are equivalent. Although we might expect minor differences in small samples, there is usually no reason to prefer one method over another. For this reason, we describe joint model imputation and fully conditional specification in this section, and we introduce model-based imputation later in the context of a random coefficient analysis. As we will see, joint model imputation uses covariance matrices and correlated residuals to preserve associations among the variables, whereas fully conditional specification uses a series of regression models.

Joint Model Imputation

Schafer's classic work outlined a joint model approach to multilevel imputation (Schafer, 2001; Schafer & Yucel, 2002), and a variety of extensions have since appeared in the literature (Asparouhov & Muthén, 2010b;

Carpenter et al., 2011; Carpenter & Kenward, 2013; Goldstein et al., 2009; Goldstein et al., 2014; Yucel, 2008, 2011). As explained previously, the joint model specifies a multivariate distribution for the incomplete variables, in this case a multivariate normal distribution. We illustrate an unconditional version of the joint model that treats all analysis variables as outcomes (Asparouhov & Muthén, 2010b; Carpenter & Kenward, 2013), and a comparable parameterization designates incomplete variables as outcomes and treats complete variables as fixed covariates (Goldstein et al., 2009; Schafer, 2001; Schafer & Yucel, 2002; Yucel, 2008).

The level-1 part of the joint imputation model expresses Y and X as the sum of a latent group mean (i.e., a random intercept) and within-cluster deviation score.

$$y_{ij} = \mu_{j(y)} + e_{ij(y)} \tag{15.6}$$

$$x_{ij} = \mu_{j(x)} + e_{ij(x)} \tag{15.7}$$

The model assumes that residuals are multivariate normal with zero means and an unrestricted covariance matrix Σ_e (e.g., a 2 by 2 matrix in our example), with the off-diagonal elements of Σ_e preserving within-cluster associations among group mean centered level-1 variables. The level-2 part of the joint model specifies latent group means and manifest level-2 variables as the sum of a grand mean and a between-cluster deviation score.

$$\mu_{j(y)} = \mu_{(y)} + r_{j(y)} \tag{15.8}$$

$$\mu_{j(x)} = \mu_{(x)} + r_{j(x)} \tag{15.9}$$

$$w_j = \mu_{(w)} + r_{j(w)} \tag{15.10}$$

The deviation scores are again assumed to follow a multivariate normal distribution with an unrestricted between-cluster covariance matrix Σ_r (e.g., a 3 by 3 matrix in our example) that preserves level-1 relations. Throughout the chapter we subscript parameters and random effects with variable names (as we do above) to link terms on the left and right sides of the equations.

The previous imputation models require four sets of unknowns: a vector of grand means, μ, within- and between-cluster covariance matrices, Σ_e and Σ_r, respectively, and a vector of latent group means for each level-2 unit, μ_j. An MCMC algorithm similar to the one described earlier in the section on Bayesian estimation for multilevel models draws these quantities from a series of probability distributions, after which it uses the resulting estimates to define the multivariate normal distribution from which it draws missing values.

$$\begin{pmatrix} y_{ij(mis)} \\ x_{ij(mis)} \end{pmatrix} \sim N_2(\dot{\mu}_j, \dot{\Sigma}_e) \quad \begin{pmatrix} \dot{\mu}_{j(y)} \\ \dot{\mu}_{j(x)} \\ w_{j(mis)} \end{pmatrix} \sim N_3(\dot{\mu}, \dot{\Sigma}_r) \tag{15.11}$$

In words, Equation 15.11 says to (a) sample level-1 imputations from a bivariate normal distribution centered at the current latent group means and (b) draw latent group means and level-2 imputations from a trivariate normal distribution. Visually, level-1 imputations are drawn from a distribution similar to Figure 15.1, with $\dot{\mu}_{j(x)}$ and $\dot{\mu}_{j(y)}$ located at the highest (center-most) contour. Importantly, the elements in Σ_e and Σ_r are free to differ, such that the within-cluster association between Y and X need not be the same as the level-2 relation between $\mu_{j(y)}$ and $\mu_{j(x)}$. Although our random intercept example does not require this complexity, joint model imputations are nevertheless appropriate for analyses that posit distinct within- and between-cluster associations (e.g., contextual effects models that incorporates level-2 group means as additional predictors; multilevel factor analysis models).

Fully Conditional Specification

van Buuren and colleagues (van Buuren, 2007, 2011; van Buuren et al., 2006; van Buuren et al., 2018) popularized fully conditional specification for single-level imputation, and analogous extensions for multilevel data have since appeared in the literature (Enders, Keller et al., 2018; Keller & Du, 2019; van Buuren, 2011, 2012). As explained previously, fully conditional specification approximates a multivariate distribution by specifying a sequence of univariate distributions with one variable conditioning on all others. The regression models that spawn these distributions are as follows.

$$y_{ij} = \beta_{0(y)} + \beta_{1(y)}(x_{ij}) + \beta_{2(y)}(\mu_{j(x)}) + \beta_{3(y)}(w_j) + b_{0j(y)} + \varepsilon_{ij(y)} \tag{15.12}$$

$$x_{ij} = \beta_{0(x)} + \beta_{1(x)}(y_{ij}) + \beta_{2(x)}(\mu_{j(y)}) + \beta_{3(x)}(w_j) + b_{0j(x)} + \varepsilon_{ij(x)} \tag{15.13}$$

$$w_j = \beta_{0(w)} + \beta_{1(w)}(\mu_{j(y)}) + \beta_{2(w)}(\mu_{j(x)}) + b_{0j(w)} \tag{15.14}$$

As before, we subscript coefficients and residual terms with variable names to emphasize that each equation requires unique estimates (as well as a supporting MCMC estimation sequence like the one from Equation 15.4 to generate the estimates).

The MCMC algorithm follows a round-robin pattern where the estimation-imputation sequence is applied to each variable in turn. After sampling the regression model parameters and random effects from their probability distributions, the algorithm uses the resulting estimates to define a univariate normal distribution of missing values for the target variable, with filled-in variables from a previous iteration or cycle functioning as predictors. The distributions of missing values for our example are as follows.

$$y_{ij(mis)}^{(t)} \sim N\left(\dot{\beta}_{0(y)} + \dot{\beta}_{1(y)}(x_{ij}^{(t-1)}) + \dot{\beta}_{2(y)}(\dot{\mu}_{j(x)}^{(t-1)}) + \dot{\beta}_{3(y)}(w_j^{(t-1)}) + \dot{b}_{0j(y)}, \dot{\sigma}_{\varepsilon(y)}^2\right) \quad (15.15)$$

$$x_{ij(mis)}^{(t)} \sim N\left(\dot{\beta}_{0(x)} + \dot{\beta}_{1(x)}(y_{ij}^{(t-1)}) + \dot{\beta}_{2(x)}(\dot{\mu}_{j(y)}^{(t)}) + \dot{\beta}_{3(x)}(w_j^{(t-1)}) + \dot{b}_{0j(x)}, \dot{\sigma}_{\varepsilon(x)}^2\right) \quad (15.16)$$

$$w_{j(mis)}^{(t)} \sim N\left(\dot{\beta}_{0(w)} + \dot{\beta}_{1(w)}(\dot{\mu}_{j(y)}^{(t)}) + \dot{\beta}_{2(w)}(\dot{\mu}_{j(x)}^{(t)}), \dot{\sigma}_{b(w)}^2\right) \quad (15.17)$$

As before, a predicted value defines the mean and the within-cluster residual variance defines spread. We attach the iteration superscript t to the variables to highlight the flow of terms from the left to the right sides of the equations. Returning to Figure 15.1, the contour plot can be viewed as the bivariate distribution of X and Y for a particular level-2 cluster j. Visually, the level-1 imputation models from Equations 15.15 and 15.16 fall along the vertical and horizontal hash marks, respectively; vertical slices represent the distribution of Y at different values of X, and horizontal slices give the distribution of X at different values of Y.

Notice that the level-2 cluster means, $\mu_{j(y)}$ and $\mu_{j(x)}$, appear as explanatory variables on the right side of the equations. In Equations 15.12 and 15.13, the group means function like contextual effects that allow the between-cluster regressions to differ from their within-cluster counterparts (Kreft et al., 1995; Lüdtke et al., 2008; Lüdtke et al., 2011; Raudenbush & Bryk, 2002). Although these group-level predictors are not necessary for our particular analysis model, including them mimics the parameterization and complexity of the joint model (Carpenter & Kenward, 2013; Mistler & Enders, 2017). In Equation 15.14, the group means play a more fundamental role by preserving the association between W and the between-cluster parts of Y and X. It is worth noting that the level-2 group means can be defined in one of three ways: as (a) within-group aggregates of the imputed level-1 scores (Enders, Keller et al., 2018; Gelman & Hill, 2007; Yucel, 2008); (b) plausible values from the psychometrics tradition (Grund et al., 2017; Mislevy, 1991); or (c) normally distributed latent variables (Keller & Du, 2019). Although the differences among these methods are often small (Grund et al., 2017; Keller & Du, 2019), the latter two options are preferable because they naturally accommodate unequal cluster sizes.

Data Analysis Example 1

Multilevel imputation schemes are available in some commercial software packages, including M*plus* (Muthén & Muthén, 1998–2017) and MLwiN (Carpenter et al., 2011). We use the R package jomo (Quartagno & Carpenter, 2018) and a standalone application called Blimp (Keller & Enders, 2019) for the data analysis examples, as both are freely available for macOS, Windows, and Linux operating systems. To provide data for the example, we used computer simulation to create an artificial data set from a random intercept population model with the following parameter values.

$$y_{ij} = 5 + 5(x_{ij}) + 5(w_j) + b_{0_j} + \varepsilon_{ij}$$

$$\sigma_b^2 = 100 \quad \sigma_\varepsilon^2 = 100$$

Although it is not possible to draw firm conclusions from a single data set, we sought to minimize contamination due to sampling error by generating a large sample of $N = 25{,}000$ comprised of 500 level-2 units and 50 level-1 observations per group. After creating the complete data, we imposed a 20% MAR missing data rate on X and W, such that missingness increased with the values of Y. The raw data files and analysis scripts for our three analysis examples are available for download from the Blimp website, www.appliedmissingdata.com/blimp. In the interest of space, we focus on the mechanics of the analysis and point readers to the software documentation for details about the scripts.

As noted previously, it is important to examine the convergence of the MCMC algorithm prior to generating imputations for analysis. Whereas maximum likelihood estimation converges to a single set of estimates, MCMC converges when the sampling steps from Equation 15.4 produce estimates that form a stable distribution (i.e., running the algorithm for more iterations does not change the mean or variance of the posterior distribution). Blimp implements a popular diagnostic measure called the potential scale reduction factor (Gelman et al., 2014; Gelman & Rubin, 1992), and the jomo package outputs parameter estimates that can be used to construct trace plots (Schafer & Olsen, 1998). The idea behind the potential scale reduction factor is to compare posterior distributions from two separate MCMC processes and determine the number of iterations required for the two distributions to achieve similar mean values (an indication that the distributions have stabilized). Lower values closer to unity (the hypothetical minimum) are better, and a common recommendation is that convergence is achieved when the index falls below 1.05 to 1.10 (Asparouhov & Muthén, 2010a; Gelman et al., 2014).

A preliminary diagnostic run in Blimp indicated that potential scale reduction factors dropped below 1.05 after approximately 200 iterations. As

TABLE 15.2 Estimates from Data Analysis Example 1

Parameter		Complete		Listwise		Joint Model		FCS	
		Est.	SE	Est.	SE	Est.	SE	Est.	SE
β_0	5	5.06	0.45	1.28	0.43	4.80	0.48	4.80	0.47
β_1	5	5.05	0.06	4.64	0.07	4.93	0.10	5.12	0.07
β_2	5	4.98	0.46	3.79	0.42	4.97	0.52	4.88	0.48
σ_b^2	100	100.27	6.03	70.20	4.63	98.34	6.76	98.97	6.41
σ_ε^2	100	99.48	0.95	89.76	1.03	99.83	1.25	98.70	0.98

we will see, models with random coefficients and interactive terms converge more slowly, so it is important to inspect these quantities prior to creating imputations for analysis. Based on the diagnostic run, we took a conservative approach that generated 20 imputations from MCMC chains with 500 burn-in and thinning iterations (i.e., data sets were saved for analysis after the first 500 iterations and every 500th iteration thereafter). We then used the R package mitml (Grund, Robitzsch et al., 2021) to fit the random intercept model from Equation 15.1 to each imputed data set and pool the resulting estimates and standard errors following Rubin's rules (Rubin, 1987).

Table 15.2 gives estimates from the complete data (artificial data prior to imposing missing values), listwise deletion, joint model imputation, and fully conditional specification. Because the data were generated from a population with known parameter values, an estimate more two standard errors from its true value can be viewed as significantly distorted. The table shows that the two imputation approaches produced estimates that were effectively equivalent and indiscernible from those of the complete-data analysis. This result is expected because the joint model and fully conditional specification are equivalent when applied to random intercept models with continuous variables. Notice that deleting observations with incomplete predictors (the standard behavior of multilevel software packages) leads to severe distortion, with estimates that differ from their true values by several standard errors.

IMPUTATION FOR RANDOM COEFFICIENT ANALYSES

Random coefficient models such as the one from Equation 15.2 are far more complex from a missing data handling perspective because they posit group-specific distributions that differ both with respect to their means and within-cluster variances and covariances. Equation 15.2 has an additional between-cluster residual term that allows the influence of the level-1 predictor to vary across level-2 units. Applied to the education example, this model allows the influence of socioeconomic status on achievement to vary across schools, with the slope for school j defined as $\beta_1 + b_{1j}$. The same collection of terms gives the slope of mood on stress for person j in the diary data context. As we will see, the product of the level-1 predictor and the random slope residual induces a non-linearity that has important implications for specifying the distributions of the explanatory variables.

Joint Model Imputation

Most formulations of the joint model assume a homogenous within-cluster covariance matrix where the variances and covariances in $\mathbf{\Sigma}_e$ are common to all level-2 units. This specification is incompatible with the random coefficient model because a common covariance matrix aligns with an analysis where the level-1 slope is fixed. Not surprisingly, applying the joint model presented in the previous section for random intercept analyses to a random coefficient model biases the resulting slope variance estimates toward zero, and other parameters may also exhibit bias (Enders, Mistler et al., 2016; Lüdtke et al., 2017).

Yucel (2011) proposed a flexible version of the joint model that incorporates random level-1 covariance matrices that vary across level-2 units. The level-1 part of Yucel's joint model changes to incorporate group-specific covariance matrices, but the level-2 model is unchanged. Applied to our analysis example, the level-1 model is

$$\begin{pmatrix} y_{ij(mis)} \\ x_{ij(mis)} \end{pmatrix} \sim N_2(\dot{\mu}_j, \dot{\Sigma}_{e_j}) \quad \dot{\Sigma}_{e_j} = \begin{pmatrix} \dot{\sigma}^2_{ej(y)} & \dot{\sigma}_{ej(yx)} \\ \dot{\sigma}_{ej(xy)} & \dot{\sigma}^2_{ej(x)} \end{pmatrix} \tag{15.18}$$

where the j subscript indicates that variances and covariances vary across level-2 units. The covariance matrices in this specification are "random" in the sense that they follow a probability distribution like other model parameters. The MCMC algorithm includes additional steps that estimate the center and spread of this distribution as well as the group-specific covariance matrices, $\mathbf{\Sigma}_{e_j}$ (Quartagno & Carpenter, 2016), but the resulting imputations can still be viewed as the sum of a latent group mean (i.e., random intercept) and a random within-cluster error term. The key change is that the magnitude of these errors as well as the strength of their correlation varies across clusters.

In principle, a joint imputation model with random covariance matrices should preserve random slope variation because it allows level-1 variances and covariances to differ across groups, just as the analysis model incorporates heterogeneity in the level-1 slopes. However, the model has a subtle feature that could introduce bias in some situations. Recall that our random coefficient example specified an unstructured covariance matrix Σ_b that allows the intercepts and slopes to covary. Importantly, the joint model components that capture intercept and slope variation—Σ_r and Σ_{e_j}, respectively—are orthogonal. Although the latent group means are drawn from a normal distribution that conditions on Σ_{e_j}, the model has no parameter that explicitly links between-cluster covariation in Σ_{e_j} to between-cluster mean variation in Σ_r. Thus, the joint model appears to align best with a random coefficient analysis where Σ_b is a diagonal matrix, and it may not be ideal for models where the covariance is of substantive interest (e.g., longitudinal growth models where initial status informs growth rates). Few empirical studies have investigated the joint model (Enders, Hayes et al., 2018; Quartagno & Carpenter, 2016), and more research is needed to understand how Yucel's (2011) method works in different applications.

Fully Conditional Specification

When the outcome is missing, the fully conditional specification imputation model has the same form as the analysis model from Equation 15.2. The distribution of missing values is the same as Equation 15.5 except that $b_{1j}(x_{ij})$ appears in the linear predictor that defines the mean parameter. A standard method for imputing a level-1 predictor is to specify a "reverse random coefficient" imputation model with X regressed on Y.

$$x_{ij} = \beta_{0(x)} + \beta_{1(x)}(x_{ij}) + \beta_{2(x)}(w_j) + b_{0j(x)} + b_{1j(x)}(x_{ij}) + \varepsilon_{ij(x)} \tag{15.19}$$

The model may also include the group means, as in Equation 15.12, but we omit them here for clarity. Consistent with the analysis model, imputation assumes that the residuals are normal with level-2 covariance matrix $\Sigma_{b(x)}$ and level-1 residual variance $\sigma^2_{\varepsilon(x)}$. After estimating the parameters and random effects from this model, the MCMC algorithm draws imputations from a normal distribution with constant variance.

$$x_{ij(mis)} \sim N\left(\dot{\beta}_{0(x)} + \dot{\beta}_{1(x)}(x_{ij}) + \dot{\beta}_{2(x)}(w_j) + \dot{b}_{Cj(x)} + \dot{b}_{1j(x)}(x_{ij}), \dot{\sigma}^2_{\varepsilon(x)}\right) \tag{15.20}$$

A number of recent papers have studied reverse random coefficient imputation (Enders, 2020; Enders, Hayes et al., 2018; Enders, Keller et al., 2018; Grund, Lüdtke et al., 2016, 2018; Kunkle & Kaizer, 2017; Lüdtke et al., 2017), and a common finding is that the method is prone to systematic biases that persist even in very large samples (e.g., random slope variation is typically underestimated by 10 to 20%). Unlike a random intercept model where reversing regression equations tends to work well (e.g., predicting Y from X and X from Y), the distribution in Equation 15.20 is misspecified because the $b_{1j}(x_{ij})$ product term in the analysis model implies that the correct distribution of X is a complex non-linear function of Y with non-constant variance (Enders et al., 2020). Because the accumulating literature is mostly critical of reverse random coefficient imputation, we cannot recommend it.

Model-Based Imputation

Model-based imputation can be thought of as two-part specification where the researcher's analysis model is paired with a second model featuring just the explanatory variables (Ibrahim et al., 2002; Ibrahim et al., 2005). Table 15.1 denotes the model-implied distribution of this approach as $p(Y|X,W)p(X,W)$, where $p(Y|X,W)$ is the distribution of Y induced by the analysis model, and $p(X,W)$ is a bivariate distribution for the predictor variables. By specifying both components separately, we can derive distributions for X and W that are compatible or consistent with the non-linear random coefficient term. As we will see, the same procedure readily accommodates interactive and other types of non-linear effects (e.g., polynomial terms). The limited research to date suggests that model-based imputation is effectively unbiased as sample size increases, and the procedure provides a substantial improvement over reverse random coefficient imputation (Enders et al., 2020; Erler et al., 2017; Erler et al., 2016; Grund et al., 2018).

Model-based imputation is built around the researcher's analysis model, in this case the random coefficient example from Equation 15.2. The analysis model predicts that, after conditioning on the predictor variables and random effects, Y is normally distributed with center and spread equal to a linear predictor and within-cluster residual variance, respectively.

$$p(Y|X,W) = N\left(\beta_{0(y)} + \beta_{1(y)}(x_{1ij}) + \beta_{2(y)}(x_{2j}) + b_{0j(y)} + b_{1j(y)}(x_{1ij}), \sigma^2_{\varepsilon(y)}\right) \quad (15.21)$$

Missing outcome scores are drawn from this distribution, with the MCMC algorithm again providing the necessary parameter values and random effects. Next, we need to specify a second model for the incomplete predictors. Borrowing from the joint model framework, a simple strategy is to specify $p(X,W)$ as a bivariate normal distribution. Recycling some earlier material, Equations 15.7 and 15.9 decompose X into within- and between-group components, and Equation 15.10 gives the between-cluster expression for W. The level-1 and level-2 parts of the normal distribution are

$$x_{ij} \sim N(\mu_{j(x)}, \sigma^2_{\varepsilon(x)}) \quad (15.22)$$

$$\begin{pmatrix} \mu_{j(x)} \\ w_j \end{pmatrix} \sim N_2\left(\begin{pmatrix} \mu_{(x)} \\ \mu_{(w)} \end{pmatrix}, \begin{pmatrix} \sigma^2_{r(x)} & \sigma_{r(xw)} \\ \sigma_{r(wx)} & \sigma^2_{r(w)} \end{pmatrix}\right) \tag{15.23}$$

where $\mu_{j(x)}$ is a normally distributed latent group mean (i.e., a random intercept).

Having specified multilevel models for all analysis variables, we can now derive appropriate distributions for the incomplete explanatory variables. We illustrate the process for X and refer interested readers to Enders et al. (2020) for details on level-2 and level-3 imputation. Applying basic axioms of probability shows that the conditional distribution of X given the other analysis variables is proportional to the product of the two distributions in which x_{ij} appears

$$\begin{aligned} p(X|Y,W) &\propto p(Y|X,W) \times p(X|W) \\ &\propto N\left(\beta_{0(y)} + \beta_{1(y)}(x_{1ij}) + \beta_{2(y)}(w_j) + b_{0j(y)} + b_{1j(y)}(x_{1ij}), \sigma^2_{\varepsilon(y)}\right) \\ &\times N(\mu_{j(x)}, \sigma^2_{e(x)}) \end{aligned} \tag{15.24}$$

where $p(X|W)$ is the distribution of X conditioning on or controlling for W. Although W does not explicitly appear in the right-most distribution, the expression is conditioning on this variable via the latent group mean $\mu_{j(x)}$, which is derived using information from level-2 predictors (Enders et al., 2020; Keller & Du, 2019).

Conceptually, the two-parts of Equation 15.24 work as follows: The right term says that X values for observation i are normally distributed around the latent mean of group j, and the left term weights each point on that normal curve (i.e., adjusts its vertical height) by the likelihood that results from substituting a particular value of X (including its product with random slope residual) into the right side of the analysis model, with higher weights assigned to scores that produce a small discrepancy between the predicted and observed values of Y. Multiplying the two normal distributions and applying some algebra shows that the random coefficient model induces a normal distribution for X with nonlinear functions in the mean and variance parameters.

$$\begin{aligned} x_{ij(mis)} &\sim N(\mu_{X|Y,W}, \sigma^2_{X|Y,W}) \\ \mu_{X|Y,W} &= \frac{\dot{\sigma}^2_{e(x)}(\dot{\beta}_{1(y)} + \dot{b}_{1j(y)})\left(y_i - \dot{\beta}_{0(y)} - \dot{\beta}_{2(y)}(w_j) - \dot{b}_{0j(y)} + \dot{\sigma}^2_{\varepsilon(y)}(\dot{\mu}_{j(x)}\right)}{\dot{\sigma}^2_{e(x)}(\dot{\beta}_{1(y)} + \dot{b}_{1j(y)})^2 + \dot{\sigma}^2_{\varepsilon(y)}} \\ \sigma^2_{X|Y,W} &= \frac{\dot{\sigma}^2_{\varepsilon(y)}\dot{\sigma}^2_{e(x)}}{\dot{\sigma}^2_{e(x)}(\dot{\beta}_{1(y)} + \dot{b}_{1j(y)})^2 + \dot{\sigma}^2_{\varepsilon(y)}} \end{aligned} \tag{15.25}$$

Equation 15.25 is instructive because it highlights that that the appropriate distribution for X is a heteroscedastic function that depends, among other things, on the random effects for level-2 unit j. The reverse random coefficient model failed to capture this complexity.

To understand how Equation 15.25 works, notice that the denominator of the mean and variance parameters increases as the difference between cluster j's random slope and the mean slope (i.e., β_1) increases. As such, the denominator can be viewed as a scaling factor that adjusts the distribution of X to compensate for the random slope. To illustrate, consider a random coefficient model with the following parameters and variables: $Y = 5$, $\beta_0 = \beta_1 = \beta_2 = 5$, and $W = 0$ (its mean). Further, assume that X is a standardized z-score variable. To minimize the difference between the predicted and observed Ys, model-based imputation must select X imputations that fall along cluster j's regression line. Figure 15.2 shows the distribution of X for three different regression lines. The solid curve is the distribution that results when b_{0j} and b_{1j} equal zero (the cluster's data fall on the average regression line). Because this cluster is situated near the center of the data, a wide range of X values generate predicted scores close to 5. The dashed curve corresponds to b_{0j} and b_{1j} values at one standard deviation above the mean, and the dotted curve corresponds to random effects at two standard deviations

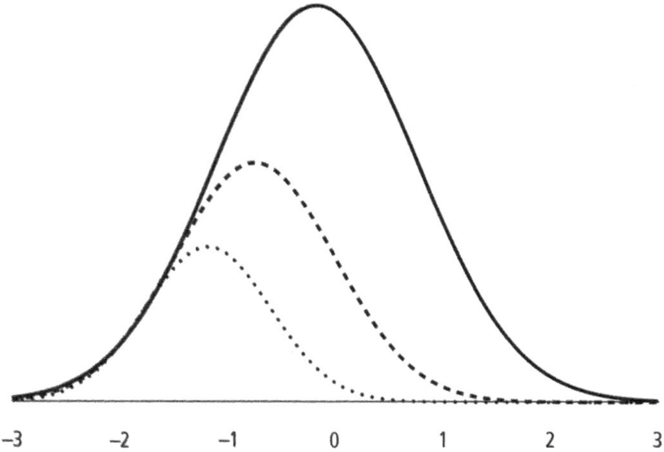

Figure 15.2 Distribution of imputations for model-based imputation. Note: The distributions were generated from a random coefficient model with the following parameters and variables: $Y = 5$, $\beta_0 = \beta_1 = \beta_2 = 5$, and $W = 0$ (its mean). The curves correspond to random effects at the mean (solid line), one standard deviation above the mean (dashed line), and two standard deviations above the mean (dotted line). As the random effects get larger, the distribution of X is rescaled to achieve a predicted value close to $Y = 5$.

above the mean As Equation 15.25 would predict, the mean and variance of X shrink as the random slope residual increases in magnitude. This makes intuitive sense—as the random effects increase, obtaining a predicted value as close as possible to $Y = 5$ can only be achieved by drawing low values of X that are situated along the leftmost side of the regression line.

The distributions of other explanatory variables such as W can be derived in a similar fashion, although the form of each distribution depends on the level at which a variable is measured as well as its role in the analysis (e.g., whether it is attached to a random coefficient). In practice, it is more straightforward to use a Metropolis-Hastings algorithm (Hastings, 1970) to sample imputations because this eliminates the need to derive distributions for a specific application. Details on this algorithm are widely available in Bayesian textbooks and in the missing data literature (Enders et al., 2020; Gelman et al., 2014; Goldstein et al., 2014; Lynch, 2007).

Data Analysis Example 2

We again use the R package jomo (Quartagno & Carpenter, 2018) and the Blimp application (Keller & Enders, 2021) to illustrate multiple imputation for a random coefficient analysis. To provide data for the example, we used computer simulation to create an artificial data set from a random coefficient population model with the following parameter values.

$$y_{ij} = 5 + 5(x_{ij}) + 5(w_j) + b_{0j} + b_{1j}(x_{ij}) + \varepsilon_{ij}$$

$$\Sigma_b = \begin{pmatrix} 100 & 25 \\ 25 & 25 \end{pmatrix} \quad \sigma_\varepsilon^2 = 100$$

Consistent with the first example, we generated a large sample of $N = 25{,}000$ comprised of 500 level-2 units and 50 level-1 observations per group, and we subsequently imposed a 20% MAR missing data rate on X and W, such that missingness increased with the values of Y.

For this illustration, we applied three missing data handling strategies to the incomplete data: joint model imputation with random (i.e., group-specific) level-1 covariance matrices, fully conditional specification with a reverse random coefficient model specification, and model-based imputation. We used jomo to implement the joint model, and we used Blimp for fully conditional specification and model-based imputation. A preliminary diagnostic run in Blimp indicated that potential scale reduction factors dropped below 1.05 after approximately 2,000 iterations. Based on the diagnostic run, we took a conservative approach that generated 20 imputations from MCMC chains with 2,500 burn-in and thinning iterations (i.e., data sets were saved for analysis after the first 2,500 iterations and every 2,500th iteration thereafter). In our experience, a longer burn-in period is often needed with model-based imputation, but the large sample size is (perhaps paradoxically) contributing to the slower convergence as well. We again used the R package mitml to fit the model from Equation 15.2 to each imputed data set and pool the resulting estimates and standard errors.

Table 15.3 gives the estimates and standard errors for each imputation approach along with those of the complete data (the deletion estimates were again severely distorted, so we omit them). Recall that the joint model does not include a parameter that links the random intercepts and slopes, but this feature did not appear to have a substantial impact on the covariance. As we might expect, the estimate is somewhat lower than its true value, but the difference is not significant. Turning to fully conditional specification, the estimates in bold typeface are significantly different from their true values. In particular, notice that the slope variance is three standard error units below its true value, and the covariance and within-cluster residual variance are similarly distorted. Finally, the model-based imputation estimates are negligibly different from those of the complete data and their true values. Taken as a whole, Table 15.3 suggests that researchers should avoid fully conditional specification and opt for the joint model with random covariance matrices or model-based imputation (the former is limited to 2-level models, whereas Blimp can be applied to 3-level models). Of course, caution is warranted when interpreting estimates from a single large data set, but the pattern of results in the table is consistent with published simulation studies (Enders et al., 2020; Enders, Hayes, et al., 2018; Enders, Keller et al., 2018; Grund, Lüdtke et al., 2016; Grund et al., 2018).

TABLE 15.3 Estimates from Data Analysis Example 2

Parameter		Complete		Joint Model		FCS		Model-Based	
		Est.	SE	Est.	SE	Est.	SE	Est.	SE
β_0	5	4.90	0.45	4.96	0.48	4.89	0.46	4.80	0.46
β_1	5	5.31	0.23	5.16	0.23	4.91	0.22	5.28	0.23
β_2	5	4.42	0.38	4.53	0.47	4.72	0.44	4.75	0.45
$\sigma^2_{b_0}$	100	96.10	6.20	98.94	7.04	97.47	6.64	94.17	6.39
$\sigma_{b_0 b_1}$	25	24.05	2.39	21.49	2.53	**19.63**	**2.32**	22.97	2.43
$\sigma^2_{b_1}$	25	24.31	1.57	23.66	1.55	**20.71**	**1.43**	24.14	1.58
σ^2_{ε}	100	99.03	0.92	101.01	1.25	**103.50**	**1.30**	99.46	1.00

ANALYSES WITH INTERACTIVE AND NON-LINEAR EFFECTS

Models that include interactive (i.e., moderation) effects are among the most popular applications of multilevel modeling, and cross-level interactions like the one in Equation 15.3 are particularly common. Applied to the education example, the model allows the level-1 influence of socioeconomic status on achievement to vary by school size, and in the daily diary context it allows the influence of stress on mood to differ as a function of neuroticism. When missing values are restricted to the outcome variable, the researcher simply needs to specify an imputation model that features the same configuration of lower-order and interactive effects as the analysis model (auxiliary variables can enter as additional predictors). However, missing data imputation for incomplete explanatory variables is more difficult because the distributions of these variables involve complex nonlinear functions. Recall that the same was true for random coefficient models. Model-based imputation readily accommodates a variety of interactive or non-linear effects, and the underlying mechanics of the procedure are virtually unchanged. A growing body of literature suggests that model-based imputation is the preferred strategy for single-level moderated regression analysis (Bartlett et al., 2015; Kim et al., 2018; Kim et al., 2015; Zhang & Wang, 2017), and the same appears to be true for multilevel regression models (Enders et al., 2020; Erler et al., 2017; Erler et al., 2016).

Fully Conditional Specification

Studies investigating imputation for interactive and non-linear effects have largely focused on single-level regression models (Enders et al., 2014; Kim et al., 2018; Kim et al., 2015; Seaman et al., 2012; van Buuren, 2012; van Buuren et al., 2018; Vink & van Buuren, 2013; von Hippel, 2009). We recommend recent papers by Kim and colleagues (Kim et al., 2018; Kim et al., 2015) for a comprehensive evaluation of several imputation strategies. Two methods that readily extend to multilevel models—passive imputation and just-another-variable imputation—can be implemented in the fully conditional specification framework (Grund et al., 2018). The literature is mostly critical of these approaches because both are prone to substantial biases, and there is no reason to expect their performance to improve when applied to multilevel data. Nevertheless, a brief discussion of these procedures is warranted so that readers can have an appreciation for how they differ from model-based imputation.

To illustrate passive and just-another-variable imputation, consider the analysis model from Equation 15.3. The imputation model for X is a reverse random coefficient model with Y, W, and their interaction as predictors.

$$x_{ij} = \beta_{0(x)} + \beta_{1(x)}(y_{ij}) + \beta_{2(x)}(w_j) + \beta_{3(x)}(y_{ij})(w_j) + b_{0j(x)} + b_{1j(x)}(y_{ij}) + e_{ij(x)} \quad (15.26)$$

Grund et al. (2018) show a similar model that decomposes the effect of Y and its interaction with W into unique within- and between-cluster components, but we omit these group effects for simplicity. We previously explained that reversing regressions is not valid when the analysis model includes non-linear terms, and the cross-level interaction in Equation 15.26 only exacerbates the misspecification.

More to the point, the two imputation strategies differ in how they treat products of incomplete variables. Passive imputation is a deterministic procedure that computes interaction terms by multiplying the most recent updates of the lower-order terms. For example, the product term in Equation 15.26 would be computed by multiplying the Y and W imputations from the most recent computational cycle, and the X imputations would similarly be multiplied by the current values of W when imputing Y. In contrast, just-another-variable imputation is a stochastic procedure that treats the product term as a variable to be imputed. This approach incorporates additional regression models where product terms are regressed on other analysis variables, and the incomplete interactions are subsequently drawn from a normal distribution—just like any other variable. Among other reasons, just-another-variable imputation is problematic because a product term is not normal even when its constituent lower-order terms are. Limited research to date suggests that just-another-variable imputation is prone to dramatic biases in a 3-level model with random coefficients and a cross-level interaction effect (Enders et al., 2020).

Model-Based Imputation

Model-based imputation adopts the same two-part specification as before, where the researcher's analysis model is paired with a second model featuring just the explanatory variables. The only change to the procedure is that $p(Y|X,W)$ now corresponds to the moderated regression from Equation 15.3. This analysis model predicts that, after conditioning on the predictor variables, their interaction, and the random effects, Y is normally distributed with center and spread equal to a linear predictor and within-cluster residual variance, respectively.

$$p(Y|X,W) = N\begin{pmatrix} \beta_{0(y)} + \beta_{1(y)}(x_{ij}) + \beta_{2(y)}(w_j) + \beta_{3(y)}(x_{ij}) \\ (w_j) + b_{0j(y)} + b_{1j(y)}(x_{ij}), \sigma^2_{\varepsilon(y)} \end{pmatrix} \quad (15.27)$$

The model for the incomplete predictors is the same bivariate normal distribution in Equations 15.22 and 15.23.

After specifying multilevel models for the analysis variables, we can derive appropriate distributions for the incomplete explanatory variables. As before, the conditional distribution of X given the other analysis variables is proportional to the product of the two distributions in which x_{ij} appears.

$$p(X|Y,W) \propto p(Y|X,W) \times p(X|W)$$
$$\propto N\begin{pmatrix}\beta_{0(y)} + \beta_{1(y)}(x_{ij}) + \beta_{2(y)}(w_j) + \beta_{3(y)}(x_{ij}) \\ (w_j) + b_{0j(y)} + b_{1j(y)}(x_{ij}), \sigma^2_{\varepsilon(y)}\end{pmatrix} \times N(\mu_{j(x)}, \sigma^2_{e(x)}) \quad (15.28)$$

Conceptually, the two-parts of Equation 15.28 work the same as before: The right term posits that X is normally distributed around its latent group mean, and the left term weights each point on the normal curve (i.e., adjusts its vertical height) by the likelihood that results from substituting a particular value of X (and its product with b_{1j} and W) into the right side of the analysis model, with higher weights assigned to scores that produce a small discrepancy between the predicted and observed values of Y. Multiplying the two normal distributions and applying some algebra shows that the interactive model induces a normal distribution for X with non-linear functions in the mean and variance parameters.

$$x_{ij(mis)} \sim N(\mu_{X|Y,W}, \sigma^2_{X|Y,W})$$

$$\mu_{X|Y,W} = \frac{\dot{\sigma}^2_{e(x)}\left(\dot{\beta}_{1(y)} + \dot{\beta}_{3(y)}(w_j) + \dot{b}_{1j(y)}\right)\left(y_i - \dot{\beta}_{0(y)} - \dot{\beta}_{2(y)}(w_j) - \dot{b}_{0j(y)}\right) + \dot{\sigma}^2_{\varepsilon(y)}(\dot{\mu}_{j(x)})}{\dot{\sigma}^2_{e(x)}\left(\dot{\beta}_{1(y)} + \dot{\beta}_{3(y)}(w_j) + \dot{b}_{1j(y)}\right)^2 + \dot{\sigma}^2_{\varepsilon(y)}} \quad (15.29)$$

$$\sigma^2_{X|Y,W} = \frac{\dot{\sigma}^2_{\varepsilon(y)}\dot{\sigma}^2_{e(x)}}{\dot{\sigma}^2_{e(x)}\left(\dot{\beta}_{1(y)} + \dot{\beta}_{3(y)}(w_j) + \dot{b}_{1j(y)}\right)^2 + \dot{\sigma}^2_{\varepsilon(y)}}$$

As before, the denominator can be viewed as an adjustment term that rescales the mean and variance parameters to account for the random slope and the interactive effect. As noted previously, a Metropolis-Hastings algorithm provides a convenient method for drawing imputations because it can approximate this distribution without the need for complicated derivations (Enders et al., 2020).

It is important to emphasize that the cross-level product term is not a variable that is imputed at any point during the procedure. Rather, model-based imputation selects replacement values for X and W that are consistent with the hypothesized interaction effect. At the beginning of each iteration, the researcher's analysis model is estimated from the filled-in data, and the product term is computed by multiplying the current values of the two variables. Similarly, the interaction is not a byproduct of imputation, and the researcher must compute the product term in each data set prior to analysis. This feature is quite convenient because the resulting imputations can be centered at their group means or their grand means (or left uncentered) prior to forming the product (Enders, 2013; Enders & Tofighi, 2007; Kreft et al., 1995).

Data Analysis Example 3

The final analysis example applies multiple imputation to the cross-level interaction model from Equation 15.3. To provide data for the example, we again used computer simulation to create a large artificial data set ($N = 25{,}000$) from a population model with the following parameter values.

$$y_{ij} = 5 + 5(x_{ij}) + 5(w_j) + 4(x_{ij})(w_j) + b_{0j} + \beta_{1j}(x_{ij}) + \varepsilon_{ij}$$

$$\Sigma_b = \begin{pmatrix} 100 & 25 \\ 25 & 25 \end{pmatrix} \quad \sigma_\varepsilon^2 = 100$$

As before, the data featured a 20% MAR missing data rate on X and W, such that missingness was a function of Y. For this illustration, we applied a fully conditional specification procedure that treats the cross-level product as a variable to be imputed (just-another-variable imputation) and model-based imputation. We used the Blimp application to implement both procedures, and model-based imputation is also available in jomo (the jomo procedure is called substantive model-compatible imputation, and an R script for the example is available with the other scripts).

A preliminary diagnostic run in Blimp indicated that potential scale reduction factors dropped below 1.05 after approximately 2,500 iterations. Based on this information, we generated 20 imputations from MCMC chains with 3,000 burn-in and thinning iterations (i.e., data sets were saved for analysis after the first 3,000 iterations and every 3,000th iteration thereafter). In our experience, the model-based imputation tends to converge slowly when applied to interactive models, and the large sample size almost certainly exacerbates this behavior. We again used the R package mitml to fit the model from Equation 15.3 to each imputed data set and pool the resulting estimates and standard errors.

Table 15.4 gives the estimates and standard errors from the analysis, with bold typeface highlighting estimates that are significantly different from their true values. The deletion estimates were again severely distorted, so we omit them. Perhaps not surprisingly, applying just-another-variable

TABLE 15.4 Estimates from Data Analysis Example 3

Parameter		Complete		FCS (JAV)		Model-Based	
		Est.	SE	Est.	SE	Est.	SE
β_0	5	4.88	0.46	5.13	0.49	5.22	0.49
β_1	5	4.70	0.23	4.70	0.24	4.74	0.25
β_2	5	5.23	0.44	**3.68**	**0.51**	4.74	0.63
β_3	4	3.89	0.23	**2.95**	**0.18**	3.89	0.26
$\sigma^2_{b_0}$	100	102.38	6.50	109.38	7.66	113.50	9.07
$\sigma_{b_0 b_1}$	25	28.45	2.65	29.59	2.73	27.15	3.55
$\sigma^2_{b_1}$	25	25.42	1.66	24.19	1.60	27.03	2.01
σ^2_{ε}	100	99.73	0.94	**103.21**	**1.44**	101.20	1.19

imputation to the product term produced substantial biases. For example, the interaction slope is nearly six standard error units from its true value. In contrast, model-based imputation estimates are negligibly different from those of the complete data and their true values. Again, caution is warranted when interpreting results from a single large data set, but the pattern of results is consistent with published simulation studies (Enders et al., 2020; Erler et al., 2016).

CONCLUSIONS

Sophisticated missing data handling methods for multilevel data are now widely available to researchers. Missing data handling is straightforward when missing values are restricted to the outcome because either maximum likelihood or multiple imputation will yield consistent estimates if the MAR assumption is satisfied. The situation changes considerably when predictors are missing because these variables are usually treated as fixed constants during estimation. This chapter focused on multiple imputation because it is currently the most flexible and mature technology for dealing with the complexities of multilevel missing data. In particular, we outlined three broad imputation frameworks: joint model imputation, fully conditional specification, and model-based (fully Bayesian) imputation.

A key take-home message is that the composition of the analysis model—particularly whether or not it includes random coefficients or interactive effects—has important implications for modeling incomplete explanatory variables. For random intercept analyses, the choice of imputation procedure rarely matters. For models with random slopes, the joint model with random level-1 covariance matrices or model-based imputation are appropriate; the former is currently restricted to 2-level analyses, and the latter is available for

3-level analyses. Finally, model-based imputation is the method of choice for 2- and 3-level analyses that include interactive or polynomial effects.

RECOMMENDED READINGS

- Enders, C. K., Du, H., & Keller, B. T. (2020). A model-based imputation procedure for multilevel regression models with random coefficients, interaction effects, and other nonlinear terms [Manuscript submitted for publication].
- Enders, C. K., Keller, B. T., & Levy, R. (2018). A fully conditional specification approach to multilevel imputation of categorical and continuous variables. *Psychological Methods, 23*(2), 298–317. https://doi.org/10.1037/met0000148
- Enders, C. K., Mistler, S. A., & Keller, B. T. (2016). Multilevel multiple imputation: A review and evaluation of joint modeling and chained equations imputation. *Psychological Methods, 21*, 222–240. https://doi.org/10.1037/met0000063
- Goldstein, H., Carpenter, J., Kenward, M. G., & Levin, K. A. (2009). Multilevel models with multivariate mixed response types. *Statistical Modelling, 9*(3), 173–197. https://doi.org/10.1177/1471082x0800900301
- van Buuren, S. (2011). Multiple imputation of multilevel data. In J. J. Hox & J. K. Roberts (Eds.), *Handbook of advanced multilevel analysis* (pp. 173–196). Routledge.

TRY THIS!

The data for this exercise (see the online appendix) mimic four repeated measurements from a clinical trial where a quality of life scale is the primary outcome. Participants were assigned to either a treatment or control condition, and background variables were collected for use as covariates. The QualityOfLife.csv data set contains the follow variables:

- ID = Participant (level-2) identifier
- Wave = Data collection wave (1, 2, 3, 4)
- Months = Months since Wave 1 assessment (Wave 1 = 0)
- QOL = Quality of Life scale score (continuous)
- Age = Participant age (continuous)
- Educ = Ordinal education attainment (1 = less than high school, 2 = high school or GED, 3 = college, 4 = graduate school)
- Txgrp = Treatment group indicator (0 = control, 1 = treatment)

The three incomplete variables—QOL, Age, and Educ—have 999 as a missing value code.

The analysis model examines whether the treatment groups exhibit differential growth trajectories (i.e., a cross-level interaction between Months and Txgroup) after controlling for age and education. The hypothesis is that the decline in quality of life scores should be less for patients in the treatment condition. The analysis model is

$$QOL_{ij} = \gamma_{00} + \gamma_{10}(Months_{ij}) + \gamma_{11}(Months_{ij})(Txgrp_j) + \gamma_{01}(Txgrp_j) + \gamma_{02}(Age_j) + \gamma_{03}(HighSchool_j) + \gamma_{04}(College_j) + \gamma_{05}(Graduate_j) + u_{0j} + u_{1j}(Months_{ij}) + \varepsilon_{ij}$$

where HighSchool, College, and Graduate are dummy codes that contrast each education group against the lowest category reference group (less than high school).

For this exercise, you will apply the skills, concepts, and approaches that we covered in this chapter to generate an appropriate set of imputations (at least 20 data sets) and subsequently analyze and pool the estimates with the software package of your choice. You will want to follow and/or consider the following issues as you complete this exercise.

1. To preserve the cross-level interaction effect, you will either need to (a) create the cross-level product term and include it in the imputation model or (b) perform imputation separately for the treatment and control groups.
2. Use an imputation model that preserves random slope variation in the incomplete quality of life scores.
3. Use an ordinal or nominal imputation model for the incomplete education variable.

AUTHOR NOTE

This work was supported by Institute of Educational Sciences award R305 D150056.

REFERENCES

Arbuckle, J. L. (1996). Full information estimation in the presence of incomplete data. In G. A. Marcoulides & R. E. Schumacker (Eds.), *Advanced structural equation modeling*. Lawrence Erlbaum Associates.

Asparouhov, T., & Muthén, B. (2010a). *Bayesian analysis using Mplus: Technical implementation*. M*plus*. https://www.statmodel.com/download/Bayes3.pdf

Asparouhov, T., & Muthén, B. (2010b). *Multiple imputation with Mplus*. M*plus*. http://www.statmodel.com/download/Imputations7.pdf

Bartlett, J. W., Seaman, S. R., White, I. R., & Carpenter, J. R. (2015). Multiple imputation of covariates by fully conditional specification: Accommodating the substantive model. *Statistical Methods in Medical Research, 24*(4), 462–487. https://doi.org/10.1177/0962280214521348

Browne, W. J., & Draper, D. (2000). Implementation and performance issues in the Bayesian and likelihood fitting of multilevel models. *Computational Statistics, 15*(3), 391–420. https://doi.org/10.1007/s001800000041

Carpenter, J. R., Goldstein, H., & Kenward, M. G. (2011). REALCOM-IMPUTE software for multilevel multiple imputation with mixed response types. *Journal of Statistical Software, 45*(5), 1–14. https://doi.org/10.18637/jss.v045.i05

Carpenter, J. R., & Kenward, M. G. (2013). *Multiple imputation and its application*. Wiley.

Enders, C. K. (2010). *Applied missing data analysis*. Guilford Press.

Enders, C. K. (2013). Centering predictors and contextual effects. In M. A. Scott, J. S. Simonoff, & B. D. Marx (Eds.), *The SAGE handbook of multilevel modeling* (pp. 89–107). SAGE.

Enders, C. K., Baraldi, A. N., & Cham, H. (2014). Estimating interaction effects with incomplete predictor variables. *Psychological Methods, 19*(1), 39–55. https://doi.org/10.1037/a0035314

Enders, C. K., Du, H., & Keller, B. T. (2020). *A model-based imputation procedure for multilevel regression models with random coefficients, interaction effects, and other nonlinear terms* [Manuscript submitted for publication].

Enders, C. K., Hayes, T. B., & Du, H. (2018). A comparison of multilevel imputation schemes for random coefficient models: Fully conditional specification and joint model imputation with random covariance matrices *Multivariate Behavioral Research, 53*(5), 695–713. https://doi.org/10.1080/00273171.2018.1477040

Enders, C. K., Keller, B. T., & Levy, R. (2018). A fully conditional specification approach to multilevel imputation of categorical and continuous variables. *Psychological Methods, 23*(2), 298–317. https://doi.org/10.1037/met0000148

Enders, C. K., Mistler, S. A., & Keller, B. T. (2016). Multilevel multiple imputation: A review and evaluation of joint modeling and chained equations imputation. *Psychological Methods, 21*(2), 222–240. https://doi.org/10.1037/met0000063

Enders, C. K., & Tofighi, D. (2007). Centering predictor variables in cross-sectional multilevel models: A new look at an old issue. *Psychological Methods, 12*(2), 121–138. https://doi.org/10.1037/1082-989x.12.2.121

Erler, N. S., Rizopoulos, D., Jaddoe, V. W., Franco, O. H., & Lesaffre, E. M. (2017). Bayesian imputation of time-varying covariates in linear mixed models. *Statistical Methods in Medical Research, 28*(2). https://doi.org/10.1177/0962280217730851

Erler, N. S., Rizopoulos, D., Rosmalen, J., Jaddoe, V. W., Franco, O. H., & Lesaffre, E. M. (2016). Dealing with missing covariates in epidemiologic studies: A comparison between multiple imputation and a full Bayesian approach. *Statistics in Medicine, 35*(17), 2955–2974. https://doi.org/10.1002/sim.6944

Gelman, A., Carlin, J. B., Stern, H. S., Dunson, D. B., Vehtari, A., & Rubin, D. B. (2014). *Bayesian data analysis* (3rd ed.). CRC Press.

Gelman, A., & Hill, J. (2007). *Data analysis using regression and multilevel/hierarchical models.* Cambridge University Press.

Gelman, A., & Rubin, D. B. (1992). Inference from iterative simulation using multiple sequences. *Statistical Science, 7,* 457–472. https://doi.org/10.1214/ss/1177011136

Goldstein, H., Bonnet, G., & Rocher, T. (2007). Multilevel structural equation models for the analysis of comparative data on educational performance. *Journal of Educational and Behavioral Statistics, 32*(3), 252–286. https://doi.org/10.3102/1076998606298042

Goldstein, H., Carpenter, J. R., & Browne, W. J. (2014). Fitting multilevel multivariate models with missing data in responses and covariates that may include interactions and non-linear terms. *Journal of the Royal Statistical Society Series a-Statistics in Society, 177*(2), 553–564. https://doi.org/10.1111/rssa.12022

Goldstein, H., Carpenter, J., Kenward, M. G., & Levin, K. A. (2009). Multilevel models with multivariate mixed response types. *Statistical Modelling, 9*(3), 173–197. https://doi.org/10.1177/1471082x0800900301

Graham, J. W., Olchowski, A. E., & Gilreath, T. D. (2007). How many imputations are really needed? Some practical clarifications of multiple imputation theory. *Prevention Science, 8*(3), 206–213. https://doi.org/10.1007/s11121-007-0070-9

Grund, S., Lüdtke, O., & Robitzsch, A. (2016). Multiple imputation of missing covariate values in multilevel models with random slopes: A cautionary note. *Behavior Research Methods, 48*(2), 640–649. https://doi.org/10.3758/s13428-015-0590-3

Grund, S., Lüdtke, O., & Robitzsch, A. (2017). Multiple imputation of missing data at level 2: A comparison of fully conditional and joint modeling in multilevel designs. *Journal of Educational and Behavioral Statistics, 43,* 316–353.

Grund, S., Lüdtke, O., & Robitzsch, A. (2018). Multiple imputation of missing data for multilevel models: Simulations and recommendations. *Organizational Research Methods, 21*(1), 111–149. https://doi.org/10.1177/1094428117703686

Grund, S., Robitzsch, A., & Lüdtke, O. (2016). *mitml: Tools for multiple imputation in multilevel modeling.* CRAN. https://cran.r-project.org/web/packages/mitml/mitml.pdf

Hastings, W. K. (1970). Monte-Carlo sampling methods using Markov chains and their applications. *Biometrika, 57*(1), 97–109. https://doi.org/10.2307/2334940

Ibrahim, J. G., Chen, M. H., & Lipsitz, S. R. (2002). Bayesian methods for generalized linear models with covariates missing at random. *Canadian Journal of Statistics-Revue Canadienne De Statistique, 30*(1), 55–78. https://doi.org/10.2307/3315865

Ibrahim, J. G., Chen, M. H., Lipsitz, S. R., & Herring, A. H. (2005). Missing-data methods for generalized linear models: A comparative review. *Journal of the American Statistical Association, 100*(469), 332–346. https://doi.org/10.1198/016214504000001844

Kasim, R. M., & Raudenbush, S. W. (1998). Application of Gibbs sampling to nested variance components models with heterogeneous within-group variance. *Journal of Educational and Behavioral Statistics, 32,* 93–116.

Keller, B. T., & Du, H. (2019). A fully conditional specification approach to multilevel multiple imputation with latent cluster means. *Multivariate Behavioral Research, 54,* 149–150.

Keller, B. T., & Enders, C. K. (2021). *Blimp user's guide* (version 3). www.appliedmissingdata.com/blimp

Kim, S., Belin, T. R., & Sugar, C. A. (2018). Multiple imputation with non-additively related variables: Joint-modeling and approximations. *Statistical Methods in Medical Research, 27*(6), 1683–1694. https://doi.org/10.1177/0962280216667763

Kim, S., Sugar, C. A., & Belin, T. R. (2015). Evaluating model-based imputation methods for missing covariates in regression models with interactions. *Statistics in Medicine, 34*(11), 1876–1888. https://doi.org/10.1002/sim.6435

Kreft, I. G., de Leeuw, J., & Aiken, L. S. (1995). The effect of different forms of centering in hierarchical linear models. *Multivariate Behavioral Research, 30*(1), 1–21. https://doi.org/10.1207/s15327906mbr3001_1

Kunkle, D., & Kaizer, E. E. (2017). A comparison of existing methods for multiple imputation in individual participant data meta-analysis. *Statistics in Medicine, 36,* 3507–3532.

Little, R. J. A., & Rubin, D. B. (1987). *Statistical analysis with missing data* (1st ed.). Wiley.

Little, R. J. A., & Rubin, D. B. (2002). *Statistical analysis with missing data.* Wiley.

Lüdtke, O., Marsh, H. W., Robitzsch, A., Trautwein, U., Asparouhov, T., & Muthén, B. (2008). The multilevel latent covariate model: A new approach to group-level effects in contextual studies. *Psychological Methods, 13*(3), 201–229.

Lüdtke, O., Marsh, H. W., Robitzsch, A., & Trautwein, U. (2011). A 2 × 2 taxonomy of multilevel latent contextual models: accuracy–bias trade-offs in full and partial error correction models. *Psychological Methods, 16*(4), 444-467. https://doi.org/10.1037/a0024376

Lüdtke, O., Robitzsch, A., & Grund, S. (2017). Multiple imputation of missing data in multilevel designs: A comparison of different strategies. *Psychological Methods, 22*(1), 141–165. https://doi.org/10.1037/met0000096

Lynch, S. M. (2007). *Introduction to applied Bayesian statistics and estimation for social scientists.* Springer.

Mislevy, R. J. (1991). Randomization-based inference about latent-variables from complex samples. *Psychometrika, 56*(2), 177–196. https://doi.org/10.1007/Bf02294457

Mistler, S. A., & Enders, C. K. (2017). A comparison of joint model and fully conditional specification imputation for multilevel missing data. *Journal of Educational and Behavioral Statistics, 42*(4), 432–466. https://doi.org/10.3102/1076998617690869

Muthén, L. K., & Muthén, B. O. (1998–2017). *Mplus user's guide* (8th ed). Muthén & Muthén.

Nezlek, J. B. (2012). Multilevel modeling analyses of diary-style data. In M. R. Mehl & T. S. Conner (Eds.), *Handbook of research methods for studying daily life* (pp. 357–383). The Guilford Press.

Quartagno, M., & Carpenter, J. R. (2016). Multiple imputation for IPD meta-analysis: Allowing for heterogeneity and studies with missing covariates. *Statistics in Medicine, 35*(17), 2938–2954. https://doi.org/10.1002/sim.6837

Quartagno, M., & Carpenter, J. (2018). *jomo: Multilevel Joint Modelling Multiple Imputation.* CRAN. cran.r-project.org/web/packages/jomo/

Raudenbush, S. W., & Bryk, A. S. (2002). *Hierarchical linear models: Applications and data analysis methods* (2nd ed.). SAGE.

Raudenbush, S. W., Bryk, A. S., & Congdon, R. (2013). HLM 7.01 for Windows. Skokie, IL.

Rubin, D. B. (1976). Inference and missing data. *Biometrika, 63*(3), 581–592. https://doi.org/10.1093/biomet/63.3.581

Rubin, D. B. (1987). *Multiple imputation for nonrespose in surveys.* Wiley.

Schafer, J. L. (1997). *Analysis of incomplete multivariate data.* Chapman & Hall.

Schafer, J. L. (2001). Multiple imputation with PAN. In A. G. Sayer & L. M. Collins (Eds.), *New methods for the analysis of change* (pp. 355–377). American Psychological Association.

Schafer, J. L., & Olsen, M. K. (1998). Multiple imputation for multivariate missing-data problems: A data analyst's perspective. *Multivariate Behavioral Research, 33*(4), 545–571. https://doi.org/10.1207/s15327906mbr3304_5

Schafer, J. L., & Yucel, R. M. (2002). Computational strategies for multivariate linear mixed-effects models with missing values. *Journal of Computational and Graphical Statistics, 11*(2), 437–457. https://doi.org/10.1198/106186002760180608

Seaman, S. R., Bartlett, J. W., & White, I. R. (2012). Multiple imputation of missing covariates with non-linear effects and interactions: An evaluation of statistical methods. *BMC Medical Research Methodology, 12,* 46. https://doi.org/10.1186/1471-2288-12-46

Shin, Y., & Raudenbush, S. W. (2007). Just-identified versus overidentified two-level hierarchical linear models with missing data. *Biometrics, 63*(4), 1262–1268. https://doi.org/10.1111/j.1541-0420.2007.00818.x

Shin, Y., & Raudenbush, S. W. (2010). A latent cluster-mean approach to the contextual effects model with missing data. *Journal of Educational and Behavioral Statistics, 35*(1), 26–53. https://doi.org/10.3102/1076998609345252

Shin, Y., & Raudenbush, S. W. (2013). Efficient analysis of *Q*-level nested hierarchical general linear models given ignorable missing data. *Int J Biostat, 9*(1), 109–133. https://doi.org/10.1515/ijb-2012-0048

Sinharay, S., Stern, H. S., & Russell, D. (2001). The use of multiple imputation for the analysis of missing data. *Psychological Methods, 6*(4), 317–329. https://doi.org/10.1037//1082-989x.6.4.317

van Buuren, S. (2007). Multiple imputation of discrete and continuous data by fully conditional specification. *Statistical Methods in Medical Research, 16*(3), 219–242. https://doi.org/10.1177/0962280206074463

van Buuren, S. (2011). Multiple imputation of multilevel data. In J. J. Hox & J. K. Roberts (Eds.), *Handbook of advanced multilevel analysis* (pp. 173–196). Routledge.

van Buuren, S. (2012). *Flexible imputation of missing data.* Chapman & Hall.

van Buuren, S., Brand, J. P. L., Groothuis-Oudshoorn, C. G. M., & Rubin, D. B. (2006). Fully conditional specification in multivariate imputation. *Journal of Statistical Computation and Simulation, 76*(12), 1049–1064. https://doi.org/10.1080/10629360600810434

van Buuren, S., Groothuis-Oudshoorn, K., Robitzsch, A., Vink, G., Doove, L., Jolani, S., . . . Gray, B. (2021). Package 'mice'. CRAN. https://cran.r-project.org/web/packages/mice/mice.pdf

van de Shoot, R., Winter, S., Ryan, O., Zondervan-Zwijnenburg, M., & Depaoili, S. (2017). A systematic review of Bayesian papers in psychology: The last 25 years. *Psychological Methods, 22*(2), 217–239. https://doi.org/10.1037/met0000100

Vink, G., & van Buuren, S. (2013). Multiple imputation of squared terms. *Sociological Methods & Research, 42*(4), 598–607. https://doi.org/10.1177/0049124113502943

von Hippel, P. T. (2009). How to impute interactions, squares, and other transformed variables. *Sociological Methodology, 39*, 265–291.

Wilkinson, L., &Taskforce on Statistical Inference. (1999). Statistical methods in psychology journals—Guidelines and explanations. *American Psychologist, 54*(8), 594–604. https://doi.org/10.1037//0003-066x.54.8.594

Yucel, R. M. (2008). Multiple imputation inference for multivariate multilevel continuous data with ignorable non-response. *Philosophical Transactions of the Royal Society A: Mathematical, Physical, and Engineering Sciences, 366*(1874), 2389–2403. https://doi.org/10.1098/rsta.2008.0038

Yucel, R. M. (2011). Random-covariances and mixed-effects models for imputing multivariate multilevel continuous data. *Stat Modelling, 11*(4), 351–370. https://doi.org/10.1177/1471082X1001100404

Zhang, Q., & Wang, L. (2017). Moderation analysis with missing data in the predictors. *Psychological Methods, 22*(4), 649–666. https://doi.org/10.1037/met0000104

CHAPTER 16

MULTILEVEL MEDIATION ANALYSIS

Nicholas J. Rockwood
Loma Linda University

Andrew F. Hayes
University of Calgary

Researchers are often interested in understanding the underlying process by which causal effects operate. For example, rather than just identifying whether participation in an employee training program leads to improved job performance, interest may be in *why* or *how* the training program is effective. Perhaps the training program leads to higher levels of employee engagement, which then results in improved performance, or maybe the training targets technological improvements that enable the employees to work more efficiently. Identification of such mechanisms can have important implications for the development and testing of theory as well as provide the employer a better understanding of which aspects of the training are effective.

Statistical mediation analysis is commonly used to identify such mechanisms of effects as well as quantify the underlying causal processes. In a

Multilevel Modeling Methods with Introductory and Advanced Applications, pages 567–597

mediation process, the independent variable (X) exerts a causal effect on the dependent variable (Y) by way of an intervening, mediator variable (M). That is, X causally influences M which, in turn, causally influences Y. Thus, the effect of X on Y occurs indirectly through M. X may also influence Y directly (i.e., not through M).

Many researchers have used statistical mediation analysis to test a mediation process. For example, Lam et al. (2014) tested the effect of neurocognition (X) on symptomatology (Y) by way of social cognition (M) in people with schizophrenia. Decker et al. (2018) were interested in identifying which cognitive skills (M) mediate the relationship between reading skill (X) and reading comprehension (Y) in grade school children. Using a sample of retail service employees, Tremblay et al. (2017) found that the relationship between leader-member exchange (X) and affective organizational commitment (Y) is mediated by person-organization fit (M).

Statistical mediation analysis is most commonly conducted by applying path-analytic methods using ordinary least squares (OLS) regression or maximum likelihood estimation via structural equation modeling (SEM) programs. Although these analytic strategies are effective for modeling "single-level" data, where all of the sampling units (e.g., participants) are independent of each other and all variables within the analyses are measured at the "level" of these units, they are not appropriate for analyzing "multilevel" data, where the sampling units are nested within higher-level units. Examples of multilevel data include employees nested within organizations, students nested within classrooms, or repeated measurements nested within individuals. Here, there may be variables corresponding to the lower-level units, such as the years of work experience of the employees, or organization-level variables, such as the size of the organization. In such data we refer to the lower-level units (e.g., employees, students, repeated measurements) as level-1 units, and the variables corresponding to these units (e.g., work experience) as level-1 variables. The higher-level units (organizations, classrooms, individuals) are referred to as level-2 units, groups, or clusters, with corresponding level-2 variables (e.g., size of the organization). Thus, level-1 units are clustered, or nested, within level-2 units, resulting in a hierarchical structure. Multilevel designs can be extended to include more than two levels, but in this chapter we restrict our attention to 2-level designs. We also make the assumption that each level-1 unit belongs to only one level-2 unit. For example, we assume that if a student attends a specific school, he or she does not simultaneously attend another school.

The reason the traditional methods for assessing mediation are not appropriate in multilevel contexts is due to the non-independence between sampling units within the same level-2 unit. For example, students within the same classroom tend to have more similar values on variables of interest than students in different classrooms, and such dependence between units violates commonly imposed regression assumptions. Some alternative modeling strategies can be implemented within these frameworks to correct for non-independence (e.g., cluster-robust standard errors), but multilevel modeling is often the chosen strategy, as it allows for the explicit modeling of hierarchical data structures. Multilevel modeling also provides more flexibility than some of the alternative methods.

Over the past few decades there has been an influx in the adaptation of statistical mediation methods to the multilevel context (e.g., Bauer et al., 2006; Kenny et al., 2003; Krull & MacKinnon, 1999, 2001; MacKinnon, 2008; Pituch et al., 2006; Preacher et al., 2010; Tofighi et al., 2013; Zhang et al., 2009; and others), and the use of multilevel mediation in applied research has also increased dramatically. Sometimes the researcher is simply interested in the causal process of individual-level variables, but individuals are nested in some higher-level unit so the researcher must account for the dependencies in the data. For example, using data collected from a construction services company, Kao et al. (2016) hypothesized that insomnia (*X*) influences workplace safety behaviors (*M*) which then influence frequency of workplace injuries (*Y*). Multilevel modeling was used because the employees were nested within supervisors. Other times, the causal process of a group-level variable is of interest, either due to the study design or because of the nature of the construct. For example, Reyes et al. (2012) tested the indirect effect of classroom emotional climate (*X*) on student language arts grades (*Y*) through student engagement (*M*). Here, classroom emotional climate is a group-level variable because it describes the level-2 unit, which is classrooms in this example, and classroom emotional climate does not vary over students within the same classroom. Lastly, multilevel modeling can be useful for testing mediation processes with longitudinal (Bolger & Laurenceau, 2013) or dyadic data (Ledermann & Kenny, 2017). Ilies et al. (2017) used daily surveys to assess the within-person indirect effect of daily work engagement (*X*) on daily family satisfaction and work–family balance (*Y*) through daily work–family interpersonal capitalization (*M*), defined as the discussion of positive work experiences with one's partner at home (Ilies et al., 2011).

In the multilevel mediation literature, mediation processes are commonly labeled in "*X-M-Y*" format by the level at which the independent (*X*), mediator (*M*), and dependent (*Y*) variables are measured. For example, in a "2-1-1" design, the independent variable *X* is measured at level-2, while the mediator *M* and dependent variable *Y* are measured at level-1. The study by Reyes et al. (2012) described above is an example of a 2-1-1 design, as classroom climate (*X*) was a level-2 variable while *M* and *Y* were level-1 variables. The most flexible 2-level mediation design is the 1-1-1 design, such as the ones presented by Kao et al. (2016) and Ilies et al. (2017). In this model, a mediation process can exist at both levels of the hierarchy. Further, the variability of the within-group mediation process across level-2 units can be estimated. And as we will elaborate on later in the chapter, multilevel mediation models for 2-1-1 and 2-2-1 data designs can be viewed as special cases of the general multilevel model for 1-1-1 designs.

The purpose of this chapter is to provide an introduction to statistical mediation analysis with multilevel data. We begin by providing a brief review of statistical mediation analysis for single-level models. Next, we present the general 2-level multilevel mediation model before detailing how to appropriately estimate all model parameters and make inferences about the effects of interest. Following, we demonstrate these methods with an example analysis. We conclude the chapter by describing ways in which the multilevel mediation model may be adapted or extended to model data from a variety of research designs.

Before delving into the details, we note that mediation is inherently a causal process, yet no statistical method can prove causality. The methods presented throughout this chapter provide an analytic framework that, when coupled with sound theory and research design, can provide support for causal mediation processes. However, the strength of these causal claims is often limited by several factors including, but not limited to, the presence of unmeasured confounding variables or a lack of temporal precedence resulting from a cross-sectional study. Several articles, books, and book chapters (including Chapter 4 of this volume, "Causal Inference in Multilevel Settings") are devoted to understanding and identifying conditions in which effects may be interpreted causally. In the context of mediation, we refer readers to Imai et al. (2010), Preacher (2015), and Hayes (2018) for an overview of such conditions. In this chapter, we instead focus on the statistical tools for multilevel mediation analysis, rather than philosophical issues regarding cause-effect relationships or research designs that can be utilized to strengthen claims of such relationships.

SINGLE-LEVEL MEDIATION

When M and Y are continuous, the single-level three-variable mediation model is often estimated as:

$$M_i = d_M + aX_i + e_{M_i} \tag{16.1}$$

$$Y_i = d_Y + c'Xi + bM_i + e_{Y_i} \tag{16.2}$$

which, as we will demonstrate, decomposes the effect of X on Y in the equation:

$$Y_i = d_{Y^*} + cX_i + e_{Y_i^*} \tag{16.3}$$

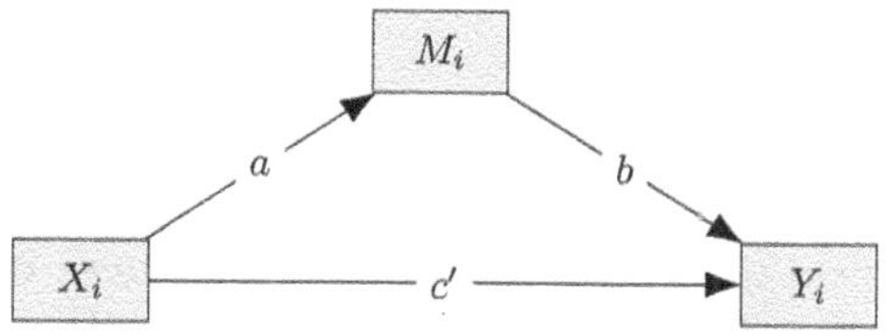

Figure 16.1 A path diagram corresponding to the single-level three variable mediation model.

The error terms, e_{M_i}, e_{Y_i}, and $e_{Y_i^*}$, are often assumed to be normally distributed random variables with means of zero and constant variances ($\sigma^2_{e_M}$, $\sigma^2_{e_Y}$, $\sigma^2_{e_{Y*}}$, respectively). A path diagram corresponding to these equations is displayed in Figure 16.1.

Throughout this chapter, we use d for intercepts, with the subscript denoting the response variable for the given equation. The *total effect* of X on Y is represented by c in Equation 16.3, which corresponds to the expected change in Y resulting from a one-unit change in X. Equations 16.1 and 16.2 partition the total effect into the *direct effect* and the *indirect effect*. The direct effect, represented as c' in Equation 16.2, quantifies the expected change in Y resulting from a one-unit change in X while holding M constant. The indirect effect, which quantifies the mediation process, is the expected change in Y through M resulting from a one-unit change in X. It can be quantified as ab, which is the product of the effect of X on M (a) and the effect of M on Y controlling for X (b). The sum of the indirect and direct effects equals the total effect, $c = ab + c'$, which can be demonstrated by substituting Equation 16.1 for M_i in Equation 16.2:

$$\begin{aligned} Y_i &= d_Y + c'X_i + b(d_1 + aX_i + e_{M_i}) + e_{Y_i} \\ &= d_Y + c'X_i + bd_M + abX_i + be_{M_i} + e_{Y_i} \\ &= \underbrace{(d_Y + bd_M)}_{d_{Y*}} + \underbrace{(ab + c')}_{c} X_i + \underbrace{(be_{M_i} + e_{Y_i})}_{e_{Y_i^*}} \end{aligned} \tag{16.4}$$

where the terms underneath Equation 16.4 denote the terms in Equation 13.3 to which that set of parentheses in Equation 16.4 corresponds. As shown, the sum of the indirect and direct effects in the second set of parentheses in Equation 16.4, $ab + c'$, is equal to the total effect, c, in Equation 16.3.

Equations 16.1–16.3 can also be expanded to include covariates, so the effects of potential confounding variables are partialled out of the estimated direct, indirect, and total effects, by simply including these covariates as additional predictors in the model. When the same covariates are included in all of the equations, the equality $c = ab + c'$ holds.

Methods of Inference

Inference about the direct and total effects is straightforward, as the confidence intervals or *p*-values provided by any standard OLS regression program can be used to test the hypothesis that these effects equal zero. On the other hand, inference about the indirect effect has been heavily debated by methodologists. In the past, inference about the indirect effect was usually conducted through a series of statistical significance tests on the individual parameters in the model (i.e., a and b; Baron & Kenny, 1986), or by comparing the ratio of the estimated indirect effect ab and its standard error to the standard normal distribution (Sobel, 1982) to obtain a *p*-value. Unfortunately, the series of significance tests only test the components of the indirect effect, rather than the indirect effect as a whole, and the ratio of the indirect effect to its standard error is generally not normally distributed (Aroian, 1947), so inference (e.g., a *p*-value) relying on the standard normal distribution is inappropriate.

Today, many methodologists recommend using other methods of inference for the indirect effect (Hayes, 2009, 2018; MacKinnon et al., 2004). The cases bootstrap (Efron & Tibshirani, 1994), which is a general resampling procedure, is often used to assess single-level mediation because no assumptions are made about the shape of the sampling distribution of the indirect effect. Many resources describe the implementation of bootstrapping for single-level mediation analysis (e.g., Hayes, 2018; Hayes & Scharkow, 2013). However, bootstrapping is not as straightforward for multilevel designs (see, e.g., Van der Leeden et al., 2008), which is the focus of this chapter. Therefore, we focus instead on the Monte Carlo method (Preacher & Selig, 2012), as several methodologists have recommended it for multilevel mediation (Bauer et al., 2006; Preacher et al., 2010).

The Monte Carlo method involves randomly simulating values from the sampling distribution of the parameters of interest. In the context of mediation, this means simulating a^* and b^* values from the joint sampling distribution of a and b to empirically estimate the sampling distribution of ab. The process begins by fitting the model defined by Equations 16.1–16.2 and obtaining estimated values of a and b, denoted by $\hat{a}$ and $\hat{b}$, respectively, with corresponding standard errors $\hat{\sigma}_{\hat{a}}$ and $\hat{\sigma}_{\hat{b}}$. Since the joint sampling distribution of the regression coefficients is multivariate normal (by assumption, which is often reasonable), we can simulate values from this joint distribution, for which we use $\hat{a}$ and $\hat{b}$ as the means, and the estimated sampling covariance matrix of $\hat{a}$ and $\hat{b}$ as the covariance matrix. Symbolically, this can be expressed as:

$$\begin{bmatrix} a^* \\ b^* \end{bmatrix} \sim MVN\left(\begin{bmatrix} \hat{a} \\ \hat{b} \end{bmatrix}, \begin{bmatrix} \hat{\sigma}^2_{\hat{a}} & \hat{\sigma}_{\hat{a},\hat{b}} \\ \hat{\sigma}_{\hat{a},\hat{b}} & \hat{\sigma}^2_{\hat{b}} \end{bmatrix} \right)$$

where "~" is read "is distributed as"; *MVN* refers to a multivariate normal distribution where the first element is the vector of means and the second element is the covariance matrix; $\hat{a}$ and $\hat{b}$ are defined as above; $\hat{\sigma}^2_{\hat{a}}$ and $\hat{\sigma}^2_{\hat{b}}$ are the corresponding estimated sampling variances (i.e., the square of the standard errors); and $\hat{\sigma}_{\hat{a},\hat{b}}$ is the sampling covariance (which is often assumed to be zero). After simulating an arbitrarily large number of a^* and b^* values (usually 1,000+), the sampling distribution of *ab* can be empirically estimated by multiplying the a^* and b^* values from each draw from the joint distribution. A 95% confidence interval for the indirect effect can be calculated using the 2.5th and 97.5th percentiles of the a^*b^* values, and a confidence interval not including zero is evidence of mediation. Although this may sound complex, software and macros developed for statistical mediation analysis (e.g., the PROCESS macro for SPSS and SAS; Hayes, 2018) can make these calculations for the user and there are also some online utilities available (e.g., Selig & Preacher, 2008) to make implementation of this method easy.

MULTILEVEL MEDIATION

We now demonstrate how methods for assessing mediation can be expanded to the multilevel context. We first present mediation equations for a 1-1-1 design, where there exists within-group variability in *X*, *M*, and *Y* (i.e., not all level-1 units within a given level-2 unit have the same values for *X*, *M*, and *Y*), as this is the most general form of the 2-level model. Next, we define within-group and between-group indirect effects and discuss inference about these effects. To help make the ideas we present throughout this section concrete, we will describe the substantive meaning of the model parameters in the context of the study by Kao et al. (2016), where employees (level-1 units) are nested within supervisors (level-2 units), insomnia is the independent variable, workplace safety behaviors is the mediator, and workplace injuries is the dependent variable. Insomnia and safety behaviors were measured using 4 and 14 items, respectively. The workplace injuries variable was constructed using a 13-item survey for which employees rated the frequency of different types of injuries on a 1–5 scale. The items for each of the variables were averaged to construct scale scores, which were treated as continuous. Henceforth, we will refer to this study simply as the "workplace study" for convenience.

The General Model

Multilevel models are often formulated using different equations for each level of analysis (i.e., level 1 and level 2). When *M* and *Y* are continuous, the level-1 equations for the three-variable multilevel mediation model can be estimated as:

$$M_{ij} = d_{M_j} + a_j(X_{ij} - \bar{X}_{.j}) + e_{M_{ij}} \tag{16.5}$$

$$Y_{ij} = d_{Y_j} + c'_j(X_{ij} - \bar{X}_{.j}) + b_j(M_{ij} - \bar{M}_{.j}) + e_{Y_{ij}} \tag{16.6}$$

where i refers to level-1 units, j refers to level-2 units/groups, and $\bar{X}_{.j}$ and $\bar{M}_{.j}$ represent the observed group means of X and M, respectively, so that $(X_{ij} - \bar{X}_{.j})$ and $(M_{ij} - \bar{M}_{.j})$ are within-group centered X and M variables. As before, we use d to refer to the intercepts, d_{M_j} and d_{Y_j}, and the errors, $e_{M_{ij}}$ and $e_{Y_{ij}}$, are assumed to be normally distributed random variables with variances σ_M^2 and σ_Y^2, i.e., $e_{M_{ij}} \sim N(0, \sigma_M^2)$, $e_{Y_{ij}} \sim N(0, \sigma_Y^2)$.

Besides the group-mean centering of X and M, the level-1 equations for the general 2-level mediation model closely resemble the single-level mediation model. Specifically, X is a predictor of M with corresponding coefficient a_j, while both X and M are predictors of Y with corresponding coefficients c'_j and b_j. Because the variables are group-mean centered, they contain no between-group variability, and so these coefficients capture the within-group relationship between the given predictor variable and the outcome.

We use j subscripts for the intercepts and slopes to denote that they may vary across the level-2 units. We can model the between-group differences in these coefficients as a deterministic function of level-2 variables and also allow the intercepts and slopes to randomly vary across level-2 units, thus resulting in what are termed *random effects*. These random effects can be viewed as level-2 residuals, or errors, for each of the coefficients, as they contain unexplained between-group variability in the coefficients.

In the workplace study, i refers to employees and j refers to supervisors. The variables $(X_{ij} - \bar{X}_{.j})$ and $(M_{ij} - \bar{M}_{.j})$ are supervisor-mean centered insomnia and workplace safety behaviors, respectively. The coefficient a_j corresponds to the within-supervisor effect of insomnia on workplace safety behaviors for supervisor j, and the coefficients b_j and c'_j are the within-supervisor effects of workplace safety behaviors and insomnia on workplace injuries for supervisor j. We can model a_j, b_j, and c'_j, as well as the intercepts, as varying across supervisors. Differences in these coefficients across supervisors may stem from observed supervisor-level variables, or it may be a result of random (i.e., unexplained) variation.

To formalize these possible differences, the coefficients themselves can be modeled in what are often referred to as the level-2 equations, as we are modeling the between-group (or level-2) variation in the coefficients. For the level-2 equations, we include the observed group means as predictors of the intercepts and allow all coefficients to randomly vary. Specifically, the upper-level equations are:

$$d_{M_j} = d_M + a_B \overline{X}_{.j} + u_{M_j} \tag{16.7}$$

$$d_{Y_j} = d_Y + c'_B \overline{X}_{.j} + b_B \overline{M}_{.j} + u_{Y_j} \tag{16.8}$$

$$a_j = a_W + u_{a_j} \tag{16.9}$$

$$b_j = b_W + u_{b_j} \tag{16.10}$$

$$c'_j = c'_W + u_{c'_j} \tag{16.11}$$

where the collection of random effects for group j (i.e., u_{M_j}, u_{Y_j}, u_{a_j}, u_{b_j}, $u_{c'_j}$) are collected into the vector $\mathbf{u}_j$, and $\mathbf{u}_j \sim MVN(\mathbf{0},\mathbf{T})$. In its most general form, $\mathbf{T}$ is an unstructured covariance matrix for the vector of random effects, where all random effect variances and covariances are model parameters. Together, Equations 16.5–16.6 (level 1) and 16.7–16.11 (level 2) define the general multilevel mediation model. Figures 16.2 and 16.3 display conceptual and statistical diagrams, respectively, of the model. Within the diagrams, variables that contain between-group variability are in the box labeled "Between" and variables that contain within-group variability are in the box labeled "Within."

The level-2 equations for the intercepts also closely resemble the single-level mediation model. The group means of X ($\overline{X}_{.j}$) are predictors of the

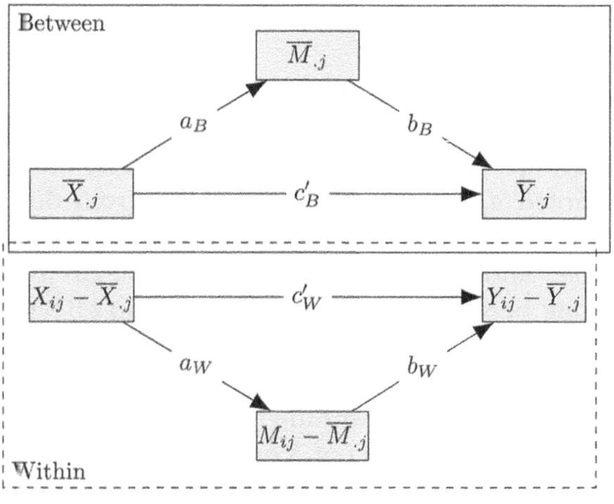

Figure 16.2 A conceptual diagram corresponding to the general multilevel mediation model. Both a between- and a within-mediation process may take place.

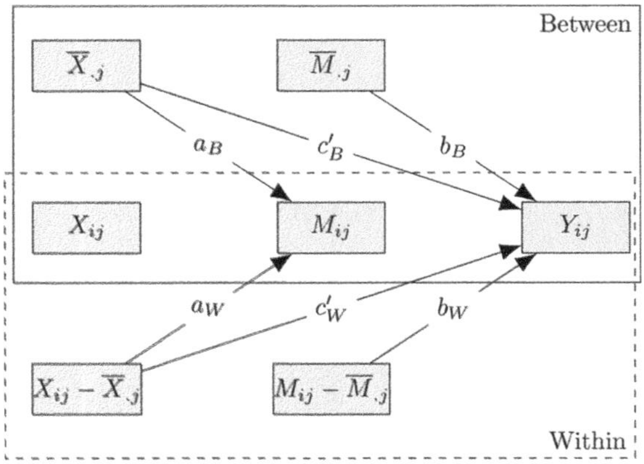

Figure 16.3 A statistical diagram corresponding to the general multilevel mediation model. M_{ij} is predicted by the within- and between- components of X_{ij} and Y_{ij} is predicted by the within- and between- components of X_{ij} and M_{ij}.

intercept M and the group means for X and M ($\overline{M}_{.j}$) are predictors of the intercept for Y. Because the group means quantify between-group differences, the corresponding coefficients quantify between-group relationships between the predictor and outcome variables. For example, two groups whose average value on X differ by one unit are expected to differ, on average, by a_B units on M. In the context of the workplace study, the average value of insomnia for all employees nested within a given supervisor is likely to differ from the average value for employees nested within a different supervisor. These differences in the average value of insomnia are captured by variability in $\overline{X}_{.j}$, and so the effect of $\overline{X}_{.j}$ on supervisor safety behaviors is quantified as a_B.

Without any level-2 predictors for a_j, b_j, and c'_j, their models correspond to a mean value plus a group-level deviation. That is, the within-group effect of X on M for group j is modeled as an overall mean within-group effect of X on M, denoted a_W, plus group j's deviation from this overall effect, denoted by u_{a_j}. Throughout this chapter we use the group-mean centered predictor variables and the corresponding group-means as additional predictors to separate between-group and within-group effects (Zhang et al., 2009). Thus, we denote between-group and within-group effects using the subscripts B and W, respectively. Therefore, a_W is the average within-group effect of X on M, $a_W + u_{a_j}$ is the within-group effect of X on M for group j, and a_B is the between-group effect of X on M.

Further clarification of the meaning of within-group and between-group effects can be made by considering the workplace study. As stated, $a_W + u_{a_j}$ can be thought of as the effect of insomnia on workplace safety behaviors for employees of supervisor j. Suppose the study only considered the employees within supervisor j, and so $a_W + u_{a_j}$ is the expected difference in workplace safety behaviors for two employees (who have the same supervisor since we are only considering supervisor j) who differ by one unit on insomnia. Now suppose that we repeated the experiment, now only considering employees from a different supervisor, say supervisor j'. Now, the predicted effect of insomnia on workplace safety behaviors (for employees of supervisor j') is $a_W + u_{a_{j'}}$, which may differ from the predicted effect in supervisor j (if $u_{a_{j'}} \neq u_{a_j}$). If we repeat the study using the employees within each supervisor and average the effects across each of these studies, we would expect to obtain a_W, the average within-supervisor effect of insomnia on workplace safety behaviors. It is a within-supervisor effect because it captures the relationship within a given supervisor, averaged across supervisors. The between-supervisor effect, on the other hand, captures the relationship among the aggregates of the variables. For this example, if employees of a given supervisor tend to be higher (on aggregate) on both insomnia and workplace safety behaviors relative to employees of a different supervisor, this relationship is captured by the between-supervisor effect.

As with the single-level mediation model, the effect of covariates can be partialled out of our estimated effects by including the covariates as additional predictors in the model. The covariates can be measured at level 1, in which they are entered into Equations 16.5–16.6, or at level 2, in which they are entered into Equations 16.7–16.8. When including a level-1 covariate, use within-group centering and include the group-mean on the covariate as an additional predictor in Equations 16.7–16.8, just as is done with X and M above. That way the within-group and between-group effects of the covariate are partialled out of our estimated within-group and between-group effects of X and M. For example, how dangerous an employee's job is most likely relates to both insomnia and workplace injuries. Therefore, we would want to control for, or partial out, the effect of job dangerousness when trying to quantify the mediation process. By including the within-supervisor centered job dangerousness, we partial out the effect of job dangerousness on the within-supervisor effects. That is, we control for potential differences in job dangerousness between employees who have the same supervisor. In addition, by including the supervisor-mean values of job dangerousness we control for differences in employees' job dangerousness across supervisors. Some supervisors will tend to have more employees with dangerous jobs relative to other supervisors, and by including the supervisor means we control for these differences.

Within-Group Indirect Effect

When within-group and between-group effects are separated, both within-group and between-group indirect effects can be calculated. A within-group indirect effect quantifies the expected difference in Y through M for two level-1 units in the *same* level-2 unit who differ by one unit on X. In the context of the workplace example, the within-group indirect effect is the effect of insomnia on workplace injuries *through workplace safety behaviors* for employees of a given supervisor. Because a_j and/or b_j may randomly vary across level-2 units, the within-group indirect effect may also randomly vary across level-2 units. That is, the indirect effect for employees of one supervisor may differ from the indirect effect for employees of a different supervisor. In this scenario, the average within-group indirect effect and the variance of the within-group indirect effect across level-2 units (e.g., supervisors) are of interest.

The average within-group indirect effect can be defined as the expected value of a_jb_j. When a_j and b_j covary, the expected value of a_jb_j is not the product of the expected values of a_j and b_j (Goodman, 1960), but, rather:

$$E(a_jb_j) = a_W b_W + \sigma_{a_j,b_j} \tag{16.12}$$

where σ_{a_j,b_j} is the covariance between a_j and b_j (Kenny et al., 2003). When $\sigma_{a_j,b_j} = 0$, which may be a result of a_j and/or b_j being fixed across level-2 units, the within-group indirect effect simplifies to $a_W b_W$. Thus, Equation 16.12 is the general form of the within-group indirect effect that holds under simplifications to Equations 16.9–16.11. The variance of the within-group indirect effect across level-2 units is (Kenny et al., 2003):

$$Var(a_jb_j) = b_W^2\sigma_{a_j}^2 + a_W^2\sigma_{b_j}^2 + \sigma_{a_j}^2\sigma_{b_j}^2 + 2a_W b_W \sigma_{a_j,b_j} + \sigma_{a_j,b_j}^2 \tag{16.13}$$

Again, this equation is for the most general model, where all slopes randomly vary and covary, but it also applies with simplifications to Equations 16.9–16.11, in which some of the terms in Equation 16.13 drop out.

Between-Group Indirect Effect

When the between-group effect of X is a result of the group aggregate of X, the between-group indirect effect quantifies the expected difference in the group aggregate of Y through the group-aggregate of M for two level-2 units that differ by one-unit in the group-aggregate of X. The effect of aggregated (across supervisors) insomnia on aggregated workplace injuries by way of aggregated workplace safety behaviors is the between-group indirect effect for the workplace study. It can be calculated as:

$$E(a_B b_B) = a_B b_B \tag{16.14}$$

Often, the between-group indirect effect in this situation can be difficult to interpret, as the meaning of the group aggregate of a variable differs from the meaning of the variable at the individual level. For instance, Preacher et al. (2010) make use of an example differentiating individual efficacy and the collective efficacy of a group. They argue that, although the aggregated individual efficacy for a given group is in fact a group-level variable (in that it only varies between groups), the focus is still at the individual level. The meaning of such a variable differs from the meaning of a variable characterizing the dynamics of the self-efficacy of the group as a collective. Therefore, caution is recommended when interpreting the meaning of between-group effects when they correspond to the group aggregate of level-1 variables.

Inference

Many of the methods of inference about indirect effects used for single-level mediation can be adopted within the multilevel context to make inferences about the within-group and between-group indirect effects. Bauer et al. (2006) derived the sampling variance of the average within-group indirect effect:

$$\begin{aligned} Var(\hat{a}_W\hat{b}_W + \hat{\sigma}_{a_j,b_j}) &= b_W^2 Var(\hat{a}_W) + a_W^2 Var(\hat{b}_W) + Var(\hat{a}_W)Var(\hat{b}_W) \\ &\quad + 2a_W b_W Cov(\hat{a}_W \hat{b}_W) + Cov(\hat{a}_W \hat{b}_W)^2 + Var(\hat{\sigma}_{a_j,b_j}) \end{aligned} \tag{16.15}$$

where *Var* and *Cov* refer to sampling variances and covariances, respectively. The sampling variance of the between-group indirect effect is (Tofighi et al., 2013):

$$Var(\hat{a}_B\hat{b}_B) = b_B^2 Var(\hat{a}_B) + a_B^2 Var(\hat{b}_B) + Var(\hat{a}_B)Var(\hat{b}_B) \tag{16.16}$$

The square root of these sampling variances correspond to the standard errors of the indirect effects, and these values are useful in indicating how precise we are in estimating the effects. However, as with single-level mediation, using the estimated sampling variances for the construction of symmetric confidence intervals around the indirect effects is inappropriate due to the non-normal sampling distribution of the indirect effects. Thus, the Monte Carlo method is often recommended for confidence interval construction for multilevel mediation (Bauer et al., 2006).

Monte Carlo confidence intervals for the within-group and between-group indirect effect can be constructed by simulating $k \times p$ values from the multivariate distribution:

$$\begin{bmatrix} \mathbf{f}^* \\ \mathbf{r}^* \end{bmatrix} \sim MVN\left(\begin{bmatrix} \hat{\mathbf{f}} \\ \hat{\mathbf{r}} \end{bmatrix}, \begin{bmatrix} \hat{\Sigma}_{\hat{f}} & 0 \\ 0 & \hat{\Sigma}_{\hat{r}} \end{bmatrix} \right),$$

where k is an arbitrarily large number, p is the total number of fixed and random parameters, $\hat{\mathbf{f}}$ is a vector containing all of the fixed effect estimates with estimated asymptotic sampling covariance matrix $\hat{\Sigma}_{\hat{f}}$, $\hat{\mathbf{r}}$ is a vector containing all of the variance/covariance parameter estimates with estimated asymptotic sampling covariance matrix $\hat{\Sigma}_{\hat{r}}$, and the sampling covariances between fixed and random effects is assumed to be zero. The average within-group and between-group indirect effects can be calculated using Equations 16.12 and 16.14 in each of the k vectors of simulated values to obtain empirical sampling distributions of the indirect effects, and $100(1-\alpha)\%$ CIs can be calculated using the $100(\alpha/2)$ and $100(1-\alpha/2)$ percentiles.

AN EXAMPLE

In this section we walk through an example to demonstrate how to conduct a multilevel mediation analysis. After providing an overview of the data and design, we fit three models of increasing complexity. The first model is a 1-1-1 mediation model with random intercepts and fixed slopes. The second model extends the first by allowing two slopes to randomly vary, and the third model includes the covariance between these random slopes as an additional model parameter. Throughout, we link the estimates obtained to their statistical meaning as described in the previous section, as well as to their substantive meaning in the context of the example. We conclude with some recommendations for choosing a final model and reporting the results.

Data and Design

For this example, we use the dataset `tutor_data.sav` (available from the online appendix), which is a simulated dataset based on an educational experiment. Suppose 48 classrooms were randomly sampled within a particular state, where each classroom is from a different school. Next, students within each classroom were randomly sampled to participate in an after-school tutoring program throughout the school year. The total number of students sampled is 450, where 223 students are assigned to the after-school tutoring program and 227 students are assigned to the control condition. The `tutor` variable in the dataset codes the assignment of each student

(0 = control, 1 = tutoring). Before completing an end-of-year mathematics post-test (`post`), the students' academic motivation (`motiv`) was measured. There is also data on the students' test scores from the previous year (`pre`).

The Model

We are interested in testing whether there is evidence that participation in the after-school tutoring program (X = `tutor`) results in higher mathematics post-test scores (Y = `post`), on average, than not participating in after-school tutoring due to an increase in student motivation (M = `motiv`). That is, tutoring is expected to increase student motivation which, in turn, is expected to result in higher test scores. Further, we are interested in whether this effect is consistent across classrooms, or whether there is between-classroom variability in the effect. Throughout, we will use the previous year's mathematics test score (Q = `pre`) as a covariate.

Here, all variables contain within-class variability, as different students were assigned to different experimental conditions within classes (i.e., not all students in the same class had the same assignment), and student motivation and test scores are also student-level variables. Further, because the proportion of students assigned to tutoring within each class is not equivalent, there is also between-class variability on X, and since the class-aggregate of student motivation and pre and post test scores differ across classrooms, there is also between-class variability on M (ICC[1] = 0.09), Q (ICC = 0.47), and Y (ICC = 0.31). Thus we group-mean center X, M, and Q to remove between-class variability and add this variability back into the model using the classroom means as predictors of the random intercepts so that within-class and between-class effects can be estimated separately. Specifically, our model is:

$$
\begin{aligned}
MOTIV_{ij} &= d_{M_j} + a_j(TUTOR_{ij} - \overline{TUTOR}_{.j}) + g_{1_j}(PRE_{ij} - \overline{PRE}_{.j}) + e_{M_{ij}} \\
POST_{ij} &= d_{Y_j} + c'_j(TUTOR_{ij} - \overline{TUTOR}_{.j}) + b_j(MOTIV_{ij} - \overline{MOTIV}_{.j}) \\
&\quad + g_{2_j}(PRE_{ij} - \overline{PRE}_{.j}) + e_{Y_{ij}} \\
d_{M_j} &= d_M + a_B\overline{TUTOR}_{.j} + g_{1_B}\overline{PRE}_{.j} + u_{M_j} \\
d_{Y_j} &= d_Y + c'_B\overline{TUTOR}_{.j} + b_B\overline{MOTIV}_{.j} + g_{2_B}\overline{PRE}_{.j} + u_{Y_j} \\
a_j &= a_W, b_j = b_W, c'_j = c'_W, g_{1_j} = g_{1W}, g_{2_j} = g_{2W}
\end{aligned}
\tag{16.17}
$$

For now, we have specified the intercepts as random and all of the within-classroom slopes as fixed, though we relax this constraint later. All level-2 random effects, as well as the level-1 residual variances, are each assumed to be normally distributed with a mean of zero and a constant variance.

The MLmed Macro

Many statistical software packages can be used to fit the multilevel mediation model. These include software designed specifically for multilevel modeling, such as HLM (Raudenbush et al., 2013) or MLwiN (Rasbash et al., 2009), software designed for structural equation modeling, such as M*plus* (Muthén & Muthén, 2015), or functions or packages within general purpose statistical software, such as the MIXED procedure in SPSS (IBM Corp., 2013), PROC MIXED in SAS (SAS Institute Inc., 2011), or the lme4 package (Bates et al., 2015) in R (R Core Team, 2016). Although there are some minor differences in the algorithms implemented within each of these options, they all tend to produce similar parameter estimates and so the choice of software becomes a matter of preference. However, some software requires less programming effort than others in the context of multilevel mediation analysis.

Throughout this example, we analyze the data using `MLmed` (Rockwood, 2017), a free SPSS macro specifically designed for multilevel mediation analysis which can be downloaded from www.njrockwood.com. `MLmed` simplifies the process by automatically group-mean centering predictor variables to separate between-group and within-group effects, stacks the equations to simultaneously estimate all model parameters (Bauer et al., 2006), estimates the parameters using the SPSS MIXED procedure (IBM Corp., 2013), calculates within-group and between-group indirect effects, and constructs Monte Carlo confidence intervals for these effects. Both a syntax version (used in this chapter) and a point-and-click interface for setting up the model are available.

The Analysis

We begin by fitting the model for the 1-1-1 data design with random intercepts in the *M* and *Y* equations, but no random slopes. This model implies that there may be some classroom-level differences in *M* and *Y*, but the relationships among the variables in the model (including the indirect effects) are the same in all classrooms. In `MLmed`, this model can be estimated using the syntax:

```
mlmed data = DataSet1
     /x = tutor
     /m1 = motiv
     /y = post
     /cov1 = pre
     /cluster = classid
     /folder = /Users/username/Desktop/.
```

where `DataSet1` is the dataset containing the variables in the analysis, `classid` is a variable identifying the classroom that each student is a member of, and the `folder` argument specifies a working folder on the user's computer. A thorough overview of all `MLmed` syntax commands can be found in the user guide that accompanies the macro.

The output containing the model fit statistics and fixed effects estimates for this model is displayed in Figure 16.4. The fixed effects are separated by

```
Model Specification
N               450
Fixed            12
Rand(L1)          2
Rand(L2)          2
Total            16

Model Fit Statistics
         Value
-2LL  3989.573
AIC   3997.573
AICC  3997.618
CAIC  4020.729
BIC   4016.729


**************************** FIXED EFFECTS ***************************

**********************************************************************
 Outcome: motiv
```

Within- Effects

		Estimate	S.E.	df	t	p	LL	UL
d_M	constant	1.2844	.5128	46.8208	2.5049	.0158	.2528	2.3160
a_W	tutor	1.3517	.0462	398.0015	29.2783	.0000	1.2610	1.4425
g_{1W}	pre	.0194	.0022	398.0015	8.7609	.0000	.0151	.0238

Between- Effects

		Estimate	S.E.	df	t	p	LL	UL
a_B	tutor	.8443	.5654	47.1328	1.4933	.1420	-.2930	1.9817
g_{1B}	pre	.0039	.0058	45.9489	.6771	.5017	-.0077	.0156

```
**********************************************************************
 Outcome: post
```

Within- Effects

		Estimate	S.E.	df	t	p	LL	UL
d_Y	constant	50.3689	7.2512	43.9803	6.9463	.0000	35.7548	64.9830
c'_W	tutor	-3.2913	1.4885	394.8792	-2.2112	.0276	-6.2176	-.3649
b_W	motiv	4.4675	.9097	394.8792	4.9112	.0000	2.6791	6.2559
g_{2W}	pre	.3746	.0440	394.8792	8.5215	.0000	.2882	.4611

Between- Effects

		Estimate	S.E.	df	t	p	LL	UL
c'_B	tutor	-24.5103	7.7505	48.0227	-3.1753	.0026	-40.1935	-9.0271
b_B	motiv	12.3961	1.9524	43.5166	6.3490	.0000	8.4600	16.3323
g_{2B}	pre	.0119	.0775	44.9123	.1532	.8789	-.1443	.1680

Figure 16.4 Excerpt MLmed output containing the model fit statistics and fixed effect estimates from the tutoring example with only random intercepts.

each equation (M and Y). Within each equation, the estimates are further separated by within-class and between-class effects. For this analysis, we are predominantly interested in the within-class effects. If wanted, we could remove the between-class effects from the model. However, for the purpose of this example, we will keep them in the model.

From the output, two students within the same classroom who differ in their tutoring assignment, but have the same pretest score, are estimated to differ by $a_W = 1.352$ units, on average, in their motivation, with the student who attended tutoring being more motivated than the student who did not attend tutoring. Further, those with a given tutoring assignment who performed higher on the pretest also tended to be slightly more motivated ($g_{1_W} = 0.019$).

Students who differ by one unit on their motivation, but have the same tutoring assignment and pretest scores, are estimated to differ by $b_W = 4.468$ units, on average, on the mathematics posttest, with more motivated students performing better. The pretest is a strong, positive predictor of the posttest ($g_{2_W} = 0.375$), and the direct effect of tutoring on the posttest is negative ($c'_W = -3.291$).

The random effect estimates, as well as the indirect effects, are displayed in Figure 16.5. The level-1 residual variances for the posttest and student

```
      *************************  RANDOM EFFECTS  **************************

      Level-1 Residual Estimates
            Estimate      S.E.    Wald Z          p         LL        UL
$\sigma^2_M$  motiv     .2256     .0160   14.1068      .0000      .1963     .2592
$\sigma^2_Y$  post    74.4739    5.3001   14.0513      .0000    64.7777   85.6213

      Random Effect Estimates
         Estimate       S.E.    Wald Z          p         LL        UL
$\sigma^2_{u_M}$  1      .1568     .0398    3.9451      .0001      .0954     .2578
$\sigma^2_{u_Y}$  2    22.7390    7.0743    3.2143      .0013    12.3582   41.8397

      Random Effect Key
      1     Int         motiv
      2     Int         post

      ********************************************************************

      ***********************  INDIRECT EFFECT(S)  ************************

      Within- Indirect Effect(s)
              E(ab)   Var(ab)    SD(ab)
      motiv   6.0388     .0000     .0000

      Within- Indirect Effect(s)
              Effect        SE         Z          p       MCLL      MCUL
$ab_W$  motiv   6.0388    1.2475    4.8408      .0000     3.6469    8.5605

      Between- Indirect Effect(s)
              Effect        SE         Z          p       MCLL      MCUL
$ab_B$  motiv  10.4665    7.2843    1.4368      .1508    -3.3670   25.7041
```

Figure 16.5 Excerpt of MLmed output containing the random and indirect effect estimates for the tutoring example with only random intercepts.

motivation are 74.474 and 0.226, respectively, and the corresponding random intercepts for each of these equations is 22.739 and 0.157.

The "Indirect Effect(s)" section contains the average within-class indirect effect, the between-class variance and standard deviation of the within-class indirect effect, an inferential test on the average within-class indirect effect, and an inferential test on the between-class indirect effect. Because neither of the components of the within-class indirect effects was specified as random, the variance of the within-class indirect effect across classrooms is zero. The within-class indirect effect is $E(a_j b_j) = a_W b_W + \sigma_{a_j,b_j} = (1.352)(4.468) + 0 = 6.039$. Thus, within a given classroom, a student who attends the after-school tutoring program is expected to perform 6.039 units higher on the mathematics posttest, *by way of increased motivation*, than a student with the same pretest score who did not participate in the tutoring program. The corresponding 95% Monte Carlo confidence interval ranges from 3.647 to 8.561. Thus, within a given classroom, there is evidence that participation in tutoring leads to higher motivation which, in turn, leads to better mathematics performance.

Although the between-class indirect effect is not of interest for this example, we walk through its calculation for the purpose of demonstration. It is calculated as $E(a_B b_B) = a_B b_B = (0.844)(12.396) = 10.467$. Since `tutor` is a dichotomous variable indicating whether student i in classroom j was assigned to tutoring, $\overline{TUTOR}_{.j}$ is the proportion of students in classroom j assigned to tutoring. The between-class indirect effect suggests that the class-average math scores for two classes that differ by one unit in the proportion of students assigned to tutoring are expected to differ by 10.467 units by way of class-average motivation. Although this effect is larger than the average within-class indirect effect, there is much more uncertainty in the estimate, and we cannot conclude that the effect is significantly different from zero using the Monte Carlo confidence interval (–3.367, 25.704). Because the between-class indirect effect is not of interest here, we will not refer to it for the remainder of the analysis.

Adding Random Slopes

Thus far, we have modeled each of the components of the within-class mediation model as fixed effects, where the effects in each class are constrained to be equal to the effects in the other classes. However, we are interested in whether the within-class indirect effect varies across classrooms. Therefore, we need to model the components as random. We can use `MLmed` to fit a model including random within-class a and b paths by adding the arguments `randx = 01` and `randm = 1` to the previously used syntax. This specifies that the effects of X on Y and M are fixed (0) and random (1), respectively, and the effect of M on Y is random (1). That is, $a_j = a_W + u_{a_j}$ and $b_j = b_W + u_{b_j}$. By default, `MLmed` assumes the covariance between these effects is zero. More details regarding model specification and defaults in MLmed can be found in the User Guide. The full syntax for this model is:

```
mlmed data = DataSet1
     /x = tutor
     /randx = 01
     /m1 = motiv
     /randm = 1
     /y = post
     /cov1 = pre
     /cluster = classid
     /folder = /Users/username/Desktop/.
```

The fixed effects for the new model (not displayed) are similar to those from the previous model. The random effects and indirect effects of this model are displayed in Figure 16.6.

```
************************** RANDOM EFFECTS ***************************

Level-1 Residual Estimates
        Estimate      S.E.    Wald Z         p         LL        UL
motiv     .1965      .0147   13.3937      .0000     .1698     .2275
post    64.2962     4.8869   13.1570      .0000   55.3974   74.6245

Random Effect Estimates
     Estimate      S.E.    Wald Z         p         LL        UL
   1    .1614     .0399    4.0459     .0001     .0995     .2620
   2  24.3558    7.1420    3.4102     .0006   13.7088   43.2718
σ²aj 3    .1043     .0387    2.6924     .0071     .0504     .2160
σ²bj 4  14.8723    5.8741    2.5318     .0113    6.8577   32.2534

Random Effect Key
1     Int         motiv
2     Int         post
3     Slope       tutor         ->       motiv
4     Slope       motiv         ->       post

*************************************************************************

************************ INDIRECT EFFECT(S) *************************

Within- Indirect Effect(s)
         E(ab)   Var(ab)    SD(ab)
motiv   5.8294   30.4889    5.5217

Within- Indirect Effect(s)
         Effect        SE         Z         p      MCLL      MCUL
motiv   5.8294    1.5014    3.8827     .0001    2.8628    8.8188

Between- Indirect Effect(s)
         Effect        SE         Z         p      MCLL      MCUL
motiv  10.4663    7.2949    1.4348     .1514   -3.4790   25.5035
```

Figure 16.6 Excerpt of MLmed output containing the random and indirect effect estimates for the tutoring example with random slopes.

The estimated variance of the within-class effects of tutoring on motivation and motivation on posttest are $\sigma^2_{a_j} = 0.104$ and $\sigma^2_{b_j} = 14.872$, respectively. Further, although the average within-class indirect effect has dropped only slightly ($E(a_j b_j) = 5.829$), and the confidence interval around the effect is still fully above zero, there is much between-class variability of this effect (variance = 30.489, SD = 5.522). That is, it is likely that in some classrooms, the indirect effect is actually negative or close to zero.

Adding Random Slope Covariance

The previous model assumed that the random slopes do not covary, which may not be reasonable as classrooms that have higher tutor-to-motivation relationships may also tend to have higher (or lower) motivation-to-posttest relationships. We can relax this assumption in `MLmed` by adding the argument `covmat = UN`, which states that the random effect covariance matrix is unstructured. Within `MLmed` this will actually just estimate the covariance between random slopes, σ_{a_j,b_j}, but not the covariance between random slopes and intercepts. These additional covariances can be estimated if desired, though. The full syntax for this new model is:

```
mlmed data = DataSet1
      /x = tutor
      /randx = 01
      /m1 = motiv
      /randm = 1
      /y = post
      /cov1 = pre
      /covmat = UN
      /cluster = classid
      /folder = /Users/username/Desktop/.
```

The random effect and indirect effect estimates for this model are displayed in Figure 16.7. In the output, the *Random Effect Key* indicates the random slopes are the 3rd and 4th random effects. The positive covariance between the within-class a and b paths of $\sigma_{a_j,b_j} = 0.328$ (found in the *Random Effect Covariance Matrix*), which translates into a correlation of 0.265,[2] means that classes that have a higher than average Tutoring → Motivation effect tend to also have a higher than average Motivation → Posttest effect. This covariance is important for the calculation of the average within-group indirect effect, $E(a_j b_j)$ (Equation 16.12). In this model, $a_W = 1.349$ and $b_W = 4.118$ (not shown), so the average within-group indirect effect is $(1.349)(4.118) + 0.328 = 5.881$, with corresponding 95% Monte Carlo confidence limits of 2.844 and 9.032.

```
**************************  RANDOM EFFECTS  **************************

Level-1 Residual Estimates
      Estimate      S.E.   Wald Z          p        LL        UL
motiv    .1965     .0147  13.3957      .0000     .1698     .2275
post   64.3516    4.8979  13.1387      .0000   55.4336   74.7042

Random Effect Estimates
 Estimate       S.E.    Wald Z          p         LL         UL
    .1614      .0399    4.0460      .0001      .0995      .2621
  24.3469     7.1418    3.4091      .0007    13.7011    43.2644
    .1046      .0388    2.6932      .0071      .0505      .2166
σaj,bj .3280  .3207    1.0227      .3064     -.3006      .9565
  14.6106     5.8470    2.4988      .0125     6.6685    32.0120

Random Effect Covariance Matrix
          1         2         3         4
1     .1614     .0000     .0000     .0000
2     .0000   24.3469     .0000     .0000
3     .0000     .0000     .1046     .3280  σaj,bj
4     .0000     .0000     .3280   14.6106
                         σaj,bj
Random Effect Correlation Matrix
          1         2         3         4
1    1.0000     .0000     .0000     .0000
2     .0000    1.0000     .0000     .0000
3     .0000     .0000    1.0000     .2653
4     .0000     .0000     .2653    1.0000

Random Effect Key
1    Int        motiv
2    Int        post
3    Slope      tutor        ->       motiv
4    Slope      motiv        ->       post

**********************************************************************

************************  INDIRECT EFFECT(S)  ************************

Within- Indirect Effect(s)
          E(ab)   Var(ab)   SD(ab)
motiv    5.8811   33.6256   5.7988

Within- Indirect Effect(s)
         Effect        SE         Z          p       MCLL      MCUL
motiv    5.8811    1.5550    3.7821      .0002     2.8442    9.0318

Between- Indirect Effect(s)
         Effect        SE         Z          p       MCLL      MCUL
motiv   10.4662    7.2948    1.4348      .1514    -3.4069   25.6182
```

Figure 16.7 Excerpt of MLmed output containing the random and indirect effect estimates for the tutoring example with random covarying slopes.

Choosing a Final Model

We have now fit three models to the data. The first contained only random intercepts for the *M* and *Y* equations, the second included random slopes for the within-class *a* and *b* paths, and the third permitted these random slopes to covary. Choosing which of these models to use can be accomplished using both theory and statistical methods. For example, one of the hypotheses of interest was about between-class variation in the within-class indirect effect. Therefore, the first model does not allow us to adequately address our hypotheses. Based on our analyses, it is also clear that the first model is an oversimplification of the underlying causal process by assuming that the causal process is consistent across classrooms.

Statistical methods can also be employed to test the relative fit of competing models to the data. For example, the likelihood ratio test can be used when the models are nested (i.e., the estimated parameters in one model are a subset of the estimated parameters in another). Here, the second model is nested within the third, and the first model is nested within the second (and the third). Although we do not walk through the details here, comparing the models using likelihood ratio tests results in the second model being chosen, as the first model is too simple, while the added complexity of the random slope covariance in the third does not adequately improve the model fit. Readers interested in more detail regarding using statistical methods, such as the likelihood ratio test, to compare competing models are referred to Chapter 3 of this volume, "Evaluation of Model Fit and Adequacy."

Although these statistical tools are useful for finding a balance between model fit and model complexity, in the context of mediation analysis our main goal is obtaining an accurate estimate of the causal process by which *X* transmits its effect on *Y* through *M*, as well as a quantification of the uncertainty in this estimate. With this goal in mind, it becomes clear that the specific model chosen in this example makes little difference on the estimated average indirect effect, as all three models produced similar estimates. Further, both the second and the third model produced similar estimates of the between-class variance of the within-class indirect effect.

ADAPTING AND EXTENDING THE MULTILEVEL MEDIATION MODEL

Here, we provide an overview of ways in which the multilevel mediation model can be adapted or extended to include alternative designs, multiple mediators within a single model, or moderators of the indirect effect(s).

Alternative Designs

The model presented in Equations 16.5–16.11 corresponds to the scenario where within-group and between-group variability exists within *X*, *M*, and *Y*. In the case where variability for one or some of these variables only exists at one level, either due to the study design or the nature of the variables, the model can be simplified. These simplifications may also be warranted to create a more parsimonious model, especially if there is not substantial variability in one or more of the variables at a given level, and the mediation process at that level is not of interest. In general, a mediation process at a given level will be more difficult to detect when the variables at that level do not contain much variability.

A common experimental design is the cluster randomized design, where level-2 units are randomly assigned to treatment conditions, and the treatment condition is usually the independent variable *X*. For example, Clements et al. (2011) randomly assigned schools to different treatment groups to better understand the impact of a new mathematics curriculum. Similarly, Dunstan et al. (2013) proposed a study in which worksites are randomly assigned to an intervention targeting reduced prolonged sitting time in office workers. Because every level-1 unit (students/office workers) within each level-2 unit (schools/worksites) receives the same treatment in both of these cluster randomized experiments, there is no within-group variability for the treatment, *X*. Equations 16.5 and 16.6 can be altered to fit this design, and any other designs in which *X* only varies across groups (i.e., *X* contains no within-group variability), by removing the (non-existent) within-group effects of *X* on *M* and *Y*. This results in what was described earlier as a 2-1-1 mediation model. It is often stated that within-group indirect effects in this context are not meaningful (Preacher et al., 2010; Zhang et al., 2009), though Pituch and Stapleton (2012) suggest that they may be under some circumstances. Nevertheless, the calculation of the between-group indirect effect remains as before.

Another common design for which multilevel mediation can be used involves repeated measurements nested within individuals who participate in multiple research conditions. This includes the popular within-participants or AB/BA crossover design, as well as extensions involving more than two conditions or repeated measurements on each participant within each condition. If the research condition is *X* and all individuals participate in all research conditions (an equal number of times), there is no between-person variability on *X* and the between-group effects of *X* can be removed from Equations 16.7 and 16.8. Now, there is no longer a between-group indirect effect, but the calculation of the within-group indirect effect remains as before.

An even simpler data design consists of *X* and *M* variables measured at level 2. That is, there is no within-group variability for *X* or *M*. With such data, the effect of *X* on *M* can be estimated using OLS regression, and the effect of *M* on *Y*, controlling for *X*, can be estimated using Equations 16.6 and 16.8 after removal of the (non-existent) within-group effects of *X* and *M*. The between-group indirect effect follows as the product of these effects.

Estimation of mediation processes that involve "upward effects," where a level-1 variable exerts its effect on a level-2 variable, are generally not possible within standard multilevel modeling software but may be estimated as a multilevel structural equation model (MSEM), as described by Preacher et al. (2010). Although the MSEM framework allows for these upward effects to be estimated and also provides some added benefits in the form of reduced bias of between-group effects based on group-aggregates, McNeish (2017) notes that most empirical studies have used 1-1-1 and 2-1-1 multilevel mediation models and most of these studies had smaller than recommended cluster-level sample sizes for MSEM. For these analyses, the multilevel mediation framework discussed throughout this chapter is recommended (McNeish, 2017).

Multiple Mediator Models

Causal effects often operate through more than one mechanism. The multilevel mediation model presented throughout this chapter can be extended to simultaneously model multiple mediation processes in a similar manner in which multiple mediator models are formulated in the single-level context (see, e.g., Preacher & Hayes, 2008). The utility of simultaneously modeling these processes is two-fold. First, if the causal effect is expected to be transmitted through multiple mediators and the mediators are correlated, failing to include one of the mediators in the model may result in a biased estimate of the indirect effect through the other mediator. And second, indirect effects through different mediators within the same model can be statistically compared, allowing for tests of the difference between the indirect effects. This can be useful for testing competing theories, where each theory posits that a different mediator is responsible for *X*'s effect on *Y*. The `MLmed` macro can model mediation processes involving up to three mediators.

Moderated Mediation/Conditional Process Models

In the multilevel mediation model, the between-group indirect effect is modeled to be the same for all groups in the study. The within-group indirect effect can be modeled as varying across groups but, as we have presented, this variation is considered random. We may expect, however, that the between-group and/or within-group indirect effect is a function of some group-level variable. For example, in the analysis we conducted in the previous section, there was considerable variability in the within-class indirect effect across classes. Suppose we believed this variability was due to differences in the class size, where the strength of the indirect effect of the tutoring program is smaller in smaller classes. Perhaps the relationship between tutoring and motivation is smaller for smaller class sizes because teachers with fewer students can more easily motivate their students, making the difference in motivation for students who did and did not participate in tutoring smaller.

We could extend the multilevel mediation model to include this level-2 moderator (class size) by including it as a predictor of the within-group a path. To test whether the within-group indirect effect changes systematically as a function of class size, we would use the *index of moderated mediation* (Hayes, 2015) adapted to the multilevel context. This index quantifies the linear relationship between the moderator and the indirect effect, and so a confidence interval around the index can provide evidence that the mediation process is moderated. The details of such an approach are beyond the scope of this chapter, but are the focus of Rockwood (2017) and Hayes and Rockwood (2019). Further, `MLmed` makes the implementation of multilevel moderated mediation fairly straightforward.

CONCLUSION

Throughout this chapter, we provided an overview of multilevel statistical mediation analysis. After reviewing the single-level mediation model, we presented the general 1-1-1 multilevel mediation model, defined the average within-group indirect effect and the between-group indirect effect, and explained how to make inferences about these effects. We also discussed estimation of the between-group variance of the within-group indirect effect, as well as its substantive meaning.

To demonstrate the statistical methods presented, we introduced and used `MLmed`, an SPSS macro designed for fitting multilevel mediation models, in an example analysis in which the within-class indirect effect of tutoring on posttest scores through motivation was quantified and assessed using three models of increasing complexity. The output obtained using `MLmed` was interpreted and mapped to the concepts and equations introduced previously in the chapter, and recommendations were made to help choose a final model.

Last, we provided an overview of some miscellaneous topics in multilevel mediation analysis. These include details on adapting the model to account for other multilevel data designs where not all variables contain within-group and between-group variability, as well as extending the model to include multiple mediators or moderators, where the strength or direction of the within-group or between-group indirect effect varies systematically as a function of a level-2 variable. These topics should help readers identify ways in which the models presented throughout this chapter can be extended to match their own research hypotheses.

RECOMMENDED READINGS

Technical

Pituch, K. A., & Stapleton, L. M. (2012). Distinguishing between cross- and cluster-level mediation processes in the cluster randomized trial. *Sociological Methods & Research, 41*(4), 630–670.

Preacher, K. J., Zyphur, M. J., & Zhang, Z. (2010). A general multilevel SEM framework for assessing multilevel mediation. *Psychological Methods, 15*(3), 209–233.

Zhang, Z., Zyphur, M. J., & Preacher, K. J. (2009). Testing multilevel mediation using hierarchical linear models: Problems and solutions. *Organizational Research Methods, 12*(4), 695–719.

Applied

Berli, C., Bolger, N., Shrout, P. E., Stadler, G., & Scholz, U. (2017). Interpersonal processes of couples' daily support for goal pursuit: The example of physical activity. *Personality and Social Psychology Bulletin, 44*(3). https://doi.org/10.1177/0146167217739264

Braun, S., Peus, C., Weisweiler, S., & Frey, D. (2013). Transformational leadership, job satisfaction, and team performance: A multilevel mediation model of trust. *The Leadership Quarterly, 24*(1), 270–283.

Brincks, A. M., Feaster, D. J., & Mitrani, V. B. (2010). A multilevel mediation model of stress and coping for women with HIV and their families. *Family Process, 49*(4), 517–529.

TRY THIS!

For this exercise, use the `tutor_data.sav` dataset described throughout the chapter (available through the online appendix). Within the chapter we presented an experimental design in which students within classrooms were randomly assigned to participate in a tutoring program or a control condition. We were interested in the within-classroom indirect effect of tutoring on student test scores through motivation. Now consider a scenario in which teachers completed a training program designed to teach a number of skills focused on engaging their students through the use of interactive real-world applications. It is thought that students who are exposed to the interactive real-world applications will see the utility of the content being taught so they will be more motivated, leading to an increase in their post-test performance, where the post-test is completed at the end of the academic year.

The variables in the dataset corresponding to this experiment are:

- classid = Classroom ID (which also corresponds to Teacher ID)
- train = Teacher Training Program (1 = completed training, 0 = control)
- motiv = Student motivation
- pre = Student pre-test score
- post = Student post-test score

Apply the skills learned throughout this chapter to test the following hypotheses:

1. Students of teachers who completed the training program were more motivated, on average, than students of teachers who were in the control group.
2. Classes in which students tend to be more motivated tended to perform better, on average, on the posttest.
3. The average amount of student motivation mediates the relationship between the teachers' completion of the training program and the average post-test score of their students.

After analyzing the data, summarize the results with an emphasis on the hypotheses. Make sure to include any information about the choices made when building the model. Before beginning the analysis, it may be useful to:

1. Identify at which level each variable was measured.
2. Using (1) above, identify which variables will have between- and/or within-class effects.
3. Using (2) above, identify at which level(s) an indirect effect can be estimated.

NOTES

1. Intra-class correlation coefficient

$$(\text{ICC}) = \frac{\sigma_B^2}{\sigma_B^2 + \sigma_W^2},$$

where σ_B^2 and σ_W^2 are the between-group and within-group variances, respectively. The ICC represents the proportion of between-group variability.
2. The correlation can be computed by dividing the covariance by the product of the standard deviations for a and b,

$$Cor(a,b) = \frac{\sigma_{a_j,b_j}}{\sqrt{\sigma_{a_j}^2}\sqrt{\sigma_{b_j}^2}} = \frac{0.328}{\sqrt{0.105}\sqrt{14.611}} = 0.265.$$

REFERENCES

Aroian, L. A. (1947). The probability function of the product of two normally distributed variables. *The Annals of Mathematical Statistics, 18*(2), 265–271.

Baron, R. M., & Kenny, D. A. (1986). The moderator–mediator variable distinction in social psychological research: Conceptual, strategic, and statistical considerations. *Journal of Personality and Social Psychology, 51*(6), 1173–1182.

Bates, D., Mächler, M., Bolker, B., & Walker, S. (2015). Fitting linear mixed-effects models using lme4. *Journal of Statistical Software, 67*(1), 1–48. https://doi.org/10.18637/jss.v067.i01

Bauer, D. J., Preacher, K. J., & Gil, K. M. (2006). Conceptualizing and testing random indirect effects and moderated mediation in multilevel models: New procedures and recommendations. *Psychological Methods, 11*(2), 142–163.

Bolger, N., & Laurenceau, J. (2013). *Intensive longitudinal methods: An introduction to diary and experience sampling research.* The Guilford Press.

Clements, D. H., Sarama, J., Spitler, M. E., Lange, A. A., & Wolfe, C. B. (2011). Mathematics learned by young children in an intervention based on learning trajectories: A large-scale cluster randomized trial. *Journal for Research in Mathematics Education, 42*(2), 127–166.

Decker, S. L., Strait, J. E., Roberts, A. M., & Wright, E. K. (2018). Cognitive mediators of reading comprehension in early development. *Contemporary School Psychology, 22*(3), 249–257.

Dunstan, D. W., Wiesner, G., Eakin, E. G., Neuhaus, M., Owen, N., LaMontagne, A. D., Winkler, E., Fjeldsoe, B. S., Lawler, S., & Healy, G. N. (2013). Reducing office workers' sitting time: Rationale and study design for the Stand Up Victoria cluster randomized trial. *BMC Public Health, 13*(1), 1–14.

Efron, B., & Tibshirani, R. J. (1994). *An introduction to the bootstrap.* CRC Press.

Goodman, L. A. (1960). On the exact variance of products. *Journal of the American Statistical Association, 55*(292), 708–713.

Hayes, A. F. (2009). Beyond Baron and Kenny: Statistical mediation analysis in the new millennium. *Communication Monographs, 76*(4), 408–420.

Hayes, A. F. (2015). An index and test of linear moderated mediation. *Multivariate Behavioral Research, 50*(1), 1–22.

Hayes, A. F. (2018). *Introduction to mediation, moderation, and conditional process analysis: A regression-based approach* (2nd ed.). The Guilford Press.

Hayes, A. F., & Rockwood, N. J. (2019). Conditional process analysis: Concepts, computation, and advances in the modeling of the contingencies of mechanisms. *American Behavioral Scientist. 64*(1), 19–54.

Hayes, A. F., & Scharkow, M. (2013). The relative trustworthiness of inferential tests of the indirect effect in statistical mediation analysis: Does method really matter? *Psychological Science, 24*(10), 1–10.

IBM Corp. (2013). IBM SPSS statistics for Windows, version 22.0 [Computer software manual]. IBM Corp.

Ilies, R., Keeney, J., & Scott, B. A. (2011). Work–family interpersonal capitalization: Sharing positive work events at home. *Organizational Behavior and Human Decision Processes, 114*(2), 115–126.

Ilies, R., Liu, X.-Y., Liu, Y., & Zheng, X. (2017). Why do employees have better family lives when they are highly engaged at work? *Journal of Applied Psychology, 102*(6), 956–970.

Imai, K., Keele, L., & Tingley, D. (2010). A general approach to causal mediation analysis. *Psychological Methods, 15*(4), 309–334.

Kao, K.-Y., Spitzmueller, C., Cigularov, K., & Wu, H. (2016). Linking insomnia to workplace injuries: A moderated mediation model of supervisor safety priority and safety behavior. *Journal of Occupational Health Psychology, 21*(1), 91–104.

Kenny, D. A., Korchmaros, J. D., & Bolger, N. (2003). Lower level mediation in multilevel models. *Psychological Methods, 8*(2), 115–128.

Krull, J. L., & MacKinnon, D. P. (1999). Multilevel mediation modeling in group-based intervention studies. *Evaluation Review, 23*(4), 418–444.

Krull, J. L., & MacKinnon, D. P. (2001). Multilevel modeling of individual and group level mediated effects. *Multivariate Behavioral Research, 36*(2), 249–277.

Lam, B. Y., Raine, A., & Lee, T. M. (2014). The relationship between neurocognition and symptomatology in people with schizophrenia: social cognition as the mediator. *BMC Psychiatry, 14*(1), 138.

Ledermann, T., & Kenny, D. A. (2017). Analyzing dyadic data with multilevel modeling versus structural equation modeling: A tale of two methods. *Journal of Family Psychology, 31*(4), 442–452.

MacKinnon, D. P. (2008). *Introduction to statistical mediation analysis.* Routledge.

MacKinnon, D. P., Lockwood, C. M., & Williams, J. (2004). Confidence limits for the indirect effect: Distribution of the product and resampling methods. *Multivariate Behavioral Research, 39*(1), 99–128.

McNeish, D. (2017). Multilevel mediation with small samples: A cautionary note on the multilevel structural equation modeling framework. *Structural Equation Modeling: A Multidisciplinary Journal, 24*(4), 609–625.

Muthén, L. K., & Muthén, B. (2015). M*plus* version 7.4 [Computer software manual].

Pituch, K. A., & Stapleton, L. M. (2012). Distinguishing between cross-and cluster-level mediation processes in the cluster randomized trial. *Sociological Methods & Research, 41*(4), 630–670.

Pituch, K. A., Stapleton, L. M., & Kang, J. Y. (2006). A comparison of single sample and bootstrap methods to assess mediation in cluster randomized trials. *Multivariate Behavioral Research, 41*(3), 367–400.

Preacher, K. J. (2015). Advances in mediation analysis: A survey and synthesis of new developments. *Annual Review of Psychology, 66,* 825–852.

Preacher, K. J., & Hayes, A. F. (2008). Asymptotic and resampling strategies for assessing and comparing indirect effects in multiple mediator models. *Behavior Research Methods, 40*(3), 879–891.

Preacher, K. J., & Selig, J. P. (2012). Advantages of Monte Carlo confidence intervals for indirect effects. *Communication Methods and Measures, 6*(2), 77–98.

Preacher, K. J., Zyphur, M. J., & Zhang, Z. (2010). A general multilevel SEM framework for assessing multilevel mediation. *Psychological Methods, 15*(3), 209–233.

R Core Team. (2016). *R: A language and environment for statistical computing* [Computer software manual]. Vienna, Austria. https://www.R-project.org/

Rasbash, J., Charlton, C., Browne, W. J., Healy, M., & Cameron, B. (2009). *MLwin version 2.1* [Computer software manual].

Raudenbush, S. W., Bryk, A. S., & Congdon, R. (2013). *HLM 7.00 for Windows* [Computer software manual]. Scientific Software International, Inc.

Reyes, M. R., Brackett, M. A., Rivers, S. E., White, M., & Salovey, P. (2012). Classroom emotional climate, student engagement, and academic achievement. *Journal of Educational Psychology, 104*(3), 700–712.

Rockwood, N. J. (2017). *Advancing the formulation and testing of multilevel mediation and moderated mediation models* [Unpublished master's thesis]. The Ohio State University, Columbus, OH.

SAS Institute Inc (2011). *SAS/STAT software version 9.3* [Computer software manual].

Selig, J. P., & Preacher, K. J. (2008). *Monte Carlo method for assessing mediation: An interactive tool for creating confidence intervals for indirect effects* [Computer software]. http://quantpsy.org/medmc/medmc.htm

Sobel, M. E. (1982). Asymptotic confidence intervals for indirect effects in structural equation models. *Sociological Methodology, 13*, 290–312.

Tofighi, D., West, S. G., & MacKinnon, D. P. (2013). Multilevel mediation analysis: The effects of omitted variables in the 1–1–1 model. *British Journal of Mathematical and Statistical Psychology, 66*(2), 290–307.

Tremblay, M., Hill, K., & Aubé, C. (2017). A time-lagged investigation of the mediating role of person–organization fit in the relationship between leader–member exchange and employee affective organizational commitment. *European Journal of Work and Organizational Psychology, 26*(1), 53–65.

Van der Leeden, R., Meijer, E., & Busing, F. M. (2008). Resampling multilevel models. In J. de Leeuw & E. Meijer (Eds.), *Handbook of multilevel analysis* (pp. 401–433). Springer.

Zhang, Z., Zyphur, M. J., & Preacher, K. J. (2009). Testing multilevel mediation using hierarchical linear models: Problems and solutions. *Organizational Research Methods, 12*(4), 695–719.

CHAPTER 17

REPORTING RESULTS OF MULTILEVEL DESIGNS

John M Ferron
University of South Florida

Yan Wang
University of Massachusetts

Zhiyao Yi
Chongqing Technology and Business University

Yue Yin
University of South Florida

Eunsook Kim
University of South Florida

Robert F. Dedrick
University of South Florida

Over the past 20 years, the number, variety, and complexity of multilevel applications in education and the social sciences have increased dramatically. These multilevel applications have included both measured and latent

Multilevel Modeling Methods with Introductory and Advanced Applications, pages 599–621

variables and a range of outcomes consisting of ordinal, nominal, continuous, count, zero-inflated, and time to event variables. Multilevel applications involving techniques such as latent class analysis (Fagginger et al., 2016), event history analysis (Chen, 2012), growth mixture modeling (Palardy & Vermunt, 2010), and intensive longitudinal (e.g., experience sampling) analysis (Nezlek et al., 2016), which were less common 20 years ago, are now increasing in use. In addition to these newer applications, there continues to be increasing use of more traditional applications of multilevel models with both cross-sectional and longitudinal data. For example, multilevel models are widely used to analyze data from cluster randomized controlled trials and multisite cluster randomized controlled trials.

The range and complexity of these applications pose challenges to researchers aiming for transparency, accuracy, completeness, and ethical reporting of their research results. To make it easier to interpret the findings from studies, the open science movement has put a premium on the completeness and accuracy of reporting. As part of this movement, pre-registration of study protocols (e.g., hypotheses/goals, design, planned sample, exclusion criteria, procedures, outcomes, analysis plan, contingency plans such as handling missing data, outliers, multicollinearity) with time stamping is now common and may be accomplished using platforms such as the Open Science Framework (https://osf.io/) and AsPredicted (https://aspredicted.org/; e.g., van 't Veer & Giner-Sorolla, 2016). The practice of pre-registration helps to make public the researchers' plan so that the evaluation of the final report can include consideration of the consistency between what was done and what was planned. To evaluate this consistency, however, the reporting of the multilevel modeling methods in both the preregistration and in the final report has to be detailed and clear.

To improve research reporting and increase reproducibility and usability of research results, general guidelines and reporting standards have been developed by various organizations to assist researchers in achieving these qualities in their reporting (e.g., American Educational Research Association's "Standards for Reporting on Empirical Social Science Research in AERA Publications" [AERA, 2006]; *Publication Manual of the American Psychological Association* [APA, 2020] and task force report on publications and communication [Appelbaum et al., 2018]; American Statistical Association's [2016] *Ethical Guidelines for Statistical Practice*; The Transparency and Openness Promotion (TOP) Committee [Nosek et al., 2015]; Consolidated Standards of Reporting Trials [CONSORT; Schulz et al., 2010]). Reporting guidelines also have been developed for various research designs such as STROBE for observational studies (von Elm et al., 2008) and PRISMA for systematic reviews and meta-analyses (Moher et al., 2009).

In addition to these general research reporting guidelines, specific reporting guidelines for multilevel modeling applications have been created by

Ferron et al. (2008) and McCoach (2010) to provide guidance on *what* to report from the multilevel analyses and "*how* to present these results using text, tables, and figures" (Ferron et al., 2008, p. 392). More recently, Kim et al. (2016) presented guidelines for reporting multilevel analyses involving latent variables used in multilevel exploratory and confirmatory factor analyses. These guidelines have been developed in the form of checklists, flow diagrams, and narrative reports to address inadequate or incomplete reporting of results. Improvements in research reporting aim to enhance the value of research by improving researchers' and consumers' ability to understand, evaluate, replicate, synthesize, and use the results of research reports.

This prior work focusing on reporting practices has provided a foundation of guidelines for reporting results from multilevel applications. In light of the dynamic field of multilevel modeling, this chapter provides updated guidelines regarding the reporting of multilevel results. These updates were informed by the general and specific guidelines noted above, as well as by other sources including journals' policy statements and instructions for authors (e.g., *Journal of the American Statistical Association*). Additionally, broader issues facing the field of research that have been raised, for example, by the open science movement, were also used to develop guidelines. These issues center on research transparency, reproducibility, replicability (Munafo et al., 2017), and what John et al. (2012) have called questionable research practices (QRPs; e.g., selective reporting of dependent variables; reporting an unexpected finding as one that was predicted *a priori*) that may result in false-positive results and impede reproducing results.

In updating reporting guidelines for multilevel modeling, we present an initial set of guidelines organized in a checklist consisting of seven sections: (a) research purposes and design type, (b) sampling, (c) variables, (d) model specification, (e) estimation and inference, (f) model evaluation, and (g) results reporting. Within these sections, we identified specific items to consider in reporting results of multilevel applications.

RESEARCH PURPOSES AND DESIGN TYPE

The reporting of research results cannot be viewed in isolation and must be considered in connection with the theoretical foundation or perspective underlying the study. McCoach (2010) emphasized the importance of having the statistical approach, such as multilevel modeling, align with the research questions or purposes of the study. This alignment can be best assessed when the written summary makes explicit both the study purposes and the rationale for the analytical method. As part of the specification of the research design, the researcher also needs to communicate information about the components of the research plan and their connections.

Information is needed about the: (a) study variables such as the level at which they are measured (e.g., student or level-1 versus school or level-2); how they are measured and their measurement quality, type of variable (observed vs. latent), and their role in the study (independent, dependent, mediator, moderator, control variable); (b) units of analysis (the number and method of selection at each level); (c) data collection procedures; and (d) specific type of design (e.g., randomized controlled trial, quasi-experiment, correlational). Information about these design components are critical to evaluating the internal and external validity of the study conclusions (Shadish et al., 2002).

SAMPLING

Sampling is critical to research in social and behavioral sciences and involves a variety of decisions, such as choosing the sampling procedure, determining sample sizes, and handling issues in the sampling process (e.g., missing data). A variety of probability-based methods (e.g., simple random, stratified random, disproportionate random) and non-probability based methods (e.g., convenient, snowball) can be used at each level in a multilevel context, and as a consequence researchers need to communicate the sampling method used for level 1 and level 2 (and higher levels if present) and justify the use of the sampling method at each level. For example, it is important to report how schools and individuals within schools are selected to be included in the study. Note that for secondary data analysis, researchers need to obtain and report information about the sampling method (e.g., referring to the technical manual). If sampling weights are used, it should be clear how data are weighted and why sampling weights are used. When multiple sets of sampling weights are available, researchers should provide the rationale for choosing certain sampling weights.

Sample sizes should be clearly communicated for the sampling process, including the number of units for each level, and the distribution of lower-level sample sizes across upper-level units. For example, in a school organizational design where students are nested within schools, the number of students and the number of schools are the sample sizes for level 1 and level 2, respectively. When the number of individuals is not the same across clusters or the number of repeated measures is not the same across individuals, researchers should consider how to report the distribution of level-1 units across level-2 units (Ferron et al., 2008). Justification for the sample sizes at each level is needed and may rely on statistical power analyses (see Chapter 5, "Statistical Power for Linear Multilevel Models," in this volume and Spybrook et al., 2016; Spybrook et al., 2011), along with budgetary and practical constraints.

Tables 17.1 and 17.2 provide an example of reporting sample size information in cross-sectional and longitudinal multilevel designs, respectively. Table 17.1 shows that there were 86 schools and 600 students in the sample, with the number of students ranging from 1 to 10 per school and most of schools having 7 or more students. Table 17.2 presents a total sample size of 140 individuals among which half has 5 repeated measures and the other half is measured one to four times. Both tables communicate the sample size at each level and the distribution of level-1 units across level-2 units, which allows readers to evaluate the precision of parameter estimates from the multilevel models.

However, it might not be feasible to include such tables when the dataset is large (e.g., many different numbers of students per school or repeated measures taken intensively or with varying time intervals). Instead,

TABLE 17.1 Example of Table for Summarizing Sample Sizes for Students Nested Within Schools in a Two-Level Design

Number of students per school	Number of schools with specified number of students	Cumulative frequency of schools	Cumulative frequency of students
1	11	11	11
2	3	14	17
3	6	20	35
4	2	22	43
5	3	25	58
6	2	27	70
7	10	37	140
8	10	47	220
9	10	57	310
10	29	86	600

TABLE 17.2 Example of Table for Summarizing Sample Sizes for Two-Level Growth Curve Design

Number of time points observed	Number of individuals	% of individuals	Cumulative frequency of individuals
1	10	7.1	10
2	20	14.2	30
3	16	11.4	46
4	21	15.0	67
5	73	52.1	140

researchers can report descriptive statistics regarding the distribution of level-1 sample sizes across level-2 units, such as the minimum, maximum, mean, and standard deviation of level-1 sample sizes per level-2 unit as well as the percent of level-2 units that are singletons. Note that with the presence of level-3 or higher-level units, additional information on sample sizes should be reported, such as the number of level-3 units and the distribution of level-2 sample sizes per level-3 unit.

In addition to reporting the sampling procedures and sample size issues in the sampling process, researchers should also clearly communicate characteristics of the final sample that will be used for the analysis. For example, if there are missing data at each level, the proportion of missingness (for both individuals and variables), the degree to which missingness is related to the variables being studied, and the method used to handle missing data (e.g., multiple imputation; see Chapter 15, "Missing Data Handling for Multilevel Data" in this volume and Enders, 2017) should be discussed. If, for example, due to the deletion of missing data or other factors not all the schools or all the students are considered in the final sample, it is important to report sample sizes for the original and final analytic samples. CONSORT (http://www.consort-statement.org/) has created a useful flow diagram for communicating the number of individuals at each phase in a randomized controlled trial: including enrollment, allocation, follow up, and analysis. Modification of this diagram can be made to fit other types of studies (e.g., correlational, meta-analytic) to communicate information about the sample.

VARIABLES

A critical part of communicating a study's research questions and subsequent results from multilevel analyses is specifying the variables used in each of the research questions. Depending on the research questions, these variables may include outcomes, predictors, control variables, mediators, or moderators. As with any research study, researchers need to provide descriptions and definitions of their variables either in the text or in a table and how they are being treated in the analyses (e.g., continuous, ordered categorical, unordered categorical, count). For example, questionnaire items that use a five-point Likert scale may be treated as continuous variables by some researchers and as ordinal, categorical variables by others. The type of coding of categorical predictor variables (e.g., dummy coding, effect coding, contrast coding) also needs to be specified because the type of coding affects the interpretation of parameter estimates of main effects and interactions involving these variables.

For continuous predictor variables, researchers need to communicate if they are centering their predictors, and if they are centering the variables, what type of centering (e.g., group-mean, grand-mean, theoretically meaningful value) is being used. This information is critical because the type of centering has implications for the interpretation of the variance components and fixed effects (Enders & Tofighi, 2007). In longitudinal studies (see Chapters 9–12 in this volume), where there are multiple options for scaling the time variable (e.g., chronological age, grade level, time in the study) communicating what time metric is used is necessary to understand how change is modeled and how the growth parameters are interpreted (Grimm et al., 2017). For example, one researcher could be comparing two groups over time with a linear growth curve model and center time such that zero corresponded to the beginning of the study, whereas another may center time such that zero corresponded to the end of the study. This centering choice influences whether the difference between the groups is estimated at the beginning of the study or the end of the study.

Researchers also need to provide information about how any derived or created scores used in the analyses were formed (e.g., mathematics scale scores were derived using a three-parameter model using item response theory) along with specific evidence to support the reliability and validity of the original variables and any derived scores (American Educational Research Association, American Psychological Association, & the National Council on Measurement in Education, 2014). Evidence of reliability and validity of the scores needs to be obtained and reported for the specific sample used in the study (Thompson & Vacha-Haase, 2000). The description, conceptualization, and psychometric analyses of study variables in multilevel analyses must use a multilevel measurement framework (see Chapter 14, "Common Measurement Issues in a Multilevel Framework" in this volume), and researchers need to provide level-specific evidence of reliability and validity (e.g., multilevel reliability; Geldhof et al., 2014) as part of the reporting process. Measurement quality of study variables is important for the modeling process because at all levels in the model measurement error can lead to biased estimates of the fixed effect parameters and standard errors (Woodhouse et al., 1996).

For example, in a 2-level organizational model (see Chapter 2, "Introduction to Multilevel Models for Organizational Data" in this volumne) in which a researcher is interested in disentangling person-level effects from contextual (compositional) effects (Raudenbush & Bryk, 2002), the researcher may group-mean center a level-1 predictor (e.g., student socioeconomic status [SES]) and create an aggregate of this predictor variable (school-average SES) and include this aggregate variable in the level-2 model. Even though this variable of SES may appear to be the same, the meaning and psychometric properties of the scores of these two variables (student-SES and school-SES) differ at the two levels of analysis. The implication of these differences is that providing reliability and validity evidence about individual-level variables may not be sufficient for making inferences for group-level variables (Rosenberg, 2009).

In a recent review of 72 multilevel factor analysis applications (exploratory and confirmatory) in journals across a range of disciplines (e.g., education, health/nursing, management, and psychology) published between 1994 and 2014, Kim et al. (2016) found that explicit discussions of how researchers conceptualized the constructs in their studies, and how the constructs (or models) were specified, estimated, and interpreted at each level were lacking. Explicit description of the type of multilevel constructs used in the study (e.g., Stapleton et al., 2016) is important in the reporting process because the conceptualization influences the type of psychometric evidence needed to support the meaningfulness of the score interpretation and how the models are specified at each level of the analysis.

MODEL SPECIFICATION

In multilevel modeling applications the researcher specifies, or defines, one or more statistical models to be estimated. Specification decisions include the choice of the dependent and predictor variables, along with assumptions about the: (a) probability distribution of the dependent variable (e.g., normal, Poisson, binomial); (b) model parameters (e.g., which vary randomly across level-2 units and which do not); and (c) covariance structure (e.g., level-2 errors correlate or they do not). For readers to critically evaluate the multilevel application, they need to know what specification decisions were made and exactly what statistical models were estimated. Without such information, it is difficult to assess the alignment between the models and the research questions, the consistency between the models and the data, or to replicate the analyses.

Among the first decisions to communicate, is the number of levels in the multilevel model. For example, a researcher's research question might involve examining how student-, classroom-, and school-level factors relate to student mathematics achievement, and therefore might specify a 3-level model (students nested in classrooms nested in schools). Communicating the number of levels in the model and the units at each level (e.g., students at level 1 and classrooms at level 2) allows the reader to assess the degree to which the levels and units in the models match the levels and units in the sampling method. This assessment of congruence is critical in evaluating the appropriateness of the standard errors, and the credibility of confidence intervals or tests of significance (Van den Noortgate et al., 2004).

As noted earlier in the variables section, researchers should be clear about how the outcome variable is being treated (e.g., ordered categorical vs. continuous variable). The choice of how to treat the dependent variable, and the underlying probability distribution, impacts how the relationships between the predictors and the outcome are indexed, and potentially the conclusions about which predictors are related to the outcome.

In addition to being transparent in the specification decisions about outcome variables, researchers need to be transparent in their specification decisions about the independent or predictor variables. It is helpful for readers to know what the predictors are at each level of the model (e.g., past achievement at level 1, and school level SES at level 2), the role of each variable (e.g., independent variable or covariate), and how and why the predictors were chosen. In addition, predictors at lower levels of the model may have coefficients that are specified to vary randomly across upper-level units, or may be specified as fixed across upper-level units. Furthermore, when there are multiple random effects at a level of the model, there are decisions to be made about the covariance structure of these effects (e.g., the covariance matrix of level-2 errors is estimated as an unstructured matrix versus a diagonal matrix). Like other specification decisions, these decisions may be made *a priori* or may be guided by the data. If guided by the data the researcher needs to communicate this to the reader and acknowledge how the use of the data to guide the analyses can negatively impact the replicability of the results (e.g., Gelman & Loken, 2014). In either case, the decisions made may influence the conclusions drawn. For example, fixing a coefficient that should vary can lead to bias in the standard errors (Van den Noortgate et al., 2004), whereas letting coefficients vary randomly that do not, can lead to difficulties in estimation (Raudenbush & Bryk, 2002).

Because some applications involve the estimation of a single specified model, whereas others are based on model comparisons, the number of models estimated may vary greatly across applications (Dedrick et al., 2009). Confirmatory data analyses that are conducted to assess the researchers' hypotheses and existing theory typically involve fewer models than exploratory research where researchers are investigating the potential relationships among variables. In confirmatory studies, the preregistration process should establish exactly which model(s) will be estimated and the final report should contain these same models, whereas preregistration of exploratory studies would indicate the analyses are exploratory and the range of models and model modifications that will be explored (see https://cos.io/prereg/; Nosek et al., 2018). For example, during preregistration, a researcher could provide the initial model specification, along with a statement like "there will be no changes to the fixed effects, but if restricted maximum likelihood estimation fails to converge or the estimated level-2 error covariance matrix is not positive definite, the covariance structure will be simplified by eliminating error terms estimated to have no variance." In the final reporting of the study, it should be clear how many models were run. Model modifications could be presented step by step, or the final model could be presented with discussion of the modifications that had been made. The focus of the results reporting (discussed in a later section of this chapter) should depend on whether the study was confirmatory or exploratory and on the number of models estimated. As the number of models increases, aspects of model interpretation (e.g., establishing Type I error control for tests of effects of interest) become more challenging and in some cases not feasible, and thus exploratory studies may avoid reporting inferences (e.g., hypothesis tests) and focus on a description of what was found in the sample.

Even when a single model is estimated, the model can be complex, and as a result, equations and matrices are typically a more straightforward and efficient way of communicating model specification than relying only on verbal descriptions. By using equations and matrices, researchers can concisely and precisely communicate the number of levels in the model, the outcome variable and how it is being treated, and the predictor variables along with the assumptions about which of their coefficients are random and the covariance structure of these random effects. An accompanying paragraph can provide the rationale for the specification decisions that were made (see Chapter 2, "Introduction to Multilevel Models for Organizational Data," and Chapter 3, "Evaluation of Model Fit and Adequacy" in this volume).

ESTIMATION AND INFERENCE

Sufficient technical details about the estimation of the multilevel models should be provided to allow readers to evaluate the validity of the statistical analyses and replicate findings. Specifically, the software program used to perform the analyses and important details about the estimation method should be discussed, because differences might exist among various software programs and estimation methods (McCoach et al., 2018). Some of the most common software programs for multilevel analyses include HLM, MLwiN, M*plus*, R, SAS, SPSS, and Stata. In addition to the software program name, the version should also be specified due to possible differences in the estimation procedures and results across versions (McCoach et al., 2018).

Some software programs have different procedures (or packages) for running multilevel models and a variety of alternative estimation methods (see Table 17.3 for some examples). The method used to estimate multilevel models should be reported. Common estimation methods for multilevel models include: maximum likelihood (ML), restricted maximum

TABLE 17.3 Common Software Programs for Multilevel Models and Estimation Methods Available

Software Program & Version	Procedure or Package	Estimation Method
HLM 7	NA	REML (default), ML, PQL (penalized quasi-likelihood), AGH (Adaptive Gauss-Hermite Quadrature), high-order Laplace approximations to ML
M*plus* 8	TYPE = TWOLEVEL	ML (default), MLR, WLSMV, Bayes, MLM, MLMV, MLF, MUML, WLS, WLSM, ULS, ULSMV, GLS
MLwiN 3.01	NA	ML, IGLS, RIGLS, Bayes
R 3.4.3	lme4, lmer, nlme	REML, ML
SAS 9.4	PROC MIXED	REML (default), ML, MIVQUE0, Type 1, Type 2, Type 3
	PROC GLIMMIX	RSPL (default), MSPL, RMPL, MMPL, LAPLACE, QUAD
SPSS 25	MIXED	REML (default), ML
Stata 15	MIXED	ML (default), REML

likelihood (REML), robust ML estimations (e.g., MLM, MLMV, and MLR), Bayesian estimation, and so on. ML and REML have been most widely used.

Across these estimation methods, sufficient details should be provided, such as the estimation algorithm (e.g., expectation-maximization or EM, Newton-Raphson, the Fisher scoring algorithm, iterative generalized least squares or IGLS, the Gibbs sampler for Bayesian estimation), and the convergence criterion. If default settings are used, researchers should explicitly report this and provide the software documentation (manual or website) available as a reference. For Bayesian estimation, the prior distributions for model parameters should be discussed and justified, even if the default priors of the software program are adopted. Additionally, the number of Markov chain Monte Carlo (MCMC) chains and thinning should be specified. If estimation details are modified (e.g., informative prior distributions are specified in M*plus* instead of the default noninformative priors), such modification should be clearly communicated, including the rationale and modified estimation settings. In addition, if estimation problems (e.g., nonconvergence) were encountered, discussions should be provided regarding how they were addressed. When the estimation details are provided and they match those that were preregistered, consumers of the research can evaluate the appropriateness of the estimation choices, while also being confident that the researchers did not estimate the model in a variety of ways and report just the most favorable set of results.

For similar reasons, researchers should also report the methods used to make inferences. The inferences may be about overall model fit (e.g., one model provides better fit than another) or a specific parameter in the model (e.g., a variance component differs significantly from zero or when a confidence interval is provided for a fixed effect regression coefficient). For example, one may test a variance component using the approximate χ^2 described by Raudenbush and Bryk (2002), bootstrapping, a likelihood ratio χ^2, or other methods of comparing model fit such as deviance, AIC, or BIC (see Chapter 3 for more details). Similarly, researchers have options for testing fixed effects. One researcher may use robust standard errors (Maas & Hox, 2004) to account for non-normality, whereas another may not. In addition to alternative ways of estimating the standard errors there are alternative ways of estimating degrees of freedom (e.g., Kenward & Roger, 1997) and these differences can substantially impact the validity of the inferences when the sample size is small (Ferron et al., 2009). Because there are a variety of methods available to make each of these inferences, researchers need to be explicit about what methods were used for each type of inference. If a study is preregistered, the researcher could specify in the analysis plan what methods would be used to evaluate model fit, estimate parameters, and make inferences.

Reporting details on estimation and inference can often be accomplished in a single sentence. For example, a researcher could state, "The multilevel models were estimated with HLM 7.02 using restricted maximum likelihood via the default settings for estimation and inference." Or another researcher could state, "Multilevel models were estimated using restricted maximum likelihood estimation with the Mixed Procedure in SAS 9.4 using default settings for estimation and inference, with the exception that the Kenward-Roger approach was used to make the treatment effect inference because of the relatively small sample size." Researchers sharing their data in open repositories may also share a link to their analysis scripts (e.g., R scripts) and code used to analyze the data. These scripts help communicate the analysis approaches used by the researcher and also facilitate other researchers' ability to reproduce the results (see Klein et al., 2018).

MODEL EVALUATION

Evaluating a model is essential because a proposed model is never a perfect representation of the phenomenon investigated in a study. Models are estimated based on a set of underlying assumptions so when there are violations of these assumptions, model evaluation becomes more complex. Sensitivity analyses provide one method of addressing the degree to which the results or conclusions of the model change with alternative assumptions (e.g., different assumed prior distributions, or alternative methods of treating missing data or outlying values).

For the evaluation of multilevel regression models, residual analysis is commonly recommended. However, O'Connell et al. (2015) reported that among 29 multilevel applications they reviewed only one application reported (raw) residual diagnostics for model evaluation. Diagnostics may include assumption checking (e.g., potential violations of assumptions of normality, homogeneity, or linearity identified though residual plots or tests) and identifying influential data points and outliers, which may impact parameter estimates (e.g., identified through Cook's D values) or unit standard errors (e.g., identified via CovRatios). Details on methods for examining residuals and software options for doing so are available in O'Connell et al. (2015) and Bell et al. (2010), and R users are referred to https://cran.r-project.org/web/packages/HLMdiag/README.html.

Importantly, because there are multiple sets of residuals in multilevel modeling, the reporting of residual analyses needs to consider each set of residuals. Accordingly, the source of model misfit (e.g., for which component at which level) needs to be clearly communicated. In reporting the residual analysis the focus should be on making transparent the degree to which the results of the residual analyses are consistent with the underlying assumption of the multilevel model. The discussion should then consider the consequences of any assumption violations. Can it be argued that the uncovered violation is not consequential based on robustness studies? Or does the violation have known consequences, such as biasing the parameter estimates or the standard errors? Without a reporting of the consistency of the residuals with the underlying assumptions the transparency goals of open science cannot be fully met.

Similarly, when researchers report the reduction in a residual variance that is the proportion explained by one or more predictors (namely, explained variance or pseudo-R-square), the residual variance involved in the computation should be specified. When the outcome variable is categorical and hierarchical generalized linear modeling is used as demonstrated in Chapters 7 and 8 of this volume ("Multilevel Logistic and Ordinal Models" and "Single and Multilevel Models for Counts"), the level-1 residuals may be heteroscedastic and the corresponding variance may depend on the predicted probability (Raudenbush & Bryk, 2002). However, the higher-level residuals are typically assumed to be normally distributed (Raudenbush & Bryk, 2002), and residual diagnostics and pseudo-R-square statistics (reduction in level-2 residual variances) are still relevant.

When model evaluation involves the comparison of multiple models, the method of model comparisons and the model selection criteria should be communicated clearly. When likelihood ratio tests are employed for the comparison of nested models, the differences in deviance (or chi-square) and degrees of freedom (or number of free parameters) need to be reported along with the associated p value. If information criteria (IC) are used for model comparisons, the selection criteria (e.g., the model with the smallest IC or cutoffs of IC) should be presented. Such communication can be facilitated with a summary report like a table of the model evaluation statistics.

REPORTING RESULTS

Because multilevel applications vary substantially in their purpose and methods, there is no single best way to organize and present results. Therefore, in this section we provide some general guidelines that hold across applications and then give some examples of tables and figures that can be useful for specific applications. First, the overall organization and reporting should be based on the purpose of the study, such that specific aspects of the results (e.g., a specific cross-level interaction effect estimate) should be explicitly connected to, and interpreted in light of, the research questions driving the study. Second, there should be a presentation of results from preliminary analyses that are appropriate to the research purposes and the models estimated. These preliminary analyses may include descriptive analyses (e.g., univariate statistics of each variable at each level of the model and bivariate analyses of associations among variables at each level of the model; Ferron et al., 2008), analyses of missing values (Peugh & Enders, 2004), and estimates of the intraclass correlation (ICC; Raudenbush & Bryk, 2002).

Third, there should be a presentation of the results from the primary analysis. What is included here varies a bit across applications. For instance, the results of a measurement study comparing several hypothesized multilevel factor analytic models will need to be reported differently than a cluster randomized controlled trial. For a measurement study that has the purpose of determining which of several multilevel factor analytic models is most appropriate for a particular measure, the focus of reporting would be on the results of the model evaluations. In the case of a cluster randomized controlled trial, the presentation of primary analyses should highlight the effect estimate, which may be presented in both raw score and standardized form and may be given separately for different groups or as a function of some moderators.

Regardless of the type of applications, the reporting of results typically includes parameter estimates from at least one multilevel model. The focus may be on specific fixed effect estimates or may be on specific variance parameter estimates. Regardless of whether answering the research questions brings focus to the fixed effects, variance components, or some combination, we recommend reporting all fixed effect and variance/covariance parameter estimates of the interpreted model, because doing so fully defines the model for the reader. If the model is complex and space limitations preclude the presentation of all parameter estimates (e.g., multilevel latent variable models with large numbers of indicator items), the complete set of estimates could be made available in supplemental materials available through the journal or author. For all parameter estimates there should be some measure of uncertainty, such as a standard error or confidence interval limits (Ferron et al., 2008). Finally, presentation of the primary results should include results from model evaluations and effect sizes when available.

Take effect sizes in a 2-level cluster randomized controlled trial as an example. In this type of application, the specific statistics chosen to index the effect will vary depending on the outcome variable (e.g., an odds ratio if the outcome is binary versus a regression coefficient and/or a standardized mean difference if the outcome is continuous). Assuming a continuous outcome, a standardized mean difference to index the size of the treatment effect can be defined using the parameters from the following multilevel model:

$$Y_{ij} = \beta_{0j} + r_{ij}$$

$$\beta_{0j} = \gamma_{00} + \gamma_{01}X_j + u_{0j}$$

where X_j is dummy coded (1 = treatment, 0 = control), $r_{ij} \sim N(0, \sigma^2)$ and $u_{0j} \sim N(0, \tau_{00})$. In principle, there are three options for the standard deviation researchers can choose to use: level-1, level-2, and the total standard deviation (Hedges, 2007). The choice of an appropriate standard deviation is based on the inference of interest to researchers. For instance, when the outcome is a level-1 construct with the individual as the unit of interest (e.g., student achievement), it would be appropriate to use the standard deviation of the population of level-1 units. Thus, the standardized mean difference could be estimated as

$$\hat{\delta}_T = \frac{\hat{\gamma}_{01}}{\sqrt{\hat{\sigma}^2 + \hat{\tau}_{00}}},$$

assuming a large enough sample size that a correction for small sample bias is not needed. However, if the outcome of interest was a cluster level construct (e.g., school climate) where the individual observations are only of interest because their aggregation defines the cluster level construct, then it would be appropriate to use the standard deviation of the population of level-2 units. In this circumstance the standardized mean difference could be estimated as

$$\hat{\delta}_B = \frac{\hat{\gamma}_{01}}{\sqrt{\hat{\tau}_{00}}},$$

again assuming a large sample size (Hedges, 2007). Similarly, various standardized effect estimates have been developed for 3-level cluster randomized experiments (Hedges, 2011) and for partially clustered designs (Hedges & Citkowicz, 2015; Lai & Kwok, 2016).

Next, we discuss the presentation of results from primary analyses using an example application. The application is a multilevel regression model with a continuous outcome variable and predictors at level-1 and level-2. Note that despite our focus on this example application, we acknowledge the need to discuss the presentation of results for other applications. Interested readers are referred to Chapter 6 of this volume, "Cross-Classified Random-Effects Models," Chapter 7 for binary and ordinal data, Chapter 8 for count data, and Chapters 9–12 for longitudinal models.

Example: Multilevel Regression Model

Suppose a researcher conducts a cluster randomized controlled trial to investigate whether sixth-grade reading ability is impacted by the provision of extra reading materials in class and if so, the extent the effect of extra reading materials is moderated by students' previous reading ability. In this multilevel regression model, the researcher may start by providing the results of the model evaluation, indicating the degree to which the data were consistent with the model assumptions. For example, the researcher may state, "Residuals at level-1 were approximately normally distributed ($|sk| < 1$; $|ku| < 1$), as were each of the level-2 residuals ($|sk| < 1$; $|ku| < 1$), with no identified outliers at either level. In addition, plots of residuals with predicted values (at each level of the model) revealed no substantial heteroscedasticity or nonlinear relationships."

The researcher may then provide parameter estimates along with measures of the precision of those estimates, such as through standard errors or confidence intervals. A common reporting practice is using a table to communicate the information. An example format is shown in Table 17.4, and additional examples are given in previous chapters of this book and in the *Publication Manual of the American Psychological Association* (American Psychological Association, 2020). In Table 17.4, the fixed effect parameter estimates (e.g., γ_{00}, γ_{10}), variance estimates (e.g., σ^2, τ_{00}), and standard errors, as well as confidence intervals for all parameter estimates, are listed. The symbols presented in the table should match the ones used in model specification equations or descriptions. In addition, the ICC is presented in the table. For better communication of the content in the table, researchers could include additional notations to give brief descriptions about the parameter estimates as shown in the example. Alternative examples for representing analysis results can be found in previous chapters in this book.

Given the focal interest of this analysis is to evaluate the effect of the reading materials, the presentation of results should bring focus to the estimated

TABLE 17.4 Example Table of Parameter Estimates for Two-Level Model of 6th Grade Reading Ability

	Unconditional model				Full model			
Parameter	Estimate	SE	*df*	95%CI	Estimate	SE	*df*	95%CI
Fixed effects								
Level 1								
Intercept (γ_{00})	34.11	0.50	59	33.1, 35.1	32.14	0.70	58	30.8, 33.5
Reading5 (γ_{10})	—	—		—	0.64	0.21	58	0.2, 1.1
Level 2								
Materials (γ_{01})	—	—		—	10.85	1.90	716	7.1, 14.6
Cross-level interaction								
Materials * Reading5 (γ_{11})	—	—		—	0.47	0.03	716	0.4, 0.5
Variance estimates								
Level 1 variance (σ^2)	7.53	0.34		6.9, 8.2	6.48	0.32		5.9, 7.1
Intercept variance (τ_{00})	1.24	0.12		1.0, 1.5	2.69	1.15		0.4, 4.9
Slope variance (τ_{11})	—	—		—	0.24	0.02		0.2, 0.3
Covariance (τ_{01})	—	—		—	0.35	0.17		0.0, 0.7
ICC	0.16							

Note: Reading5 is 5th grade (or prior year) reading ability grand-mean centered; Materials is coded 1 for students whose teachers provided extra reading materials and 0 for those whose teachers did not; level-1 sample size equals 836; level-2 sample size equals 60. ICC = Intraclass correlation coefficient.

effect of the reading materials (γ_{01}), which depends on how each variable was centered or scaled. In this example, the fifth-grade reading ability was grand-mean centered, and thus the interpretation of γ_{01} would be that students with an average level of fifth-grade reading ability who were provided with extra reading materials in class are predicted to score 10.85 points higher than similar students if they did not get extra reading materials.

In multilevel models, interactions may manifest within a certain level (e.g., an interaction between two variables at the lower level or at the upper level) or between different levels (e.g., an interaction between a lower-level predictor and an upper-level predictor). In this example, the cross-level interaction effect (γ_{11}) indicates the effect of being exposed to extra reading materials on students' sixth-grade reading ability would vary depending on the students' last year reading ability. For displaying interactions, using graphs is more efficient and concrete than relying solely on verbal description. The graphical example (see Figure 17.1) that was constructed based on the scenario above shows that students with higher levels of prior reading ability benefit more from additional reading materials. For an alternative graphical approach for displaying interactions see Tate (2004).

Researchers may also report a standardized mean difference, as we have previously discussed. Although standardization of regression coefficients can be misleading (Willett et al., 1998) and thus avoided by some researchers, others prefer a standardized measure of effect because it can help with comparisons to the results of other studies. In addition, reporting and interpreting alternative indices, such as pseudo-R-square (Singer & Willett, 2003) or R-square measures (Rights & Sterba, 2019), may aid in communicating the

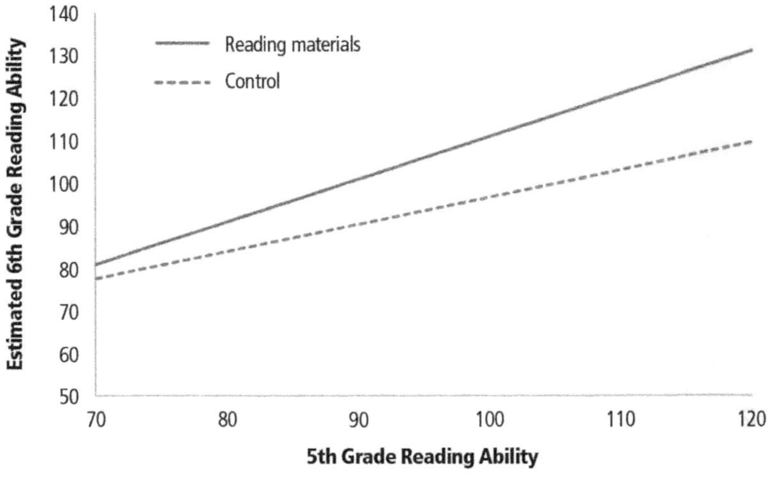

Figure 17.1 Graphical illustration of the effect of the extra reading materials on sixth-grade reading ability as a function of fifth-grade reading ability.

size of the effect. Researchers should recognize there are advantages and limitations of various standardized effect measures, and clearly communicate the particular methods applied for estimating standardized effects.

CHAPTER SUMMARY

To summarize suggestions across these sections, a checklist (see Table 17.5) has been developed where items that should be communicated with appropriate level of detail and justification are listed. This checklist is an updated version of the one provided previously (Ferron et al., 2008). Readers are encouraged to use this updated checklist as a reference in reporting multilevel analyses or reviewing the reporting practices of multilevel analyses. We view this checklist as providing a foundation for reporting of multilevel analyses that can be expanded as new techniques and applications of multilevel model emerge.

TABLE 17.5 Checklist for Multilevel Analysis Reporting

Could the reader answer the following questions?	Yes	No
Could the reader answer the following questions about Research Purposes and Design Type?		
What were the research purposes for the study?		
What was the rationale for the multilevel model(s)?		
What type of research design was used?		
Could the reader answer the following questions about Sampling?		
What sampling method was used at each level?		
How many units were at each level of the sampling?		
How many units were at each level of analysis?		
If any, how were missing data treated?		
What was the distribution of lower-level units across upper-level units?		
Could the reader answer the following questions about Variables?		
What was the meaning of each variable at each level?		
What was assumed about the nature of each variable (e.g., continuous, binary, count)?		
How was each variable scaled (e.g., coding or centering)?		
What was the reliability evidence for each applicable variable?		
What was the validity evidence for each applicable variable?		
Could the reader answer the following questions about Model Specification?		
How many models were estimated?		
How many levels were in each estimated model?		

(continued)

TABLE 17.5 Checklist for Multilevel Analysis Reporting (continued)

Could the reader answer the following questions?	Yes	No
What was the outcome and its assumed distribution (e.g., normal, Poisson, binomial)?		
What were the predictors/IVs? Did they have interactions?		
What was the assumed variance/covariance structure at each level?		
Could the reader answer the following questions about Estimation and Inference?		
What software and version were used (e.g., HLM 7.03; SAS 9.4 Proc Mixed)?		
What method of estimation was used (e.g., REML; Bayesian)?		
What methods of implementing the estimation were used (e.g., software defaults)?		
What methods were used to make inferences (e.g., *t*-test with Satterthwaite *df*)?		
Could the reader answer the following questions about Model Evaluation?		
What methods were used to evaluate the overall model (e.g., AIC, BIC)?		
What methods were used to evaluate the lowest level (e.g., level-1 residual analysis)?		
What methods were used to evaluate upper levels (e.g., level-2 residual analysis)?		
Could the reader answer the following questions about Results Reporting?		
What were the connections between research purposes and specific results?		
What were the results of preliminary analyses?		
What were the parameter estimates for each interpreted model?		
What were the CIs or standard errors for each parameter estimate?		
What were the results of model evaluations?		

REFERENCES

American Educational Research Association. (2006). Standards for reporting on empirical social science research in AERA publications. *Educational Researcher, 35*(6), 33–40.

American Educational Research Association, American Psychological Association, & National Council on Measurement in Education. (2014). *Standards for educational and psychological testing.* American Psychological Association.

American Psychological Association. (2020). *Publication manual of the American Psychological Association* (7th ed.). American Psychological Association.

American Statistical Association. (2016). *Ethical guidelines for statistical practice.* http://www.amstat.org/asa/files/pdfs/EthicalGuidelines.pdf

Appelbaum, M., Cooper, H., Kline, R. B., Mayo-Wilson, E., Nezu, A. M., & Rao, S. M. (2018). Journal article reporting standards for quantitative research in psychology: The APA publications and communications board task force report. *American Psychologist, 73*, 3–25.

Bell, B. A., Schoeneberger, J., Morgan, G., Kromrey, J., & Ferron, J. (2010). Fundamental diagnostics for two-level mixed models: The SAS macro MIXED DX. *Proceedings of SAS Global Forum 2010 Conference.* SAS Institute Inc.

Chen, R. (2012). Institutional characteristics and college student dropout risks: A multilevel event history analysis. *Research in Higher Education, 53,* 487–505. https://doi.org/10.1007/s11162-011-9241-4

Dedrick, R. F., Ferron, J. M., Hess, M. R., Hogarty, K. Y., Kromrey, J. D., Lang, T. R., Niles, J. D., & Lee, R. S. (2009). Multilevel modeling: A review of methodological issues and applications. *Review of Educational Research, 79,* 69–102.

Enders, C. K. (2017). Multiple imputation as a flexible tool for missing data handling in clinical research. *Behavior Research and Therapy, 98,* 4–18.

Enders, C. K., & Tofighi, D. (2007). Centering predictor variables in cross-sectional multilevel models: A new look at an old issue. *Psychological Methods, 12,* 121–138.

Fagginger Auer, M. F., Hickendorff, M., Van Putten, C. M., Béguin, A. A., & Heiser, W. J. (2016). Multilevel latent class analysis for large-scale educational assessment data: Exploring the relation between the curriculum and students' mathematical strategies. *Applied Measurement in Education, 29,* 144–159. https://doi.org/10.1080/08957347.2016.1138959

Ferron, J. M., Bell, B. A., Hess, M. R., Rendina-Gobioff, G., & Hibbard, S. T. (2009). Making treatment effect inferences from multiple-baseline data: The utility of multilevel modeling approaches. *Behavior Research Methods, 41,* 372–384.

Ferron, J. M., Hogarty, K. Y., Dedrick, R. F., Hess, M. R., Niles, J. D., & Kromrey, J. D. (2008). Reporting results from multilevel analyses. In A. O'Connell & B. McCoach (Eds.), *Multilevel modeling of educational data* (pp. 391–426). Information Age Publishing.

Geldhof, G. J., Preacher, K. J., & Zyphur, M. J. (2014). Reliability estimation in a multilevel confirmatory factor analysis framework. *Psychological Methods, 19,* 72–91.

Gelman, A., & Loken, E. (2014). The statistical crisis in science. *American Scientist, 102,* 460–465.

Grimm, K. J., Ram, N., & Estabrook, R. (2017). *Growth modeling: Structural equation and multilevel modeling approaches.* The Guilford Press.

Hedges, L. V. (2007). Effect sizes in cluster randomized designs. *Journal of Educational and Behavioral Statistics, 32,* 341–370.

Hedges, L. V. (2011). Effect sizes in three-level cluster-randomized experiments. *Journal of Educational and Behavioral Statistics, 36,* 346–480.

Hedges, L. V., & Citkowicz, M. (2015). Estimating effect size when there is clustering in one treatment group. *Behavior Research Methods, 47,* 1295–1308.

John, L. K., Loewenstein, G., & Prelec, D. (2012). Measuring the prevalence of questionable research practices with incentives for truth telling. *Psychological Science, 23,* 524–532. https://doi.org/10.1177/0956797611430953

Kenward, M. G., & Roger, J. H. (1997). Small sample inference for fixed effects from restricted maximum likelihood. *Biometrics, 53,* 983–997.

Kim, E. S., Dedrick, R. F., Cao, C., & Ferron, J. M. (2016). Multilevel factor analysis: Reporting guidelines and a review of reporting practices. *Multivariate Behavioral Research, 51,* 881–898. https://doi.org/10.1080/00273171.2016.1228042

Klein, O., Hardwicke, T. E., Aust, F., Breuer, J., Danielsson, H., Hofelich Mohr, A., IJzerman, H., IJzerman, H., Nilsonne, G., Vanpaemel, W., & Frank, M. C.

(2018). A practical guide for transparency in psychological science. *Collabra: Psychology*. https://doi.org/10.1525/collabra.158

Lai, M. H. C., & Kwok, O. (2016). Estimating standardized effect sizes for two- and three-level partially nested data. *Multivariate Behavioral Research, 51*, 740–756.

Maas, C. J. M., & Hox, J. J. (2004). Robustness issues in multilevel regression analysis. *Statistica Neerlandica, 58*, 127–137.

McCoach, D. B. (2010). Hierarchical linear modeling. In G. R. Hancock & R. O. Mueller (Eds.), *The reviewer's guide to quantitative methods in the social sciences* (pp. 123–140). Routledge.

McCoach, D. B., Rifenbark, G. G., Newton, S. D., Li, X., Kooken, J., Yomtov, D., Gambino, A. J., & Bellara, A. (2018). Does the package matter? A comparison of five common multilevel modeling software packages. *Journal of Educational and Behavioral Statistics, 43*, 594–627,

Moher, D., Liberati, A., Tetzlaff, J., Altman, D. G., & PRISMA Group. (2009). Preferred reporting items for systematic reviews and meta-analyses: The PRISMA statement. *Annals of Internal Medicine, 151*, 264–269.

Munafo, M. R., Bosek, B. A., Bishop, D. V. M., Button, K. S., Chambers, C. D., du Sert, N. P., Simonsohn, U., Wagenmakers, E.-J., Ware, J. J., & Ioannidis, P. A. (2017). A manifesto for reproducible science. *Nature of Human Behavior, 1*, 1–9.

Nezlek, J. B., Holas, P., Rusanowska, M., & Krejtz, I. (2016). Being present in the moment: Event-level relationships between mindfulness and stress, positivity, and importance. *Personality and Individual Differences, 93*, 1–5. https://doi.org/10.1016/j.paid.2015.11.031

Nosek, B. A., Alter, G., Banks, G. C., Borsboom, D., Bowman, S. D., Breckler, S. J., Buck, S., Chambers, C. D., Chin, G., Christensen, G., Contestabile, M., Dafoe, A., Elich, E., Freese, J., Glennerster, R., Goroff, D., Green, D. P., Hesse, B. Humphreys, M., . . . Yarkoni, T. (2015). Promoting an open research culture. *Science, 348*, 1422–1425.

Nosek, B. A., Ebersole, C. R., DeHaven, A. C., & Mellor, D. T. (2018). The preregistration revolution. *PNAS, 115*(11), 2600–2606.

O'Connell, A. A., Yeomans-Maldonado, G., & McCoach, D. B. (2015). Residual diagnostics and model assessment in a multilevel framework: Recommendations toward best practice. In J. R. Harring, L. M. Stapleton, & S. N. Beretvas (Eds.), *Advances in multilevel modeling for educational research: Addressing practical issues found in real-word applications* (pp. 97–138). Information Age Publishing.

Palardy, G. J., & Vermunt, J. K. (2010). Multilevel growth mixture models for classifying groups. *Journal of Educational and Behavioral Statistics, 35*, 532–565. https://doi.org/10.3102/1076998610376895

Peugh, J. L., & Enders, C. K. (2004). Missing data in educational research: A review of reporting practices and suggestions for improvement. *Review of Educational Research, 74*, 525–556.

Raudenbush, S. W., & Bryk, A. S. (2002). *Hierarchical linear models: Applications and data analysis methods* (2nd ed.). SAGE Publications.

Rights, J. D., & Sterba, S. K. (2019). Quantifying explained variance in multilevel models: An integrative framework for defining R-squared measures. *Psychological Methods, 24*(3), 309–338. https://doi.org/10.1037/met0000184

Rosenberg, S. L. (2009). *Multilevel validity: Assessing the validity of school-level inferences from student achievement test data* [Doctoral dissertation, University of North Carolina]. ProQuest Dissertations Publishing. (3352691)

Schulz K. F., Altman, D. G., Moher D., for the CONSORT Group (2010). CONSORT 2010 Statement: Updated guidelines for reporting parallel group randomised trials. *Journal of Clinical Epidemiology, 63*(8), 834–840. https://doi.org/10.1016/j.jclinepi.2010.02.005

Shadish, W. R., Cook, T. D., & Campbell, D. T. (2002). *Experimental and quasi-experimental designs for generalized causal inference.* Houghton Mifflin.

Singer, J. D., & Willett, J. B. (2003). *Applied longitudinal data analysis: Modeling change and event occurrence.* Oxford University Press.

Spybrook, J., Kelcey, B., & Dong, N. (2016). Power for detecting treatment by moderator effects in two- and three-level cluster randomized trials. *Journal of Educational and Behavioral Statistics, 41,* 605–627.

Spybrook, J., Raudenbush, S. W., Liu, X.-F., Congdon, R., & Martinez, A. (2011). *Optimal design plus empirical evidence: Documentation for the "optimal design" software.* http://hlmsoft.net/od/od-manual-20111016-v300.pdf

Stapleton, L. M., Yang, J. S., & Hancock, G. R. (2016). Construct meaning in multilevel settings. *Journal of Educational and Behavioral Statistics, 41,* 481–520. https://doi.org/10.3102/1076998616646200

Tate, R. (2004). Interpreting hierarchical linear and generalized linear models with slopes as outcomes. *Journal of Experimental Education, 73,* 71–95.

Thompson, B., & Vacha-Haase, T. (2000). Psychometrics is datametrics: The test is not reliable. *Educational and Psychological Measurement, 60,* 174–195.

Van den Noortgate, W., Opdenakker, M.-C., & Onghena, P. (2004). The effects of ignoring a level in multilevel analysis. *Journal School Effectiveness and School Improvement, 16,* 281–303.

van 't Veer, A. E., & Giner-Sorolla, R. (2016). Pre-registration in social psychology—A discussion and suggested template. *Journal of Experimental Psychology, 67,* 2–12.

von Elm, E., Altman, D. G., Egger, M., Pocock, S. J., Gøtzsche, P. C., & Vandenbroucke, J. P. (2008). The strengthening the reporting of observational studies in epidemiology (STROBE) statement: Guidelines for reporting observational studies. *Journal of Clinical Epidemiology, 61*(4), 344–349. https://doi.org/10.1016/j.jclinepi.2007.11.008

Willett, J. B., Singer, J. D., & Martin, N. C. (1998). The design and analysis of longitudinal studies of development and psychopathology in context: Statistical models and methodological recommendations. *Development and Psychopathology, 10,* 395–426.

Woodhouse, G., Yang, M., Goldstein, H., & Rasbash, J. (1996). Adjusting for measurement error in multilevel analysis. *Journal of the Royal Statistical Society—A, 159,* 201–212.

ABOUT THE CONTRIBUTORS

Bethany A. Bell is an associate professor in the research, statistics, and evaluation program and chair of the Department of Leadership, Foundations, and Policy at the University of Virginia's School of Education and Human Development. Her research is both applied and methodological with a focus on expanding the methodological knowledge base of complex statistical procedures to advance health and education equity research and to make complex analyses more accessible to applied researchers. To achieve these goals, she conducts Monte Carlo simulation studies, develops SAS macros and "how to" papers, and examines social determinants of various health outcomes, using both primary and secondary data sources.

Aarti P. Bellara is a senior lecturer in the quantitative research, evaluation, and measurement program in the College of Education and Human Ecology (EHE) at the Ohio State University. She teaches courses in educational research, statistics, and measurement. Her research focuses on assessment literacy, measurement theory, instrument design, and the use and misuse of propensity score modeling with nested data.

S. Natasha Beretvas is the senior vice provost for faculty affairs at The University of Texas at Austin and the John L. and Elizabeth G. Hill Centennial Professor in the quantitative methods program in the Educational Psychology Department. Her research focuses on use of multilevel modeling for handling complex data and on meta-analysis—for single-case and group-comparison experimental designs' data.

Multilevel Modeling Methods with Introductory and Advanced Applications, pages 623–630

Nivedita Bhaktha is a senior researcher in the Department of Survey Design and Methodology at GESIS Leibniz-Institut für Sozialwissenschäften. She did her PhD in quantitative research, evaluation, and measurement from the Ohio State University. Her research interests are in the areas of categorical data analysis, latent variable modeling, and exploratory data analysis.

Shelley Blozis is a professor of psychology at the University of California, Davis, where she has been since 2001. She earned her PhD in psychology from the University of Minnesota. Her research concerns the development and application of statistical methods for psychological data. Recent project topics include the estimation of nonlinear mixed-effects models and nonlinear latent curve models for longitudinal data.

Robert Dedrick is a professor in the educational measurement and research program in the Department of Educational and Psychological Studies at the University of South Florida. He teaches courses in research design and measurement and has research interests that focus on the development and validation of educational and psychological measures.

Nianbo Dong is associate professor of quantitative methods in the School of Education, University of North Carolina at Chapel Hill. Dr. Dong's research program centers on developing and applying rigorous quantitative methods to evaluate educational policies, programs, and practice. His current interests in quantitative methodology focus on power analyses of the main, moderation, and mediation effects in multilevel experiments and causal inference. His substantive research focuses on the evaluations of the effectiveness of teacher and principal training programs and early child education programs. His work has been supported by the U.S. Department of Education's Institute of Education Science and National Science Foundation (NSF). Dr. Dong received the NSF Faculty Early CAREER award in 2017.

Craig Enders is a professor in the Department of Psychology at UCLA where he is a member of the quantitative program area. Enders teaches graduate-level courses in missing data analyses, multilevel modeling, and longitudinal modeling. The majority of his research focuses on analytic issues related to missing data analyses and multilevel modeling. His book, Applied Missing Data Analysis, was published with Guilford Press in 2010 and is currently under revision.

John Ferron is a professor in the educational measurement and research program in the Department of Educational and Psychological Studies at the University of South Florida. He teaches courses in educational statistics and has research interests that focus on the development and application of statistical methods in educational research.

Holmes Finch is the George and Frances Ball Distinguished Professor of educational psychology at Ball State University in Muncie, Indiana. His research interests are in latent variable modeling, with particular interest in differential item functioning, multidimensionality assessment, and statistical methods for small samples.

Brian F. French is the Berry Family Distinguished Professor in the College of Education at Washington State University in Pullman, Washington. His research interests are in the area of fair and accurate score use for making decisions about individuals and groups. His work sits at the intersection of applied and methodological studies to explore educational and psychological measurement issues.

Anthony J. Gambino earned his PhD in educational psychology from the research methods, measurement, and evaluation program at the University of Connecticut. He currently works as a postdoctoral research associate for the National Center for Research on Gifted Education at the University of Connecticut. His areas of expertise/research include causal inference, research/evaluation methodology, multilevel and structural equation modeling, and measurement theory.

Jessica Goldstein is the principal of Boulevard Research Partners LLC, which she founded after a decade as research faculty at the University of Connecticut. Dr. Goldstein's current work focuses on assessment literacy, early childhood evaluation, and strategic planning for businesses, nonprofits, and state agencies focused on education.

Jeffrey Harring is a professor of measurement, statistics, and evaluation in the Human Development and Quantitative Methodology Department at the University of Maryland, College Park. His research focuses on longitudinal methods of linear and nonlinear repeated measures data, finite mixture modeling, multilevel modeling with nonlinear effects, and statistical computing.

Andrew F. Hayes is Distinguished Research Professor at the Haskayne School of Business at the University of Calgary. He conducts research on and develops statistical methods for quantifying and assessing the mechanisms by which causal effects operate, their boundary conditions, and the contingencies of the mechanisms generating those effects. He is the author of *Introduction to Mediation, Moderation, and Conditional Process Analysis: A Regression-Based Approach* (2018) as well as *Regression Analysis and Linear Models: Concepts, Applications, and Implementation* (2017, with Richard B. Darlington), both published by The Guilford Press.

Timothy Hayes is an assistant professor of quantitative psychology at Florida International University, with research interests in structural equation models (SEMs), multilevel models, and missing data. His recent work has focused on: (a) extending factor score regression (FSR) methods to models with correlated residuals and cross-loadings, (b) developing approaches to calculating R-squared change in SEMs with latent variables and missing data, and (c) evaluating and describing novel approaches to multilevel multiple imputation.

Donald Hedeker is a professor of biostatistics in the Department of Public Health Sciences at the University of Chicago. Don's main expertise is in the development of methods for clustered and longitudinal data, with emphasis on mixed-effects models. With Robert Gibbons, Don is the author of the text *Longitudinal Data Analysis* published by Wiley in 2006. He is also the primary author of several freeware computer programs for mixed-effects analysis. Don was named a fellow of the American Statistical Association in 2000, and he is an associate editor for *Statistics in Medicine.*

Hsien-Yuan Hsu is an assistant professor of research in education and evaluation at the University of Massachusetts Lowell. His research interests include multilevel structural equation modeling, latent growth curve modeling, and the applications of these models in educational and psychological research.

Ben Kelcey is professor of quantitative research methodologies in the College of Education, Criminal Justice, and Human Services, University of Cincinnati. His research interests center on the development and application of quantitative and measurement methodologies within the context of multilevel and multidimensional settings such as schools, hospitals, and neighborhoods.

Eunsook Kim is an associate professor in the educational measurement and research program in the Department of Educational and Psychological Studies at the University of South Florida. She has a broad interest in research methodology in social sciences including structural equation modeling, multilevel modeling, factor mixture modeling, and propensity score analysis. Her focal research interests include measurement invariance testing in multilevel and longitudinal data.

Minjung Kim is an assistant professor of quantitative research, evaluation, and measurement at the Department of Educational Studies at the Ohio State University. She received her PhD in research, measurement, and statistics program at Texas A&M University in 2012. Her research interests are quantitative methods including structural equation modeling (SEM) and multilevel modeling. Dr. Kim is also interested in applying the advanced statistical models to real data in social science research, including education and psychology.

Oi-Man Kwok is professor of research, measurement, and statistics in the Department of Educational Psychology at Texas A&M University. He received his PhD in 2005 from Arizona State University. Dr. Kwok's research interests include examining the methodological issues of both multilevel models and structural equation models, and the applications of these models in different educational and psychological research.

Audrey J. Leroux is an associate professor in the research, measurement, and statistics program in the Department of Educational Policy Studies at Georgia State University. She teaches courses in introductory and advanced multilevel modeling, among other quantitative methods courses. Her research work evaluates, using Monte Carlo simulations, innovative models as well as their applications in multilevel modeling and computerized adaptive testing. More specifically, her multilevel modeling research examines the effect of different multilevel models (cross-classified and multiple membership random effects models) for handling complex data structures such as individual mobility across clusters, particularly with longitudinal data.

Eva Yujia Li received her PhD in educational psychology from the research method, measurement and evaluation program at the University of Connecticut. She is currently a statistician in the Office of Institutional Research at Yale University, where her research focuses on applying quantitative methods to studying topics related to college admissions, student experiences, outcomes, and diversity.

Meng-Ting Lo is a research fellow at National Yang Ming Chiao Tung University. She earned a doctoral degree in the quantitative research, evaluation, and measurement program at the Ohio State University in 2020. Her research work is in multilevel modeling with a focus on modeling approaches for the analysis of categorical nested data. Her research interests also include data mining techniques and structural equation modeling, especially as they pertain to educational research.

D. Betsy McCoach is professor of research methods, measurement, and evaluation in the Department of Educational Psychology at the University of Connecticut. Dr. McCoach is the co-author of *Introduction to Modern Modelling Methods* (2021, SAGE, with Dakota Cintron) and is the founder and program chair of the Modern Modeling Methods Conference, held at UCONN. Dr. McCoach is co-principal investigator for the National Center for Research on Gifted Education. Dr. McCoach's research interests include latent variable modeling, multilevel modeling, longitudinal modeling, instrument design, and gifted education.

Robin Mermelstein is distinguished professor, Department of Psychology, and is the director of the Institute for Health Research and Policy at the University of Illinois at Chicago. Her research has focused on understanding the etiology and progression of tobacco use in adolescents and young adults, in developing interventions to reduce tobacco use and increase cessation in both adolescents and in young adults, and in translating research into practice and policy. She has a special interest and expertise in the use of ecological momentary assessments (EMA or real time data capture) for understanding the contexts surrounding tobacco use.

Sarah D. Newton is postdoctoral research associate in the University of Connecticut's Department of Educational Psychology. She completed her PhD and MA in educational psychology (with a concentration in research methods, measurement, and evaluation) from the same institution. Sarah also earned an MS in criminal justice and a BA in criminology, with completed course requirements in psychology, from Central Connecticut State University. Her research focuses include: model/data fit and model adequacy as complementary tools for multilevel model evaluation and selection, information criteria performance in multilevel modeling contexts, latent variable modeling; instrument design, reliability and validity theory, and quantitative research methodology.

Ann A. O'Connell is professor of quantitative research, evaluation and measurement in the College of Education and Human Ecology (EHE) at the Ohio State University. Her primary areas of specialization are generalized linear and mixed models, analysis of clustered and complex sample data, and methods for assessing health and education interventions. She is a former Fulbright Scholar to Addis Ababa University in Ethiopia from 2013–2014 and is committed to efforts that improve capacity for quantitative research, data use, and statistical methods locally and abroad.

C. Y. Joanne Peng is adjunct professor at National Taiwan University and formerly at Indiana University. She and co-authors have researched definitions and applications of ESs for group and single-case designs, including: ES reporting, alternative ESs, treating trends, missing data, ties in SCD, and computing tools.

Christopher Rhoads received his PhD in statistics from Northwestern University and is currently an associate professor in the research methods, measurement, and evaluation program in the Department of Educational Psychology at the Neag School of Education at the University of Connecticut, where he teaches classes in research design and quantitative methods. His research focuses on improving causal inference in quantitative social science research, particularly on the design and analysis of large field studies for the purposes of evaluating educational policies and practices.

Nicholas J. Rockwood is an assistant professor in the School of Behavioral Health at Loma Linda University. His research focuses on latent variable modeling (i.e., multilevel, structural equation, and item response theory) and statistical mediation and moderation analysis.

H. Jane Rogers is associate professor emeritus in the research methods, measurement, evaluation, and assessment program in the Neag School of Education at the University of Connecticut. Her research interests are in applications of item response theory, assessment of differential item functioning, and educational statistics.

Jason Schoeneberger has more than 20 years of experience in applied education research, measurement, and evaluation methods, having worked for several large, urban school districts and a large-scale assessment organization. He has conducted numerous mixed-methods evaluations of educational programs and policies including Teach for America and strategic staffing initiatives, and has participated as part of a RCT research team studying Math for All. Dr. Schoeneberger specializes in multilevel modeling, multiple imputation procedures, regression discontinuity methods, statistical programming, and large-scale data management.

Jessaca Spybrook is professor of evaluation, measurement, and research in the Department of Educational Leadership, Research, and Technology in the College of Education and Human Development at Western Michigan University. Her research focuses on improving the design of causal inference studies, particularly in education.

Laura M. Stapleton is interim dean of the College of Education at the University of Maryland and professor in measurement, statistics, and evaluation in the Department of Human Development and Quantitative Methodology. Her research focuses on analysis of administrative data and survey data obtained under complex sampling designs, multilevel latent variable models, and tests of mediation within a multilevel framework. She currently serves as director of the National Science Foundation quantitative research methods scholars program at the University of Maryland.

Scott L. Thomas is the John P. "Jack" Ellbogen Dean of the College of Education and professor at the University of Wyoming. His substantive research interests include higher education policy, science and technology, and the stratification of postsecondary opportunity. In addition to his scholarship in these substantive areas, Thomas has a related line of interest and work focusing on methodological issues, including multilevel statistical modeling, social network analysis, and spatial analytics.

Thao Vo is currently a doctoral student in educational psychology at Washington State University. Her research broadly examines equity in measurement at the intersection of critical theories and quantitative research methods.

Yan Wang is an assistant professor in the Department of Psychology at the University of Massachusetts Lowell. She teaches statistics courses and is interested in the application of statistical models (e.g., structural equation modeling, longitudinal data analysis) to psychology, education, and health.

Zhiyao Yi is an instructor in the School of Mathematics and Statistics at Chongqing Technology and Business University. Her research interests include structural equation modeling, multilevel modeling, models for intensive longitudinal data and measurement quality of psychological instruments.

Yue Yin received his PhD in the educational measurement and research program in the Department of Educational and Psychological Studies at the University of South Florida. He has research interests that focus on the development and application of statistical methods and psychometrics in educational research.

Jing Zhang is an assistant professor of human development and family studies at Kent State University. Her research focuses on mechanisms linking contextual risks and behavioral and health outcomes of marginalized youth and families and related preventive interventions. Dr. Zhang is interested in applying multilevel modeling, structural equation modeling, and mixture modeling to the research of adolescent health behaviors.

Printed and bound by CPI Group (UK) Ltd, Croydon, CR0 4YY

15/09/2026

14962337-0002